INTERNATIONAL DOMAIN NAME LAW

The Domain Name System (DNS), which matches computer addresses to human-friendly domain names, has given rise to many legal issues. Two important issues are the institutional arrangements for governing the DNS and the use of trade marks as domain names. This book is the first complete statement of this rapidly-evolving area of the law. In particular, the book includes a comprehensive statement of decisions under the Uniform Domain Name Dispute Resolution Policy (UDRP), the international system for resolving disputes between trade mark owners and domain name registrants. In this path-breaking work the author examines the extent to which principles of national trade mark law have been used in UDRP decisions. It will be essential reading for anyone, whether academic or practitioner, interested in internet law, intellectual property and e-commerce law.

International Domain Name Law

ICANN and the UDRP

DAVID LINDSAY

• HART •
PUBLISHING
OXFORD AND PORTLAND, OREGON
2007

Published in North America (US and Canada) by
Hart Publishing
c/o International Specialized Book Services
920 NE 58th Avenue, Suite 300
Portland, OR 97213-3786
USA
Tel: +1 503 287 3093 or toll-free: (1) 800 944 6190
Fax: +1 503 280 8832
E-mail: orders@isbs.com
Website: www.isbs.com

Hart Publishing, 16C Worcester Place, OX1 2JW
Telephone: +44 (0)1865 517530 Fax: +44 (0)1865 510710
E-mail: mail@hartpub.co.uk
Website: http://www.hartpub.co.uk

British Library Cataloguing in Publication Data
Data Available

ISBN-13: 978-1-84113-584-7

Typeset by Hope Services, Abingdon
Printed and bound in Great Britain by
TJ International Ltd, Padstow, Cornwall

Foreword

The authorisation of commercial activity on the Internet in 1995 changed both commerce and the Internet. For commerce, the 'bright lattices of logic'[1] unfolding invisibly across national boundaries opened a global marketplace and distributional channels of unprecedented power and reach. For the Internet, the age of innocence passed. Its traffic would no longer be confined to the public goods of research and defence, but would thenceforth concern also the interaction of private persons and enterprises and private goods.

Any enterprise wishing to trade in the new global marketplace needed, however, an address at which its offerings could be located and viewed. The Internet provided a ready answer in the form of the domain name system, which permitted human-friendly names or alphanumeric strings to be used in place of the Internet Protocol numbers which enabled computers to connect to each other, but which were so easy to misread, mistype or forget. The adoption of domain names as commercial addresses, however, caused a mutation. Domain names no longer served only the function of technical addresses connecting computers, but became also brands or signals of commercial presence that entered the world of publicity and advertising as part of the standard identification apparatus of enterprises.

Thus was born a seemingly intractable legal problem, the conflict between two systems of identifiers designed for different purposes and different contexts. On the one hand, there was the well established system of trademarks, legal titles granted pursuant to a regulated procedure designed to avoid confusion with similar marks or legally respected commercial indicators. They were territorially limited and had effect only in commerce. On the other hand, there was the new breed of domain names, technical addresses accorded following a largely unregulated application procedure that did not seek to eliminate confusing similarities. They had global effect and, while being able to be used in commerce, were equally available for social, cultural, scientific and political communication

For various reasons, traditional legal processes offered little hope of providing a completely satisfactory means of dealing with the conflict between these two systems of identifiers. The speed of Internet activity was one such reason. A domain name could be acquired in a matter of seconds and gave an immediate global presence, with the potential for considerable damage to be inflicted on a pre-existing trademark before litigation could be commenced and successfully terminated. Another reason was the international nature of the Internet and its users. A domain name could be held by anyone, anywhere in the world and, thus, possibly in a jurisdiction on the opposite side of the world from the country in which a similar trademark was held. The inconvenience of litigation in a foreign language and in a foreign legal system and country was manifest. Furthermore, the normal instrument of providing for a means of resolving such international issues, the treaty, which usually takes several years to negotiate and many more to enter into force over a geographically extensive range, was woefully inadequate in the circumstances of the very rapidly expanding use of the Internet.

[1] William Gibson, *Neuromancer* (Victor Gollanz, 1984) 11.

The Uniform Domain Name Dispute Resolution Policy (UDRP) was the novel solution invented to overcome the short-comings of traditional legal processes and to provide a partial, but very effective, solution to the interface between trademarks and domain names. It was the outcome of a very expeditious, but nevertheless extensive, international process conducted by the World Intellectual Property Organization (WIPO), which was in turn adopted by the Internet Corporation for Assigned Names and Numbers (ICANN) and given effect through the contractual web that links ICANN, registrars and domain name holders.

David Lindsay's work provides a comprehensive and invaluable guide to the engaging story of the evolution of the domain name system, ICANN and the UDRP for students, practitioners and the interested public. It is a story that has intrinsic merit, but also one that may provide lessons for other legal problems that may arise in the future as global technologies become increasingly commonly deployed in a world that remains governed essentially by national legal systems.

Francis Gurry
Deputy Director General
World Intellectual Property Organization (WIPO)

Preface

This book explains and analyses what, by any measure, has been a remarkable and unprecedented achievement: the emergence of the complex international legal framework dealing with the governance of, and disputes relating to, Internet domain names. In doing so, the book aims to provide a resource for practitioners, as well as to make a contribution to our understanding of this relatively new and extraordinary area of the law.

The book addresses three main areas. First, Chapter 1 provides an historical account of the development of domain names and the domain name system (DNS), and a technical explanation of the operation of the DNS. This chapter establishes the necessary technical and historical foundations for the later sections of the book. It introduces essential concepts, which are necessary to understand key issues, such as the importance of decisions regarding the introduction of new generic Top Level Domains (gTLDs).

Secondly, in Chapter 2, the book explains the history of, and the legal structures governing, the institutional arrangements for making decisions regarding domain names and the DNS. An understanding of the convoluted history behind DNS decision-making is necessary to appreciate the complex legal arrangements by which the most significant decisions regarding domain names are made by a unique, non-profit-making Californian corporation, the Internet Corporation for Assigned Names and Numbers (ICANN). In explaining the legal arrangements governing DNS decision-making, the book places the emergence of ICANN within the wider context of broad changes in public policy decision-making, such as the greater role of private sector organisations and the involvement of interdependent networks of actors in policy-making, which have been associated with the term 'governance'. Chapter 2 also puts DNS governance within the broader context of Internet governance, outlining recent developments in this area, including the World Summit for the Information Society (WSIS) and the Internet Governance Forum (IGF). The material in the book dealing with the DNS and domain name governance is, as far as possible, accurate to September 2007.

Thirdly, the remainder of the book, in Chapters 3 through 7, deals with the *sui generis* international system for resolving disputes between trade mark owners and domain name holders, known as the Uniform Domain Name Dispute Resolution Policy (UDRP). Chapter 3 sets the stage for the remaining chapters by introducing the problem of 'cybersquatting' and explaining the historical development of the UDRP. An adequate appreciation of the decisions made under the UDRP depends, first, upon an understanding of the *travaux préparatoires* for the UDRP and, secondly, an understanding of the distinction between traditional trade mark law, on the one hand, and disputes involving the use of trade marks as confusingly similar domain names, on the other. A major underlying theme of the book is that, as the objectives of trade mark law and domain name dispute resolution are quite different, the legal principles that are applied to resolving these different kinds of legal disputes should be tailored accordingly.

Chapters 4 to 7 of the book deal with decisions that have been made under the UDRP. The UDRP procedural rules are dealt with in Chapter 4, while the three substantive elements that must be established for a trade mark owner to bring a successful complaint against the registered holder of a disputed domain name under the UDRP are covered in Chapters 5 to 7 respectively. In relation to each of the key issues that have arisen in the application of the UDRP, the book endeavours to refer to the most important UDRP decisions, and to provide illustrations of the operation of the principles by copious examples drawn from relevant panel decisions. The material dealing with UDRP decisions endeavours to be accurate to the end of February 2007.

The attempt to bring some cohesion to UDRP decisions has involved sifting through, and analysing, a vast, and ever-developing, reservoir of primary material. According to the most recent available statistics, from the introduction of the UDRP, in December 1999, through September 2007, a total of 11,727 UDRP or UDRP-based disputes had been filed in relation to 21,435 domain names with the World Intellectual Property Organization (WIPO) Arbitration and Mediation Center, which is but one of the UDRP dispute resolution service providers.[1] As of December 2006, a total of 7,328 decisions had been given by WIPO in UDRP disputes,[2] with an additional 1,410 disputes dealt with in 2007 through to late September 2007. Importantly, while some predicted that there would be, over time, a reduction in the number of domain name disputes because, for example, of the near-universal reliance on search engines, experience has proved to the contrary. For instance, the number of domain name disputes lodged with WIPO in 2006 increasing by 25 per cent as compared with 2005, and UDRP disputes filed in 2007 have, if anything, increased.

There are three main explanations for the persistent rate of domain name disputes. First, there has been the progressive introduction of new gTLDs, which invariably creates scope for those seeking to capitalise on trade marks by registering variants as domain names. In this respect, the momentum gathered throughout 2007 by ICANN processes for introducing new gTLDs is likely to result in the introduction of further new gTLDs in 2008. Secondly, recent years have seen a boom in the domain name industry that is comparable to the domain name 'gold rush' of the late 1990s. For example, according to news reports, in mid-2007 the domain name industry was valued at US$2 billion, with an average 90,000 domain names being registered each day, most in the .com gTLD.[3] Thirdly, and most importantly, there is the seemingly limitless capacity of those seeking to register variants of trade marks as domain names to develop strategies in an attempt to profit from the use of marks as domain names. In this respect, we have seen the emergence of a range of practices that take advantage of the use of software for automatically registering expired domain names. Such practices include domain name 'parking', involving the registration of large numbers of domain names which are 'parked' and used for 'click-through' advertising, and domain name 'tasting' (or 'domain kiting'), involving the registration of large numbers of domain names by professional speculators during the five day grace period granted by most registries during which refunds will be granted.[4] The continuing interest in new gTLDs, and the continuing development of new practices seeking to exploit marks used as domain names, suggests that disputes involving the use of trade marks as domain names are unlikely to diminish, at least so long as Internet addressing remains tied to the DNS.

One of the great challenges of writing in any area relating to the Internet is the daunting pace of change. This book has been several years in the making, during which time large

[1] See <http://www.wipo.int/amc/en/domains/statistics>.

[2] WIPO, 'Cybersquatting Remains on the Rise with further Risk to Trademarks from New Registration Practices', Media Release, 12 March 2007, at <http://www.wipo.int/pressroom/en/articles/2007/article_0014.html>.

[3] National Arbitration Forum (NAF), (2007) 8 *Domain News* (31 August 2007).

[4] See, eg, ICANN GNSO, *GNSO Issues Report on Domain Tasting* (14 June 2007).

sections have required major rewrites. As a respected colleague said when I described my work on the project, 'It must be like trying to catch a river'. The relentless stream of UDRP decisions has meant that I have had to establish a cut-off point, after which I have not referred to any new panel decisions. I have therefore quite arbitrarily set this cut-off point at the end of February 2007.

The rapid development of information technologies and, in particular, Internet-related technologies, has been associated with a 'cult of the new', in which there is an intense focus on each new development and issue, with attention shifting quickly to the next new issue that emerges. When the problem of 'cybersquatting' was first recognised in the 1990s, and when the UDRP was implemented in 1999, there was intense interest in the area by academics, legal practitioners, policy-makers and other interested parties. But that interest waned, as new developments—commons-based intellectual creation, web 2.0 or virtual worlds—caught the attention. Nevertheless, in the early stages of a major legal innovation, such as the UDRP, there is a limit to what can be said with any degree of certainty. It is only with experience accrued over time, in applying legal rules to a variety of disputes in a variety of contexts, that a reasoned assessment of the rules becomes possible.

When I commenced work on this project, I envisaged (and hoped) that it would have taken much less time than it has. In retrospect, the passage of time has been a blessing in disguise: the accumulated weight of UDRP decisions has thrown light on issues that were either recognised as important in the early analyses but imperfectly understood, or given rise to new issues that were unforeseen. The chapters of the book that deal with the UDRP are therefore able to draw on a richer body of decisions than earlier analyses of the UDRP, in a way that hopefully adds some coherence to this unique body of legal decisions.

Throughout the sections of the book that deal with the UDRP, I have, above all, been interested in exploring the relationship between trade mark law, which remains essentially a product of national legal systems, and disputes involving the registration of domain names that are identical or confusingly similar to trade marks which, in the form of the UDRP, is effectively an international dispute resolution system. An important part of this analysis has been tracking, and explaining, the use of traditional trade mark legal principles in UDRP decisions. Over and above this, however, there are important questions of the extent to which the uncritical application of principles drawn from traditional trade mark law to disputes involving the use of marks as domain names is necessary or desirable. In some circumstances, such as determining what amounts to a trade mark and, in particular, in determining what constitutes an unregistered (or 'common law') mark, recourse to national trade mark laws is necessary. Given the quite different objectives of trade mark law and domain name dispute resolution, however, there are other areas—such as the comparison of marks with domain names, or the application of principles of freedom of expression to the registration of marks as domain names—where the application of national principles of trade mark law is unhelpful or counter-productive. This book therefore attempts to distinguish the objectives of domain name dispute resolution from the objectives of trade mark law, to specify where the wholesale application of trade mark law is inappropriate, and to suggest the principles that should apply where this is the case. That said, it is likely that the aims I have set for the more strictly analytical sections of the book have not been fully realised, and no doubt further examination of these issues is warranted.

From the scope of the material covered in this book, it should be clear that domain name law is a truly vast area. Indeed, given the sheer scale of the law relating to domain names, which has developed in such a short period of time, I have had to exclude some important topics from the book. First, I have not tried to deal exhaustively with the decisions of courts involving disputes between trade mark owners and domain name holders under national trade mark laws. Instead, I have focused my attention on the UDRP, the system under which the great majority of domain name disputes are resolved. Consequently, the book, in general, deals with the decisions of courts in matters involving disputes between trade marks and domain names only insofar as they are relevant to the UDRP. Secondly, I have not dealt exhaustively with the decisions of courts under the one national law that is expressly directed at resolving disputes between trade mark owners and domain names holders, the US Anti-Cybersquatting Consumer Protection Act (ACPA). As with other decisions of national courts, I have confined myself to explaining only those implications of the decisions of US courts under the ACPA that are most relevant to the UDRP.

I have many people to thank for their assistance in the production of this book. My debts are explained in the Acknowledgements to this book. I, of course, bear sole responsibility for any and all errors of fact, law and analysis that have made their way into the book.

David Lindsay
Melbourne
September 2007

Acknowledgements

I have many people to thank for their assistance in the production of this book.

The initial research for the book was undertaken when I was fortunate enough to be a Research Fellow at the Centre for Media and Communications Law at Melbourne Law School (CMCL). For their support for this project, I owe a debt of gratitude, both financial and otherwise, to Andrew Kenyon, Director of the CMCL, and Andrew Christie, Director of the Intellectual Property Research Institute of Australia (IPRIA), which also supported the research. I, for one, appreciate the difficulties and challenges associated with being a director of an academic research institute and I hope that this book, in part, repays the support of the two research institutes.

While I was at Melbourne Law School, I benefited greatly from the intellectual stimulation, and companionship, of the law faculty members, especially those working in the areas of intellectual property and technology law. In particular, I would like to thank David Brennan, Dan Hunter, Janice Luck, Megan Richardson and Kimberlee Weatherall (now at the University of Queensland), who have all been valuable colleagues, and contributed to my understanding of intellectual property issues. The assistance of Sam Ricketson, whose advice is always valuable, requires special acknowledgement.

In 2005, I accepted a position with the faculty of law at Monash University. Monash Law School has provided a most supportive and stimulating environment, which has allowed me to complete the greater part of the research and writing for this book. I could not hope for a more talented and collegiate group of academic experts in the areas of intellectual property and technology law than my colleagues at Monash. I would like, in particular, to acknowledge my debts to Mark Davison, Melissa DeZwart, Ann Monotti, Moira Paterson and Sharon Rodrick for their cooperation, friendship, support and intellectual insights.

In the comparatively short history of domain name law, the contribution of Australians, I believe, merits some general acknowledgement. At the international level, Francis Gurry, the Deputy Director General of the World Intellectual Property Organization (WIPO) was instrumental in the introduction of the UDRP, and in overseeing WIPO's important subsequent work in dispute resolution and other domain name issues. Paul Twomey, an Australian, has been President and CEO of ICANN since March 2003, successfully navigating the organisation through some difficult issues. Bruce Tonkin of Melbourne IT has been especially active in the important work undertaken by ICANN's GNSO. In researching this book, I have also been particularly struck by the care and attention taken by Australian panellists in writing UDRP decisions. Although it is unfair to single out the decisions of particular panellists, I have benefited from reading the decisions of Australian panellists such as Philip Argy, Neil Brown QC, Andrew Christie, Dan Hunter, Alan Limbury, Sam Ricketson and Des Ryan. Of course, the diligence and care of the great majority of UDRP panellists must be commended.

I would like to thank Richard Hart of Hart Publishing for his patience in waiting for the final manuscript, and for his encouragement. I would also like to acknowledge the assistance of the editorial team at Hart, especially Mel Hamill, who has been extraordinarily diligent and patient, and Kate Elliott, who made many helpful suggestions.

I am grateful to ICANN for permission to reproduce the material that appears in the Appendicies to the book, including the diagram of ICANN's structure, and the text of the UDRP and the UDRP Rules.

Last, but far from least, I am grateful to my partner, Robyn Parker, who lived through the writing of this book, and who has been a continual source of support.

Contents

List of Abbreviations

ASO	Address Supporting Organisation
ACPA	Anti-cybersquatting Consumer Protection Act
ARPA	Advanced Research Project Agency
ADNRC	Asian Domain Name Resolution Centre
ALAC	At-Large Advisory Committee
BIND	Berkeley Internet Name Domain
BWG	Boston Working Group
CERN	European Centre for Nuclear Research
ccNSO	country-code Names Supporting Organisation
ccTLDs	country-code Top-Level Domains
DARPA	Defense Advanced Research Project Agency
DDN-NIC	Defense Data Network-Network Information Center
DHCP	Dynamic Host Configuration Protocol
DISA	Defense Information Systems Agency
DOC	Department of Commerce (US)
DeC	Disputes.org/eResolution Consortium
DNS	Domain Name System
DNSO	Domain Name Supporting Organisation
DNSSEC	DNS Security Extensions
ERDRP	Eligibility Requirements Dispute Resolution Policy
ERP	Eligibility Reconsideration Policy
ETSI	European Telecommunication Standards Institute
FNC	Federal Networking Council
GAC	Government Advisory Committee
GNSO	Generic Name Supporting Organisation
gTLDs	generic Top-Level Domains
HKIAC	Hong Kong International Arbitration Centre
IAB	Internet Activities Board/Internet Architecture Board
IAHC	International Ad Hoc Committee
IANA	Internet Assigned Numbers Authority
ICANN	Internet Corporation for Assigned Names and Numbers
ICCB	Internet Configuration Control Board
IETF	Internet Engineering Task Force
IGF	Internet Governance Forum
IFWP	International Forum on the White Paper
IGOs	International Intergovernmental Organisations
INNs	International Non-proprietary Names
INTA	International Trademark Association
INWG	International Network Working Group
IP	Internet Protocol
IPDRCP	Intellectual Property Defensive Registration Challenge Policy
IRP	Independent Review Panel

IRTF	Internet Research Task Force
ISI	Information Sciences Institute
ISO	International Standardization Organization
ISOC	Internet Society
ITAG	IANA Transition Advisors Group
ITU	International Telecommunication Union
KIDRC	Korean Internet Address Dispute Resolution Committee
NAF	National Arbitration Forum
NCP	Network Control Program
NCSA	National Center for Supercomputing Applications
NIC	Network Information Center
NLAs	Next-Level Aggregators
NRO	Number Resource Organisation
NSF	National Science Foundation
NSI	Network Solutions, Inc
NTEPPTF	New TLD Evaluation Planning Task Force
NTIA	National Telecommunication and Information Administration
NWG	Network Working Group
ORSC	Open Root Server Confederation
PDPs	Policy-Development Processes
PIR	Public Interest Registry
PSO	Protocol Supporting Organisation
QCP	Qualification Challenge Policy
RALOs	Regional At-Large Organisations
RDRP	Restrictions Dispute Resolution Policy
RFCs	Requests for Comments
RFP	Request for Proposals
SAC	Security and Stability Advisory Committee
SAIC	Science Applications International Corporation
SLAs	Site-Level Aggregators
SMTP	Simple Mail Transfer Protocol
SRI	Stanford Research Institute
sTLDs	sponsored Top-Level Domains
STOP	Start-up Trade Mark Opposition Policy
TCP	Transmission Control Protocol
TCP/IP	Transmission Control Protocol/Internet Protocol
TDRP	Transfer Dispute Resolution Policy
TLAs	Top-Level Aggregators
TLG	Technical Liaison Group
TRIPS	Agreement on Trade-Related Aspects of Intellectual Property Rights
UDRP	Uniform Domain Name Dispute Resolution Policy
URLs	Uniform Resource Locators
uTLDs	unsponsored Top-Level Domains
W3C	World Wide Web Consortium
WGIG	Working Group on Internet Governance
WHO	World Health Organization
WIPO	World Intellectual Property Organisation
WSIS	World Summit on the Information Society

Table of Cases and Domain Name Decisions

Australia

European Court of Justice

New Zealand

United Kingdom

United States of America

UDRP Panel Decisions

STOP (Start-up Trade Mark Opposition Policy) Panel Decisions

Table of Legislation, Agreements and Policies

LEGISLATION

Norway

Spain

Turkey

United Kingdom

United States of America

International

World Summit on the Information Society

DOMAIN NAME CONSTITUTIONAL DOCUMENTS, AGREEMENTS AND POLICIES

ICANN's Constitution

ICANN Internet Coordination Policies (ICPs)

Major ICANN Agreements

ICANN gTLD Agreements

Other Domain Name Dispute Resolution Policies

UDRP Supplementary Rules

1

The Domain Name System

The Internet

[1.1] The Internet

The Internet is a global network of interconnected packet-switched networks that uses the TCP/IP (Transmission Control Protocol/Internet Protocol) suite of protocols.[1] The Internet (with an upper-case 'I') refers to the network of interconnected networks resulting from research on computer networking, which commenced in the 1960s and was funded by the United States Advanced Research Project Agency (ARPA).[2] The research was first implemented in an experimental network, which became known as ARPANET. The original ARPANET, initially linking just four nodes, was established in late 1969. The first public demonstration of ARPANET took place at the First International Conference on Computer Communications held in Washington DC in October 1972.

ARPANET was based on the use of an innovative concept in computer communications, known as packet-switching. Packet-switching involves dividing messages into small digital message units, known as 'packets'. The packets are then transmitted, over potentially diverse routes, across multiply redundant links to their destination. Packet-switching differs from the form of communications traditionally used in voice telephony, known as circuit-switching, in that it does not require a continuous physical link between the source of a message and its destination.

Another important element of the ARPANET research involved the development of network protocols (or agreements). Protocols establish the rules governing communications between computers. The first protocols were developed by a group of graduate students,

[1] In Oct 1995 the Federal Networking Council (FNC), the body responsible for coordinating communications networks among US federal agencies, unanimously passed the following resolution defining the 'Internet':

> 'The Federal Networking Council (FNC) agrees that the following language reflects our definition of the term "Internet":
>
> "Internet" refers to the global information system that—
> (i) is logically linked together by a globally unique address space based on the Internet Protocol (IP) or its subsequent extensions/follow-ons;
> (ii) is able to support communications using the Transmission Control Protocol/Internet Protocol (TCP/IP) suite or its subsequent extensions/follow-ons, and/or other IP-compatible protocols; and
> (iii) provides, uses or makes accessible, either publicly or privately, high level services layered on the communication and related infrastructure described herein.'

Federal Networking Council (FNC), FNC Resolution: Definition of 'Internet' (24 Oct 1995), available at www.nitrd.gov/fnc/Internet_res.html.

[2] On the history of the Internet see K Hafner and M Lyon, *Where Wizards Stay Up Late* (New York, Simon & Schuster, 1996); J Abbate, *Inventing the Internet* (Cambridge, Mass, The MIT Press, 1999); J Naughton, *A Brief History of the Future* (London, Phoenix, 1999).

mostly from UCLA, known as the Network Working Group (NWG). The NWG operated according to an informal, consensus-based decision making process. In 1970 the NWG developed the first protocol for setting up connections between ARPANET host computers, known as the Network Control Program (NCP).

The NCP protocol reflected the extent to which the original ARPANET was a relatively homogeneous and reliable network. In particular, the host computers connected to the ARPANET were not concerned with ensuring the safe transmission of packets, but simply assumed that the packets would be reliably delivered. From 1973, ARPA researchers began investigating problems associated with sending communications across incompatible networks that were much less reliable than ARPANET. This research became known as the 'internetworking project'. The problems of 'internetworking' were addressed by a group of international researchers coordinated by the International Working Group (INWG). The group agreed on a single host protocol, known as the Transmission Control Protocol (TCP), which was responsible for establishing connections between host computers, verifying the arrival of packets, error correction and data flow management.

The TCP protocol was based on two important innovations. First, the problem of incompatible networks was dealt with by computers known as gateways (later known as routers), which were responsible for passing packets between networks. Secondly, the packets were encapsulated in electronic envelopes (or datagrams), which wrapped the messages in common headers. This meant that the routers needed to be concerned only with the envelope and not with the contents of the message.

[1.2] TCP/IP

The initial TCP protocol required the routers to be involved with reliability functions, including error correction and data flow control. This placed considerable demands on the routers. In 1978, an important decision was taken to simplify the operations involved in internetworking by splitting the TCP protocol into two: a host-to-host protocol known as TCP, and an internetworking protocol, known as the internet protocol (IP). The revised TCP protocol became responsible for dividing messages into packets, putting the messages into envelopes (attaching headers), reassembling the packets at the destination, error correction and retransmitting packets in the event of errors in transmission. The IP protocol became responsible for passing packets across networks, essentially performing addressing and routing functions. Splitting the protocol in this way meant that routers only needed to run the IP protocol, and were no longer required to handle error detection and correction. In other words, the host computers at the ends of the network became responsible for the most complex networking functions. The main Internet protocols, which continue to evolve, became known as the TCP/IP suite of protocols. In January 1983, all host computers connected to ARPANET were required to discontinue use of the NCP protocol and adopt the TCP/IP protocol.

With the implementation of the TCP/IP protocol, the term 'Internet' became the standard term for referring to the network of networks that resulted from ARPA's internetworking research. The term 'internet' (with a lower case 'i'), on the other hand, refers to any network of interconnected networks that use common networking protocols.[3]

[3] As noted in RFC 1206: 'You will often see "internet" with a small "i". This could refer to any network built based on TCP/IP, or might refer to networks using other protocol families that are composites built of smaller networks': GS Malkin and AN Marine, 'FYI on Questions and Answers: Answers to Commonly asked "New Internet User" Questions' RFC 1206 (Feb 1991) para [3].

[1.3] Internet Standards

The NWG pioneered an informal, practically-oriented, consensus-based process for determining Internet technical standards. The tone of the process for determining standards is captured by the name given to the documents that constitute Internet standards, which are known as Requests for Comments (RFCs). RFCs comprise proposed standards, standards, and other statements and comments.[4]

The institutional arrangements for determining Internet standards have evolved over time. The NWG was disbanded when ARPANET became operational in the early 1970s. The role of standards development was then taken over by the Internet Program, which established an advisory group, known as the Internet Configuration Control Board (ICCB). In 1983, standards development became the responsibility of the Internet Activities Board (IAB), which consisted of a 10-member board, each member being responsible for a task force. Membership of IAB task force working groups was completely open.

In the 1980s, the United States' National Science Foundation (NSF) invested substantial amounts in building a national computer network, designed mainly to link its supercomputer centres. This programme resulted in an internet, known as NSFNET, which eventually adopted the TCP/IP protocols. In the late 1980s, ARPANET was gradually dismantled and NSFNET took over as the Internet 'backbone' network. Associated with the growth of the Internet, technical problems became more complex and there was an increased interest in standard making. In 1989, one of the IAB task forces was renamed the Internet Engineering Task Force (IETF), which assumed responsibility for protocol development and immediate technical problems. Another task force, known as the Internet Research Task Force (IRTF) assumed responsibility for long term technical planning.

In the mid 1990s NSFNET was privatised, with NSFNET being decommissioned in 1995 and Internet access services being henceforth provided by commercial Internet Service Providers (ISPs). The growing commercialisation and internationalisation of the Internet were associated with a perceived need for greater separation of technical standard-making from the United States government. In 1992, the Internet Society, a private, non-profit-making body was formed as an umbrella organisation for the IAB (which became known as the Internet Architecture Board) and the IETF. While the informal standard-making process continued, however, the formal organisational arrangements remained imprecise.

Internet Addressing and the DNS

[1.4] Internet Naming and Addressing

Like any communications system required to transmit information from one point to another point, including traditional postal and telephone systems, the Internet relies on an

[4] RFCs may be defined as 'the body of literature comprising Internet protocols, standards, research questions, hot topics, humor (especially those dated 1 April), and general information': G Kessler and S Shepard, 'A Primer on Internet and TCP/IP Tools and Utilities' RFC 2151 (June 1997) para [9.1]. The authoritative collection of RFCs is available at www.ietf.org/rfc.html.

addressing system.[5] The Internet naming and addressing system evolved with the institutional and technical evolution of the Internet.

A basic distinction is drawn between names and addresses.[6] A name is simply an identifier that identifies an entity, such as a person or computer. An address, on the other hand, is an identifier that also reveals information about either the physical or logical location of the entity. Host computers attached to the Internet have both names and addresses.

An early decision was taken to assign a human-readable name to each host computer connected to the ARPANET, in addition to a computer-readable numerical address.[7] There are two sound reasons for the decision to name computers.[8] First, names are easier for humans to remember and use than long strings of numbers. In other words, the names assigned to computers act as mnemonics. Secondly, names are more stable than numerical addresses. In a dynamic system such as the Internet numerical addresses may be subject to reassignment if, for example, there is a need to expand the numerical address space, or a change in the organisation responsible for a block of numerical addresses. Consequently, the use of domain names allows for the portability of Internet addresses.

Early in the development of ARPANET, the Stanford Research Institute (SRI) was given a contract to maintain a directory resource known as the Network Information Center (NIC). Amongst its activities, the NIC became responsible for maintaining the RFCs. Moreover, all applications for names were sent to the NIC. The NIC vetted the names to avoid duplication before adding the new name to a file called <hosts.txt>. Each host computer was required to keep a copy of the <hosts.txt> file. Whenever a new host was connected to the ARPANET, the <hosts.txt> file needed updating, and copies of the new file were required to be sent to all hosts. With the transformation of ARPANET into the Internet, it was clear that the original address system was unsustainable. First, the <hosts.txt> file grew to unwieldy proportions. Secondly, distributing the updated files to the host computers threatened to impose an unmanageable burden on the interconnected networks. In other words, the original address system did not scale.

The current Domain Name System (DNS) was designed to overcome the problems associated with the original ARPANET address and naming system.

[1.5] IP Addresses

Computers are machines for processing information in binary form. For messages to be transmitted across computer networks, information regarding the logical location of the destination computer is required. If error correction is dealt with by re-transmission of packets, as is the case with the TCP protocol, information regarding the logical location of the source computer is also required. For information to be successfully sent across a computer network from source to destination, each address must be unique. Functions associated with Internet addressing are dealt with by the IP protocol.

[5] For a comprehensive technical, institutional and political analysis of Internet addressing and naming see M Mueller, *Ruling the Root* (Cambridge, Mass., The MIT Press, 2002). For a practical, technical overview of Internet addressing and naming see P Albitz and C Liu, *DNS and BIND*, 4th edn (Sebastopol, CA, O'Reilly & Associates, 2001).

[6] See U Black, *Internet Architecture: An Introduction to IP Protocols* (Upper Saddle River, NJ, Prentice Hall PTR, 2000) 19.

[7] The concept of allocating human-readable names to host computers appears to have been first formally suggested by Peggy Karp in 1971: P Karp, 'Standardization of Host Mnemonics 1' RFC 226 (20 Sept 1971).

[8] See Mueller, above n 5, 39–40.

Network addressing entails two distinct functions.[9] First, an address space must be defined and, secondly, addresses must be assigned to network entities. The current IP address space, known as IPv4, was originally specified by RFC 790 in 1981. The IPv4 address space consists of 32-bit numbers, giving a theoretical maximum of 4.3 billion unique numbers. The IP address has two main parts: the network prefix (or network address), which identifies the network a computer is attached to; and the host ID (or host address), which identifies the logical location of the host computer on the network. IP addresses are conventionally represented in dotted decimal form divided into four binary octets. For example, the IP address for the Internet Corporation for Assigned Names and Numbers (ICANN) may be represented as 192.0.34.65.

As explained at [1.2] above, the TCP protocol divides messages into packets. When each packet is created, a header is attached to the packet. The header includes the IP addresses of both the source and destination computers. The division of an IP address into two parts aids the efficient transmission of messages across the Internet. Routers, which are responsible for directing packets to their destinations, contain files of network information, called routing tables. When a packet reaches a router, the router compares the destination address with the information in the routing table. As each router is responsible only for directing the packet to the next router, routing tables need only include network prefixes, not complete IP addresses.

Blocks of IP addresses are assigned by Internet address registries. Address registries assign blocks of addresses based on network prefixes, with assignees assuming responsibility for assigning host addresses. Originally, IPv4 addresses were defined in accordance with rigid classes based upon the number of bits allocated to the network prefix. The main classes used the following formats:

- Class A addresses, used for networks with large numbers of host computers, assigned 7 bits to the network prefix and 24 bits to the host ID;
- Class B addresses, used for intermediate size networks, assigned 14 bits to the network prefix and 16 bits to the host ID; and
- Class C addresses, used for networks with relatively few host computers, assigned 21 bits to the network prefix.

When the IPv4 address space was originally specified, it was impossible to anticipate the growth of the Internet. Consequently, from the early 1990s it became apparent that the IPv4 address space would not be able to meet the demand for new addresses. Some of the problems derived from the rigid class-based allocation system. For example, class A addresses were far from fully utilised, whereas class B addresses were running out. Moreover, the growth in routing tables was becoming unsustainable. Steps were therefore taken to deal with the impending shortage of IP addresses.

Within the IPv4 address space, two main methods were introduced to conserve the limited address space. First, policies for assigning address blocks became more restrictive. Thus, assignees of address blocks were required to indicate their requirements and fees were introduced for address blocks. Secondly, new protocols were introduced to deal with inefficiencies in the assignment of address blocks, especially the rigid class-based allocation system. In particular, a new, more flexible, system of address allocation, known as Classless Inter-Domain Routing (CIDR—pronounced 'cider'), was introduced. CIDR replaced the

[9] *Ibid*, 15–19.

rigid class-based designations by allowing the network prefix to be of a flexible length, from 13 to 27 bits. The introduction of CIDR was associated with hierarchical address (or route) aggregation. With address aggregation, networks are grouped together logically, sharing parts of the IP network prefix. Hierarchical aggregation of logically grouped addresses means that routers need only refer to the common element of the network prefix, thereby reducing the size of routing tables.

Two further mechanisms introduced to conserve the IP address are dynamic address allocation and subnet addressing. Dynamic allocation provides for the assignment of addresses for limited periods and automatic reuse of addresses that are no longer needed. The framework for dynamic addressing is specified by the Dynamic Host Configuration Protocol (DHCP). Subnet addressing allows multiple networks to be identified by one address. This means that fewer addresses need to be included in routing tables.

Although the above techniques for conserving the IPv4 address space were regarded as temporary measures, their success has apparently inhibited more fundamental reforms to the address space, especially the adoption of a new address space, known as IPv6.

[1.6] IPv6

The obvious deficiencies with IPv4 led to steps to define a new address space. In 1992 the IETF established a working group to develop a new address space, known as IPv6. The original protocols for IPv6 were specified in 1995.

IPv6 is intended to extend the IP address space and make it more efficient. In place of the 32-bit IPv4 address space, IPv6 provides for a 128-bit address space. The large address space was based on estimates of population growth to 2020.[10] IPv6 addresses are represented as eight 16-bit integers separated by colons, for example:

0123:4567:89ab:cdef:0123:4567:89ab:cdef

Like IPv4 addresses, IPv6 addresses are hierarchical. While IPv4 addresses essentially have two elements, however, an IPv6 address consists of a prefix, which determines the format of the address, and five hierarchical subfields. The subfields consist of top-level aggregators (TLAs—which connect directly to the Internet backbone), next-level aggregators (NLAs), site-level aggregators (SLAs) and the Interface ID (the host address).

The IETF delegated parts of the IPv6 address space to address registries in 1999. During the transition from IPv4 to IPv6, the two address systems will co-exist. The IETF has published RFCs that specify techniques for the transition to IPv6.

[1.7] The Domain Name System (DNS)

In addition to a computer-readable IP address, each host computer connected to the Internet is assigned a human-readable name, known as a domain name. Attaching both a name and an address to a single computer creates a need for a system to correlate, or 'map', human-friendly domain names to computer-friendly IP addresses. The main purpose of the Domain Name System (DNS) is to map domain names to IP addresses.

[10] Black, above n 6, 174.

The original ARPANET address system, described at [1.4], was based on a flat name space. This means that each host computer connected to ARPANET was identified by a distinct name in the <hosts.txt> file. As explained, however, the <hosts.txt> system did not scale. There were two main problems. First, as each host computer requires a unique name, the more hosts connected to the network, the more difficult it was to assign unique names. Secondly, the registration of names, and the maintenance and distribution of the <hosts.txt> file, was highly centralised. As the network grew, especially with the transition from the ARPANET to the Internet, the resources expended on the <hosts.txt> system threatened to grow to unmanageable proportions. The DNS was designed to overcome these problems. The solution to the problems involved dividing the flat name space into a number of smaller domains. This meant establishing a hierarchical name space, which enabled a decentralisation of Internet naming and address functions.

From the late 1970s the concepts underlying the DNS were discussed by researchers associated with the internetworking project.[11] Researchers based at the Information Sciences Institute (ISI), an affiliate of the University of Southern California, were particularly active in relation to naming and addressing issues. The ISI was established in 1972 as a non-profit-making research centre with the objective of obtaining funding for computer network research from ARPA. Researchers based at the ISI included Steve Crocker, Jon Postel and Paul Mockapetris. Crocker and Postel had been members of the NWG. Crocker was responsible for coining the term 'Request for Comments' (RFCs) to refer to documents relating to Internet, and had written the first RFC. Postel became responsible for editing the RFCs and for the administration of unique number assignments for ports and protocols.[12] When Postel moved from UCLA to ISI, he took these functions with him. In 1982 Zaw-Sing Su and Jon Postel published some of the basic concepts underlying the DNS as RFC 819.[13] The conceptual framework of the DNS and technical details relating to the implementation of the DNS were completely specified by Mockapetris in RFCs 882 and 883, which were released in 1983.[14]

The DNS is a distributed, hierarchical database of Internet name, address and associated information. It is a separate Internet standard. The original specifications for the DNS, published as RFCs 882 and 883, have been superseded by the current standards, RFCs 1034 and 1035.[15] Mockapetris released RFCs 1034 and 1035 in 1987. The specifications in RFCs 1034 and 1035 have been supplemented and expanded by a considerable number of subsequent RFCs.

As specified in RFC 1034, the DNS has three major components:

- a hierarchical domain name space (see [1.8]ff) and associated resource records (see [1.18]);
- name servers (see [1.16]–[1.17]); and
- name resolvers (see [1.18]).

[11] See Mueller, above n 5, 77–8. The original hierarchical concept of the DNS was conceived by David Mills from COMSAT Laboratories in RFC 799, issued in Sept 1981: see D Mills, 'Internet Name Domains' RFC 799 (Sept 1981).

[12] See, eg, J Postel, 'Proposed Standard Socket Numbers', RFC 349 (30 May 1972).

[13] Zaw-Sing Su and J Postel, 'The Domain Naming Convention for Internet User Applications', RFC 819 (Aug 1982).

[14] P Mockapetris, 'Domain Names—Concepts and Facilities', RFC 882 (Nov 1983); P Mockapetris, 'Domain Names—Implementation and Specification', RFC 883 (Nov 1983).

[15] P Mockapetris, 'Domain Names—Concepts and Facilities', RFC 1034 (Nov 1987); P Mockapetris, 'Domain Names—Implementation and Specification', RFC 1035 (Nov 1987).

[1.8] The Domain Name Space

The Domain Name Space (DNS) is organised hierarchically around a root and tree structure. At the top of the DNS is a single root, which is unnamed. The DNS root is conventionally written as a single dot (.). Connections between each level in the DNS hierarchy are also conventionally represented by a single dot (.). The DNS tree has a maximum limit of 127 levels.

Below the DNS root, the DNS tree is hierarchically organised in subtrees or branches. A domain is a subtree of the DNS. The branches connect nodes, which are the points at which the branches intersect. Each node has a name (or label) that can be up to 63 octets (or characters) in length. The name for the DNS root is reserved and given a null, or zero length, label. The hierarchical nodes are also referred to as parents and children, the parent being the node closer to the root than the child. The DNS specifies that nodes that are children of the same parent must have different names.

The domain name of a node is simply the list of names on the path from the node to the DNS root.[16] Conventionally, domain names are read from left to right, with the most specific (the farthest from the root) on the left, and the most general (the closest to the root) on the right. The name of each node on the path is conventionally separated by a dot (.). The maximum length of a domain name is limited to 255 octets (or characters). A domain is defined as that part of the domain name space that is at or below the domain name that identifies the domain.[17]

It is customary to divide the DNS into the following hierarchical levels:

- *Top-Level Domains,* or TLDs. A TLD is a child of the root;
- *Second-Level Domains,* or 2LDs. A 2LD is a child of a TLD;
- *Third-Level Domains,* or 3LDs. A 3LD is a child of a 2LD;
- *Nth-Level Domain.* Meaning the name of the host computer.

The hierarchical levels of the DNS therefore continue until the name of the host computer is reached. The domain name is then read from the name of the host computer on the left to the TLD on the right, for example:

<law.monash.edu.au> or
<law.harvard.edu>.

The main purpose of the DNS was to decentralise the administration of Internet naming and address functions. This was made clear by Jon Postel and Joyce Reynolds in RFC 920, which states:

> Domains are administrative entities. The purpose and expected use of domains is to divide the name management required of a central administration and assign it to sub-administrations.[18]

The hierarchical arrangement of domain names reflected this objective. Given the hierarchical arrangement, responsibility for assigning and managing names at different levels in the hierarchy could be delegated to different entities.

[16] P Mockapetris, 'Domain Names—Concepts and Facilities', RFC 1034 (Nov 1987) para [3.1].
[17] *Ibid.*
[18] J Postel and J Reynolds, 'Domain Requirements', RFC 920 (Oct 1984).

Top-level Domains (TLDs)

[1.9] Top-level Domains (TLDs)

Top-level Domains (TLDs) are necessarily included in all domain names. Unsurprisingly, the specification of TLDs has been one of the most controversial issues associated with the DNS.

The first TLD was .arpa. The .arpa TLD was first used to assist in the transition from the ARPANET <hosts.txt> system to the DNS. As specified in RFC 881, which was issued in November 1983, each name in the <hosts.txt> file was allocated a .arpa suffix.

Apart from .arpa, the initial TLDs were specified by Postel and Reynolds in RFC 920, which was issued in October 1984.[19] RFC 920 defined two distinct kinds of TLDs, which later became known as generic TLDs (gTLDs) and country code TLDs (ccTLDs).

[1.10] Original Generic Top-level Domains (gTLDs)

Following some controversy,[20] RFC 920 defined five gTLDs (other than .arpa)—.com. .edu, .gov, .mil and .org—and gave an indication of the uses to which it was envisaged they would be put. Guidelines on the establishment and administration of domain names were established by RFC 1032, issued in November 1987.[21] RFC 1032 stated that a .net gTLD had been introduced for 'various network-type organizations', including information centres and operations centres. A further gTLD, .int, was created for international treaty organisations in 1988. RFC 1591, issued by Postel in March 1994, listed the seven initial gTLDS (other than .arpa) and their intended uses. As the intended uses of the gTLDs had changed over time, RFC 1591 clarified the uses. The original gTLDs and intended uses are set out in Table 1.1.

Table 1.1 Original gTLDs

gTLD	Use
.com	intended for commercial entities, that is companies
.edu	originally intended for all educational institutions, but subsequently limited to US educational institutions that are 4 year colleges or universities
.gov	originally intended for any kind of government office or agency, but subsequently limited to US federal government agencies
.int	limited to organisations established by international treaties, or international databases
.mil	limited to the US military
.net	intended to hold only the computers of network providers, meaning organisations providing network infrastructure. Restrictions on registration were removed in 1996.
.org	originally intended as a miscellaneous TLD for organisations that did not fit anywhere else, although often identified as being for non-commercial organisations. Restrictions on registration were removed in 1996.

[19] *Ibid.*
[20] Mueller, above n 5, 78–81.
[21] M Stahl, 'Domain Administrators Guide', RFC 1032 (Nov 1987).

The original gTLDs are divided into open and closed gTLDs. The .com, .net and .org gTLDs are known as open gTLDs, as there are no restrictions on eligibility for registration. Although restrictions were originally imposed on the .net and .org gTLDs, all such restrictions were removed from 1996. The .edu, .gov and .mil gTLDs are known as closed gTLDs, as registrations are limited to US educational institutions, US federal government organisations and US military organisations, respectively.

To a large extent, the choice of gTLDs is arbitrary. Early controversies relating to the choice of gTLDs concerned the meanings, if any, to be given to gTLDs. In order to discuss DNS policy issues, in 1983 Jon Postel began a mailing list known as the *namedroppers* list. Before the release of RFC 920, the issue of how to determine which gTLD an organisation should be registered under was canvassed on the *namedroppers* mailing list. It was, for example, suggested that names could be used as a form of directory assistance to aid in locating host computers or entities associated with host computers. In an important posting to the mailing list, Postel was careful to define the functions of the DNS in extremely narrow terms. In particular, Postel maintained that the DNS was a system for naming computers, primarily in order to decentralise naming and address administration, and was not a form of directory assistance. Thus, Postel stated:

> One concept that seems to have been widely misunderstood is that this is a naming system, not a general directory assistance system.
>
> This is supposed to be a system for finding specific types of information about exactly named things ... This is not intended to be a system for finding partially described things ...
>
> The whole point of domains is to subdivide the name assignment problem. To try to preserve some higher level uniqueness would require the very central coordination we are trying to eliminate! It should be clear that there is no hope of such uniqueness in the long run, so let's not make any plans based on such false hope.[22]

Postel therefore simply proclaimed that the problem of logically grouping entities according to some meaningful system was something that either was not, or could not be, addressed by the DNS. In practice, of course, people did attribute meanings to the gTLDs. The natural tendency for people to attribute meanings to domain names, instead of regarding them as simply a convention for naming host computers attached to the Internet, has been the source of many of the difficulties subsequently encountered with the DNS, including the difficulties experienced by trade mark owners. These underlying issues emerged again in debates associated with the approval of new gTLDs in the late 1990s, discussed below at [1.12]–[1.13].

[1.11] Country Code Top-level Domains (ccTLDs)

Given the extent of international collaboration on the internetworking project, interest in the DNS was not confined to computer scientists based in the United States. Among the international community there was a view that TLDs should be based on geographical or national boundaries.[23] The first TLD associated with a country, the .uk TLD, was designated before the release of RFC 920.

[22] J Postel, 'Comments on Domain Requirements', original available at namedroppers@SRI-NIC, 20 May 1984, archived at http://bebas.vlsm.org/v01/internet/ietf/00/0131.txt.

[23] Mueller, above n 5, 79.

In order to avoid predictable debates about the proper designation of countries, Postel subsequently decided to use an established list of abbreviations for countries. The International Standardization Organization (ISO) had determined a standard known as 'Codes for the Representation of Names of Countries' (ISO-3166), which assigned two-letter alphabetic codes to countries and other political entities for the purposes of postal, transport and communication authorities. RFC 920 incorporated the ISO-3166 list of two-letter country codes as TLDs, simply stating that country-based TLDs would consist of:

> The English two letter code (alpha-2) identifying a country according to the ISO Standard for "Codes for the Representation of Names of Countries".[24]

There is, nevertheless, one exception to the use of ISO-3166 codes for country-based TLDs. The ISO-3166 code for Great Britain is 'gb' and not 'uk'. As the .uk TLD had already been designated, this was retained as the TLD for Great Britain.

The TLDs representing countries, based on ISO-3166, are known as country code TLDs, abbreviated to ccTLDs.

[1.12] Commercialisation of the Internet

In the early to mid-1990s the Internet was transformed from a predominantly academic research network to a commercial network with widespread appeal. When ARPANET was replaced as the Internet backbone by NSFNET in the late 1980s, the NSF implemented an Acceptable Use Policy that prohibited commercial activities. In the late 1980s and early 1990s, however, commercial networks based on the TCP/IP protocol were established and began to grow. In 1991, the NSF first permitted commercial traffic on NSFNET subject to certain restrictions. In 1993 a decision was taken to privatise the Internet backbone and, when NSFNET was decommissioned in 1995, the Internet was fully commercialised.

At the same time as the Internet was being privatised, an application for locating documents on computer networks by the use of hypertext, known as the World Wide Web, was being developed. In 1990 the first World Wide Web software, including the first web browser, was installed at CERN (the European Centre for Nuclear Research), based in Berne, Switzerland. The software, created by Tim Berners-Lee and Robert Cailliau, was made available by means of the Internet in 1991. In 1993 a graphical browser, known as Mosaic, was developed at the National Center for Supercomputing Applications (NCSA) at the University of Illinois, and released to the public. The user-friendly World Wide Web was instrumental in the Internet becoming a mass medium.

The commercialisation of the Internet and the success of the World Wide Web placed considerable strains on the DNS. In particular, the popularity of the World Wide Web created a significant increase in demand for new domain names. Moreover, the World Wide Web was responsible for a transformation in the function of domain names. The World Wide Web is an application built on the TCP/IP protocol. The World Wide Web, however, has its own address system, called Uniform Resource Locators (URLs). As explained at [1.10] above, domain names were originally intended to function as names for host computers connected to the Internet. A URL, on the other hand, is designed to locate resources on the web, such as documents, images and other files. Just as the World Wide Web is built

[24] J Postel and J Reynolds, 'Domain Requirements 1', RFC 920 (Oct 1984) 1.

on the Internet architecture, URLs incorporate domain names in web addresses. A typical URL uses the following format:

<http://www.icann.org/general/fact-sheet.html>

The first part of the URL (http) indicates that the World Wide Web communications protocol, Hypertext Transfer Protocol is used to access the resource. The second part of the URL (www) indicates that it is a World Wide Web resource. The third part of the URL (<icann.org>) is the domain name, which identifies the host computer on which the resource is stored. The fourth and subsequent parts of the URL (/general/fact-sheet.htm/) identify the directories, sub-directories and file name of the resource. The fourth and subsequent parts of the URL can include an unlimited number of directories and sub-directories.

The use of domain names in URLs meant that domain names became used for locating content on the Internet. Moreover, as the World Wide Web began to be used to promote oganisations and products, domain names became used to locate organisations and products.[25] At the same time, new users began to type domain names into web browsers in order to find web pages. Despite the intentions of those who designed the DNS, domain names became used as a form of directory for locating material stored on the Internet.

[1.13] New Generic Top-level Domains (gTLDs)

As explained at [1.4], the NIC, based at SRI, was originally responsible for registering IP addresses and domain names under contract with the US Defense Department. In 1983, partly because of security concerns, the military part of ARPANET was separated from the rest of ARPANET to form a distinct network, known as MILNET. In 1991, the Defense Department decided to seek tenders for the provision of registration services for the military part of the network. The military contract was awarded to Government Systems, Inc (GSI), which sub-contracted the registry functions to Network Solutions, Inc (NSI). In 1992, the NSF opened the non-military registration and information functions of the NIC to tender. In early 1993, NSI was awarded the contract for providing civilian registration services. NSI, in turn, sub-contracted the registration functions to the University of Southern California's ISI, where Jon Postel and his colleague, Joyce Reynolds, were based. A commercial company, NSI, therefore became responsible for registering IP addresses and domain names just as the commercialisation of the Internet and the impending growth of the World Wide Web placed considerable pressures on the DNS. The number of registrations handled by NSI escalated significantly between 1993 and 1995. Consequently, in September 1995, NSF authorised NSI to charge registration fees for domain name registrations in the .com, .net and .org gTLDs. At the same time, the demand for domain names made it administratively impossible for NSI to maintain restrictions on registration in the .net and .org gTLDs, which became open gTLDs.

The NSF decision to impose charges for domain name registration led influential members of the Internet technical community, including Jon Postel, to favour competition in domain name registration via the introduction of new top-level domains. In May 1996, Postel issued a draft RFC, entitled 'New Registries and the Delegation of International Top-Level Domains', which proposed the immediate introduction of 50 new registries and 150

[25] As Mueller explains, 'Domain names began to refer to content resources rather than just network resources': Mueller, above n 5, 108.

new TLDs.[26] In part, because of debates concerning the legitimacy of any organisation making decisions regarding the DNS, Postel's proposal to introduce large numbers of new TLDs failed to gain acceptance.

As explained at [2.11], the Internet Corporation for Assigned Names and Numbers (ICANN), a non-profit-making corporation, assumed responsibility for the DNS in February 1999, after being officially recognised by the US Department of Commerce. ICANN referred consideration of the expansion of gTLDs to its Domain Name Supporting Organisation (DNSO), which, in July 1999, established a working group, known as 'Working Group C'. In April 2000, following a report from the working group, the DNSO adopted a *Statement on New gTLDs,* which recommended the introduction of a limited number of new TLDs.[27] The ICANN board adopted the DNSO recommendations on 16 July 2000, authorising the ICANN President to call for proposals for new gTLDs.[28] On 14 November 2000, after an application process, in which applicants for establishing and managing new TLDs were required to satisfy guidelines and pay an application fee of US$50,000, the ICANN board approved seven new gTLDs from among 44 proposals.[29] The ICANN board authorised the seven new gTLDs as a 'proof of concept', essentially meaning that a cautious approach was adopted towards the introduction of new gTLDs due to concerns regarding the potential effects of new gTLDs on the technical operation of the DNS, as well as concerns from trade mark owners regarding the need for new gTLDs. Following negotiations, ICANN eventually entered into registry agreements with the selected new registry operators.

The new gTLDs are divided into two main categories: sponsored TLDs (sTLDs or sgTLDs) and unsponsored TLDs (or uTLDs). An sTLD is a specialised TLD that has a sponsor representing the community that is most affected by it. A sponsor is an organisation that has defined delegated policy-formulation responsibilities regarding the way in which the sTLD is operated. The sTLDs operate within an agreed charter that defines who can be a registrant and how policy decisions will be made within the relevant community, which is known as the sponsored TLD community. The sponsor is usually responsible for selecting the registry operator. The new sTLDs—.museum, .aero and .coop—were delegated to non-profit-making organisations that are accountable to a defined community. The remaining four new gTLDs—.biz, .info, .pro and .name—are known as uTLDs. Three of the new uTLDs—.biz, .pro and .name—are known as restricted TLDs, as registration is restricted to organisations that satisfy defined criteria for eligibility. The remaining uTLD, .info, is an unrestricted or open gTLD, meaning that there are no criteria restricting registrations.

The new gTLDs approved by ICANN, the status of the TLD (whether it is sponsored or unsponsored), the registry operators and sponsoring organisations (where applicable), and the eligibility criteria for registration are set out in Table 1.2.

[26] J Postel, 'New Registries and the Delegation of International Top-Level Domains', Internet Draft, available at http://tools.ietf.org/html/draft-postel-iana-itld-admin-00. (May 1996). Mueller points out that this later became known as 'draft-postel'.

[27] 'DNSO Names Council Statement on New gTLDs', 19 Apr 2000, available at http://www.dnso.org/dnso/notes/20000419.NCgtlds-statement.html.

[28] ICANN regular board meeting, 16 July 2000, available at http://www.icann.org/tlds/new-tld-resolutions-16jul00.htm#00.460.

[29] See documentation at http://www.icann.org/yokohama/new-tld-topic.htm.

Table 1.2 New gTLDs

gTLD	Sponsored/ Unsponsored	Registry Operator/ Sponsoring Organisation	Eligibility criteria
.biz	Unsponsored.	Neulevel (a joint venture of Neustar, Inc. and Melbourne IT Ltd).	bona fide business or commercial purposes.
.info	Unsponsored.	Afilias (a consortium of registrars and investors, including DNS Investment GmbH, CORE, Corporate Domains, Domain Bank, Domain Registration Services, Global Media Online (GMO), NS Holding Company, Register.com, Schlund+Partner AG, Sitename.com LLC and Tucows International).	open.
.name	Unsponsored.	Global Name Registry, Ltd (a subsidiary of GNR).	individuals.
.museum	Sponsored.	Sponsoring organisation: Museum Domain Management Association (MuseDoma). Registry operator: Internet Council of Registrars (CORE).	museums recognised by the International Council of Museums (ICOM).
.coop	Sponsored.	Sponsoring organisation: DotCooperation LLC (DCLCC or DotCoop) (National Cooperative Business Association (NCBA is sole member of DCLCC). Registry operator: Poptel.coop (run by Poptel Ltd).	organisations operating according to cooperative principles.
.aero	Sponsored.	Sponsoring organisation: Société Internationale de Télécommunications Aéronautiques (SITA) (wholly owned and operated by the Air Transport Community (ATC)). Registry operator: SITA Information Networking Computing BV (SITA INC).	air transport industry.
.pro	Unsponsored.	RegistryPro (a subsidiary of Hostway Corporation).	Recognised professionals, particularly those in the legal, medical and accountancy fields.

[1.14] New Sponsored TLDs (sTLDs)

When ICANN selected the seven new gTLDs in November 2000, it also convened a New TLD Evaluation Planning Task Force (NTEPPTF). In its final report, issued in July 2002, the task force recommended that ICANN initiate a process for evaluating the new gTLDs and consider planning for additional gTLDs.[30] In August 2002, the ICANN board requested the President of ICANN to prepare a recommendation about how ICANN might initiate a process for introducing a limited number of new TLDs. In November 2002, the ICANN President issued

[30] New TLD Evaluation Planning Task Force (NTEPPTF), *Final Report* (31 July 2002), available at http://www.icann.org/committees/ntepptf/final-report-31jul02.htm.

a document in response to the request, entitled 'A Plan for Action Regarding New gTLDs' (the 'Plan for Action').[31] The Plan for Action recommended that ICANN solicit proposals for new sTLDs as an extension of the cautious, 'proof of concept' underlying the new gTLDs. The Plan for Action further recommended that any action in relation to new uTLDs await a full evaluation of those uTLDs that had been approved by ICANN in November 2000.

At its annual meeting in December 2002, the ICANN board accepted the recommendation to proceed with soliciting proposals for a limited number of new sTLDs. The board directed the preparation of a draft Request for Proposals (RFP) for new sTLDs and, in June 2002, directed that the draft be made available for public comment. An important issue in the public consultations was whether the RFP should be limited to applicants for sTLDs that had been unsuccessful in November 2000. In October 2003, following a recommendation from the Generic Name Supporting Organisation (GNSO),[32] the ICANN board directed the ICANN President to prepare a final RFP, which was not restricted to previous unsuccessful applicants, for a limited number of new sTLDs. The RFP for new sTLDs, which was released on 15 December 2003, included selection criteria in relation to technical standards, business/financial standards and sponsorship requirements.[33] In summary, the approved selection criteria were:

- The technical standards included 'evidence of ability to ensure stable registry operation', 'evidence of ability to ensure that the registry conforms with best practice technical standards for registry operations', 'evidence of a full range of registry services', and 'assurance of continuity of registry operation in the event of business failure of the proposed registry'.
- The business plan had to 'demonstrate the applicant's methodology for introducing a new sTLD and the ability of the organization to implement a robust and appropriately resourced organization'. The financial model had to 'outline the financial, technical and operational capabilities of the organization'.
- The sponsorship information had to include a 'definition of sponsored TLD community', 'evidence of support from the Sponsoring Organization', 'appropriateness of the Supporting Organization and the policy formulation environment', and 'level of support from the Community'. In addition, the criterion of 'community value' had to be demonstrated by the 'addition of new value to the Internet name space', protection for 'the rights of others', 'assurance of charter-compliant registrations and avoidance of abusive registration practices', 'assurance of adequate dispute-resolution mechanisms' and 'provision of ICANN-policy compliant WHOIS service'.[34]

Before the application period closed on 16 March 2004, ICANN received applications for the following nine proposed new sTLDs: .asia, .cat, .jobs, .mail, .mobi, .post, .tel, .travel and .xxx. Two different applicants submitted applications for the proposed .tel sTLD. Following an initial period of public comment, ICANN appointed a project manager, Summit Strategies International LLC, to coordinate evaluation of the new sTLD applications. The applications were submitted to an independent panel of experts, with expertise in the technical, business/financial and sponsorship/community areas. Following the submission of the reports of the expert teams to ICANN in July 2004, ICANN decided to allow each proposal

[31] S Lynn, 'A Plan for Action Regarding New gTLDS' (18 Oct 2002, posted 8 Nov 2002), available at http://www.icann.org/committees/ntepptf/new-gtld-action-plan-18oct02.htm.

[32] The GNSO, which is dealt with at [2.24.2], succeeded the DNSO.

[33] See http://www.icann.org/tlds/new-stld-rfp/.

[34] See ICANN, *Status Report on the sTLD Evaluation Process* (28 Nov 2005; updated 3 Dec 2005), available at http://www.icann.org/tlds/stld-apps-19mar04/stld-status-report.pdf.

to progress on its own timetable. This process provided the opportunity for applicants to furnish ICANN with further information and clarifications in response to the expert reports.

Table 1.3 sets out the status of the new sTLDs, the sponsoring organisations (SO) and registry operators (RO), and the eligibility criteria for registration.

Table 1.3 New sTLDs

sTLD	Status	Sponsoring Organisation (SO)/ Registry Operator (RO)	Eligibility criteria
.asia	6 December 2006: ICANN and DotAsia signed registry agreement. 2 May 2007: .asia added to the root.	SO: DotAsia Organisation Limited ('DotAsia'). RO: DotAsia.	Open.
.cat	23 September 2005: ICANN and puntCAT signed registry agreement. 16 September 2006: .cat added to the root.	SO: Fundació Privada (a Catalonia private foundation) (puntCAT). RO: puntCAT.	Catalan linguistic & cultural community.
.jobs	5 May 2005: ICANN and Employ Media signed registry agreement. 9 September 2005: .jobs added to the root.	SO: The Society for Human Resource Management (SHRM). RO: Employ Media LLC (Employ Media).	Open, but only company names accepted for registration.
.mobi	10 July 2005: ICANN and mTLD Top Level Domain Ltd signed registry agreement. 20 October 2005: .mobi added to the root.	SO: DotMobi, Ltd. RO: mTLD Top Level Domain Ltd.	Open.
.post	31 July 2004: ICANN informed applicant that it satisfied selection criteria and it was ready to commence negotiations.	SO: Universal Postal Union (UPU). RO: UPU.	Universal Postal Union (UPU) members.
.tel (Telnic)	30 May 2006: ICANN and Telnic Limited signed registry agreement. 2 March 2007: .tel added to the root.	SO: Telname Limited. RO: Telnic Limited.	Open.
.travel	5 May 2005: ICANN and Tralliance Corporation signed registry agreement. 21 July 2005: .travel added to the root.	SO: The Travel Partnership Corporation (TPPC). RO: Tralliance Corporation.	People, organisations, associations, and private, governmental and non-governmental agencies in the travel and tourism industry.
.xxx	10 May 2006: majority of ICANN board revoked approval. 7 January 2007 and 16 February 2007: revised proposals posted for public comment. 30 March 2007: majority of ICANN board rejected application.	SO: The International Foundation for Online Responsibility (IFFOR). RO: ICM Registry LLC.	Online adult-entertainment industry.

In relation to two of the applications, .mail and the .tel application submitted by Pulver.com, the ICANN board determined that the applicants had failed to provide adequate or sufficient information for negotiations to be authorised. After considerable controversy, on 30 March 2007 a majority of the ICANN board voted to reject the .xxx application, largely on the basis that it failed to meet the sponsorship criteria of the RFP for the new sTLDs.

[1.15] Policy on Introducing New gTLDs

The process for developing a coherent policy for introducing new gTLDs has been protracted and contentious.

The November 2002 President's 'Plan for Action', referred to at [1.14], recommended a two-phase process for the introduction of new gTLDs: first, that steps should be taken towards approval of a limited number of new sTLDs; and, secondly, that the question of the future evolution of gTLDs be referred to the GNSO. In December 2002, the ICANN board adopted the two-phase process recommended by the 'Plan for Action', requesting the President to prepare a draft Request for Proposals (RFP) for new sTLDs, while directing the GNSO to provide a recommendation on the future of the gTLDs namespace. In May 2003, the GNSO Council responded to this direction by determining that:

- Expansion of the gTLD namespace should be a bottom-up approach with names proposed by the interested parties to ICANN. Expansion should be demand-driven.
- Furthermore, there should be a set of objective criteria to be met in any future expansion.
- The development of this set of objective criteria should be the subject of a new policy-development process (PDP).

The ICANN policy-development processes (PDPs) are dealt with at [2.24.2].

In October 2003, the ICANN board directed the President to begin an 'expeditious and targeted development of strategy and policy leading to a streamlined process for the introduction of new gTLDs'. At the same time, the board directed the President to seek community input into the development of a new gTLD strategy. In September 2004, ICANN staff produced a draft document setting out a 'strategy', in very general terms, for introducing new gTLDs, that was made available for public comment.[35]

The development of a policy for the introduction of new gTLDs occurred in parallel with the evaluation of the new gTLDs introduced in November 2000. The NTEPPTF report, referred to at [1.14], formulated 19 questions to be addressed in evaluating the new gTLDS, covering technical, business, legal and procedural aspects of introducing the new gTLDs. In July 2004, a comprehensive report on the policy and legal questions surrounding the introduction of the new gTLDs was prepared for ICANN.[36]

In late 2005, the GNSO Council determined that work should be completed on the evaluation of the new gTLDs and that a new gTLD policy should be developed. Accordingly, in September 2005, the GNSO Council resolved to request ICANN staff to prepare an issues paper on key policy areas relating to the introduction of new gTLDs as a basis for a policy development process (PDP). In November 2005, the GNSO Council initiated a PDP on the

[35] ICANN, *Strategy: Introduction of New Generic Top-Level Domains* (30 Sept 2004), available at http://www.icann.org/tlds/new-gtld-strategy.pdf. .

[36] Summit Strategies International, *Evaluation of the New gTLDs: Policy and Legal Issues* (10 July 2004), available at http://icann.org/tlds/new-gtld-eval-31aug04.pdf.

introduction of new gTLDs. As part of this process, the GNSO released an *Issues Report* in December 2005 and, following public comment, an *Initial Report* was released in July 2006. In August 2007, the GNSO released its *Final Report*, which included sets of proposed principles, recommendations and implementation guidelines for the introduction of new gTLDS. The proposed principles, recommendations and guidelines are set out in Appendix 4. In expectation of a board decision in favour of introducing new gTLDs, in September 2007 ICANN posted a notice soliciting the development of a request for proposals (RFP) for a process and documentation for introducing new gTLDs.

Operation of DNS

[1.16] Name Servers

The DNS operates according to a client-server model. In a client-server model, one program, known as the client, makes a request for services or resources from another program, known as the server. Although not invariably the case, a client program is commonly located on an end user's computer, while a server is commonly located elsewhere on the network. DNS server programs are known as name servers.

The functions and technical operation of name servers are set out in RFC 1034, which states:

> NAME SERVERS are server programs which hold information about the domain tree's structure and set information. A name server may cache structure or set information about any part of the domain tree, but in general a particular name server has complete information about a subset of the domain space, and pointers to other name servers that can be used to lead to information from any part of the domain name tree. Name servers know the parts of the domain tree for which they have complete information; a name server is said to be an AUTHORITY for these parts of the name space. Authoritative information is organized into units called ZONEs, and these zones can be automatically distributed to the name servers which provide redundant service for the data in the zone.[37]

A name server is therefore a server program that stores information about a part of the hierarchical domain name space. The part of the domain name space about which the name server stores information is known as a zone, which is simply a section of the DNS database. A zone is not the same as a domain. All TLDs, and some second- and lower-level domains, are split into smaller units, which are known as zones.

RFC 1034 introduced the two important concepts of authority and delegation. As specified by RFC 1034, the data that describe a zone have four major parts:

- Authoritative data for all nodes within the zone.
- Data that define the top node of the zone (can be thought of as part of the authoritative data).
- Data that describe delegated subzones, i.e., cut around the bottom of the zone.
- Data that allow access to name servers for subzones (sometimes called 'glue' data).[38]

[37] P Mockapetris, 'Domain Names—Concepts and Facilities', RFC 1034 (Nov 1987), para [2.4].
[38] *Ibid*, para [4.2.1].

A name server has authority in relation to a zone for which it stores authoritative data. A name server may, however, be authoritative in relation to multiple zones. Responsibility for subzones may be delegated to other organisations, meaning that responsibility for the subzone is assigned to different name servers. When responsibility for a subzone is delegated, the original zone does not include the subzone, but name servers that store information for the original zone must include pointers to name servers that store information for the subzone.

The primary function of a name server is to answer queries about information concerning zones in relation to which it is authoritative. Name servers have two modes of answering queries, known as recursive and non-recursive (or iterative). As the main purpose of the DNS is to decentralise administration of Internet address and naming functions, any given name server does not have complete information regarding addresses and names of all hosts connected to the Internet. In other words, a given name server can only be relied upon directly to provide information in relation to a zone for which it is authoritative.

Recursive and non-recursive modes of operation refer to two different ways a name server deals with queries concerning zones in relation to which the server is not authoritative. The mode of operation of the name server depends upon whether the query is recursive or non-recursive. If a query is recursive, the name server must respond by returning either the requested data or an error stating that the requested data, or requested domain name, does not exist. A name server cannot respond to a recursive query by referring the query to another name server. This means that, if the name server is not authoritative in relation to the zone for which the data are sought, it must query other name servers in the hierarchy of name servers in order to retrieve the required information. In querying other name servers, the original name server may make either a recursive or non-recursive query. A recursive query would pass on responsibility for retrieving the requested data to the queried name server.

Under most current implementations of the DNS, name servers send non-recursive queries.[39] The recursive mode of operation is the simplest mode for the client software, but the most complex for the server. If a name server receives a non-recursive (or iterative) query, the server can respond by returning the requested data, returning an error stating that the requested data do not exist, or by referring the client to another name server. In other words, the name server responds solely on the basis of the data stored on the server, and is not responsible for sending its own queries to other name servers. If a name server responds to a non-recursive query with a referral to another name server, the client is responsible for sending the query to the other name server. The non-recursive mode of operation is the simplest for the name server, but the most complex for the client software.

RFC 1034 specifies two types of name server: primary or master name servers and secondary or non-master name servers (often referred to as slaves).[40] The distinction between primary and secondary name servers is based on the relationship of the server to data relating to the zone for which the server is authoritative. A primary name server acquires data for the zone in relation to which it is authoritative from a master file stored on its host. Any changes in data for the zone are made to the master file. Secondary name servers, on the other hand, acquire data from primary servers. Secondary name servers are required because it is important to have redundant copies of zone data and because multiple servers promote more efficient network operations. Secondary servers are required periodically to update zone information from primary servers. When zone data is sent from a primary server to a secondary server the process is known as a zone transfer. A secondary server may

[39] Albitz and Liu, above n 5, 29.
[40] *Ibid*, 24–5.

also acquire zone data from other secondary servers, for example, if the primary server is unavailable or to improve the transfer process. When a secondary server acquires zone data from another secondary server the process is also known as a zone transfer.

[1.17] Root Name Servers and the Root Zone File

The DNS is a hierarchical, distributed database of Internet address and naming information. At the top of the hierarchy is a single, globally unique root, which is conventionally represented by a dot (.). Immediately underneath the root are the TLDs.

The root zone file is the master file that contains a list of the delegations of the TLDs, associated records required by the DNS protocol, and pointers to primary and secondary name servers for each TLD. The root name servers are authoritative in relation to data stored in the root zone file. Operational requirements for root name servers, including security requirements, are specified in RFC 2870, issued in June 2000.[41] A technical comment on the need for a unique DNS root was issued by the Internet Architecture Board as RFC 2826 in May 2000.[42] Data in the root zone file is made available to the root zone servers either in-band by zone transfers as defined by RFC 1034, or out-of-band by means of the FTP (file transfer protocol) protocol.[43] As explained in RFC 2826:

> There is one specific technical respect in which the root zone differs from all other DNS zones: the addresses of the name servers for the root zone come primarily from out-of-band information.[44]

To an extent, the problems in transferring root zone information are dealt with by the requirement, pursuant to RFC 2870, that in-band zone transfers use the DNS Security Extensions (DNSSEC) for zone transfers of root data.[45] As RFC 2826 points out, however, even with DNSSEC an out-of-band configuration problem arises in relation to the cryptographic signature key to the root zone.[46]

At present, there are 13 root name servers. The domain names of the root servers take the form letter.root-servers.net, with the *letter* ranging from A to M. The current details of the 13 root name servers, including the root server operators, locations and IP addresses, are set out in Table 1.4.[47]

The authoritative copy of the root zone file is stored on the 'A' root server, which is operated by VeriSign Global Registry from Dulles, Virginia. The 12 secondary root servers acquire copies of the root zone file from the 'A' root server. The root servers all use some vari-

[41] R Bush, D Karrenberg, M Kosters and R Plzak, 'Root Name Server Operational Requirements', RFC 2870 (June 2000).

[42] Internet Architecture Board, 'IAB Technical Comment on the Unique DNS Root', RFC 2826 (May 2000). This RFC is for informational purposes only, and is not an Internet standard. To the same effect, see also ICANN, *A Unique, Authoritative Root for the DNS*, Internet Coordination Policy 3 (ICP-3, 9 July 2001), available at http://www.icann.org/icp/icp-3.htm..

[43] Black, above n 6, 57.

[44] Internet Architecture Board, 'IAB Technical Comment on the Unique DNS Root', RFC 2826 (May 2000) para [1.3].

[45] DNS Security Extensions (DNSSEC) are specified by RFC 2535: see D Eastlake, 'Domain Name System Security Extensions', RFC 2535 (Mar 1999).

[46] Internet Architecture Board, 'IAB Technical Comment on the Unique DNS Root', RFC 2826 (May 2000) para [1.3]. RFC 2870 provides: '[t]he root zone MUST be signed by the Internet Assigned Numbers Authority (IANA) in accordance with DNSSEC, see [RFC2535] or its replacements': R Bush, D Karrenberg, M Kosters and R Plzak, 'Root Name Server Operational Requirements', RFC 2870 (June 2000) para [3.3.1].

[47] See http://www.root-servers.org/.

Table 1.4 Root Name Servers

Server	Operator	Locations	IP Address	Home ASN
A	VeriSign Naming and Directory Services	Dulles VA.	198.41.0.4	19836
B	Information Sciences Institute	Marina Del Rey CA.	IPv4: 192.228.79.201 IPv6: 2001:478:65::53	*tba*
C	Cogent Communications	Distributed using anycast. Herndon VA; Los Angeles: New York City; Chicago.	192.33.4.12	2149
D	University of Maryland	College Park MD.	128.8.10.90	27
E	NASA Ames Research Center	Mountain View CA.	192.203.230.10	297
F	Internet Systems Consortium, Inc.	Distributed using anycast. 40 sites.	IPv4: 192.5.5.241 IPv6: 2001:500::1035	3557
G	US DOD Network Information Center	Columbus OH	192.112.36.4	568
H	US Army Research Lab	Aberdeen MD.	IPv4: 128.63.2.53 IPv6: 2001:500:1::803f:235	13
I	Autonomica/ NORDUnet	Distributed using anycast. 29 sites	192.36.148.17	29216
J	VeriSign Naming and Directory Services	Distributed using anycast. 22 sites	192.58.128.30	26415
K	Réseaux IP Européens–Network Coordination Centre (RIPE NCC)	Distributed using anycast. 17 sites	IPv4: 193.0.14.129 IPv6: 2001:7fd::1	25152
L	Internet Corporation for Assigned Names and Numbers (ICANN)	Los Angeles	198.32.64.12	20144
M	WIDE Project	Tokyo; Seoul (KR); Paris (FR); San Francisco CA.	IPv4: 202.12.27.33 IPv6: 2001:dc3::35	7500

ant of the UNIX operating system, although there is considerable diversity in the hardware and software used by the root servers.[48]

Control of the root zone file and root zone servers confers considerable de facto control over the domain name space. If a client requests information regarding a domain name with a TLD that is not contained in the root zone file, the client will be unable to retrieve information concerning that domain name. In other words, the domain name will simply

[48] Black, above n 6, 57.

not appear to the end user. There is, nevertheless, no technical or legal requirement that name servers point to the root name servers. Any given name server could direct requests to different name servers that store information relating to different TLDs. A number of alternative root servers have been established.[49] As a matter of convention, however, all name servers that matter point to the authoritative root servers.

Root name-servers C, F, I, J and K are based in multiple locations on different continents. These servers are implemented in clusters of machines that use anycast announcements to provide a decentralised service. Anycast is a scheme for network addressing and routing in which data are routed to the 'nearest' or 'best' destination. The use of anycast to provide an authoritative DNS service is specified in RFC 3258.[50] The adoption of anycast transmissions means that most of the physical root servers are now located outside the United States.

[1.18] Name Resolvers

DNS client programs are known as name resolvers. The functions and technical operation of resolvers are set out in RFC 1034, which provides:

> RESOLVERS are programs that extract information from name servers in response to client requests. Resolvers must be able to access at least one name server and use that name server's information to answer a query directly, or pursue the query using referrals to other name servers. A resolver will typically be a system routine that is directly accessible to user programs; hence no protocol is necessary between the resolver and the user program.[51]

A name resolver is therefore a program that performs the following functions: sends queries to name servers; interprets responses from name servers; and returns information to the client program (such as a web browser) that requested the information. As explained at [1.16], a query from a resolver may be recursive or non-recursive (iterative), the nature of the query determining, in part, the division of responsibility between the resolver and the name server. There are different types of resolvers, with differing degrees of functionality. In the main software implementation of the DNS, the Berkeley Internet Name Domain (BIND) (see [1.22]), name resolvers have limited functionality, consisting mainly of sending queries to name servers and re-sending queries in the event of an error.[52] RFC 1034 refers to resolvers with such limited functionality as stub resolvers.[53] A stub resolver sends recursive requests to a name server, thus delegating significant responsibility for finding information regarding the domain name space to the name server.

[1.19] Resource Records

Each node on the DNS tree has resource information that is associated with that node, and is composed of separate resource records. RFCs 1034 and 1035 specify the format of resource records passed across the network, but not the format of resource records stored in local file systems. The format of resource records stored in local file systems is specified by the particular implementation of the DNS used by the name server.[54] As explained at [1.22]

49 See, eg http://www.open-rsc.org.
50 T Hardie, 'Distributing Authoritative Name Servers via Shared Unicast Addresses', RFC 3258 (Apr 2002).
51 P Mockapetris, 'Domain Names—Concepts and Facilities', RFC 1034 (Nov 1987), para [2.4].
52 Albitz and Liu, above n 5, 26.
53 P Mockapetris, 'Domain Names—Concepts and Facilities', RFC 1034 (Nov 1987), para [5.3.1].
54 See, eg, B Greenblatt, *Internet Directories* (Upper Saddle River, NJ, Prentice Hall PTR, 2001) 35.

the most common implementation of the DNS for name servers is BIND, but DNS servers are not required to use the BIND storage format.

RFCs 1034 and 1035 specify different types of standardised resource record, which correspond to different kinds of data. The main types of resource record specified in RFCs 1034 and 1035 and an explanation of the types of record are summarised below.

A

The 'A' record contains the host address, which in IPv4, is the 32 bit IP address of the host.

CNAME

The 'CNAME' record, which is short for 'canonical name' record, is used when there is more than one host name for a single physical node (IP address). Where there is more than one domain name that corresponds to an IP address, one authoritative name is usually chosen as the canonical name, and the other names are referred to as aliases.[55] The 'CNAME' record maps aliases to the canonical name. This means that a query for any of the domain names will return the same IP address.

HINFO

The 'HINFO' record identifies the central processing unit (CPU) and operating system (OS) used by the host computer.

MX

The 'MX' record, which is short for 'Mail eXchange' record, is specified for the purpose of routing and delivery of electronic mail using the Simple Mail Transfer Protocol (SMTP). It contains information regarding electronic mail exchanges for the domain that will be used in the delivery of electronic mail.

NS

The 'NS' record contains the 'name server' resource records. The 'NS' records contain the host names of name servers that are authoritative for the relevant domain.

PTR

The 'PTR' record is used for address-to-name mappings. For this purpose, 'PTR' records specify the domain name in the 'in-addr.arpa' name space, which corresponds to the canonical domain name. As explained at [1.20], the 'in-addr.arpa' name space includes IP addresses in a domain name with the 'in-addr.arpa' suffix. The 'in-addr.arpa' name space is used for inverse IP address to host name lookups.

SOA

The 'SOA' record, which is the 'start of authority' resource record, contains information concerning the naming authority that is responsible for the relevant zone of the DNS, including the name of the primary master server for the zone and the e-mail address of the person responsible for the zone.

Each type of resource record also includes a TTL field, which is the 'Time to Live' for the record. The TTL field is a 32 bit integer which specifies how long a record can be cached before it should be discarded and the source of the information consulted again. Each type

[55] As RFC 2181 explains, there is no technical requirement for a single, authoritative canonical name: R Elz and R Bush, 'Clarifications to the DNS Specification', RFC 2181 (July 1997), para [10].

of resource record also includes an RDATA ('resource data') field, which is a string of octets that describes the resource and which varies according to the type of data. When a DNS message is sent, it must contain a header, which includes a fixed part, that indicates the type and TTL of the resource record, and variable part, that contains the RDATA field, which varies with the type of record.

The scope of the information included in the resource records means that the DNS can be used for much more than simply mapping domain names to IP addresses. RFC 2181 explains this point in the following terms:

> Occasionally it is assumed that the Domain Name System serves only the purpose of mapping Internet host names to data, and mapping Internet addresses to host names. This is not correct, the DNS is a general (if somewhat limited) hierarchical database, and can store almost any kind of data, for almost any purpose.[56]

[1.20] Mapping Addresses to Domain Names

As explained at [1.7], the purpose of the DNS is to map domain names to IP numbers. That is, in a standard DNS query, a domain name is entered and a name server resolves the name into an IP number. The DNS, however, also supports a mechanism, known as inverse resolution, for mapping IP numbers to domain names. In order for the DNS to map numbers to names, part of the domain name space has been allocated to IP numbers. The part of the domain name space allocated to IP addresses is the 'in-addr.arpa' domain. This means that whenever an IP address is allocated, a domain name corresponding to that IP address is established in the 'in-addr.arpa' name space. The IP address in the 'in-addr.arpa' name space is, however, represented in the reverse order from that in which it is conventionally read. For example, the IP address for ICANN, which is 192.0.34.65., is represented in the 'in-addr.arpa' name space as:

<65.34.0.192.in-addr.arpa>

The reason for representing IP addresses in reverse order in the 'in-addr.arpa' name space is that IP addresses, like domain names, are organised hierarchically. As conventionally represented, IP addresses read from the more general (on the left) to the more specific (on the right). Thus, in our example, '192' is the more general component of the IP address, just as the TLD is the more general component of a domain name. By reversing the order, the address is made to conform to domain name conventions, thereby facilitating inverse resolution.

Some Internet applications, such as file transfer protocol (ftp) servers and mail servers, use inverse resolution to confirm the legitimacy of a communication. In other words, if a communication does not reverse resolve then it may not be legitimate, and may be rejected by the application.

As explained at [1.19], PTR-type resource records specify the IP address that corresponds to the canonical domain name in the form of the reversed IP number suffixed with 'in-addr.arpa'.

[1.21] The WHOIS Directory Service

Early in the development of ARPANET, the Stanford Research Institute—Network Information Center (SRI-NIC) established a centralised database of information about host

[56] R Elz and R Bush, 'Clarifications to the DNS Specification', RFC 2181 (July 1997), para [11].

computers and systems connected to ARPANET, including e-mail addresses of the users of connected systems. The original protocol for the directory service based on the host and user information was issued as RFC 812 in March 1982.[57] RFC 812 requested that individuals connected to an ARPANET host and who were capable of passing traffic across ARPANET send the following information to SRI-NIC: full name, middle initial, mailing address, telephone number and e-mail address. RFC 812 was replaced by RFC 954, which was issued in October 1985.[58] As RFCs 812 and 954 make clear, the original WHOIS service consisted of a database of look up information stored on WHOIS servers operated by the SRI-NIC. RFC 954 even specified the IP addresses of the SRI hosts to be contacted in order to use the WHOIS directory service. The RFCs do not, however, prohibit other Internet hosts from running WHOIS servers. There are now many WHOIS servers run on a variety hosts. Most TLD administrators, for example, run their own WHOIS server. Users can search for WHOIS servers at http://www.allwhois.com.

The WHOIS service may be formally defined as 'a TCP transaction based query/response server, running on a few specific central machines, that provides netwide directory service to Internet users'.[59] The WHOIS service is used to find the names and contact details of individuals or organisations that have registered domain names. The service is also used to find other records provided at the time a domain name is registered, including the names of network and system administrators, system and network points-of-contact and contact information of other registered individuals.[60] Significantly, the WHOIS service allows for searches of the WHOIS database on the basis of different properties of registered individuals, including name and e-mail address.[61] In order to ensure that distinct records in the WHOIS database are properly identified, each record must have a unique identifier, known as a handle.[62] The WHOIS++ service is a more highly structured version of the WHOIS service, that provides greater functionality to users than the basic WHOIS service. Information regarding the WHOIS++ service, including the types of data records to be stored on WHOIS++ servers, is provided by RFC 1834, issued in August 1995.[63]

The initial purpose of establishing the WHOIS database was to provide information in order to resolve technical problems associated with the operation of ARPANET. As the Internet grew and became commercialised, however, other uses of the WHOIS service emerged.[64] For example, the WHOIS service was used to track the source of abusive practices, such as spamming or hacking attacks. The WHOIS service was also used to determine who had registered a domain name by people who were, for example, interested in purchasing the domain name. Furthermore, the WHOIS service was used by trade mark owners to

[57] K Harrenstien and V White, 'NICNAME/WHOIS', RFC 812 (1 Mar 1982). RFC 812 stated that 'NICNAME' had been chosen as the global name for the user program, but that sites could choose to use the more familiar name of 'WHOIS'.

[58] K Harrenstein, M Stahl and E Feinler, 'NICNAME/WHOIS', RFC 954 (Oct 1985).

[59] J Gargano and K Weiss, 'Whois and Network Information Lookup Service Whois++', RFC 1834 (Aug 1995), para [I].

[60] See G Kessler and S Shepard, 'A Primer on Internet and TCP/IP Tools and Utilities', RFC 2151 (June 1997), para [5.1]. As Greenblatt puts it, '[t]he intention of Whois is to provide a way to access information about people that is collected through the DNS registration process': Greenblatt, above n 51, 129.

[61] *Ibid*, 129–51.

[62] See, eg, J Gargano and K Weiss, 'Whois and Network Information Lookup Service Whois++', RFC 1834 (Aug 1995), para [IV].

[63] *Ibid*. For further information on WHOIS++, including the structure of WHOIS++ information, see Greenblatt, above n 51, 133–51.

[64] Mueller, above n 5, 235–8.

search for potentially infringing domain name registrations, and to discover information regarding the registrants of such domain names.

The expanded uses to which the WHOIS service has been put over time has resulted in the service becoming controversial. In particular, concerns have arisen in relation to the privacy rights of individuals whose information is included on WHOIS databases.

[1.22] BIND

The DNS protocol must be implemented in software that is responsible for operating name servers and name resolvers. The DNS protocol was first implemented in a program known as JEEVES, which was written by Paul Mockapetris. The most common software implementation of DNS is currently known as BIND (the Berkeley Internet Name Domain). BIND is an open source software implementation, which was written by Kevin Dunlap for Berkeley's 4.3 BSD Unix operating system. BIND is currently used to run over 90 per cent of name servers.[65] The Internet Software Consortium is responsible for maintaining BIND.[66] BIND is made available free of charge: it is usually shipped with the standard TCP/IP Unix networking software, and is otherwise freely available from the Internet Software Consortium's web site.[67]

The particular implementation of the DNS may be of considerable importance to the operation of the DNS. For example, the BIND name server software is set by default to point to the IP addresses of the authoritative root servers, thereby ensuring that alternative root directories are not available to name servers that run BIND. This means that those responsible for maintaining BIND have some potential influence over DNS policy-making. For example, as Mueller recounts, in 1996 Paul Vixie, who was responsible for maintaining BIND, threatened to reconfigure the BIND software to point to alternative root directories unless action was taken to end the NSI monopoly in relation to IP address and naming functions for the gTLDs.[68] Although nothing came of this threat, it serves to illustrate the extent to which stability of the DNS necessarily depends, to an extent, upon the conventions adopted by those responsible for maintaining and using DNS implementations, especially the software used to run name servers.

[65] Mueller, above n 5, 43.
[66] For information regarding the Internet Software Consortium see http://www.isc.org.
[67] Albitz and Liu, above n 5, 38.
[68] Mueller, above n 5, 141–2.

2

DNS Governance and ICANN

Internet and DNS Governance

[2.1] Governance

The term 'governance' has been applied to the wide range of issues associated with policy development for, and management of, the Internet. It is, however, widely acknowledged that the term is imprecise, with no single accepted definition. As Rhodes has put it:

> The term 'governance' refers to a change in the meaning of government, referring to a *new* process of governing. There are many uses of governance: for example, it refers (to) the minimal state; corporate governance; and the new public management . . . It has too many meanings to be useful.[1]

Given the widespread use of the term 'governance' it is, nevertheless, important to attempt to give some meaning to the term. Rhodes suggests that the various meanings attributed to 'governance' share the following characteristics:

1. Interdependence between organizations. Governance is broader than government, covering non-state actors. Changing the boundaries of the state meant the boundaries between public, private and voluntary sectors became shifting and opaque.
2. Continuing interactions between network members, caused by the need to exchange resources and negotiate shared purposes.
3. Game-like interactions, rooted in trust and regulated by rules of the game negotiated and agreed by network participants.
4. A significant degree of autonomy from the state. Networks are not accountable to the state; they are self-organizing. Although the state does not occupy a sovereign position, it can indirectly and imperfectly steer networks.[2]

As Rosenau has explained, the sense in which 'governance' refers to a shift from centralised control to the more indirect 'steering' of relatively autonomous actors is related to the etymological roots of the term; 'governance' being derived from the Greek 'kybernan', which means 'to steer', and 'kybernetes', which means 'pilot or helmsman'.[3] As illustrated in this chapter, the complexity of the arrangements for managing the DNS, and the involvement of a range of public and private sector actors in DNS decision-making, mean that 'governance' is an especially apt term for DNS institutional arrangements.

[1] RAW Rhodes, *Understanding Governance* (Buckingham, Open University Press, 1997) 15 (emphasis in original).
[2] *Ibid*, 53.
[3] J Rosenau, 'Governance in a Globalizing World' in D Held and A McGrew (eds), *The Global Transformations Reader* (Cambridge, Polity Press, 2000) 182.

[2.2] Internet Governance and the WSIS

What precisely is meant by Internet governance, and the appropriate role of national governments in Internet governance, have proven to be controversial questions. These questions have recently arisen within the context of the World Summit on the Information Society (WSIS), held between 2003 and 2005, and are explained in more detail at [2.30].

The first phase of the WSIS, held in 2003, resulted in the Geneva Declaration of Principles. It stated:

> The Internet has evolved into a global facility available to the public and its governance should constitute a core issue of the Information Society agenda. The international management of the Internet should be multilateral, transparent and democratic, with the full involvement of governments, the private sector, civil society and international organizations. It should ensure an equitable distribution of resources, facilitate access for all and ensure a stable and secure functioning of the Internet, taking into account multilingualism.[4]

The Declaration of Principles further indicated that the Secretary-General of the United Nations would be asked to establish a working group on Internet governance to investigate and make proposals for action on Internet governance in time for the second phase of the WSIS, which was held in November 2005.[5]

The terms of reference for the working group, as set out in the Geneva Plan of Action, were to:

i) develop a working definition of Internet governance;
ii) identify the public policy issues that are relevant to Internet governance;
iii) develop a common understanding of the respective roles and responsibilities of governments, existing intergovernmental and international organisations and other forums as well as the private sector and civil society from both developing and developed countries.[6]

The Working Group on Internet Governance (WGIG), established pursuant to the WSIS mandate, published a report in June 2005.[7] The WGIG report adopted the following working definition of Internet governance:

> Internet governance is the development and application by Governments, the private sector and civil society, in their respective roles, of shared principles, norms, rules, decision-making procedures, and programmes that shape the evolution and use of the Internet.[8]

The report adopted a broad approach to the public policy issues relevant to Internet governance, identifying the following four clusters of issues:

(a) Issues relating to infrastructure and the management of critical Internet resources, including administration of the domain name system and Internet protocol addresses (IP addresses), administration of the root server system, technical standards, peering and interconnection,

[4] World Summit on the Information Society, *Declaration of Principles. Building the Information Society: a global challenge in the new Millennium,* Document WSIS-03/GENEVA/DOC/4-E (12 Dec 2003), para 48.

[5] *Ibid,* para 50.

[6] World Summit on the Information Society, *Plan of Action,* Document WSIS-03/GENEVA/DOC/5-E (12 Dec 2003), para 13(b).

[7] Working Group on Internet Governance (WGIG), *Report on Internet Governance* (June 2005), available at http://www.wgig.org/docs/WGIGREPORT.pdf. The report was supplemented by a more detailed background report: WGIG, *Background Report* (June 2005).

[8] WGIG, *Report on Internet Governance* (June 2005), para 10.

telecommunications infrastructure, including innovative and convergent technologies, as well as multilingualization. These issues are matters of direct relevance to Internet governance and fall within the ambit of existing organizations and responsibility for these matters.

(b) Issues relating to the use of the Internet, including spam, network security and cybercrime. While these issues are directly related to Internet governance, the nature of global cooperation required is not well defined.

(c) Issues that are relevant to the Internet but have an impact much wider than the Internet and for which existing organizations are responsible, such as intellectual property rights (IPRs) or international trade.

(d) Issues relating to the development aspects of Internet governance, in particular capacity-building in developing countries.[9]

In accordance with its mandate, the WGIG report included a number of proposals for action to be considered by the second phase of the WSIS, which was held in Tunis in November 2005. The most important proposal was for the establishment of a new international forum, linked to the UN, for consideration of Internet governance issues.[10]

The Tunis Agenda for the Information Society reaffirmed the principles set out in the Geneva Declaration of Principles and adopted the WGIG working definition of Internet governance. It added:

> We recognise that Internet governance, carried out according to the Geneva principles, is an essential element for a people-centred, inclusive, development oriented and non-discriminatory Information Society. Furthermore, we commit ourselves to the stability and security of the Internet as a global facility and to ensuring the requisite legitimacy of its governance, based on the full participation of all stakeholders, from both developed and developing countries, with their respective roles and responsibilities.[11]

The Tunis Agenda included a substantial number of principles on Internet governance, including the following:

> 58. We recognise that Internet governance includes more than Internet naming and addressing. It also includes other significant public policy issues such as, *inter alia,* critical Internet resources, the security and safety of the Internet, and development aspects and issues pertaining to the use of the Internet.
> 59. We recognise that Internet governance includes social, economic and technical issues including affordability, reliability and quality of service.
> 60. We further recognise that there are many cross-cutting international public policy issues that require attention and are not adequately addressed by the current mechanisms.[12]

In accordance with the WGIG proposal for action, the Tunis Agenda asked the UN Secretary-General to convene a new forum for Internet governance, to be known as the Internet Governance Forum (IGF), with the mandate to:

> a. Discuss public policy issues related to key elements of Internet governance in order to foster the sustainability, robustness, security, stability and development of the Internet;
> b. Facilitate discourse between bodies dealing with different cross-cutting international public policies regarding the Internet and discuss issues that do not fall within the scope of any existing body;

[9] *Ibid,* para 13.

[10] *Ibid,* paras 40–47.

[11] World Summit on the Information Society, *Tunis Agenda for the Information Society,* Document WSIS-05/TUNIS/DOC/6(Rev.1)-E (18 November 2005), para 31.

[12] *Ibid,* paras 58–60.

c. Interface with appropriate inter-governmental organisations and other institutions on matters under their purview;
d. Facilitate the exchange of information and best practices, and in this regard make full use of the expertise of the academic, scientific and technical communities;
e. Advise all stakeholders in proposing ways and means to accelerate the availability and affordability of the Internet in the developing world;
f. Strengthen and enhance the engagement of stakeholders in existing and/or future Internet governance mechanisms, particularly those from developing countries;
g. Identify emerging issues, bring them to the attention of the relevant bodies and the general public, and, where appropriate, make recommendations;
h. Contribute to capacity building for Internet governance in developing countries, drawing fully on local sources of knowledge and expertise;
i. Promote and assess, on an ongoing basis, the embodiment of WSIS principles in Internet governance processes;
j. Discuss, *inter alia,* issues relating to critical Internet resources;
k. Help to find solutions to the issues arising from the use and misuse of the Internet, of particular concern to everyday users;
l. Publish its proceedings.[13]

The IGF was convened in 2006 and held its first meeting in Athens from 30 October to 3 November 2006.

[2.3] The Problem of DNS Governance

DNS governance issues represent an important, distinct sub-set of the wider set of Internet governance issues. As the WGIG Background Report pointed out, there has been a spectrum of views regarding the appropriate scope and mechanisms of Internet governance:

> At one end of the spectrum, some actors took a relatively narrow view of these questions in which only the management of the Internet's core resources (eg IP addresses, domain names, the root zone) needed special governance arrangements, and in which contracts were the principal governance mechanism. At the other end of the spectrum, some other actors took a much broader view of the scope of Internet governance and included its social, economic, cultural and political impacts in addition to its technical and logical infrastructure, and envisaged the use of a correspondingly wider range of governance mechanisms, including treaty instruments.[14]

The WSIS process, referred to at [2.2] and discussed further at [2.30], adopted a broad view of what amounts to Internet governance. Nevertheless, even on the narrowest view, management of the DNS falls firmly within the scope of Internet governance. At the same time, it is important for the understanding of DNS governance issues to see them within the broader context of the problem of creating appropriate governance structures for the Internet. In fact, given the unchallenged need for decisions to be made, and policies developed, for the management of the DNS, the broader issues of legitimate Internet governance structures initially, and inevitably, arose in debates regarding DNS governance. Consequently, DNS policy-making has been characterised, almost from the beginning, by underlying questions concerning the proper role of nation states in making rules for a global

[13] World Summit on the Information Society, *Tunis Agenda for the Information Society,* Document WSIS-05/TUNIS/DOC/6(Rev.1)-E (18 November 2005), para 72.

[14] WGIG, above n 7, para 29.

communications system; the respective roles and competences of interdependent public sector and private sector actors; and the difficulty of developing workable accountability mechanisms for a global system.

It is first necessary to understand the broader issues raised by Internet governance. The Internet is a global, decentralised communications system that was designed to ensure the reliable delivery of packets of information. In performing this task, the identity or location of parties involved in a communication and the content of the communication are notionally irrelevant. The central features that characterise the Internet as an unparalleled system for delivering information necessarily create difficulties for sovereign nation states that seek to impose laws, or other forms of regulation, on the Internet. The difficulties relate both to the practicalities of imposing and enforcing national laws, and to the legitimacy of doing so. In relation to the practicalities of enforcing territorially-based laws, it may be difficult to identify or locate parties to Internet communications. Moreover, an entity that has no physical presence within the borders of a nation state may not be readily amenable to the laws of that nation state. In relation to the legitimacy of applying territorially-based laws to a global communications system, the legitimacy of applying national laws to extra-territorial entities or conduct must, in general terms, depend upon establishing some form of accepted legal nexus between the territorial state, on the one hand, and the entity or conduct sought to be regulated, on the other.

There are inevitably tensions involved in any attempt by a territorially-based nation state to make general laws for a global communications system that is organised on the basis of logical connections between machines, and not according to any necessary connection with a sovereign, territorially-based political entity, such as a nation state. The Internet therefore presents challenges to accepted notions that the territorially-based nation state is the pre-eminent source of legitimate law-making power. In this sense, the Internet is but one of many contemporary challenges to the received model of sovereignty, traditionally traced back to the 1648 Peace of Westphalia, in which legitimate authority is seen to derive entirely from sovereign nation states with exclusive control over a geographically-bounded territory.[15] Consequently, it is important that debates concerning Internet governance are related to the problems of establishing an acceptable model of legitimacy for a global system in which territorially-bounded nation states are not the sole source of authority.[16] The tensions between the established DNS governance structures and the WSIS process, dealt with at [2.30], should be seen within this context.

How, then, do DNS governance issues fit within the broader context? The DNS is a hierarchical, distributed database of Internet address, naming and associated information. Nevertheless, the technical requirement that each domain name be unique, together with the hierarchical organisation of the domain name space, establishes the need for some form of centralised coordination.[17] Control of the hierarchical domain name space ultimately depends upon control of the apex of the hierarchy, namely the root zone file and the authoritative root zone servers. As control of the root zone file and the root zone servers determines what will (or will not) be accepted as TLDs, this confers control over the text strings that can, in practice, be used as domain names. It also confers *de facto* control over who can

[15] See eg SD Krasner, 'Compromising Westphalia' in Held and McGrew, above n 3, 124–35.

[16] See eg J Weinberg, 'ICANN and the Problem of Legitimacy' (2000) 50 *Duke Law Journal* 187.

[17] As Michael Froomkin puts it, '[d]espite being famously decentralized and un-hierarchical, the Internet relies on an underlying centralized hierarchy built into the domain name system (DNS)': AM Froomkin, 'Wrong Turn in Cyberspace: Using ICANN to Route Around the APA and the Constitution' (2000) 50 *Duke Law Journal* 17, 20.

allocate domain names that end in authorised TLDs.[18] Moreover, before any entity can appear on the Internet, the entity must first be allocated a name and an address. In other words, in order to appear on the Internet, an entity must register a domain name.

The need to obtain a domain name confers an ability on entities responsible for allocating domain names to impose terms and conditions on registration. In turn, control of the root zone file and root zone servers confers control over the policies that entities responsible for allocating domain names must adopt in determining the terms and conditions of registration. This means that control of the root zone file and root zone servers carries considerable discretionary power in relation to policies to be adopted concerning the DNS, and potentially concerning more general use of the Internet. At the same time, in practical terms, the ability to impose policies in relation to the DNS is constrained by the extent to which name servers continue to point to the authoritative root servers, which is entirely a matter of convention.

To the extent that control of the root zone file and root zone servers confers the ability to make policy decisions, and to the extent that the ability has been used to make policy decisions, questions have inevitably arisen about the legitimacy of entities responsible for such policy decisions. As the centralised coordination of the DNS has presented the main opportunity for making centralised policy decisions relating to the Internet, it is unsurprising that, to date, the most controversial issues in Internet governance have been issues of DNS governance. The remainder of this chapter explains the main governance structures that have arisen for managing the DNS and making DNS policy-decisions, including the historical development of these structures, before returning to questions of the legitimacy of the existing structures in the context of the WSIS process.

History of DNS Governance

[2.4] Early History

The history of DNS governance is extraordinarily complex and convoluted. It is, however, impossible to understand the current institutional structure for DNS governance without some reference to this history.[19] The early arrangements relating to administrative authority for the DNS centred on particular individuals, specifically Jon Postel and his colleagues. The arrangements were ad hoc, idiosyncratic and sometimes less than transparent. Nevertheless, the relatively informal administrative arrangements proved adequate so long as the Internet remained an experimental network. As the Internet grew, however, the informal arrangements were incapable of dealing with the demands of managing an address and name space for a network of unprecedented scale and complexity. Much of the history of DNS governance has been characterised by attempts to retain elements of the early informal arrangements in the new, more formal structures that were increasingly necessary as the DNS became the focus of complex social, economic and legal demands. The history has,

[18] As Froomkin explains, 'a domain name that cannot be resolved into an IP number by the vast majority of users is of very limited value on the Internet': *ibid*, 47.

[19] For comprehensive treatments of the history of DNS governance see: Froomkin, above n 17; Weinberg, above n 16; M Mueller, *Ruling the Root* (Cambridge, Mass., The MIT Press, 2002).

however, also been characterised by a reticence among most of the major actors to accept complete responsibility for important policy decisions made in relation to DNS governance.

As explained at [1.7], the DNS was essentially developed by Jon Postel and his colleagues at the University of Southern California's ISI in the early 1980s. The activities of the ISI were funded by the Defense Advanced Research Project Agency (DARPA, which was the re-named ARPA) under a long-term research contract which specified a number of functions, including number assignment and editing RFCs.[20] Formal responsibility for managing the domain name space was given to the ISI by the Defense Information Systems Agency (DISA) in 1985.[21] As explained at [1.4] above, the SRI NIC had been responsible for address and naming functions for the ARPANET. The NIC was funded under a contract with the Defense Communications Agency, and became known as the Defense Data Network—Network Information Center (DDN-NIC). With the introduction of the DNS, the DDN-NIC assumed responsibility for the root name servers and for registering TLDs and 2LDs.[22] In 1987, the ISI functions relating to IP addresses, which had been performed by Postel and Reynolds, were transferred to the DDN-NIC.[23]

As explained at [1.3], the Internet Activities Board (IAB) was established in 1983 to assume responsibility for Internet standards development. The IAB was established as a DARPA committee, but the institutional arrangements were informal. The first chair of the IAB was Vint Cerf, who had been instrumental in the development of the TCP/IP protocol. Cerf was responsible for selecting other IAB board members. The IAB later became the focus for an attempt to claim a source of authority for DNS activities independent from US government agencies.

[2.5] Internet Assigned Numbers Authority (IANA)

In 1988, DARPA entered into a new contract with the ISI, which listed a number of RFC and DNS-related functions.[24] At the same time, the ISI was internally reorganised and began to refer to its RFC and DNS functions by the name 'Internet Assigned Numbers Authority' (IANA).[25] The IANA was first officially mentioned in RFC 1083, a description of the Internet standardisation process written by Jon Postel, that was released in December 1988.[26] RFC 1083 simply stated that 'protocol standards are managed for the IAB by the Internet Assigned Numbers Authority'.[27] The contract with DARPA, however, did not refer to the IANA. On the other hand, the RFCs released during this period made no reference to the contract with DARPA.

Subsequent RFCs claimed that the IANA derived its authority over Internet names and addresses from the IAB and, furthermore, that the IANA had delegated operational authority for registering names and IP numbers to the DDN-NIC. For example, RFC 1174, an informational RFC released by Vint Cerf in August 1990, included the following statements:

[20] Mueller, above n 19, 81.
[21] See Froomkin, above n 17, 53.
[22] The NICs functions were recognised in RFC 1032: see Mary Stahl, 'Domain Administrators Guide', RFC 1032 (Nov 1987).
[23] Mueller, above n 19, 82.
[24] *Ibid*, 279, fn 24.
[25] Froomkin, above n 17, 53.
[26] Jon Postel, 'IAB Official Protocol Standards', RFC 1083 (Dec 1988).
[27] *Ibid*, para [7.2].

> Throughout its entire history, the Internet system has employed a central Internet Assigned Numbers Authority (IANA) for the allocation and assignment of various numeric identifiers needed for the operation of the Internet. The IANA function is performed by USC Information Sciences Institute. The IANA has discretionary authority to delegate portions of this responsibility and, with respect to numeric network and autonomous system identifiers, has lodged this responsibility with an Internet Registry (IR). This function is performed by SRI International at its Network Information Center (DDN-NIC).[28]

This version of the administrative arrangements for the DNS was elaborated by a general informational RFC, RFC 1206, which was released in February 1991, and which stated:

> The task of coordinating the assignment of values to the parameters of protocols is delegated by the Internet Activities Board (IAB) to the Internet Assigned Numbers Authority (IANA) . . . Internet addresses and Autonomous System numbers are assigned by the Network Information Center at SRI International. This responsibility has been delegated by the IANA to the DDN NIC, which serves as the Internet Registry. The IANA is located at USC/Information Sciences Institute.[29]

Later, in RFC 1700, released by Postel and Reynolds in October 1994, a further attempt was made to establish another source of authority for the IANA. RFC 1700 stated:

> The Internet Assigned Numbers Authority (IANA) is the central coordinator for the assignment of unique parameter values for Internet protocols. The IANA is chartered by the Internet Society (ISOC) and the Federal Network Council (FNC) to act as the clearinghouse to assign and coordinate the use of numerous Internet protocol parameters.[30]

As explained at [1.3], the Internet Society—a private, non-profit-making organisation—was formed in 1992 as an umbrella organisation for the IAB and the IETF. According to Mueller, the major motivation for forming the Internet Society came from the perceived need for IETF members to be insured against liability for legal actions.[31] Other factors leading to the formation of the Internet Society included the need to raise money to support standard-making and DNS functions, given diminishing support from DARPA, and a need for greater legitimacy in the international standard-making arena. As Froomkin points out, there were major problems with the attempt to derive the IANA's authority over the DNS from the Internet Society, including that the IANA was established before the Internet Society and that, even if the Internet Society had any claims to authority over the DNS, there was never any actual decision to delegate authority to the IANA.[32]

The implausibility of the accounts of the institutional structures for administering the DNS recorded in the RFCs of the late 1980s and early 1990s raises questions relating to the purpose sought to be achieved by these statements. Mueller suggests the following explanation:

> These descriptions reflected the technical community's growing conception of itself as an autonomous, self-governing social complex. Explicit claims on the right to manage name and address assignment were being made by an authority structure that existed solely in Internet RFCs and lacked any basis in formal law or state action. The authority claims nevertheless had significant legitimacy within the technical community. Not only was Postel known, respected, and trusted

[28] Vint Cerf, 'IAB Recommended Policy on Distributing Internet Identifier Assignment and IAB Recommended Policy Change to Internet "Connected" Status', RFC 1174 (Aug 1990).

[29] GS Malkin and AN Marine, 'FYI on Questions and Answers: Answers to Commonly asked "New Internet User" Questions', RFC 1206 (Feb 1991) para [7]. Mueller erroneously refers to this RFC as RFC 1207: see Mueller, above n 19, 280, fn 41.

[30] J Postel and J Reynolds, 'Assigned Numbers', RFC 1700 (Oct 1994).

[31] Mueller, above n 19, 94–5.

[32] Froomkin, above n 17, 57.

within the IETF and the supporting government agencies but the RFC series was, for both old and new participants in IETF, the way reality was defined on the Internet.[33]

[2.6] Privatisation of the Root

As explained at [1.13], in 1983 the military part of ARPANET was separated from the civilian part to form a separate network, known as MILNET. Under informal arrangements, the military permitted civilian agencies to use the services of the DDN-NIC in return for other services provided by agencies such as the NSF.[34] Nevertheless, there remained underlying tensions between military and civilian interests. Largely in order to deal with these tensions, in 1990 the NSF established a coordinating committee, known as the Federal Networking Council (FNC), which consisted of representatives from US federal agencies. Later in 1990, the FNC decided that the NSF should be responsible for paying for civilian name and address registration services. This decision led to the separation of registration functions for civilian and military interests.

As further explained at [1.13], in 1991 the US Defense Department awarded a contract to operate the DDN-NIC to a private corporation, GSI, which sub-contracted the registry functions to Network Solutions, Inc (NSI). In 1992, the NSF solicited proposals for operating central administrative functions of the civilian Internet. In January 1993, NSF entered into agreements with three separate corporations: an agreement with NSI to perform name and number registration services; an agreement with AT&T to perform directory and database services; and an agreement with General Atomics to provide information services.[35] In awarding the contracts, the NSF required the successful corporations to revise their proposals to form a collaborative project, which became known as the InterNIC.

The agreement between the NSF and NSI required NSI to provide registration services to non-military Internet users that were previously provided by the DDN-NIC, including domain name registration and network number assignment.[36] Interestingly, the agreement appeared to require NSI to follow policy directions from the IANA, Article 3 of the agreement stating that NSI was to 'provide registration services in accordance with the provisions of RFC 1174'.[37] As explained at [2.5], RFC 1174 stated that the IANA had authority to delegate responsibility for registering names and numbers to an Internet Registry. The agreement seemed to confirm these arrangements by also providing that NSI could delegate its functions to other registries, but only in cooperation with the IANA.[38] As Froomkin points out, however, the account of the authority structure for the DNS given by RFC 1174 was clouded by the fact that Postel and the IANA were funded by the US Department of Defense, which presumably meant that the IANA remained subject to policy directions from the Department of Defense.[39]

33 Mueller, above n 19, 93.
34 *Ibid*, 100.
35 *Ibid*, 101–2.
36 See *NSF Cooperative Agreement No NCR-9218742* (1 Jan 1993), available at http://www.icann.org/nsi/coopagmt-01jan93.htm.
37 *Ibid*, Art 3C.
38 *Ibid*, Art 3D.
39 Froomkin, above n 17, 58.

[2.7] The Crisis in Governance

The result of the multiple agreements between US federal government agencies and those directly responsible for DNS governance and administration, and the lack of clarity regarding DNS authority structures, was that in the early 1990s formal legal authority for the DNS was muddled. At the same time, however, the Internet was developing into much more than a research network. As explained at [1.12], from 1991 to 1995, when NSFNET was formally decommissioned, restrictions on the commercial use of the Internet were progressively removed. As further explained, the early 1990s saw the emergence of the World Wide Web as a popular Internet application. By the mid-1990s, these developments led to a significant increase in Internet use and in demand for domain names.[40] The chaotic DNS authority structure was incapable of dealing with the pressures generated by the commercialisation and growth of the Internet.

The demand for new domain names, especially in the .com gTLD, quickly overwhelmed the resources allocated to InterNIC under the NSF cooperative agreement. The first response of InterNIC was to cut costs by automating the registration of domain names to the greatest extent possible. Thus, any attempt to impose restrictions on registrations in the .com gTLD were abandoned, and distinctions between .com, .net and .org were effectively removed. Registration on a 'first-come-first-served' basis, without any vetting of registrations, coupled with the absence of a fee for registration, resulted in a rush for registration of domain names in the open gTLDs.

The second response to the demand for domain names was to introduce charges for domain name registration. In December 1994, an independent review of InterNIC operations recommended that NSI commence charging for registrations in the .com gTLD and eventually charge for all registration services.[41] In September 1995 the NSF cooperative agreement was amended to allow NSI to charge an annual fee of US$50 for the registration of 2LDs in the .gov, .edu, .com, .net and .org gTLDs.[42] Seventy per cent of the fee was to go to NSI and thirty per cent to an infrastructure fund operated by the NSF. Before the agreement was amended to allow charging, in March 1995, NSI had been taken over by a large defence contractor, Science Applications International Corporation (SAIC).

The decisions to allocate registration functions to NSI, then to allow charges to be imposed for registrations, gave NSI considerable potential earning power, eventually resulting in a successful initial public offering on the NASDAQ exchange in September 1997.[43] The success of NSI led to opposition from two sources. First, the decision to allow charging created opposition from the traditional technical community, which was still attached to a non-commercial model for Internet registrations. Secondly, the potential profits from charging for registration resulted in demands for others to be allowed into the potentially lucrative market for domain name registration services. In both cases, concerns relating to the NSI monopoly over domain name registrations led to proposals for introducing new TLDs in order to provide competition to NSI. For instance, in an email sent to the Internet

[40] Mueller notes that InterNIC 2LD registrations handled in one month went from 16,114 in July 1994 to 263,760 in Feb 1996: Mueller, above n 17, 110.

[41] *InterNIC Midterm Evaluation and Recommendations*, A Panel Report to the National Science Foundation (Dec 1994), available at http://www.dnso.org/dnso/dnsocomments/comments-gtlds/Arc00/msg00006.html.

[42] *Amendment 4 to Cooperative Agreement Between NSI and U.S. Government* (13 Sept 1995), available at http://www.icann.org/nsi/coopagmt-amend4-13sep95.htm.

[43] The NASDAQ listing generated US$350 million: see Mueller, above n 19, 124.

Society in September 1995, Jon Postel said: 'I think this introduction of charging by the Intenic [*sic*] for domain registrations is sufficient cause to take steps to set up a small number of alternate top level domains managed by other registration centers. I'd like to see some competition between registration services to encourage good services at low prices'.[44]

In response to these concerns, in May 1996 Postel issued an Internet-Draft, which proposed a new regime that in the first year of operation would authorise 50 new registries to compete with NSI and introduce 150 new TLDs to be managed by the new registries.[45] Postel proposed that the new registries pay an application fee and an annual fee to the Internet Society, which would be used to fund IANA. Postel's Internet-Draft also proposed to authorise 10 new registries and introduce 30 new TLDs in each year after the first year of the new regime.

Although there was nothing to prevent anyone, including Jon Postel, from floating proposals for introducing new registries and TLDs, precisely who had the authority to introduce a new regime was completely unclear. More than anything else, Postel's proposal brought the unresolved questions relating to final authority over the DNS to a head. In February 1995, Postel had circulated a final draft proposal for a new charter for IANA in which, on the model set out in the RFCs referred to at [2.6], it would obtain its authority from the Internet Society. In return, in June 1996, the Internet Society supported Postel's proposals for introducing new registries and TLDs.[46] Against this, the DARPA representative on the FNC claimed that the US Defense Department had delegated policy control over the DNS to the FNC.[47] The position was complicated by non-US registries, which had become responsible for registrations in some ccTLDs, and which naturally resisted claims that they were subject to policy control by US government agencies. The position was further complicated by alternative root registries, which resented the IANA's centralised control over the DNS and proposed removing all restrictions on the introduction of new TLDs. Finally, trade mark owners, and especially the International Trademark Association (INTA), had become alarmed by various practices involving the registration of trade marks as domain names. Concerns relating to the potential for the registration of domain names to undermine the value of trade marks meant that trade mark interests, including INTA, opposed the introduction of new TLDs.

[2.8] The International Ad Hoc Committee

The lack of a clear authority structure over DNS policy-making and the diverse positions of the various stakeholders meant that Postel's proposal to introduce new TLDs resulted in an impasse. This impasse directly led to an attempt to establish a more certain authority structure for the DNS.

In October 1996, the Internet Society established an international panel, known as the International Ad Hoc Committee (IAHC), to develop a new DNS governance structure. The IAHC included representatives of the Internet Society, the IAB, IANA, the FNC, the

[44] J Postel, 'Re: ISOC Statement on Domain Name Fees' (15 Sept 1995), available at http://www.dnso.org/clubpublic/ga-full/ARC04/msg00082.html; see also Mueller, above n 19, 128.

[45] J Postel, 'New Registries and the Delegation of International Top Level Domains', Internet-Draft (May 1996), available at http://tools.ietf.org/html/draft-postel-iana-itld-admin-00 (known as 'draft-postel').

[46] See Mueller, above n 19, 135–6.

[47] *Ibid*, 137.

INTA, the World Intellectual Property Organization (WIPO) and the International Telecommunication Union (ITU). Significantly, NSI was not represented on the IAHC.

The IAHC released its final report in February 1997.[48] The IAHC process was a complete departure from the informal, open processes that, until then, had characterised the development of RFCs and Internet governance structures. Although the IAHC solicited public comments, its meetings were closed and its minutes were not published. Nevertheless, the IAHC process represented the first serious attempt to deal with the problems of DNS governance in a systemic manner. The IAHC final report attempted to address the inherent problems involved with establishing a legitimate DNS governance structure, referred to at [2.1]–[2.3], by means of a framework agreement, known as the Generic Top-level Domain Memorandum of Understanding (gTLD-MoU). The gTLD-MoU framework was based on the following underlying principle:

> the Internet top level domain space is regarded as a public resource and is subject to the public trust. Therefore, any administration, use and/or evolution of the Internet TLD space is a public policy issue and must be carried out in an open and public manner in the interests and service of the public.[49]

The main features of the governance structure proposed by the IAHC report were:

- A Policy Oversight Committee (POC) to be responsible for gTLD policy development, including decisions relating to the introduction of new gTLDs. The POC would consist of representatives of IANA, the IAB, the ITU, the Internet Society, INTA, WIPO and domain name registrars.
- A Policy Advisory Body (PAB), which would advise the POC on gTLD policy matters. The PAB would consist of signatories to the gTLD-MoU comprising governmental organisations, non-governmental organisations, industry and Internet organisations.
- A Council of Registrars (CORE), to be established by the POC and to consist of all domain name registrars. CORE was to be a non-profit-making association incorporated in Geneva and subject to Swiss law. CORE was intended to coordinate the activities of registrars, including the operation of the DNS database, which would be operated by an independent third party.
- The ITU would act as the official depository for the gTLD-MoU.

According to the IAHC report, the gTLD-MoU would come into effect upon obtaining the signatures of the IANA and the Internet Society. Nevertheless, it was envisaged that the gTLD-MoU would be signed by all major public and private sector stakeholders.

A number of elements that would later become incorporated into DNS governance structures appeared for the first time in the IAHC final report, including:

- The separation of the functions involved with the operation of the database for registering domain names for a gTLD from the functions involved in registering domain names for customers. For the first time, the entity responsible for operating the database was referred to as a 'registry' and entities responsible for dealing with customers, including the sale of domain names, were referred to as 'registrars'.

[48] International Ad Hoc Committee (IAHC), *Final Report of the International Ad Hoc Committee: Recommendations for Administration and Management of gTLDs* (4 Feb 1997), available at http://www.iahc.org/draft-iahc-recommend-00.html.

[49] *Ibid*, para [5.1.1].

- The introduction of multiple, competing registrars that would have shared access to gTLD registry databases. The IAHC report recommended the introduction of competition by means of 28 regionally based registrars, to be selected by lottery.
- The phased introduction of a small number of new gTLDs. In place of the substantial numbers of new gTLDs proposed in Postel's Internet-Draft, the IAHC report recommended the initial introduction of seven new gTLDs.
- The introduction of a procedure for dealing with the concerns of trade mark owners. Under the procedure proposed in the IAHC report, registration of second-level domain names in the gTLDs would be subject to expedited arbitration by domain name 'challenge panels' to be administered by WIPO. The IAHC report envisaged that decisions of 'challenge panels' would have no effect on the ability of parties to domain name disputes to bring actions before national courts.

The IAHC final report can be seen as a series of compromises between the major stakeholders represented on the IAHC. It was an attempt to deal with the problems of DNS governance by establishing a contract-based authority structure, with signatories including prominent public and private sector entities with an interest in the DNS. At the same time, it sought to retain a significant role for those who had traditionally been directly involved in DNS governance, in the form of the IANA and the Internet Society. It also represented an attempt to deal with the concerns of organisations that had come to have an interest in DNS governance, including the ITU, and trade mark interests, as represented by INTA and WIPO.

In March 1997, Jon Postel, representing the IANA, and a representative of the Internet Society signed the gTLD-MoU.[50] The Internet Society and the ITU organised a formal signing ceremony for the gTLD-MoU to be held in Geneva in May 1997. The underlying problem, however, remained that all of the parties, including the IANA and Internet Society, lacked clear authority to establish a new governance structure for the DNS. Moreover, the IAHC coalition did not include some prominent interest groups. In particular, the IAHC excluded NSI, which stood to lose its privileged position of monopoly control over registrations in the .com gTLD if the IAHC proposals were implemented. Nevertheless, under the agreement with the US Defense Department referred to at [2.6], NSI retained control of the authoritative A root server.

The extent to which control of the A root server and the authoritative root zone file conferred control over the introduction of new gTLDs was tested in early 1997, when an alternative root registry requested that NSI include new TLDs in the authoritative DNS and brought an action against NSI under US antitrust law.[51] NSI initially referred the request to Postel and IANA, which denied any authority in the absence of consensus from the 'Internet community'. NSI then referred the request to the NSF, which under the Cooperative Agreement with NSI directed that NSI refrain from adding new gTLDs until an internal review of DNS policy was completed. The NSF was therefore forced to acknowledge some responsibility in relation to the introduction of new gTLDs precisely during the period in which the IAHC was attempting to assert authority through the gTLD-MoU process. The NSF was, nevertheless, uncomfortable with assuming this responsibility and had, in fact, already agreed in principle to an early termination of the Cooperative Agreement with NSI.[52] The NSF's attempt to remove itself from the increasingly controversial debates

[50] See eg Mueller, above n 19, 146.
[51] *Name.Space, Inc v Network Solutions, Inc*, 202 F 3d 573, 579 (2000). The action was ultimately unsuccessful.
[52] See Mueller, above n 19, 155.

concerning DNS governance was, however, derailed by the direct intervention of the US government. Meanwhile, the cooperative agreement with NSI was extended to 30 September 1998, and the US Department of Commerce took over administration of the NSI agreement from the NSF.

Origins of ICANN

[2.9] The NTIA Green Paper

In December 1995, the Clinton administration established an Interagency Task Force responsible for policy on electronic commerce, headed by Ira Magaziner. The Task Force was largely responsible for the Clinton–Gore policy statement, entitled *Framework for Global Electronic Commerce,* which was released in July 1997.[53] The Clinton–Gore Statement included the following general statement of principle:

> For electronic commerce to flourish, the private sector must continue to lead. Innovation, expanded services, broader participation, and lower prices will arise in a market-driven arena, not in an environment that operates as a regulated industry.[54]

The Statement also included the following statements relating to DNS governance:

> The Administration supports private efforts to address Internet governance issues including those related to domain names and has formed an interagency working group under the leadership of the Department of Commerce to study DNS issues . . . The group will consider, in light of public input, (1) what contribution government might make, if any, to the development of a global competitive, market-based system to register Internet domain names, and (2) how best to foster bottom-up governance of the Internet.[55]

Magaziner had established the interagency working group earlier in 1997, with responsibility for the group being subsequently given to the National Telecommunications and Information Administration (NTIA), which is part of the Department of Commerce (DoC). In conjunction with the release of the Clinton–Gore Statement, the President directed the Secretary of Commerce to privatise the DNS, to increase competition in the DNS and to facilitate international participation in the management of the DNS. Immediately following the Presidential direction, the NTIA initiated a formal proceeding, in the form of a Request for Comments (RFC), soliciting public comments on DNS administration.[56]

By mid-1997, then, the confusion surrounding authority for the DNS had resulted in two completely separate processes aimed at establishing governance structures for the DNS based on different versions of who had authority for DNS policy-making. The gTLD-MoU process, which had obtained the support of some influential public and private sector organisations, was ultimately based on the view that the IANA and Internet Society could make DNS policy. The US government, however, as represented by the interagency working

[53] President William J Clinton and Vice President Albert Gore Jr, *Framework for Global Electronic Commerce* (1 July 1997), available at http://www.technology.gov/digeconomy/framewrk.htm. .

[54] *Ibid,* 2.

[55] *Ibid,* 11.

[56] US Department of Commerce, *Requests for Comments in the Matter of Registration and Administration of Internet Domain Names,* Doc No 970613137-7137-01 (1 July 1997).

group, decided against supporting the gTLD-MoU process. The NTIA process, on the other hand, was based on the view that US government agencies had, at the very least, a central role to play in setting DNS policy.

In practice, the policy authority claimed by the US government was largely derived from the ability to give directions to NSI, which retained control of the authoritative A root server and root zone file. It is hardly surprising, therefore, that the tensions between the two processes ultimately resulted in a conflict over control of the authoritative root zone file. In January 1998, before the release of the NTIA draft report, Postel sent an e-mail to those responsible for the authoritative root servers instructing them to point to the B root server, which was under the control of the IANA, instead of NSI's A root server.[57] Some of those responsible for the root servers modified their zone files to point to the IANA's B server, but others, including NSI, did not. Upon learning of Postel's actions, Magaziner directed him to return the root servers to their previous configuration, thereby restoring NSI's A server to the top of the DNS hierarchy. Postel complied with the direction, and later referred to his request to redirect the root servers as a 'test'. Nevertheless, the attempt at redirecting the root servers illustrated the fragility of any attempt to assert authority over the DNS, given the dependence of the DNS upon decisions made by those in charge of the root servers.

The NTIA interagency working group released a report, in the form of a Notice of Proposed Rulemaking, based on its initial public consultations in January 1998.[58] The report became known as the Green Paper, in recognition of its status as a draft report soliciting further public comment. The Green Paper set out four principles to guide DNS governance: stability, competition, private bottom-up coordination, and representation. In relation to the proposed privatisation of DNS governance, the report stated:

> Certain technical management functions require coordination. In these cases, responsible, private-sector action is preferable to government control. A private coordinating process is likely to be more flexible than government and to move rapidly enough to meet the changing needs of the Internet and of Internet users. The private process should, as far as possible, reflect the bottom-up governance that has characterized development of the Internet to date.[59]

The main substantive recommendations of the Green Paper were as follows:

- A private, non-profit-making corporation should be created to perform the necessary DNS coordination functions, including setting policy for the allocation of IP number blocks, overseeing the operation of an authoritative root server system, overseeing policy for determining the circumstances under which new TLDs are added to the root system and coordinating the development of other relevant technical parameters.
- The US government should, as soon as possible, transfer existing IANA functions to the new corporation. It should, however, participate in policy oversight to assure stability until the new corporation is established and stable.
- The new corporation would be based in the United States and incorporated as a non-profit-making corporation under US law. The board of directors for the new corporation should represent the main DNS stakeholders.
- As in the IAHC final report, a distinction was drawn between registries and registrars. The Green Paper defined a 'registry' as being 'responsible for maintaining a TLD's zone files,

[57] See eg Mueller, above n 19, 161; Froomkin, above n 17, 64.

[58] National Telecommunications and Information Administration (NTIA), *A Proposal to Improve the Technical Management of Internet Names and Addresses*, 63 Fed. Reg 8,825 (1998).

[59] *Ibid*, para VC.

which contain the name of each SLD in that TLD and each SLD's corresponding IP number'.[60] It defined a 'registrar' as 'an interface between domain-name holders and the registry, providing registration and value-added services'.[61]

- Competition should be introduced in relation to registrar functions and there should be 'limited experimentation' with competing registries.[62]
- In the transition to private management of the DNS, up to five new gTLDs should be introduced to be managed by five new registries. IANA would be responsible for establishing criteria for selecting the new registries.
- Each gTLD registry should establish minimum dispute resolution procedures relating to trade mark considerations.
- The Cooperative Agreement with NSI would be terminated by the end of September 1998, but NSI would continue to operate as the registry for the .com, .net and .org gTLDs. NSI would, nevertheless, be required to separate its registry functions from its functions as a registrar.

The Green Paper met with considerable opposition from a variety of interest groups including, but not confined to, supporters of the gTLD-MoU. Most opposition focussed on the central role that the US government had assumed in forming policy for a global communications system and on the recommendation to subject the new corporation proposed to be responsible for DNS policy-making to US law. Opposition to the dominant role assumed by the US government effectively united the European Commission, other interested national governments (including Australia) and supporters of the gTLD-MoU. A further source of opposition focussed on concerns that the new process was abandoning the traditions of Internet self-governance as practised, for example, by the IETF. The main opposition to the Green Paper therefore centred on the proposed process for establishing a new DNS governance regime rather than on the substantive recommendations relating to the nature of the new regime, which bore some similarities to the proposed gTLD-MoU regime.[63] Progress in reconciling the two processes—the gTLD-MoU process and the NTIA process—depended upon some understanding being reached concerning the role of the US government, which would include concessions made both by those involved with the gTLD-MoU process and by the US government. A new organisation, the IANA Transion Advisors Group (ITAG), which was established in February 1998 to advise Jon Postel on the transition to a new governance regime, joined negotiations between the two camps. Largely as a result of compromises reached between those involved with the two separate processes, there were significant changes between the recommendations made in the Green Paper and those made in the final NTIA report, which is known as the White Paper.

[2.10] The NTIA White Paper

The final NTIA report, commonly known alternatively as the White Paper or Magaziner Report, was released in June 1998.[64] The first significant difference between the White Paper

[60] National Telecommunications and Information Administration (NTIA), *A Proposal to Improve the Technical Management of Internet Names and Addresses,* 63 Fed. Reg 8,825 (1998), para VIB.

[61] *Ibid.*

[62] *Ibid.*

[63] As Mueller puts it, '[o]pposition to the Green Paper had more to do with who would be in control of the transition than the actual policies that would be adopted': Mueller, above n 19, 168.

[64] National Telecommunications and Information Administration (NTIA), *Management of Internet Names and Addresses* 63 Fed Reg 31,741 (1998).

and the Green Paper was the form taken by the final report. Whereas the Green Paper was a proposed rulemaking by a US federal government agency, the White Paper was a non-binding statement of policy. This change in form effectively avoided the difficult legal question of the authority of the US Department of Commerce to make binding rules in relation to the administration of the DNS. Accompanying this change in form, the recommendations made in the White Paper were much less prescriptive than those in the Green Paper. The White Paper explained the changes from the draft report in the following terms:

> In light of the public comment and the changes to the proposal made as a result, as well as the continued rapid technological development of the Internet, the Department of Commerce has determined that it should issue a general statement of policy, rather than define or impose a substantive regulatory regime for the domain name system. As such, this policy statement is not a substantive rule, does not contain mandatory provisions and does not itself have the force and effect of law.[65]

The main recommendation of the White Paper remained the privatisation of DNS governance by transferring DNS coordination functions to a private, non-profit-making corporation. The attempt to accommodate supporters of the gTLD-MoU, however, was reflected by the extent to which important policy decisions were left to be determined by the proposed new corporation rather than specified in the White Paper itself. The conciliatory attitude expressed towards the established Internet technical community, which had been instrumental in promoting the gTLD-MoU process, was summed up by the following comments concerning IANA:

> IANA has functioned as a government contractor, albeit with considerable latitude, for some time now. Moreover, IANA is not formally organized or constituted. It describes a function more than an entity, and as such does not currently provide a legal foundation for the new corporation. This is not to say, however, that IANA could not be reconstituted by a broad-based, representative group of Internet stakeholders or that individuals associated with IANA should not themselves play important foundation roles in the formation of the new corporation. We believe, and many commenters also suggested, that the private sector organizers will want Dr. Postel and other IANA staff to be involved in the creation of the new corporation.[66]

The most significant recommendations made by the White Paper were as follows:

- The four principles guiding the management of the DNS remained those set out in the Green Paper, namely: stability; competition; private, bottom-up coordination; and representation.
- A new, private, non-profit-making corporation should be established to manage and perform functions related to coordination of the DNS including: setting policy for and directing allocation of IP number blocks; overseeing operation of the authoritative root server system; overseeing policy for determining the circumstances under which new TLDs are added to the root system; and coordinating the assignment of other Internet technical parameters as needed to maintain universal connectivity on the Internet.
- The new corporation should be based in the United States and incorporated under US law. National governments and international organisations acting as representatives of national governments should not participate in the management of Internet names and addresses.

[65] *Ibid*, 13.
[66] *Ibid*, 7.

- The board of directors of the new corporation should include representatives of IP number registries, domain name registries, domain name registrars, the technical community, Internet Service Providers (ISPs), and Internet users from around the world. The White Paper did not, however, set mandatory limits for board representation, effectively leaving this to the proposed new corporation.
- Competition should be introduced to the performance of registrar functions. While the White Paper followed the Green Paper in supporting limited competition in the performance of registry functions, criteria for authorising new gTLD registries were left to the proposed new corporation.
- While the Green Paper recommended that the US government introduce up to five new gTLDs during the transition to a new regime, the White Paper recommended that the proposed new corporation be responsible for decisions relating to the introduction of new gTLDs.
- The US government would request WIPO to initiate a process to: develop recommendations for a uniform approach to resolving trade mark/domain name disputes; recommend a process for protecting famous trade marks in the gTLDs; and evaluate the effects of adding new gTLDs and related dispute resolution procedures on trade mark and intellectual property holders.
- The proposed new corporation should adopt policies which require domain name registrants to submit to the authority of a court of law in the jurisdiction in which the registry, registry database, registrar or the A root servers are located. Domain name registrants should also be required to agree, at the time of registration, to submit to and be bound by alternative dispute resolution systems determined by the new corporation to apply to trade mark/domain name disputes. Moreover, the proposed new corporation should not limit the rights available to trade mark owners or domain name registrants under national laws.
- NSI would be required to act in a manner consistent with the policies set out in the White Paper, including taking action to permit the development of competition in domain name registration and recognising the role of the proposed new corporation in setting DNS policy.

Given that the four guiding principles have become criteria for determining the success of the privatisation of DNS management functions, they are worth setting out in full. The four guiding principles as set out in the White Paper are:

1. Stability

During the transition and thereafter, the stability of the Internet should be the first priority of any DNS management system.

2. Competition

Where possible, market mechanisms that support competition and consumer choice should drive the management of the Internet because they will lower costs, promote innovation, encourage diversity and enhance user choice and satisfaction.

3. Private Sector, Bottom-Up Coordination

A private coordinating process is likely to be more flexible than government and to move rapidly enough to meet the changing needs of the Internet and of Internet users. The private process should, as far as possible, reflect the bottom-up governance that has characterized development of the Internet to date.

4. Representation

Management structures should reflect the functional and geographic diversity of the Internet and its users. Mechanisms should be established to ensure international participation in decision making.[67]

As is clear from the recommendations, the White Paper addressed the concerns of opponents of the Green Paper mainly by modifying the suggested role of the US government in the transition to the proposed new DNS governance regime. Thus, the White Paper included no specific recommendations regarding the introduction of new gTLDs, the recognition of new registries or the structure and composition of the proposed new corporation. At the same time, the essential policy directions set out in the Green Paper, including privatisation of DNS governance functions and the introduction of competition, were retained.

An important feature of the White Paper was the increased emphasis given to trade mark/domain name disputes. For example, the White Paper included express provision for a role for WIPO in developing recommendations for a new dispute resolution policy, whereas WIPO had not been mentioned in the Green Paper. Moreover, while the Green Paper had recommended that each gTLD registry establish minimum dispute resolution procedures, the White Paper supported a uniform approach to resolving domain name disputes. Although WIPO was incorporated into the proposed new regime, however, the other international organisation that had been involved with the gTLD-MoU process, the ITU, was effectively excluded. This appears to reflect the extent to which business interests that had been active in consultations regarding the Green Paper were concerned to ensure adequate protection of trade mark interests, but saw no clear value in DNS policy-making incorporating what was perceived to be a highly bureaucratic international treaty-based organisation.[68] Apart from the ITU, the other organisation obviously disadvantaged by the recommendations made in the White Paper was NSI, which was to be made subject to policy direction from the proposed new corporation in addition to being subject to competition.

The White Paper therefore established the conditions for a reconciliation between supporters of the gTLD-MoU process, including IANA and the Internet Society, and the US government, which had initiated the NTIA process. The grounds for reconciliation with opponents of the Green Paper were established by the extent to which the White Paper provided for a less prescriptive role for the US government in the transition to the new governance regime than had been envisaged in the Green Paper. At the same time, a much more influential policy role in the transition to a new regime was envisaged for the proposed new corporation. For example, the proposed new corporation would be responsible for decisions relating to the introduction of new gTLDs and the introduction of competition in the performance of relevant DNS functions. The new corporation would also be largely responsible for determining its own structure and composition. Moreover, although the White Paper identified some desirable features of the proposed new corporation, the US government would not take an active part in establishing the new corporation. In this respect, the White Paper simply stated:

Internet stakeholders are invited to work together to form a new, private, not-for-profit corporation to manage DNS functions.[69]

[67] *Ibid*, 24–5.

[68] On the coalition of business interests that formed to promote compromises between the gTLD-MoU process and the NTIA process see Mueller, above n 19, 168–72.

[69] See NTIA, above n 64, 14.

The role of the US government, as set out in the White Paper, was therefore largely confined to recognising the proposed new corporation once it had been formed by relevant interest groups, including the Internet technical community. Thus, the White Paper stated:

> the U.S. Government is prepared to recognize, by entering into agreement with, and to seek international support for, a new, not-for-profit corporation formed by private sector Internet stakeholders to administer policy for the Internet name and address system. Under such agreement(s) or understanding(s), the new corporation would undertake various responsibilities for the administration of the domain name system now performed by or on behalf of the U.S. Government or by third parties under arrangements or agreements with the U.S. Government.[70]

The White Paper envisaged that most significant decisions relating to the transition to the proposed new regime would be made by an interim board of directors appointed once the new corporation was formed. Attention therefore focussed on the process for selecting the proposed new corporation, which became known as 'NewCo'. Although the White Paper represented a reconciliation between the US government and supporters of the gTLD-MoU, the process for selecting NewCo was characterised by a further split. This time the split was predominantly between those who had been part of the deal embodied in the White Paper—essentially, policy 'insiders'—and those who attempted to coordinate an open process for establishing NewCo —the policy 'outsiders'.

[2.11] Formation of ICANN

The White Paper called on private sector stakeholders to establish the proposed new corporation, but provided no guidance in relation to the process for selecting NewCo. The invitation to establish NewCo resulted in two separate processes which, in large measure, mirrored a struggle for control over the future of DNS policy-making between insiders and outsiders.

The White Paper had clearly envisaged a role for IANA and Jon Postel in the proposed new regime. Moreover, Postel had, over a long period, accumulated substantial personal prestige among the Internet technical community. On the advice of the ITAG (referred to at [2.9] above), Postel hired a prominent antitrust lawyer, Joe Sims. Sims and Postel drafted articles of incorporation and bylaws for a new corporation which were designed to be compatible with the desirable features of NewCo identified in the White Paper. The corporate structure chosen by Sims was a non-profit-making public benefit corporation, incorporated under Californian law. Sims and Postel decided that the interim board should consist of distinguished individuals who had not been actively involved in the disputes relating to DNS governance. The articles of incorporation drafted by Sims and Postel provided that half of the NewCo board would be appointed by the interim board and half appointed by stakeholders, to be known as Supporting Organisations.

Meanwhile, a process involving public meetings, known as the International Forum on the White Paper (IFWP), had been initiated by organisations, including NSI, which had not been part of the informal coalition that had been influential in shaping the White Paper. As it progressed, however, the public process obtained the support of other groups, including representatives of Internet Service Providers (ISPs) and organisations from outside the United States, such as the European Commission. In contrast to the Sims–Postel model, the

[70] See NTIA, above n 64, 13.

IFWP process came to support a corporate structure based on board members selected to represent DNS interest groups.[71] The IFWP process, however, failed to obtain the support of most of the Internet technical community, especially the IANA and the Internet Society. Given their historic role in administration of the DNS, the participation of the IANA and Internet Society was clearly an essential element of any future DNS governance regime. The final meetings organised by IFWP participants were eventually cancelled when representatives of the IANA refused to attend.

Although the closed IANA process coordinated by Postel and Sims had derailed the IFWP process, in practice NSI retained significant political leverage, especially as it retained control of the authoritative A root server and root zone file. Magaziner therefore encouraged the IANA and NSI to enter into negotiations. As a result of these negotiations, the parties released draft articles of incorporation and bylaws for the proposed new corporation on 17 September 1998. For the first time, these draft documents referred to the proposed corporation as the Internet Corporation for Assigned Names and Numbers (ICANN).[72]

The deadline set for the formation of NewCo was designed to coincide with the termination of the cooperative agreement between the US government and NSI, which was due to expire on 30 September 1998. The draft incorporation documents agreed to between IANA and NSI, however, proved unpopular with the Internet technical community. Consequently, at the end of September 1998, Sims and Postel submitted revised draft articles of incorporation and bylaws for the proposed new corporation to the US Department of Commerce, which nevertheless retained the name 'ICANN' for the new corporation.

ICANN was formally incorporated under Californian law on 30 September 1998. In submitting ICANN's constitutional documents, Sims and Postel claimed that they reflected 'the consensus judgment of the global Internet community as to how to form a corporation that will include the IANA function'.[73] Sims and Postel also submitted a list of names of nine interim board members, who had been chosen following private negotiations involving the IANA, the Internet Society, business interests, the European Commission and the Australian government. Jon Postel was designated as chief technical officer. At the same time, two alternative proposals for the new corporation were submitted to the Department of Commerce: one by some remaining representatives of the IFWP process, known as the Boston Working Group (BWG), and one from representatives of organisations promoting alternative root servers, known as the Open Root Server Confederation (ORSC).

Given the breakdown of negotiations between the IANA and NSI, and the imminent expiry of the NSI cooperative agreement, the US Department of Commerce was required to come to terms with NSI before the new DNS governance regime could be established. Consequently, on 7 October 1998, the Department of Commerce entered into a new agreement with NSI, known as Amendment 11 to the NSI Cooperative Agreement.[74] Amendment 11 was the first time that the proposed new corporation was formally referred to as 'NewCo'. As further explained at [2.12] below, the new agreement involved a compromise, which extended NSI's role as both registry and registrar for the .com, .net and .org gTLDs, and in managing the root zone file, until 30 September 2000. In return, NSI agreed to develop a shared registry system (SRS), which would allow competition among multiple

[71] See Mueller, above n 19, 177.
[72] See M Mueller, 'ICANN and Internet Governance' (1999) 1 *info* 497, 507.
[73] See Weinberg, above n 16, 209.
[74] Amendment 11 to Cooperative Agreement Between NSI and US Government (7 Oct 1998), available at http://www.icann.org/nsi/coopagmt-amend11-07oct98.htm.

registrars in the .com, .net and .org gTLDs. Furthermore, NSI agreed to comply with a price cap on the fees it could charge registrars for its registry services. Importantly, NSI also agreed to enter into an agreement with NewCo under which it would recognise NewCo's authority over the DNS pursuant to a proposed agreement between the US government and NewCo.

The support for the Sims-Postel (ICANN) model among the Internet technical community could, in large measure, be attributed to Postel's personal prestige within that community. The process of establishing a new DNS governance regime therefore suffered a set-back when Jon Postel died from a heart attack on 18 October 1998. By this stage, however, the final proposals for NewCo had been submitted to the Department of Commerce and the new NSI cooperative agreement had been entered into. The Department of Commerce solicited public comments on the various proposals, but there never appears to have been any doubt that ICANN would eventually be selected.[75] Nevertheless, as a result of the considerable number of public submissions that were critical of the ICANN model, the interim board of the new corporation was requested to enter into negotiations with those responsible for other proposals. Following these negotiations, the Bylaws of the new corporation were amended slightly to provide for a more representative board and greater transparency.

On 25 November 1998, the US Department of Commerce entered into a Memorandum of Understanding with ICANN.[76] Later, on 26 February 1999, the NTIA notified NSI that ICANN had been designated as the private, non-profit-making organisation referred to as NewCo in Amendment 11 of the NSI cooperative agreement.[77] These steps paved the way for the interlocking contractual arrangements between the US Department of Commerce, ICANN and NSI that, in essence, continue to provide the legal framework for DNS governance.

Contractual Basis of DNS Governance

[2.12] The 1999 Agreements

The formal legal structure for DNS governance consists of interlocking contracts between the US Department of Commerce, ICANN and NSI (now VeriSign, Inc). The precise source of legal authority for each of the parties, including the Department of Commerce, to enter into agreements concerning the management of the DNS is surprisingly uncertain. In large measure, the difficulties in defining legal authority over the DNS derive from the inherent problems involved in establishing a legitimate basis for DNS governance referred to at [2.1]–[2.3]. Authority over the DNS is therefore defined mainly by means of the agreements

[75] Mueller has summed up the process of selecting the new corporation in the following terms: '[p]laced in historical context, IANA's unilateral Board selection and its unwillingness to participate meaningfully in the IFWP were unsurprising. Behind the scenes, Magaziner and the Commerce Department had already tacitly anointed it as the nucleus of the new corporation': Mueller, above n 72, 507.

[76] *Memorandum of Understanding Between the US Department of Commerce and Internet Corporation for Assigned Names and Numbers* (25 Nov 1998), available at http://www.icann.org/general/icann-mou-25nov98.htm.

[77] Letter from J Beckwith Burr to David Graves , *Re: Internet Corporation for Assigned Names and Numbers (ICANN) designated as 'NewCo' for certain purposes under Cooperative Agreement No NCR-9218742, as amended* (26 Feb 1999), available at http://www.icann.org/newco.html.

between the parties rather than by reference to any independent, external source of legal authority, such as an international treaty.

The legal relationship between the US Department of Commerce and ICANN was initially defined by the Memorandum of Understanding (MOU) first entered into by the parties on 25 November 1998. The MOU, which is also known as the 'Joint Project Agreement', committed the parties to develop mechanisms, methods and procedures to achieve the transition to private management of the DNS without disrupting the operation of the Internet. The MOU nominated ICANN as the private, non-profit-making corporation envisaged by the White Paper. It referred to ICANN in the following terms:

> DOC has determined that this project can be done most effectively with the participation of ICANN. ICANN has a stated purpose to perform the described coordinating functions for Internet names and addresses and is the organization that best demonstrated that it can accommodate the broad and diverse interest groups that make up the Internet community.[78]

Under the MOU, the Department of Commerce asserted authority to oversee DNS management functions in order to ensure continuity and stability in the performance of technical management of the DNS in the transition to a private management regime. Meanwhile, in December 1998 a separate agreement was entered into between the University of Southern California and ICANN for the transfer of the IANA functions performed by ISI to ICANN.[79]

The original MOU with the Department of Commerce was due to terminate on 30 September 2000, but was amended six times after it was originally entered into, with the expiration date being progressively extended to 30 September 2006. The amendments, which increased the Department of Commerce's supervision of ICANN, are dealt with at [2.14]. In September 2006 a new Joint Project Agreement was entered into which made major amendments to the then existing MOU. The new agreement effectively increased the day-to-day autonomy of ICANN, while maintaining a supervisory role for the Department of Commerce and extending the agreement to 30 September 2009. Details of the MOU are dealt with at [2.13] and the 2006 agreement is explained at [2.17].

As explained at [2.11], Amendment 11 to the cooperative agreement between the Department of Commerce and NSI, entered into on 7 October 1998, extended NSI's role as registry and registrar for the open gTLDs until 30 September 2000. In return, NSI agreed to develop a shared registration system and to enter into a contractual relationship with the private, non-profit-making organisation designated as NewCo. The agreement included a timetable, which specified deadlines for NSI to meet targets in relation to the development of a shared registration system. NSI also agreed to provide accredited registrars with equal access to its registry services and provide safeguards to ensure that it would not use its role as a registry to disadvantage competitive registrars. As further explained, on 26 February 1999, the NTIA notified NSI that ICANN had been designated as the private, non-profit-making organisation. Amendment 11 to the cooperative agreement is notable as it represents the first time that the US Government formally asserted 'policy authority' over the DNS root zone file. In this respect, the amendment stated:

[78] *Memorandum of Understanding Between the U.S. Department of Commerce and Internet Corporation for Assigned Names and Numbers* (25 Nov 1998), available at http://www.icann.org/general/icann-mou-25nov98.htm, para II B

[79] *University of Southern California/ICANN Transition Agreement* (24 Dec 1998), available at http://www.icann.org/general/usc-icann-transition-agreement.htm.

> While NSI continues to operate the primary root server, it shall request written direction from an authorized USG official before making or rejecting any modifications or deletions to the root zone file.

Subsequent amendments to the cooperative agreement between the Department of Commerce and NSI (now VeriSign) are dealt with at [2.19].

Upon being recognised by the Department of Commerce, the first substantive issue addressed by ICANN's management and staff was NSI's continued role as monopoly provider of registry services for the open gTLDs. It was decided that, in addition to licensing registries, ICANN would be directly responsible for authorising registrars operating in the gTLDs. Draft guidelines for the accreditation of registrars were prepared before NSI was notified of ICANN's designation and approved by ICANN's board in March 1999.[80] Under the accreditation guidelines, registrars were required to pay ICANN a fixed fee of US$5,000 and a fee of US$1 per domain name registered per year. The guidelines also included requirements aimed at protecting trade mark interests, including the domain name registrant's consent to jurisdiction. Following the issue of the registrar accreditation guidelines, Amendment 13 to the cooperative agreement between the Department of Commerce and NSI set a cap for the wholesale price NSI could charge for the registration of domain names at US$9 per name per year.

In the first months of ICANN's existence relations between ICANN and NSI were tense. Matters were not helped when, following an inquiry by the US Congress in which NSI had opposed ICANN's attempts to impose charges on registrars, the Department of Commerce required ICANN to abandon its plans to charge US$1 per domain name registered. This created a short-term funding crisis for ICANN, which had to be resolved by soliciting donations from private corporations. In ICANN's first status report, presented to the Department of Commerce in June 1999, ICANN identified the relationship with NSI as the most important immediate challenge it faced. The first status report stated:

> NSI has to date refused to accept the community-consensus registrar accreditation policies adopted by ICANN after public notice and comment, and has even asserted that it should not have to comply with the same accreditation standards that apply to all other registrars. Obviously, full and fair competition requires that all have the same opportunities, and to the extent that there are consumer protection or other requirements, that all meet them equally. Thus, it is critical to accomplishing the White Paper objective of maximizing competition that (1) NSI's registry and registrar functions be fully separated, so that NSI as a registrar does not have any structural advantage over its registrar competitors; (2) NSI accept community consensus policies relating to registrars, as reflected in ICANN's accreditation standards; and (3) the relationship between NSI as registry and all registrars does not allow NSI to impair or adversely affect the development of competition because of its continuing monopoly position as registry operator.[81]

ICANN was therefore in the position where it wished to assert policy control over registrars by means of its accreditation guidelines, but NSI, the sole registry for the open gTLDs and, until then, the sole registrar, refused to enter into an agreement. NSI's authority to operate as sole registry and as registrar for the open gTLDs derived from the cooperative agreement with the Department of Commerce, not from ICANN. While NSI refused to enter into an agreement with ICANN, it continued to have de facto control over registra-

[80] See Mueller, above n 19, 293, fn 1.

[81] ICANN, *First Status Report to the Department of Commerce* (15 June 1999), available at http://www.icann.org/general/statusreport-15june99.htm.

tions in the open gTLDs, subject only to the cooperative agreement with the Department of Commerce. Nevertheless, the new regime of competitive registrars could not effectively commence operations until NSI accepted ICANN's role in setting DNS policy.

The impasse between ICANN and NSI came to an end with four interlocking agreements between the Department of Commerce, ICANN and NSI that were announced on 28 September 1999. These four agreements, as amended by their successors, continue to be the fundamental legal documents that establish the structure for DNS governance. After a short period for public comment, the final agreements were approved on 4 November, and entered into force on 10 November 1999. The agreements and the main features of the bargain struck between the parties are as follows:

Amendment 19 to Cooperative Agreement Between NSI and US Government[82]

—The cooperative agreement between the Department of Commerce and NSI was amended to require NSI to recognise ICANN and to define the relationship between ICANN and NSI. The most important elements of Amendment 19 were:
—NSI agreed to enter into a registry agreement and a registrar accreditation agreement with ICANN.
—If ICANN terminated the registry agreement or registrar accreditation agreement, the Department of Commerce would terminate NSI's authority to provide registry or registration services under the cooperative agreement.
—NSI agreed only to accept registrations in the open gTLDs from ICANN accredited registrars.
—The Department of Commerce agreed to ensure that the authoritative root would point to a root server designated by NSI until the earlier of the termination of the cooperative agreement or the registry agreement with ICANN.
—NSI would remain the registry for the open gTLDs until the earlier of the termination of the cooperative agreement or the registry agreement with ICANN.
—NSI agreed not to deploy an alternative DNS root server.

ICANN-NSI Registry Agreement[83]

Pursuant to Amendment 19 to the cooperative agreement, NSI entered into a registry agreement with ICANN. The main elements of the ICANN-NSI Registry Agreement were as follows:

—ICANN agreed that NSI would continue to operate as the registry for the open gTLDs for the term of the agreement.
—ICANN agreed to ensure that the authoritative root would point to a root server designated by NSI for the term of the agreement.
—NSI agreed to comply, in its operation as a registry, with all consensus policies adopted by ICANN, including policies relating to the allocation of 2LDs and domain name dispute resolution.
—NSI agreed to pay a registry fee to ICANN.
—If the Department of Commerce terminated the cooperative agreement with NSI, ICANN would comply with a request from the Department of Commerce to terminate the NSI Registry Agreement.

[82] *Amendment 19 to Cooperative Agreement Between NSI and US Government* (approved 4 Nov 1999; signed 10 Nov 1999), available at http://www.icann.org/nsi/coopagmt-amend19-04nov99.htm.

[83] *ICANN-NSI Registry Agreement* (approved 4 Nov 1999; signed 10 Nov 1999), available at http://www.icann.org/nsi/nsi-registry-agreement-04nov99.htm.

—The NSI Registry Agreement was subject to the cooperative agreement between NSI and the Department of Commerce.

ICANN-NSI Registrar Transition Agreement[84]

ICANN and NSI entered into a Registrar Transition Agreement, which provided the conditions for NSI to be accredited as a registrar under ICANN's Registrar Accreditation Agreement. The main elements of the Registrar Transition Agreement were:

—ICANN agreed to accept NSI's application for accreditation as a registrar for the open gTLDs.
—NSI agreed to pay registrar accreditation fees to ICANN.

Amendment 1 to ICANN/DOC MOU[85]

The MOU between the Department of Commerce and ICANN was amended, essentially to provide for Department of Commerce supervision of the relationship between ICANN and NSI. The main elements of Amendment 1 to the MOU were:

—The Department of Commerce approved the ICANN-NSI Registry Agreement and ICANN agreed not to modify registry agreements relating to the .com, .net and .org gTLDs without the prior approval of the Department of Commerce.
—ICANN agreed not to enter into a registry agreement for the open gTLDs (.com, .net and .org) with a registry other than NSI without the prior approval of the Department of Commerce.
—ICANN agreed to terminate the NSI Registry Agreement in the event that the Department of Commerce terminated the NSI cooperative agreement.

In essence, then, under the interlocking agreements the parties agreed to the following mutually supportive arrangements:

- The Department of Commerce and ICANN agreed to recognise NSI's continued operation as the registry for the open gTLDs, and to ensure that the authoritative root would point to a root server designated by NSI;
- NSI agreed to be subject to ICANN registry and registrar agreements; and
- ICANN agreed not to modify the NSI Registration Agreement, or to authorise the operation of other registries in the open gTLDs (.com, .net and .org), without prior approval from the Department of Commerce.

The interlocking agreements between the parties are supplemented by a number of other agreements. In particular, in February 2000, the NTIA entered into a contract with ICANN, in the form of a purchase order, for the performance of the IANA technical function previously undertaken by the ISI at the University of Southern California.[86] The contract for performance of the IANA function between the US Government and ICANN is dealt with at greater length at [2.18].

[84] *ICANN-NSI Registrar Transition Agreement* (approved 4 Nov 1999; signed 10 Nov 1999), available at http://www.icann.org/nsi/icann-nsi-transition-04nov99.htm.

[85] *Amendment 1 to ICANN/DOC Memorandum of Understanding* (approved 4 Nov 1999; signed 10 Nov 1999), available at http://www.icann.org/nsi/amend1-jpamou-04nov99.htm.

[86] *ICANN/US Government Contract for the IANA Function* (9 Feb 2000), available at http://www.icann.org/general/iana-contract-09feb00.htm.

[2.13] The ICANN/DOC MOU

As explained at [2.12], the continuing legal relationship between the US Department of Commerce and ICANN is defined by the MOU, which was first entered into on 25 November 1998. A new Joint Project Agreement was entered into on 29 September 2006, which made major amendments to the MOU. The 2006 Joint Project Agreement is explained at [2.17].

The MOU committed the Department of Commerce and ICANN to collaborate on a project, referred to as the 'DNS Project' or the 'Joint Project', designed to provide for the eventual full transition of DNS management functions to the private sector. In this regard, the MOU states:

> the Parties will jointly design, develop, and test mechanisms, methods, and procedures that should be in place and the steps necessary to transition management responsibility for DNS functions now performed by, or on behalf of, the U.S. Government to a private-sector not-for-profit entity. Once testing is successfully completed, it is contemplated that management of the DNS will be transitioned to the mechanisms, methods, and procedures designed and developed in the DNS Project.[87]

The MOU adopted wholesale the four guiding principles for DNS governance originally established by the Green Paper and later incorporated in the White Paper, namely: stability; competition; private sector, bottom-up coordination; and representation.[88] The four guiding principles, as articulated in the White Paper, are set in full out at [2.10].

The general scope of the DNS Project, as defined by the MOU, consists of the following management functions:

a. Establishment of policy for and direction of the allocation of IP number blocks;
b. Oversight of the operation of the authoritative root server system;
c. Oversight of the policy for determining the circumstances under which new top level domains would be added to the root system;
d. Coordination of the assignment of other Internet technical parameters as needed to maintain universal connectivity on the Internet; and
e. Other activities necessary to coordinate the specified DNS management functions, as agreed by the Parties.[89]

Like the four guiding principles, the statement of management functions was adopted directly from the White Paper.

The transition period was always considered necessary to ensure that ICANN possessed the capabilities and resources to undertake the DNS management functions.[90] During the transition period, the Department of Commerce described itself as 'the steward of critical elements of the DNS', whereas ICANN was regarded as being responsible for 'day-to-day DNS management'.[91]

[87] *Memorandum of Understanding Between the U.S. Department of Commerce and Internet Corporation for Assigned Names and Numbers* (25 Nov 1998), available at http://www.icann.org/general/icann-mou-25nov98.htm, para IIB.

[88] *Ibid*, para IIC

[89] *Ibid.*

[90] See Department of Commerce, *Statement Regarding Extension of Memorandum of Understanding with ICANN* (20 Sept 2002), available at http://www.ntia.doc.gov/ntiahome/domainname/agreements/docstatement-09192002.htm.

[91] See Department of Commerce, *Statement Regarding Extension of Memorandum of Understanding with ICANN* (16 Sept 2003), available at http://www.ntia.doc.gov/ntiahome/domainname/agreements/sepstatement-09162003.htm.

The MOU initially provided for the agreement to expire on 30 September 2000.[92] After the agreement was originally entered into, however, the MOU was amended six times, which included four extensions to the term of the agreement. The main amendments to the MOU are explained at [2.1]. Following the sixth amendment, the agreement was set to expire on 30 September 2006.[93] Prior to the new Joint Project Agreement, which was entered into on 29 September 2006, the amendments to the MOU mainly concerned specific tasks that ICANN was required to perform and milestones that ICANN was required to meet.

[2.14] Department of Commerce Supervision

The amendments made to the MOU over the 1999–2006 period embodied a degree of supervision of ICANN by the Department of Commerce. In particular, the amendments modified the core tasks to be performed by ICANN and set relatively prescriptive, date-specific milestones required to be achieved by ICANN.

The original MOU set out nine core tasks, known as the 'transition tasks', for ICANN to complete as part of the DNS Project.[94] In summary, the original transition tasks that ICANN agreed to perform were to:

- Provide expertise and advice on private sector functions related to technical management of the DNS such as the policy and direction of the allocation of IP number blocks and coordination of the assignment of other Internet technical parameters as needed to maintain universal connectivity on the Internet.
- Collaborate on the design, development and testing of procedures by which members of the Internet community adversely affected by decisions that are in conflict with the bylaws of the organisation can seek external review of such decisions by a neutral third party.
- Collaborate on the design, development and testing of a plan for the introduction of competition in domain name registration services.
- Collaborate on written technical procedures for operation of the primary root server including procedures that permit modifications, additions or deletions to the root zone file.
- Collaborate on a study and process for making the management of the root zone server system more robust and secure.
- Collaborate on the design, development and testing of a process for affected parties to participate in the formulation of policies and procedures that address the technical management of the Internet. This process will include methods for soliciting, evaluating and responding to comments in the adoption of policies and procedures.
- Collaborate on the development of additional policies and procedures designed to provide information to the public.

[92] *Memorandum of Understanding Between the U S Department of Commerce and Internet Corporation for Assigned Names and Numbers* (25 Nov 1998), available at http://www.icann.org/general/icann-mou-25nov98.htm, para VII.

[93] *Memorandum of Understanding Between the U S Department of Commerce and Internet Corporation for Assigned Names and Numbers, Amendment 6* (16 Sept 2003), available at http://www.icann.org/general/amend6-jpamou-17sep03.htm.

[94] *Memorandum of Understanding Between the US Department of Commerce and Internet Corporation for Assigned Names and Numbers* (25 Nov 1998), available at http://www.icann.org/general/icann-mou-25nov98.htm, para VC1-9.

- Collaborate on the design, development and testing of appropriate membership mechanisms that foster accountability to and representation of the global and functional diversity of the Internet and its users, within the structure of private sector DNS management organisation.
- Collaborate on the design, development and testing of a plan for creating a process that will consider the possible expansion of the number of gTLDs.

In addition to the nine specific tasks, the original MOU included a catch-all provision requiring that ICANN '[c]ollaborate on other activities as appropriate to fulfill the purpose of this Agreement, as agreed by the Parties'.[95]

The tasks that ICANN was originally required to perform under the MOU were redefined and replaced by subsequent amendments to the agreement, which also included certain date-specific milestones. The amendments to the MOU and the main changes made to the original agreement were as follows:

- *Amendment 1*

  ICANN submitted its First Status Report[96] to the Department of Commerce in June 1999. As explained at [2.12] above, the report identified the relationship with NSI as the most significant challenge facing ICANN. As further explained at [2.12], the impasse between ICANN and NSI was resolved by the four interlocking agreements between the Department of Commerce, ICANN and NSI entered into in November 1999. An integral part of the interlocking agreements was Amendment 1 of the MOU, which essentially provided for Department of Commerce approval and supervision of the Registry Agreement between ICANN and NSI in relation to the .com, .net and .org gTLDs.

- *Amendment 2*

  ICANN submitted its Second Status Report to the Department of Commerce in June 2000.[97] The Second Status Report indicated that ICANN had completed, or made substantial progress towards completing, most of the original nine tasks. Nevertheless, the report concluded that work remained to be completed in two main areas: coordination of the root server system, which remained under the practical control of NSI; and relations with ccTLD operators. The report therefore recommended extending the term of the MOU. Following the Second Status Report, Amendment 2 of the MOU was entered into in September 2000. Amendment 2 replaced the original nine specific tasks with seven specific tasks, reflecting the extent to which work on the original tasks had been completed. The seven specific tasks emphasised the importance of ICANN establishing stable relationships with ccTLD operators and Regional Internet Registries (RIRs), and of taking steps to enhance the root server system, including root server security. At the same time, Amendment 2 extended the termination date of the agreement from 30 September 2000 to 30 September 2001.

[95] *Ibid*, para VC 10.

[96] ICANN, *Status Report to the Department of Commerce* (15 June 1999), available at http://www.icann.org/general/statusreport-15june99.htm.

[97] ICANN, *Second Status Report to the Department of Commerce* (30 June 2000), available at http://www.icann.org/general/statusreport-15june99.htm.

- *Amendment 3*

  In May 2001 the ICANN-NSI Registry Agreement was replaced by three separate registry agreements between ICANN and VeriSign, Inc.[98] VeriSign had previously taken over NSI. The three new agreements with VeriSign were in relation to the operation of registries for the .com, .net and .org gTLDs. The agreements were designed to place VeriSign, as far as practicable, in the same position as the registry operators for other gTLDs. Amendment 3 of the MOU, which was entered into on 25 May 2001, altered the MOU to reflect the new registry agreements between ICANN and VeriSign, Inc.

- *Amendment 4*

  ICANN submitted its Third Status Report to the Department of Commerce in July 2001.[99] The report indicated that considerable progress had been made towards completing the transitional tasks that had been specified in Amendment 1. At the same time, the report concluded that, as important tasks were not yet completed, the term of the Memorandum should be extended. Amendment 4 to the MOU, entered into in September 2001, implemented this recommendation by extending the term of the agreement until 30 September 2002.

- *Amendment 5*

  In February 2002, as a result of persistent criticisms of ICANN and the difficulties experienced by ICANN in performing its functions, the then President of ICANN, Dr Stuart Lynn, issued a report which included proposals for reforming ICANN. Following a period of discussion, the proposals resulted in significant structural reform. The process of structural reform is dealt with at [2.20].

  ICANN submitted its Fourth Status Report to the Department of Commerce in August 2002.[100] The report referred to the process of structural reform and acknowledged that progress in completing the transition tasks had been slow. The report concluded that priority should be given to ICANN's efforts at structural reform and recommended that the MOU be amended to reflect the new priorities.

  Amendment 5 to the MOU was entered into in September 2002. Amendment 5 contained the most significant changes to the MOU since the agreement was entered into. In addition to the substantive changes made to the terms of the MOU, the term of the agreement was extended to 30 September 2003. In agreeing to extend the agreement, the Department of Commerce expressed its disappointed with ICANN's slow progress.[101] The Department also indicated that ICANN's reform efforts should focus on the following areas: the scope of ICANN's mission; the transparency and accountability of its decision-making; ICANN's responsiveness to Internet stakeholders; defining an effective advisory role for governments; and security of Internet functions.[102]

[98] See *.com Registry Agreement with VeriSign, Inc* (25 May 2001), available at http://icann.org/tlds/agreements/verisign/registry-agmt-com-25may01.htm; *.net Registry Agreement with VeriSign, Inc* (25 May 2001), available at http://icann.org/tlds/agreements/verisign/registry-agmt-net-25may01.htm; *.org Registry Agreement with VeriSign, Inc* (25 May 2001), available at http://icann.org/tlds/agreements/verisign/registry-agmt-org-25may01.htm;.

[99] ICANN, *Third Status Report to the Department of Commerce* (3 July 2001), available at http://www.icann.org/general/statusreport-03jul01.htm.

[100] ICANN, *Fourth Status Report to the Department of Commerce* (15 Aug 2002), available at http://www.icann.org/general/status-report-15aug02.htm.

[101] Department of Commerce, *Statement Regarding Extension of Memorandum of Understanding with ICANN* (20 Sept 2002), available at http://www.ntia.doc.gov/ntiahome/domainname/agreements/docstatement-09192002.htm.

[102] *Ibid.*

The changes made to the agreement by Amendment 5 reflected the new focus on structural reform by revising and adding to ICANN's transition tasks, while requiring the Department of Commerce to engage in specified activities to assist ICANN in completing the transition tasks. Moreover, to improve ICANN's accountability to the Department, the MOU was amended to require ICANN to submit quarterly reports on progress towards completing the transition tasks and on implementing structural reforms.

- *Amendment 6*

  Pursuant to the new accountability arrangements introduced by Amendment 5 ICANN submitted quarterly reports on its progress in completing the transition tasks and on implementing structural reform.[103] In addition, ICANN submitted its *Eighth Status Report,* an annual report on progress in meeting the transition tasks, in August 2003.[104]

  Amendment 6 to the MOU was entered into in September 2003. The amendment extended the term of the agreement for an additional three years, until 30 September 2006. In extending the agreement, the Department of Commerce reiterated its commitment to privatising the DNS management functions.[105] While recognising the progress made by ICANN, especially in relation to structural reform, the Department indicated that ICANN still faced considerable work in completing the transition tasks.

  In addition to extending the term of the MOU, Amendment 6 further refined and expanded on the transition tasks defined by Amendment 5, and the Department's supervision of ICANN, by adding to the tasks and by introducing milestones for ICANN to meet in completing the tasks. Amendment 6 also replaced the requirement for quarterly status reports with an obligation for ICANN to provide the Department with six-monthly reports.

[2.15] US Government Principles and the EU

On 30 June 2005, Michael Gallagher, Assistant Secretary of Commerce for Communications and Information and Administrator of the NTIA, issued a statement, entitled *U.S. Principles on the Internet's Domain Name Addressing System.*[106] The statement set out the following general principles:

- The US Government intends to preserve the security and stability of the Internet's Domain Name and Addressing System (DNS).
- Governments have legitimate interest in the management of their country code top level domains (ccTLD).

[103] See ICANN, *Fifth Status Report to the Department of Commerce* (8 Jan 2003), available at http://www.icann.org/general/status-report-08jan03.htm; ICANN, *Sixth Status Report to the Department of Commerce* (31 Mar 2003), available at http://www.icann.org/general/status-report-31mar03.htm; ICANN, *Seventh Status Report to the Department of Commerce* (30 June 2003), available at http://www.icann.org/general/status-report-30jun03.htm. .

[104] ICANN, *Eighth Status Report to the Department of Commerce* (1 Aug 2003), available at http://www.icann.org/general/status-report-01aug03.htm.

[105] See Department of Commerce, *Statement Regarding Extension of Memorandum of Understanding with ICANN* (16 Sept 2003), available at http://www.ntia.doc.gov/ntiahome/domainname/agreements/sepstatement-09162003.htm.

[106] NTIA, *U.S. Principles on the Internet's Domain Name Addressing System* (30 June 2005), available at http://www.ntia.doc.gov/ntiahome/domainname/USDNprinciples_06302005.htm.

- ICANN is the appropriate technical manager of the Internet DNS.
- Dialogue related to Internet governance should continue in relevant multiple fora.

The statement of principles elaborated on the US government's intention to retain a central role in DNS governance, including the following important statements:

- '. . . the United States is committed to taking no action that would have the potential to adversely impact the effective and efficient operation of the DNS and will therefore maintain its historic role in authorizing changes or modifications to the authoritative root zone file'; and
- 'The United States will continue to provide oversight so that ICANN maintains its focus and meets its core technical mission'.

The NTIA statement was a reaction to the WSIS process, which is explained at [2.30] below, and to proposals emerging from the WGIG to reform DNS governance. In particular, the statement can be seen as a response to concerns expressed by the European Union, including the following statement made by the EU at a WGIG meeting in April 2005:

> the EU believes that a new cooperation model is needed in order to concretize the provisions in the WSIS Declaration of Principles regarding the crucial role of all actors within Internet governance.

The US Congress, which has a role in supervising the Department of Commerce, also responded to the WSIS process, and to the EU concerns, by expressing support for maintaining US control over ICANN. Thus, on 16 November 2005, the US House of Representatives unanimously passed a resolution stating that:

> the authoritative root zone server should remain physically located in the United States and the Secretary of Commerce should maintain oversight of ICANN so that ICANN can continue to manage the day-to-day operation of the Internet's domain name and addressing system well, remain responsive to all Internet stakeholders worldwide, and otherwise fulfill its core technical mission.[107]

On 18 November 2005, the US Senate called on the President to 'continue to oppose any effort to transfer control of the Internet to the United Nations or any other international entity'.[108]

Meanwhile, as explained at [2.30], the second phase of the WSIS was convened in Tunis from 16 to 18 November 2005. In anticipation of this, the European Commission issued a statement outlining the European Union's position on Internet governance.[109] The statement supported the privatisation of DNS governance and the role of ICANN, stating that 'one should not try to change this successful example of management in private hands'. At the same time, the statement emphasised the importance of the Internet as a 'global resource', adding that this prompted 'many to question whether one government alone should supervise such an important part of the infrastructure'. These concerns led to the EU advocating the complete privatisation of the day-to-day management of the Internet by 'phasing out the oversight functions of the US Department of Commerce over ICANN'.

To date, the concerns of the EU and others have had a limited influence on US government policy. As explained at [2.17], a new agreement between the Department of Commerce and ICANN was entered into in September 2006, which retained US govern-

[107] US House of Representatives, Expressing the sense of the Congress regarding oversight of the Internet Corporation for Assigned Names and Numbers, H Con Res 268 (16 Nov 2005).

[108] US Senate, Expressing the sense of the Senate that the United Nations and other international organizations should not be allowed to exercise control over the Internet, S Res 323 (18 Nov 2005).

[109] European Commission, *World Summit on the Information Society in Tunis, Internet Governance: Frequently asked questions*, Memo/05/428 (Brussels, 15 Nov 2005).

ment supervision of ICANN while increasing ICANN's autonomy in relation to day-to-day management of the DNS.

[2.16] 2006 NTIA Consultation

In expectation of the expiry of the MOU in September 2006, the NTIA initiated a public consultation process on the future of DNS governance in May 2006.[110] The consultation requested comments on the following matters:

- Whether the four guiding principles established by the NTIA White Paper, and set out in full at [2.10], were still relevant?
- Whether ICANN had made sufficient progress in achieving its core tasks and date-specific milestones?
- Whether the core tasks and date-specific milestones were still relevant?
- Whether all relevant stakeholder groups were effectively involved in the ICANN process?
- Is participation of stakeholders in ICANN supporting organisations and committees meeting the needs of key stakeholders and the Internet community?
- What methods and processes should be considered to encourage greater efficiency and responsiveness to governments and ccTLD managers in processing root management requests?
- How can information exchange, collaboration and enhanced cooperation among public and private organisations with roles in DNS governance be enhanced?

The NTIA consultation process generated a considerable number of public comments, the majority in favour of relaxing, or removing, the Department of Commerce's supervision of ICANN.

[2.17] The September 2006 Agreement

On 29 September 2006 the Department of Commerce and ICANN entered into a new Joint Project Agreement. The new agreement redefined the relationship between the parties by amending the original MOU and extended the term of the agreement to 30 September 2009.[111]

The new agreement retains a supervisory role for the Department of Commerce, while relaxing the more prescriptive aspects of that supervision, such as the use of date-specific milestones. In this respect, it should be noted that some of the milestones set in previous agreements had not been met by ICANN.

The main elements of the new agreement are as follows:

- The agreement reiterates the four guiding principles of DNS governance originally set out in the NTIA White Paper (see [2.10]), namely: stability and security; competition; bottom-up coordination; and representation.

[110] National Telecommunications and Information Administration (NTIA), *The Continued Transition of the Technical Coordination and Management of the Internet Domain Name and Addressing System*, 71 Fed Reg 30,388 (2006).

[111] *Joint Project Agreement Between the U.S. Department of Commerce and the Internet Corporation for Assigned Names and Numbers* (29 Sept 2006), available at http://www.icann.org/general/JPA-29sep06.pdf.

- ICANN is given more day-to-day autonomy from the Department of Commerce, with prescriptive tasks and targets being replaced by a more general statement of ICANN's responsibilities, which is set out in an annexure to the agreement.
- ICANN's obligations periodically to report directly to the Department of Commerce have been removed. Instead, ICANN is required to publish an annual report setting out its progress against the criteria set out in its Bylaws, the Affirmation of Responsibilities and its Strategic and Operating Plans. The obligation formally to report to the Department of Commerce has been replaced by a requirement for regular meetings between the Department of Commerce and ICANN senior management.

The more prescriptive elements of the Department of Commerce's supervision have therefore been largely replaced by accountability in accordance with objectives set through ICANN's own processes, such as those set out in ICANN's strategic plan.

The most important feature of the Joint Project Agreement is known as the Affirmation of Responsibilities, which was approved by the ICANN board on 25 September 2006, and which is annexed to the agreement. The Affirmation of Responsibilities establishes the following 10 ICANN responsibilities:

1. *Security and Stability*—to coordinate the global Internet's system of unique identifiers, and in particular to ensure the stable and secure operation of the Internet's unique identifier systems.
2. *Transparency*—to continue to develop, test and improve processes and procedures to encourage improved transparency, accessibility, efficiency and timeliness in the consideration and adoption of policies related to technical coordination of the Internet DNS, and funding for ICANN operations.
3. *Accountability*—to continue to develop, test, maintain and improve on accountability mechanisms to be responsive to global Internet stakeholders in the consideration and adoption of policies related to the technical coordination of the Internet DNS.
4. *Root Server Security and Relationships*—to continue to coordinate with the operators of root name servers and other appropriate experts with respect to the operational and security matters, both physical and network, relating to the secure and stable coordination of the root zone.
5. *TLD Management*—to maintain and build on processes to ensure that competition, consumer interests and Internet DNS stability and security issues are identified and considered in TLD management decisions, including the consideration and implementation of new TLDs and the introduction of Internationalised Domain Names (IDNs).
6. *Multi-stakeholder Model*—to maintain and improve the multi-stakeholder model and the global participation of all stakeholders, including conducting reviews of its existing advisory committees and supporting organisations, and to continue to further the effectiveness of the bottom-up policy development processes.
7. *Role of Governments*—to work with the Government Advisory Committee (GAC) members to review the GAC's role within ICANN so as to facilitate effective consideration of GAC advice on the public policy aspects of the technical coordination of the Internet.
8. *IP Addressing*—to continue to work collaboratively on a global and regional level so as to incorporate Regional Internet Registries' policy-making activities into the ICANN processes while allowing them to continue their technical work.

9. *Corporate Responsibility*—to maintain excellence and efficiency in operations, including good governance, organisational measures to maintain stable, international private sector organisation, and to maintain relevant technical and business experience for members of the Board of Directors, executive management and staff.
10. *Corporate Administrative Structure*—to conduct a review of, and to make necessary changes in, corporate administrative structure to ensure stability.

For the most part, the Affirmation of Responsibilities is pitched at a high level of generality. Presumably this gives ICANN a greater degree of flexibility in determining how it shall comply with its responsibilities than it had under the MOU before the new Joint Project Agreement. At the same time, the Affirmation of Responsibilities does incorporate a limited number of more specific obligations. The most contentious of these obligations, included in ICANN's responsibility for TLD Management, relates to the WHOIS policy and provides that:

> ICANN shall continue to enforce existing policy relating to WHOIS, such existing policy requires that ICANN implement measures to maintain timely, unrestricted and public access to accurate and complete WHOIS information, including registrant, technical, billing and administrative contact information.

[2.18] IANA Function Contract

Prior to the formation of ICANN, IANA technical functions were performed by the ISI at the University of Southern California under a contract between the university and DARPA for the performance of what was known as the 'Teranode Network Technology' project. In December 1998, the University of Southern California and ICANN entered into a transition agreement which provided for the transfer of IANA functions, responsibilities, assets and personnel from ISI to ICANN.[112] In February 2000, the NTIA entered into a contract with ICANN, essentially consisting of a purchase order for the performance of the IANA technical function.[113] Under the performance contract, the US government approved the transfer of functions and responsibilities from the University of Southern California to ICANN. Since being entered into, the IANA performance contract has been modified, extended and renewed.[114]

The current version of the performance contract (or purchase order) was entered into in August 2006, with the initial term expiring in September 2007, but with options to extend the term to September 2011.[115] The agreement defines the IANA functions in the following terms:[116]

[112] *University of Southern California/ICANN Transition Agreement* (24 Dec 1998), available at http://www.icann.org/general/usc-icann-transition-agreement.htm .

[113] *ICANN/U.S. Government Contract for the IANA Function* (9 Feb 2000), available at http://www.icann.org/general/iana-contract-09feb00.htm.

[114] See *Modification 0001 to ICANN/U.S. Government Contract for the IANA Function* (6 Sept 2000), available at http://www.icann.org/general/iana-contract-modification1-06sep00.htm; *ICANN/U.S. Government Contract for the IANA Function* (21 Mar 2001), available at http://www.icann.org/general/iana-contract-modification1-06sep00.htm; *ICANN/U.S. Government Contract for the IANA Function* (17 Mar 2003), available at http://www.icann.org/general/iana-contract-17mar03.htm.

[115] *ICANN/US Government Contract for the IANA Function* (11 Aug 2006), available at http://www.icann.org/general/iana-contract-14aug06.pdf.

[116] *Ibid*, para C2. 2. 1.

- *Coordinate the assignment of technical protocol parameters*—This function involves the review and assignment of unique values to various parameters (eg, operation codes, port numbers, object identifiers, protocol numbers) used in various Internet protocols. This function also includes the dissemination of the listings of assigned parameters through various means (including on-line publication) and the review of technical documents for consistency with assigned values.
- *Perform administrative functions associated with root management*—This function addresses facilitation and coordination of the root zone of the domain name system, with 24 hour-a-day/seven days-a-week coverage. It includes receiving requests for and making routine updates of the country code top level domain (ccTLD) contact (including technical and administrative contacts) and nameserver information. It also includes receiving delegation and redelegation requests, investigating the circumstances pertinent to those requests, and making recommendations and reporting actions undertaken in connection with processing such requests. All requests must be processed in accordance with procedures set out in an appendix to the agreement. The procedures require ICANN to issue a report to the Department of Commerce regarding whether the proposed changes should be authorised. ICANN is also required to develop and implement a process for consulting with the relevant governments and ccTLD managers to encourage greater efficiency and responsiveness in processing ccTLD requests.
- *Allocate Internet numbering resources*—This function involves overall responsibility for allocated and unallocated IPv4 and IPv6 address space and Autonomous System Number space. It includes the responsibility for delegation of IP address blocks to regional registries for routine allocation, typically through downstream providers, to Internet end-users within the regions served by those registries. This function also includes reservation and direct allocation of space for special purposes.
- *Other services*—ICANN is required to perform other IANA functions and implement modifications in performance of the IANA functions as needed upon mutual agreement of the parties.

The performance contract is careful to exclude matters covered by other agreements entered into by the Department of Commerce.[117] First, the performance contract does not authorise modifications, additions or deletions to the root zone file, which remain the responsibility of VeriSign pursuant to the Cooperative Agreement between the Department of Commerce and VeriSign. Secondly, the performance contract does not authorise ICANN to make material changes to established methods for performing the IANA functions, as ICANN policy development is dealt with pursuant to the MOU between the Department of Commerce and ICANN.

[2.19] VeriSign Agreement

The Cooperative Agreement between the Department of Commerce and VeriSign defines the relationship between the parties in relation to the DNS root and VeriSign's operation of the .com and .net registries. As explained at [2.12], in November 1999, Amendment 19 to Cooperative Agreement between NSI and the Department of Commerce amended the

[117] *ICANN/US Government Contract for the IANA Function* (11 Aug 2006), available at http://www.icann.org/general/iana-contract-14aug06.pdf, para C4.

agreement to require NSI to recognise ICANN and to define the relationship between ICANN and NSI. As further explained, the US Government asserts authority to give directions to NSI/VeriSign because of its purported authority over the root zone file, which was first formally expressed in Amendment 11 to the NSI cooperative agreement.

In June 2000, NSI was acquired by VeriSign, Inc., a Californian public corporation. In May 2001, Amendment 24 to the cooperative agreement substituted VeriSign for NSI and extended the expiration date of the agreement from 10 November 2003 to 10 November 2007. Amendment 24 restated VeriSign's obligations to enter into registry agreements with ICANN, and to comply with Department of Commerce policies and directives regarding the provision of registry services (as distinct from day-to-day administration of registries). Amendment 24 also restated the obligation of the Department of Commerce to ensure that the authoritative root would continue to point to the TLD zone servers designated by VeriSign for the registries for which it is responsible.

As explained at [2.12]–[2.13] above, Amendment 1 to the ICANN/DOC MOU provides that amendments to, or substitutes for, the .com registry agreement are subject to prior approval by the Department of Commerce. As explained at [2.27] below, on 29 November 2006 the Department of Commerce approved a new .com registry agreement between ICANN and VeriSign, which had extended the expiration of the term of the registry agreement from 10 November 2007 to 30 November 2012. At the same time, Amendment 30 to the cooperative agreement extended the term of the cooperative agreement to match the .com registry agreement, so that the cooperative agreement now expires on 30 November 2012.

Amendment 30 to the cooperative agreement also introduced new provisions relating to renewal of the .com registry agreement, and to approval of any renewal by the Department of Commerce. First, VeriSign agreed not to enter into a renewal of the .com registry agreement, or a new agreement, without the prior approval of the Department of Commerce. Secondly, in determining whether to approve a renewal or a new agreement, the Department of Commerce is required to apply a public interest test. Thus, the Department of Commerce must grant approval if it is satisfied that renewal will serve the public interest in:

- the continued security and stability of the DNS and the operation of the .com registry, including:

—consideration of VeriSign's compliance with Consensus Policies and technical specifications and its service level agreements; and
—the investment associated with improving the security and stability of the DNS; and
—the provision of registry services offered at reasonable prices, terms and conditions.

Thirdly, there is a presumption that the Department of Commerce will grant approval provided that the public interest test is satisfied and that VeriSign is entitled to renewal under the .com Registry Agreement. Fourthly, the Department of Commerce has a discretion whether or not to approve changes in the registry agreement relating to VeriSign's obligations to provide registry services to registrars, and to the regulation of prices charged for registry services. Fifthly, if VeriSign fails to comply with the terms of the registry agreement relating to the provision of registry services and price regulation, the Department of Commerce has a right to bring an action for specific performance of such terms.

ICANN

[2.20] ICANN's Structural Reform Process

In February 2002 the then President of ICANN, Dr Stuart Lynn, presented a paper on the need to reform ICANN entitled *ICANN: A Case for Reform.*[118] The paper argued that ICANN's performance had been disappointing, and that the organisation's future was jeopardised, mainly because of an incomplete structure and inadequate funding. In particular, the paper identified the following three systemic problems:

- Too little participation by critical entities, including operators of ccTLDs, root name server operators, Internet infrastructure providers, national governments and major Internet users, especially members of the business community. The paper argued that the original concept of a purely private sector body had proved to be impracticable, and emphasised the importance of active participation by national governments to the future success of ICANN.
- Too much focus on process. The paper argued that there had been an inordinate emphasis on ICANN's processes, and especially on attempts to establish a form of representative structure, at the expense of establishing a workable, effective and stable organisation.
- Too little funding to provide quality services. The paper concluded that ICANN was significantly under funded, meaning that it was unable to adequately perform its DNS management tasks.

The paper maintained that these problems could only be addressed by a fundamental overhaul of ICANN's structure, processes and funding arrangements.

In March 2002, the ICANN board responded to the President's paper by giving its Evolution and Reform Committee (ERC) the task of making recommendations relating to structural reform of ICANN. After a process that elicited a considerable number of public comments, the ERC published its report, *ICANN: A Blueprint for Reform,* in June 2002.[119] The ERC report refined some of the proposals made in the President's paper, modified other proposals and discarded some.

Upon receiving the ERC report, it was adopted by ICANN's board, which instructed the ERC to oversee the reform process. Eventually, in December 2002, the new ICANN structure came into effect when the ICANN board adopted new Bylaws. At the same time, the Board adopted a transition plan to guide the transition to the new structure. Since then, ICANN's Bylaws have been further amended, as the reform process has been refined in practice. The ICANN structure is outlined at [2.24]ff.

[2.21] ICANN's Constitution

ICANN is a non-profit-making public benefit corporation incorporated under the California Nonprofit Public Benefit Corporation Law. ICANN's constitution consists of its Articles of Incorporation and its Bylaws.

[118] President's Report: ICANN—The Case for Reform (24 Feb 2002), available at http://www.icann.org/general/lynn-reform-proposal-24feb02.htm.

[119] Committee on ICANN Evolution and Reform (CIER), *ICANN: A Blueprint for Reform* (20 June 2002), available at http://www.icann.org/committees/evol-reform/blueprint-20jun02.htm.

The Articles of Incorporation define ICANN's objects. The objects of ICANN, as defined in Article 3 of the Articles, are:

> to pursue the charitable and public purposes of lessening the burdens of government and promoting the global public interest in the operational stability of the Internet by (i) coordinating the assignment of Internet technical parameters as needed to maintain universal connectivity on the Internet; (ii) performing and overseeing functions related to the coordination of the Internet Protocol ("IP") address space; (iii) performing and overseeing functions related to the coordination of the Internet domain name system ("DNS"), including the development of policies for determining the circumstances under which new top-level domains are added to the DNS root system; (iv) overseeing operation of the authoritative Internet DNS root server system; and (v) engaging in any other related lawful activity in furtherance of items (i) through (iv).

The Articles of Incorporation also prescribe the way in which ICANN is to perform its objects. In this respect, Article 4 provides that ICANN:

> shall operate for the benefit of the Internet community as a whole, carrying out its activities in conformity with relevant principles of international law and applicable international conventions and local law and, to the extent appropriate and consistent with these Articles and its Bylaws, through open and transparent processes that enable competition and open entry in Internet-related markets. To this effect, the Corporation shall cooperate as appropriate with relevant international organizations.

As explained at [2.20], ICANN's structural reform process culminated in a new set of Bylaws that were approved by ICANN's Board in December 2002. As further explained, the Bylaws have since been amended, as the structural reform process has been refined. The Bylaws define ICANN's internal structure and its processes. In particular, the Bylaws deal with the following matters: ICANN's mission and core values; ICANN's structure, including ICANN's board of directors, supporting organisations and advisory committees; and ICANN's processes, including transparency, accountability and review.

Pursuant to Article XIX of the ICANN Bylaws, the Articles of Incorporation or Bylaws may only be altered, amended or repealed, or new Articles of Incorporation or Bylaws adopted, by a two-thirds vote of members of the ICANN board.

[2.22] ICANN's Mission

The scope of ICANN's responsibilities and functions has been one of the most controversial areas of DNS governance. In particular, the first years of ICANN's operations were characterised by debates over whether ICANN was a purely technical coordination body or whether it was primarily a policy-making body. The scope of ICANN's mission received considerable attention during ICANN's structural reform process, with many claiming that ICANN had acted beyond its core technical functions. The ERC report, *ICANN: A Blueprint for Reform,* reached the eminently practical conclusion that ICANN was necessarily involved with policy development and implementation, but that its policy role 'should be limited to those policy areas that are reasonably related to ICANN's technical mission'.[120] The ERC recommended a new mission statement that was adopted by ICANN's board and incorporated into ICANN's Bylaws.

[120] *Ibid.*

ICANN's mission is set out in Article I.1 of the Bylaws, which provides that the mission is:

> to coordinate, at the overall level, the global Internet's systems of unique identifiers, and in particular to ensure the stable and secure operation of the Internet's unique identifier systems. In particular, ICANN:
>
> 1. Coordinates the allocation and assignment of the three sets of unique identifiers for the Internet, which are
>   a. Domain names (forming a system referred to as "DNS");
>   b. Internet protocol ("IP") addresses and autonomous system ("AS") numbers; and
>   c. Protocol port and parameter numbers.
> 2. Coordinates the operation and evolution of the DNS root name server system.
> 3. Coordinates policy development reasonably and appropriately related to these technical functions.

[2.23] ICANN's Core Values

In addition to a new mission statement, the ERC report recommended that ICANN adopt a set of core values to provide guidance for ICANN in the performance of its mission. Article I.2 of the Bylaws sets out the following 11 core values, or guiding principles, which were refined from the recommendations made in the ERC report:

1. Preserving and enhancing the operational stability, reliability, security and global interoperability of the Internet.
2. Respecting the creativity, innovation and flow of information made possible by the Internet by limiting ICANN's activities to those matters within ICANN's mission requiring or significantly benefiting from global coordination.
3. To the extent feasible and appropriate, delegating coordination functions to or recognising the policy role of other responsible entities that reflect the interests of affected parties.
4. Seeking and supporting broad, informed participation reflecting the functional, geographic and cultural diversity of the Internet at all levels of policy development and decision-making.
5. Where feasible and appropriate, depending on market mechanisms to promote and sustain a competitive environment.
6. Introducing and promoting competition in the registration of domain names where practicable and beneficial in the public interest.
7. Employing open and transparent policy development mechanisms that (i) promote well-informed decisions based on expert advice, and (ii) ensure that those entities most affected can assist in the policy development process.
8. Making decisions by applying documented policies neutrally and objectively, with integrity and fairness.
9. Acting with a speed that is responsive to the needs of the Internet while, as part of the decision-making process, obtaining informed input from those entities most affected.
10. Remaining accountable to the Internet community through mechanisms that enhance ICANN's effectiveness.

11. While remaining rooted in the private sector, recognising that governments and public authorities are responsible for public policy and duly taking into account governments' or public authorities' recommendations.

The Bylaws recognise that it may not always be possible fully to comply with all of the core values and that, in making decisions, ICANN may sometimes need to find a balance between competing values.

ICANN's Structure

[2.24] ICANN's Structure

ICANN's structure is set out diagrammatically in Appendix 1. It consists of the following main components:

- A board of directors and a nominating committee ('nomco') that is responsible for selecting some members of the board;
- Three supporting organisations, namely the Address Supporting Organisation (ASO), Country-Code Names Supporting Organisation (ccNSO) and Generic Names Supporting Organisation (GNSO); and
- Four advisory committees, namely the Governmental Advisory Committee (GAC), Security and Stability Advisory Committee (SAC), Root Server System Advisory Committee (RSSAC) and At-Large Advisory Committee (ALAC).

This paragraph deals with each of these components. It also explains ICANN's external advisory mechanisms.

[2.24.1] Board of Directors

Since its inception, the selection of members of ICANN's board of directors has been controversial. As explained at [2.11] above, in 1998 the initial board members were selected by Jon Postel following private negotiations involving IANA, the Internet Society, business interests, the European Commission and the Australian government.[121] Moreover, as further explained at [2.11], following negotiations with the Boston Working Group (BWG) and the Open Root Server Confederation (ORSC), ICANN's original Bylaws were amended to provide for a more representative board. In particular, the original Bylaws provided for an open membership structure that would directly elect nine directors, to be known as the 'at-large directors'. The commitment to direct election of directors could be seen as reflecting one of the four guiding principles set out in the *White Paper*, and referred to at [2.10], namely 'representation', meaning that mechanisms should be established to ensure international participation in decision-making.[122] Following a board decision, the number of directly elected directors was reduced to five, each to represent a separate geographic region.

[121] See Mueller, above n 19, 180.

[122] On the other hand, it has been argued that, properly interpreted, the White Paper advocated a form of 'deliberative' policy-making, not a directly representative structure: A McLaughlin, 'The Virtues of Deliberative Policymaking' (Dec 2003), available at http://cyber.law.harvard.edu/mclaughlan/mclaughlan-response-publicparticipation.html.

In September–October 2000 ICANN conducted an on-line election, which was funded by the Markle Foundation. More than 76,000 people registered to vote in the elections, although only about 34,000 votes were valid.[123] The five elected directors served one term, which expired in December 2002. The ICANN election has been the subject of a number of studies, including a report from ICANN's At Large Membership Study Committee[124] and a report by the US-based NGOs, the Center for Democracy and Technology and Common Cause.[125] In his February 2002 paper on the need to reform ICANN, referred to at [2.20], Dr Stuart Lynn lamented the time and resources expended on the direct elections, concluding that:

> the concept of At Large membership elections from a self-selected pool of unknown voters is not just flawed, but fatally flawed, and that continued devotion of ICANN's very finite energy and resources down this path will very likely prevent the creation of an effective and viable institution.[126]

In its review of ICANN's structure, the ERC considered various options for selecting board members, and also rejected the use of direct elections.[127] Instead, the ERC recommended that the board consist of 15 members, with eight to be selected by a nominating committee ('nomco'), two to be selected from each of ICANN's supporting organisations, and the final member to be the President. The ERC recommendations were adopted by ICANN's board and incorporated in the Bylaws.

Following from the ERC recommendations, Article VI of the Bylaws provides that the ICANN board shall consist of 15 voting members, constituted as follows:

- eight to be selected by the Nominating Committee;
- two each to be selected by each of the three supporting organisations; and
- the President ex officio.

The terms of board members are generally set at three years.[128]

Article XIII.2 provides for the President of ICANN to be elected annually by the board upon the recommendation of the Chairman. Under Article VII.2, the nominating committee is to consist of the following:

- A non-voting Chair, appointed by the ICANN board;
- The immediately previous Chair, as a non-voting advisor;
- Five voting delegates selected by the ALAC;
- Two voting delegates selected by the Business Users Constituency of the GNSO;
- One voting delegate selected by each of the following entities:

  —The gTLD Registry Constituency of the GNSO;
  —The gTLD Registrars Constituency of the GNSO;
  —The Council of the ccNSO;
  —The Internet Service Providers Constituency of the GNSO;
  —The Intellectual Property Constituency of the GNSO;
  —The Council of the ASO;

[123] The election results were monitored by the Carter Center: see http://members.icann.org/carter.htm.
[124] See http://www.atlargestudy.org/final_report.shtml.
[125] See http://www.naisproject.org/report/final/.
[126] See President's Report, above n 118.
[127] See CIER, above n 119.
[128] Bylaws for Internet Corporation for Assigned Names and Numbers (as amended effective 28 Feb 2006), Art VI.8.

—An entity designated by the board to represent academic and similar organisations;
—Consumer and civil society groups, selected by the Non-commercial Users Constituency of the GNSO;
—The Internet Engineering Task Force (IETF); and
—The ICANN Technical Liaison Group (TLG);

- Three non-voting liaisons appointed by each of three advisory committees, namely the RSSAC, SAC and GAC;
- A non-voting Associate Chair appointed at the discretion of the Chair.

The Bylaws endeavour to ensure that board members have a diversity of background and experience. Thus, Articles VI.2.2 and VI.2.3 require that, in selecting directors, the nominating committee and supporting organisations shall 'seek to ensure that the ICANN Board is composed of members who in the aggregate display diversity in geography, culture, skills, experience, and perspective' by applying the criteria for selection set out in Article VI.3 of the Bylaws. Article VI.3 provides that the criteria for selection of ICANN directors are that the directors are to be:

- Accomplished persons of integrity, objectivity and intelligence, with reputations for sound judgement and open minds, and a demonstrated capacity for thoughtful group decision-making;
- Persons with an understanding of ICANN's mission and the potential impact of ICANN decisions on the global Internet community, and committed to the success of ICANN;
- Persons who will produce the broadest cultural and geographical diversity on the Board consistent with meeting the other criteria set forth in this Section;
- Persons who, in the aggregate, have personal familiarity with the operation of gTLD registries and registrars; with ccTLD registries; with IP address registries; with Internet technical standards and protocols; with policy-development procedures, legal traditions, and the public interest; and with the broad range of business, individual, academic, and non-commercial users of the Internet;
- Persons who are willing to serve as volunteers, without compensation other than the reimbursement of certain expenses; and
- Persons who are able to work and communicate in written and spoken English.

Articles VI.2.2, VI.2.3 and VI.5 further provide that the nominating committee and supporting organisations must, in selecting directors, ensure that each of the ICANN geographic regions are represented and that no one region is over-represented. Article VI.5 defines five ICANN geographic regions, namely Europe; Asia/Australia/Pacific; Latin America/Caribbean Islands; Africa; and North America.

[2.24.2] Supporting Organisations

Since its establishment, an important component of ICANN's structure has been the supporting organisations, which are designed to be the main source of policy recommendations within the scope their specialised areas of expertise. The supporting organisations are intended to 'help to promote the development of Internet policy and encourage diverse and

[129] See http://www.icann.org/general/support-orgs.htm.

international participation in the technical management of the Internet'.[129] In other words, the supporting organisations are designed to provide for bottom-up, detailed policy development that will result in recommendations to the ICANN board. Despite this objective, it has been suggested that it remains difficult reliably to assess the extent of public participation in ICANN through the supporting organisations.[130]

Before the December 2002 structural reforms, there were three supporting organisations: the Domain Name Supporting Organisation (DNSO); the Protocol Supporting Organisation (PSO); and the Address Supporting Organisation (ASO). The DNSO was responsible for policy recommendations relating to the DNS. Mueller argues that, although the DNSO was originally designed to be broadly representative, it was captured by large stakeholders, with seven of the DNSO constituencies representing business interests.[131] Mueller also points out that, although the DNSO was intended to be the source of policy recommendations regarding the DNS, important policy decisions, including the registrar accreditation agreement with NSI, were entered into before the DNSO was formally constituted.[132] The PSO was responsible for policy recommendations relating to the assignment of unique parameters for Internet protocols, and essentially consisted of standards development organisations. The ASO was responsible for policy recommendations relating to the system of IP addresses, and retains that responsibility.

Following the December 2002 structural reforms, the ICANN supporting organisations are:

- The Address Supporting Organisation (ASO), which retains responsibility for the system of IP addresses;
- The Country-code Names Supporting Organisation (ccNSO), which is concerned with policy issues relating to ccTLDs; and
- The Generic Names Supporting Organisation (GNSO), which is responsible for policy recommendations relating to the gTLDs.

Address Supporting Organisation (ASO)

The ASO is established by Article VIII of the ICANN Bylaws. Article VIII.1.1 provides that the function of the ASO is to:

> advise the Board with respect to policy issues relating to the operation, assignment, and management of Internet addresses.

Article VIII.1.2 further provides that the ASO is to be the entity established by the memorandum of understanding originally entered into between ICANN and the regional Internet registries (RIRs) in October 1999. In October 2004 a revised Memorandum of Understanding (MOU) was entered into between the Number Resource Organisation (NRO), the four regional RIRs and ICANN, which replaced the October 1999 MoU.[133] Amongst other things, the MOU, which took effect on 1 January 2005, defines the composition and responsibilities of the ASO Address Council, and a global policy development process for the ASO, including how the ICANN board will receive and deal with global policy recommendations from the ASO.

[130] See J Palfrey, C Chen, S Hwang and N Eisenkraft, *Public Participation in ICANN: A Preliminary Study* (Dec 2003), available at http://cyber.law.harvard.edu/icann/publicparticipation/.

[131] Mueller, above n 19, 198.

[132] *Ibid*, 186.

[133] *ICANN Address Supporting Organization (ASO) MoU* (29 Oct 2004), available at http://www.icann.org/aso/aso-mou-29oct04.htm.

Country-code Names Supporting Organisation (ccNSO)

As identified in the President's report and the ERC report referred to at [2.20] above, managing the relationship between ICANN and ccTLD managers has been one of the most difficult of ICANN's transition tasks. When ICANN was first established, the ccTLD registries were treated as a DNSO constituency. During ICANN's structural reform process, however, it became clear that the policy issues relating to ccTLDs and ccTLD managers were sufficiently distinct from those relating to the gTLDs to merit the establishment of a separate supporting organisation. The precise details of the ccNSO arrangements set out in the Bylaws were largely the results of the work of the ccNSO Assistance Group, that was formed in September 2002 to assist the ERC in making recommendations regarding the ccNSO. The ccNSO was formally established in June 2003, following the ICANN board adopting a relevant amendment to the ICANN Bylaws, and formally constituted in March 2004.

The ccNSO is established by Article IX of the ICANN Bylaws. Article IX.1 provides that the ccNSO is to be responsible for:

- developing and recommending to the Board global policies relating to ccTLDs;
- nurturing consensus across the ccNSO's community, including the name-related activities of ccTLDs; and
- coordinating with other ICANN supporting organisations, committees and constituencies.

The Bylaws further provide that the ccNSO may engage in other activities, including seeking to develop voluntary best practices for ccTLD managers, assisting in skills building within the global community of ccTLD managers, and enhancing operational and technical cooperation among ccTLD managers.

The Bylaws establish a membership structure for the ccNSO. Under Article IX.4 the members of the ccNSO consist of ccTLD managers that have agreed in writing to be members in accordance with the procedures set out in Article IX.4.2.

In addition to ccNSO members, the ccNSO has a ccNSO Council. Pursuant to Article IX.2, the ccNSO Council is responsible for managing the ccNSO policy development process. Article IX.3.8 further provides that the role of the ccNSO Council is 'to administer and coordinate the affairs of the ccNSO . . . and to manage the development of policy recommendations'. Under Article IX.3, the ccNSO Council consists of: three members selected by ccNSO members from within each of ICANN's five geographic regions; three members selected by the ICANN nominating committee referred to at [2.24.1] above; and ccNSO Council liaisons and observers. The method of selecting ccNSO Council members from the geographic regions is specified in Articles IX.4.7–9 of the Bylaws. Essentially, the ccNSO Council members are to be selected by the nomination of ccNSO members in the relevant geographic regions and, if necessary, by election. If an election is necessary, the Council members shall be selected on the basis of those receiving a majority of votes cast by ccNSO members within the relevant geographic region. Importantly, the ccNSO Council includes non-voting liaisons appointed by the Governmental Advisory Committee (GAC) and the At-Large Advisory Committee (ALAC).

The ccNSO policy-development process (ccPDP) is set out in Annex B to the ICANN Bylaws. The ccPDP starts with a request for an issue report, which may be made by the ccNSO Council, the ICANN Board, a recognised ccTLD regional organisation, an ICANN supporting organisation or advisory committee or the members of the ccNSO. After receiving the request, the ccNSO Council is to appoint an issue manager, who is to be responsible for preparing an issue report.

Once it receives the issue report, the ccNSO Council must vote on whether to initiate the ccPDP. If the ccNSO Council decides to initiate the ccPDP, it must decide whether to appoint a task force to address the issue. Whether or not a task force is established, there is to be public notification of the process, including a period for public comment and active solicitation of information from relevant organisations. At the end of the period for public comment, the issue manager must prepare a final report and submit it to the ccNSO Council. In deciding whether to make a recommendation on the issue, the ccNSO Council is required to seek to act by consensus. A recommendation may, nevertheless, be made if it is supported by 14 or more Council members.

If the ccNSO Council makes a recommendation, the issue manager must prepare a report to ccNSO members, who must be given an opportunity to vote on the recommendation. In the event that more than 66 per cent of the votes are in favour of the recommendation, it is to be sent to the ICANN board. The ICANN board is required to adopt the ccNSO recommendation unless, by a vote of more than 66 per cent, it determines that the policy is not in the best interest of the ICANN community or of ICANN. If the board rejects a recommendation, then the ccNSO Council is entitled to make a supplemental recommendation. Should ICANN's board reject the supplemental recommendation then, assuming it falls within the ccNSO's scope, the board is not entitled to set policy on the issue until the ccNSO makes a further recommendation that is accepted by the board.

The scope of the ccNSO is dealt with in Annex C to ICANN's Bylaws. Annex C attempts to define the scope of the ccNSO with regard to developing policies. Defining the scope of the ccNSO necessarily entails difficult issues relating to the relative responsibilities of ICANN, on the one hand, and ccTLD managers, on the other hand, in relation to policy development for ccTLDs. The demarcation of responsibilities set out in Annex C was largely the result of the work of the ccNSO Assistance Group.

The ccNSO was chartered in March 2004, when the ccNSO Launching Group announced the establishment of the ccNSO in accordance with Article XX of ICANN's Bylaws.[134]

Generic Names Supporting Organisation (GNSO)

The GNSO is established by Article X of the ICANN Bylaws. Article X.1 provides that the GNSO is responsible:

> for developing and recommending to the ICANN board substantive policies relating to gTLDs.

The GNSO has a constituency structure. Accordingly, Article X.2 provides that the GNSO is to consist of constituencies that are to represent particular stakeholders, and a GNSO Council. Under Article X.5, the six GNSO constituencies recognised by the Bylaws are:

- gTLD registries (representing all gTLD registries under contract to ICANN);
- registrars (representing all registrars accredited by and under contract to ICANN);
- Internet service and connectivity providers (essentially meaning telecommunications carriers and ISPs);
- commercial and business users (representing both large and small commercial users of the Internet);
- non-commercial users; and
- intellectual property interests (representing the full range of trademark and other intellectual property interests relating to the DNS).

[134] ICANN, *Ninth Status Report to the Department of Commerce* (7 Apr 2004), available at http://www.icann.org/general/status-report-07apr04.pdf, 9.

The registries, registrars and Internet service and connectivity providers' constituencies are provider constituencies, while the business, non-commercial users and intellectual property constituencies are user constituencies. The ICANN board may create new GNSO constituencies either of its own motion or in response to a petition for recognition from any group of individuals or entities. In deciding whether or not to recognise a new constituency, the board must take into account the views of the GNSO Council.

Article X.3.4 provides that the GNSO Council is responsible for managing the policy development process of the GNSO, including managing open forums, such as mailing lists, for the participation of all those willing to contribute to the work of the GNSO. Under Article X.3.1, the GNSO Council consists of: two representatives selected by each of the GNSO constituencies; three members selected by the ICANN nominating committee referred to at [2.24.1]; two non-voting ccNSO Council liaisons appointed by the GAC and ALAC; and observers. No two representatives selected by any one constituency can be citizens of the same country or of countries located in the same ICANN geographic region.

The GNSO policy-development process (PDP) is set out in Annex A to the ICANN Bylaws. The PDP starts with an issue being raised for consideration by the ICANN board, the GNSO Council (by a vote of at least 25 per cent of Council members) or by an ICANN advisory committee. After an issue is properly raised for consideration, the GNSO staff manager must prepare an issue report and distribute it to the full GNSO Council.

Once it receives the issue report, the GNSO Council may initiate the PDP. If the ICANN board raises the issue for consideration, the GNSO Council must initiate the PDP. If the issue is otherwise raised, then the Council must vote on whether to initiate the PDP. In that case, the PDP must be initiated if more than 33 per cent of Council members are in favour of initiating the PDP.

If the GNSO Council decides to initiate the PDP, it must decide whether to appoint a task force to address the issue. Whether or not a task force is established, there is to be public notification of the process, including a period for public comment and active solicitation of information from relevant GNSO constituencies. At the end of the period for public comment, the staff manager must prepare a final report and submit it to the GNSO Council chair, who must distribute the report to all Council members and convene a Council meeting. After the GNSO Council meeting, the staff manager must prepare a report to the ICANN board, incorporating the views of the GNSO Council.

If a supermajority (meaning more than 66 per cent) of the GNSO Council is in favour of a recommendation, the ICANN board is required to adopt the recommendation unless, by a vote of more than 66 per cent, the board determines that the policy is not in the best interest of the ICANN community or of ICANN. If the board rejects such a recommendation, then the GNSO Council is entitled to make a supplemental recommendation. ICANN's board must adopt the supplemental recommendation unless it determines, by a vote of more than 66 per cent, that the policy is not in the best interest of the ICANN community or of ICANN. If the GNSO Council is not able to reach a supermajority, the ICANN board can decide on the recommendation on the basis of a simple majority.

[2.24.3] Advisory Committees

In addition to the three supporting organisations, ICANN consists of four advisory committees. Whereas the supporting organisations are responsible for detailed policy development leading to recommendations to the ICANN board, the advisory committees are responsible for providing advice to the ICANN board in their respective areas of expertise.

Article XI.2 of the Bylaws establishes the following four specific advisory committees:

- Governmental Advisory Committee (GAC), which is responsible for providing advice on the activities of ICANN as they relate to the concerns of governments;
- Security and Stability Advisory Committee (SAC), which is responsible for providing advice on matters relating to the security and integrity of the Internet's naming and address allocation systems;
- Root Server System Advisory Committee (RSSAC), which is responsible for providing advice about the operation of the root name servers; and
- At-Large Advisory Committee (ALAC), which is responsible for providing advice on the activities of ICANN, insofar as they relate to the interests of individual Internet users.

Article XI.1 provides that the ICANN board may create advisory committees in addition to those specified in the Bylaws. It further provides that the advisory committees are to have no legal authority to act for ICANN, but are merely to report their findings and recommendations to the board.

Government Advisory Committee (GAC)

The February 2002 President's report, dealt with at [2.20], maintained that an important component of structural reform was the development of a new public–private partnership between national governments and ICANN. In the report, the then President stated:

> I am now convinced that the original desire to avoid a totally governmental takeover of the IANA functions led to an overreaction—the choice of a totally private model. With three years' experience, it is clear that model is simply not workable. It is not workable because it leaves ICANN isolated from the real-world institutions—governments—whose backing and support are essential for any effective global coordinating body to accomplish its assigned tasks. ICANN currently has an advisory committee to channel governmental input, but that mechanism has not effectively integrated the views or the influence of governments; we must find a better way.[135]

In accordance with these views, the ICANN structural reform process signalled a more active role for the GAC. The desirability of a greater role for national governments in ICANN processes was reflected in the ERC report, also dealt with at [2.20] above, which made the following recommendation:

> To strengthen the GAC's integration into ICANN and to strengthen representation of the public interest, the GAC should appoint (a) a non-voting liaison to the Board (b) one delegate to the Nominating Committee, and (c) non-voting liaisons to each of the SO Councils and . . . [the advisory committees] . . . In each case, the liaisons should have sufficient expertise to participate effectively in each of these bodies.[136]

Under Article XI.2.1 of the Bylaws, the GAC is established to:

> consider and provide advice on the activities of ICANN as they relate to concerns of governments, particularly matters where there may be an interaction between ICANN's policies and various laws and international agreements or where they may affect public policy issues.

[135] President's Report, above n 118.

[136] See CIER, above n 119.

Membership of the GAC is open to all national governments. Moreover, on the invitation of the GAC Chair, membership is also open to 'distinct economies' as recognised in international fora, and to multinational governmental organisations and treaty organisations. The GAC currently has 70 members (of which only about 30 are active members) including national governments and two treaty organisations, namely the ITU and World Intellectual Property Organisation (WIPO).[137]

Each member of the GAC is obliged to appoint one accredited representative, who must hold a formal official position with the member's public administration. The ICANN board is required to notify the GAC of any proposal that raises public policy issues, including proposals before supporting organisations or advisory councils, on which public comment is sought, and must take into account any GAC response to such proposals. Otherwise, the GAC may put issues to the board by way of comment, advice or recommendations. If the board decides to take action that is not consistent with GAC advice, it must inform the GAC and provide reasons. The board and GAC must then attempt, 'in good faith and in a timely and efficient manner', to find a solution. In the event that a mutually acceptable solution is not arrived at, the board must state the reasons for not following the GAC's advice in its final decision.

Security and Stability Advisory Committee (SAC)

The SAC was established in 2002, following an ICANN board meeting in late 2001 that focussed on DNS security issues in the wake of the 11 September 2001 terrorist attacks.

Under Article XI.2.2 of the Bylaws, the SAC is established to:

> advise the ICANN community and board on matters relating to the security and integrity of the Internet's naming and address allocation systems.

In particular, the SAC has the following responsibilities:

- To develop a security framework for Internet naming and address allocation services that defines key focus areas, and identifies where the responsibilities for each area lie;
- To communicate on security matters with the Internet technical community and the operators and managers of critical DNS infrastructure services as events and developments dictate;
- To engage in continuous threat assessment and risk analysis of the Internet naming and address allocation services to assess where the principal threats to stability and security lie, and to advise the ICANN community accordingly;
- To communicate with those who have direct responsibility for Internet naming and address allocation security matters to ensure that its advice on security risks, issues and priorities is properly synchronised with existing standardisation, deployment, operational and coordination activities;
- To report periodically to the board on its activities; and
- To make policy recommendations to the ICANN community and board.

The SAC members, and the SAC chair, are appointed by the ICANN board. The members of the SAC are drawn from operators of Internet infrastructure and other security experts.

[137] See http://www.gac-icann.org/web/about/gac-outreach_English.htm.

Root Server System Advisory Committee (RSSAC)

Under Article XI.2.3 of the Bylaws, the RSSAC is established to:

> advise the board about the operation of the root name servers of the domain name system.

In performing this function, the RSSAC is required to:

- consider and provide advice on the operational requirements of root name servers, including host hardware capacities, operating systems and name server software versions, network connectivity and physical environment;
- examine and advise on the security aspects of the root name server system; and
- review the number, location and distribution of root name servers considering the total system performance, robustness and reliability.

The members of the RSSAC consist of each operator of an authoritative root name server, as set out at [1.17], and such other members as appointed by the ICANN board. The RSSAC usually meets in conjunction with meetings of the IETF. One of the main tasks of the RSSAC has been to work on documenting the root server system as it currently exists, focussing on issues relating to security, system performance, robustness and reliability. The RSSAC has also been examining the implications of new technologies, including IPv6 referred to at [1.6], for the root server system.

At-large Advisory Committee (ALAC)

Neither the President's report nor the ERC report, dealt with at [2.20], envisaged an at-large advisory committee. Mainly as a result of public comments received during the structural reform process, however, in June 2002 the ICANN board instructed the ERC to 'consider the creation of an At Large Advisory Committee as a potential vehicle for informed participation in ICANN by the broad user community'.[138] Following further consideration, the ICANN Bylaws, adopted in December 2002, included provision for an ALAC that would provide a mechanism for receiving and incorporating input from the general Internet user community into ICANN policy-making. In large measure, then, the ALAC should be seen as a response to concerns raised about the abandonment of direct elections for at-large directors during ICANN's structural reform process.

Under Article XI.2.4 of the Bylaws, the ALAC is established to:

> consider and provide advice on the activities of ICANN, insofar as they relate to the interests of individual Internet users.

The Bylaws create a structure for the ALAC based around Regional At-large Organisations (RALOs). In this respect, the Bylaws provide for there to be five RALOs, one for each of the five ICANN geographic regions, namely Europe, Asia/Australia/Pacific, Latin America/Caribbean islands, Africa and Northern America. The Bylaws further provide that an entity will be recognised as the RALO for its geographic region upon entering a memorandum of understanding with ICANN that deals with: the process for selecting ALAC members; requirements of openness, participatory opportunities, transparency, accountability and diversity; and criteria and standards for the RALO's constituent structures. It is envisaged that the RALOs will manage outreach and public involvement of Internet users, and will be the main forum for public input to ICANN in each region.

[138] ICANN, *Fourth Status Report to the Department of Commerce* (15 Aug 2002).

The Bylaws provide for the ALAC to consist of 15 members, comprising two members selected from each of the RALOs and five members selected by the nominating committee referred to at [2.24.1]. The five members selected by the nominating committee must comprise one citizen of a country within each of ICANN's five geographic regions.

In June 2003 the ICANN board approved a framework for the formation of local and regional At-large groups, known as 'at-large structures', to promote the involvement and participation of the individual Internet user community in ICANN.[139] The framework establishes minimum criteria for at-large structures; a certification process for at-large structures; guidelines for the form of the memorandum of understanding to be entered into between ICANN and each RALO; and procedures for the ICANN board to review to certify or de-certify an at-large structure. The framework invites interested, qualified groups in each ICANN geographic region to apply for certification as an 'at-large structure'. Once sufficient at-large structures have been certified in a geographic region, and a Memorandum of Understanding has been entered into with ICANN, the group is recognised as the RALO for its region.

[2.24.4] External Advisory Mechanisms

In addition to ICANN's advisory committees, dealt with at [2.24.3], the Bylaws establish mechanisms for obtaining external expert advice, including advice on issues of public policy. Article XI-A provides for two mechanisms for receiving expert advice: expert advisory panels and the Technical Liaison Group (TLG).

Expert Advisory Panels

Article XI-A.1.2 provides that the ICANN board may, either on its own initiative or at the suggestion of an ICANN supporting organisation or advisory committee, appoint an expert advisory panel. If the advice sought concerns issues of public policy, then the board must first consult the GAC on the appropriate source from which to seek advice and the arrangements for requesting and obtaining that advice. The board may, moreover, refer any issues of public policy relating to ICANN's mission to a multinational governmental or treaty organisation. Any requests for advice from such bodies must, however, first be transmitted to the GAC, with the suggestion that the GAC transmit the request to the relevant multinational governmental or treaty organisation. ICANN's supporting organisation and advisory committees must have an opportunity to comment upon any external advice prior to any board decision.

Technical Liaison Group (TLG)

Both the President's report and the ERC report, dealt with at [2.20], envisaged a Technical Advisory Committee (TAC), which would be responsible for providing technical advice and guidance to the board and to other organisations within ICANN. During ICANN's structural reform process, it was decided that this role could be performed by an external Technical Liaison Group (TLG).

Pursuant to Article XI-A.2.2 of the Bylaws, the TLG consists of the following four organisations: the European Telecommunications Standards Institute (ETSI); the International

[139] ICANN, *At-Large Framework Formation* (June 2003), available at http://alac.icann.org/framework.htm.

Telecommunications Union's Telecommunication Standardization Sector (ITU-T); the World Wide Web Consortium (W3C); and the Internet Architecture Board (IAB).

Article XI-A.2.3 of the Bylaws provides that the TLG's role is:

> to channel technical information and guidance to the Board and to other ICANN entities.

The Bylaws further provide that this role has two components: a responsive component and an active 'watchdog' component. The responsive component requires the TLG, in response to a request for information regarding specific technical questions, to connect the board or other ICANN body with appropriate sources of technical information. Where information is requested regarding a technical standard for which one of the TLG organisations is responsible, the request must be directed to that organisation. The 'watchdog' component requires the TLG to advise the board of the relevance and progress of technical developments, in areas falling within each of the TLG organisation's scope, which could affect ICANN board decisions or other actions. The TLG must also draw attention to global technical standards issues that affect policy development within the scope of ICANN's mission. The Bylaws require each TLG organisation to designate two technical experts who are familiar with technical standards issues that are relevant to ICANN's activities. The eight experts are to be available to determine, through an exchange of e-mail messages, where to direct technical questions from ICANN. The Bylaws, however, specifically provide that the TLG is to act mainly as a conduit for technical advice. In particular, the Bylaws provide that the TLG is not to have officers or hold meetings, and is not to provide policy advice to the ICANN board as a committee.

ICANN's Processes

[2.25] ICANN's Processes

The President's report, dealt with at [2.20], identified an undue focus on process at the expense of substance and effectiveness as one of the major problems facing ICANN. While rejecting the concept of directly electing at-large directors and criticising the attempt to establish an elaborate reconsideration process, the report was concerned to reaffirm a commitment to openness and public participation. In this respect, the report stated:

> I believe strongly in ICANN's core values of openness and participation. An ICANN that is insulated from input and involvement by individuals across the global diversity of the Internet would be a failed ICANN. Though some will doubtless try to characterize it otherwise, my conclusion about the unworkability of At Large membership elections is NOT a criticism of the concept of participation by individuals. On the contrary, I believe that a reform of ICANN must result in greater openness, wider diversity, and clearer, more meaningful avenues for individual and organizational participation.[140]

The ERC report, also dealt with at [2.20], recommended mechanisms for implementing the core values of participation, openness and transparency.[141] In relation to public participation, the ERC report recommended encouraging policy development through bottom-up

[140] *President's Report—ICANN: The Case for Reform* (24 Feb 2002).

[141] See CIER, above n 119.

consensus processes. In relation to the values of openness and transparency, the ERC report followed the President's report in recommending new accountability mechanisms, including the establishment of an Ombudsman and amending ICANN's reconsideration process. This paragraph explains ICANN's decision-making processes.

[2.25.1] Policy-development Processes

The ERC report, dealt with at [2.20], stated:

> A bottom-up, consensus-driven approach to policy development is preferable wherever such approaches do not prevent the Board from carrying out it's ultimate responsibility for ensuring policies are developed, approved, and implemented as necessary to accomplish ICANN's mission. That is, wherever practical, the development or modification of policies would benefit from undergoing policy development in the appropriate policy development body acting with appropriate community review and input.[142]

As explained at [2.24.2], the detailed, bottom-up development of ICANN policies takes place through ICANN's supporting organisations. At [2.24.2] it was further explained that the ICANN Bylaws require the ccNSO Council and GNSO Council to comply with policy-development processes (PDPs), which are set out in the Bylaws, in the development of recommendations to the ICANN board. The PDPs embody public processes, including the publication of reports, public notification of processes and solicitation of comments. As explained at [2.24.3], further input into the development of ICANN policies takes place through advice provided to the ICANN board by advisory committees, including the ALAC.

[2.25.2] Transparency

Article III of the Bylaws requires ICANN and the ICANN board to act in as open a manner as possible. In this respect, Article III.1 sets out the following general principle:

> ICANN and its constituent bodies shall operate to the maximum extent feasible in an open and transparent manner and consistent with procedures designed to ensure fairness.

Article III further provides for public notice of board meetings, the publication of minutes of board meetings and meetings of supporting organisations, and timely publication of preliminary reports of actions taken by the board following board meetings. In particular, ICANN is required to make a range of specified information, including financial information, available by means of a publicly-accessible Web site. Furthermore, ICANN is required to create a staff position, known as the Manager of Public Participation, to be responsible for coordinating public participation in ICANN, including means of communicating with and receiving input from the general community of Internet users.[143]

[2.25.3] Accountability and Review

The Bylaws establish four main mechanisms that are designed to ensure that ICANN remains accountable for its actions, decisions and policies:

- a reconsideration process;
- independent review of board actions;

[142] *Ibid.*

[143] The ERC report had recommended that ICANN establish a staff position to encourage public participation in ICANN: CIER, above n 119.

- periodic review of ICANN structure and operations; and
- an ombudsman.

Article IV.1 of the Bylaws sets out the following statement of general principle concerning ICANN accountability:

> In carrying out its mission as set out in these Bylaws, ICANN should be accountable to the community for operating in a manner that is consistent with these Bylaws, and with due regard for the core values set forth in Article 1 of these Bylaws.

ICANN's core values are set out at [2.23]. In relation to accountability, core value 10 is especially important. Core value 10 is:

> Remaining accountable to the Internet community through mechanisms that enhance ICANN's effectiveness.

Reconsideration Process

Article IV.2 of the Bylaws requires ICANN to have a process by which any person or entity materially affected by an action of ICANN may request review or reconsideration of that action. A person or entity may submit a reconsideration request if he or she has been adversely affected by:

- a staff action, or inaction, that contradicts established ICANN policies; or
- an action or inaction of the board that has been taken without consideration of material information (except where the person or entity making the request could have submitted the relevant information to the board, but did not).

Reconsideration requests are submitted to a committee of the board, known as the reconsideration committee. The reconsideration committee is responsible for: evaluating the request; determining whether to stay the contested action pending resolution of the request; conducting a factual investigation; requesting additional written submissions; and making recommendations to the board. All reconsideration requests must be submitted to an e-mail address designated by the reconsideration committee within 30 days of the date on which information about the challenged board action is first published; the date on which the requesting party first became aware of the challenged staff action; or the date on which the affected person reasonably concluded that action would not be taken in a timely manner.

All reconsideration requests are required to be posted to the ICANN Web site. The reconsideration committee is required to act promptly and, within 30 days of receiving the request, indicate its intention either to decline the request or to consider it. The reconsideration committee may, however, dismiss a reconsideration request where the request is repetitive, frivolous, non-substantive or otherwise abusive, or where the affected party had notice and opportunity to, but did not, participate in a public process relating to the contested action. The reconsideration committee may also dismiss a request where the requesting party does not show that it will be affected by ICANN's action.

The reconsideration committee is required to make a final recommendation to the board within 90 days of receiving the request, unless this is impractical, in which case the committee must report to the board detailing the reasons for the delay. The board is not, however, required to follow the recommendations of the reconsideration committee. The reconsideration committee must submit an annual report to the board containing information relating to reconsideration requests for the preceding calendar year.

Independent Review of Board Actions

Article IV.3 requires ICANN to have a process for independent third-party review of board actions that are alleged to be inconsistent with ICANN's Articles of Incorporation or Bylaws. Any person materially affected by a decision or action of ICANN's board that is alleged to be inconsistent with the Articles of Incorporation or Bylaws may submit a request for independent review of that decision or action.

The Bylaws require ICANN to establish an Independent Review Panel (IRP) to be operated by an international arbitration provider appointed by ICANN, known as the IRP Provider. The IRP Provider will, subject to board approval, establish operating rules and procedures. Requests for independent review are to be referred to the IRP, which has the authority to: request additional written submissions; declare whether an action or inaction was inconsistent with the Articles of Incorporation or Bylaws; and recommend a stay of action until the board reviews and acts on the IRPs opinion.

The IRP is required, to the maximum extent feasible, to conduct its proceedings by e-mail. Moreover, the IRP is required to base its declaration solely on the documentation, supporting materials and arguments submitted by the parties. IRP declarations are required to be in writing. Where feasible, ICANN's board is required to consider the IRP declaration at the next board meeting.

In May 2004, ICANN entered into an agreement with the Independent Centre for Dispute Resolution (ICDR), an arbitration service, to provide the functions of the IRP.[144]

Review of ICANN Structure and Operations

In addition to the overall supervision by the Department of Commerce under the September 2006 Joint Project Agreement, which is dealt with at [2.17] above, Article IV.4 of the Bylaws requires the board to conduct periodic reviews of the performance and operation of each supporting organisation, each supporting organisation council, each advisory committee (other than the GAC) and the nominating committee. The reviews, which, if feasible, are to occur every three years, are to be conducted by an entity that is independent of the organisation under review. The board is required to consider the results of any such review no later than the second scheduled board meeting after the results of the review have been posted to the Web site for 30 days. The Bylaws provide for the GAC to establish its own review mechanisms.

Ombudsman

Article V of the Bylaws establishes an Office of the Ombudsman, which is responsible for receiving complaints about unfair or inappropriate actions by the ICANN board or staff, and the attempted resolution of those complaints. The Ombudsman is a full-time position, with an initial two-year appointment.

Pursuant to Article V.2, the charter of the Ombudsman is to act as a neutral dispute resolution practitioner for matters where requests for reconsideration or review have not been submitted under the reconsideration process or independent review policy. The Bylaws provide that the principal function of the Ombudsman is:

[144] ICANN, *Ninth Status Report to the Department of Commerce* (7 Apr 2004), available at http://www.icann.org/general/status-report-07apr04.pdf, 6; ICANN, *Tenth Status Report to the Department of Commerce* (7 Oct 2004), available at http://www.icann.org/general/mou-status-report-07oct04.htm, 6.

to provide an independent internal evaluation of complaints by members of the ICANN community who believe that the ICANN staff, board or an ICANN constituent body has treated them unfairly.

The Ombudsman is to seek to clarify the issues and resolve complaints using conflict resolution tools such as negotiation, facilitation and 'shuttle diplomacy'.

It is important to note that the Ombudsman cannot consider complaints that have become the subject of ICANN's reconsideration or independent review processes. Moreover, the Ombudsman cannot deal with complaints relating to ICANN employees or vendors/suppliers of ICANN. ICANN employees and board members are obliged to direct members of the ICANN community who voice problems, concerns or complaints about ICANN to the Ombudsman, who is to advise complainants of the options available for dealing with such problems, concerns or complaints. The first ICANN Ombudsman took office on 1 November 2004.

Registry and Registrar Agreements

[2.26] gTLD Registry Agreements

As explained at [2.8] above, the 1997 report of the International Ad Hoc Committee (IAHC) was the first official document to propose separating the functions involved with the operation of the database for registering domain names for a gTLD, known as the registry functions, from the functions involved in registering domain names for customers, known as the registrar functions. This fundamental distinction between registries and registrars was retained in the NTIA Green Paper, dealt with at [2.9] above, and the NTIA White Paper, dealt with at [2.10] above, which led to the formation of ICANN.

As explained at [2.11], Amendment 11 to the Cooperative Agreement between the Department of Commerce and NSI, entered into in October 1998, extended NSI's pre-existing role as both registry and registrar for the .com, .net and .org gTLDs until September 2000, in return for NSI agreeing to develop a shared registry system (SRS), to allow competition among registrars and to recognise ICANN. The SRS allows competing ICANN-accredited registrars to register domain names using a shared, central registry maintained by a single registry operator. As explained at [2.12], under Amendment 19 to the Cooperative Agreement, entered into in November 1999, NSI agreed to enter into a registry agreement and registrar accreditation agreement with ICANN, and only to accept registrations in the open gTLDs from ICANN accredited registrars. As further explained at [2.12], at the same time, ICANN and NSI entered into a registry agreement under which ICANN agreed that NSI would continue to operate as the registry for the open gTLDs and NSI agreed to comply with all consensus policies adopted by ICANN.[145] The original ICANN-NSI Registry Agreement was scheduled to expire four years after the agreement came into effect, unless it was extended pursuant to NSI divesting itself of its registrar operations.[146]

[145] http://www.icann.org/general/mou-status-report-07oct04.htm.

[146] *Ibid,* cl 23.

As explained at [2.14], in May 2001 the ICANN-NSI Registry Agreement was replaced by three separate registry agreements between ICANN and VeriSign, Inc. (which had, by then, taken over NSI) for the operation of the .com, .net and .org gTLDS. The agreements were designed to place VeriSign in a position that was, as much as possible, equivalent to the registry operators for other gTLDs. The revised agreements were reflected in Amendment 3 to the ICANN/Department of Commerce MOU. Under the revised agreements, VeriSign agreed to discontinue its role as registry for the .org gTLD in exchange for a presumptive right of renewal of the .com registry agreement and for removal of VeriSign's obligation to divest its registrar operations. Following this agreement, the .org registry was reassigned from VeriSign to Public Interest Registry (PIR), a wholly-owned subsidiary of the Internet Society (ISOC), in January 2003.

As explained at [1.13], the gTLDs are divided into sponsored TLDs (sTLDs), meaning that the TLD is a specialised TLD that has a sponsor representing the particular community most affected by the TLD, and unsponsored TLDs (uTLDs). The current ICANN gTLD registry agreements, including the identity of the registry operators, the dates that the agreements came into effect, the initial terms of the agreements and whether the agreements are sponsored or unsponsored, are set out in Table 2.1. Contract negotiations concerning renewal of the .aero, .coop and .museum sponsorship agreements were protracted due to disagreements about the ability of the registries to make policies.

[2.27] .com Registry Agreement

The .com Registry Agreement defines the relationship between ICANN and VeriSign in relation to registry operations for the .com gTLD. The current agreement resulted from the settlement of legal disputes between Verisign and ICANN. As explained at [2.12]–[2.13], Amendment 1 to the ICANN/DOC MOU provides that amendments to, or substitutes for, the .com registry agreement are subject to prior approval by the Department of Commerce.

In February 2004, VeriSign brought an action against ICANN in the US District Court for the Central District of California alleging antitrust violations and breach of contract, resulting from ICANN decisions to restrict VeriSign from offering controversial services, including the proposed SiteFinder (or 'wildcard' feature) Service, which redirected Web browsers to a search service when users attempted to access a non-existent .com or .net domain name. In September 2004, the court entered its final judgment dismissing VeriSign's complaint on the basis that it did not sufficiently allege that ICANN was controlled by VeriSign's competitors,[147] whereupon VeriSign appealed. In October 2005, ICANN announced the terms of a proposed settlement of the litigation with VeriSign, which required VeriSign to abandon the SiteFinder Service and included a new .com registry agreement. The proposed new .com agreement was made available for public comment and, following amendments, a revised agreement was posted for public comment in January 2006. On 28 February 2006, a majority of ICANN's board approved the final settlement agreements, including a new .com registry agreement, with the new .com registry agreement being entered into on 1 March 2006. Following a nine-month-long review, the Department of Commerce approved the agreement on 29 November 2006. As explained at [2.19] above, at the same time, the cooperative

[147] *VeriSign Inc v ICANN,* Case No. CV 04-1292 AHM (CTx), Final Judgment, 22 Sept 2004.

Table 2.1 gTLD Registry Agreements

gTLD Agreement	Registry Operator	Date of agreement	Initial term*	Sponsored/ Unsponsored
.aero TLD Sponsorship Agreement	Société Internationale de Télécommunications Aéronautiques SC, (SITA)	17 December 2001 (renewal agreement posted for public comment: 21 December 2005)	5 years	Sponsored
.asia TLD Registry Agreement	DotAsia Organisation Limited	6 December 2006	10 years	Sponsored
.biz Registry Agreement	Neulevel	8 December 2006	Until 31 December 2012	Unsponsored
.cat TLD Sponsorship Agreement	Fundació puntCAT	23 September 2005	10 years	Sponsored
.com Registry Agreement	VeriSign, Inc.	1 March 2006 (Approved by Department of Commerce: 29 November 2006)	Until 30 November 2012	Unsponsored
.coop TLD Sponsorship Agreement	Dot Cooperation LLC	1 July 2007	10 years	Sponsored
.info Registry Agreement	Afilias	8 December 2006	Until 31 December 2012	Unsponsored
.jobs Registry Agreement	Employ Media LLC	5 May 2005	10 years	Sponsored
.mobi TLD Sponsorship Agreement	mTLD Top Level Domain, Ltd.	10 July 2005	10 years	Sponsored
.museum TLD Sponsorship Agreement	Museum Domain Management Association (MuseDoma)	17 October 2001 (renewal agreements posted for public comment: 21 December 2005; 2 March 2007; 31 August 2007)	5 years	Sponsored
.name Registry Agreement	Global Name Registry, Ltd	1 August 2001 (as amended 8 August 2003) (renewal agreement posted for public comment: 29 June 2007)	5 years	Unsponsored
.net Registry Agreement	VeriSign, Inc.	29 June 2005	6 years	Unsponsored
.org Registry Agreement	Public Interest Registry (PIR)	8 December 2006 (as-amended 13 February 2007)	Until 30 June 2013	Unsponsored

.pro Registry Agreement	RegistryPro, Inc.	3 May 2002	5 years	Unsponsored
.tel Registry Agreement	Telnic Limited	30 May 2006	10 years	Sponsored
.travel Registry Agreement	Tralliance Corporation	5 May 2005	10 years	Sponsored

* There is considerable variation among the agreements concerning the date of the commencement of the term of the agreement.

agreement between the Department of Commerce and VeriSign was amended. The new .com Registry Agreement should be read together with the amended cooperative agreement.

The new .com Registry Agreement extended the term of the agreement to 30 November 2012. The main obligations of VeriSign under the agreement are:

- to comply with and implement all temporary specifications or policies established by ICANN that are necessary to maintain the stability or security of the Internet or DNS.
- to comply with and implement all ICANN 'Consensus Policies'. 'Consensus Policies' are specifications or policies established pursuant to the procedures set out in ICANN's Bylaws, especially the Policy-Development Process (see [2.25.1] above), and that relate to:
  —issues for which uniform or coordinated resolution is reasonably necessary to facilitate interoperability, security and/or stability of the Internet or DNS;
  —functional and performance specifications for the provision of registry services;
  —security and stability of the registry database for the .com gTLD;
  —registry policies reasonably necessary to implement consensus policies relating to registry operations or registrars; or
  —resolution of disputes regarding the registration of domain names (as opposed to the use of such domain names).
- to establish a data escrow or mirror site policy for .com registry data, to be mutually agreed between ICANN and VeriSign.
- to notify registrars of the purposes for which personal data submitted by registrars is collected, the intended recipients and mechanisms for access and correction of such data.
- to provide ICANN with bulk access to the zone files for the .com registry on a continuous basis in the manner reasonably specified by ICANN and to provide bulk access to third parties on the terms set out in an agreement reasonably established by ICANN.
- to provide a monthly report to ICANN.
- to reserve and not register domain names that include reserved TLD strings. The reserved TLD strings, identified in Appendix 6 to the agreement, include ICANN-related names, such as 'icann' and 'gnso', and IANA-related names, such as 'ripe' and 'iana'.
- to comply with the functional and performance specifications set out in Appendix 7 to the agreement, and to allow for ICANN to audit technical and operational records.
- to comply with the process for consideration of proposed registry services. Under this process, VeriSign must notify ICANN of proposed changes in registry services, including sufficient information for ICANN to make determinations. Following notification, ICANN has 15 days to make a preliminary determination whether the registry service requires further consideration because it could raise significant security, stability or competition issues. If ICANN determines that the registry service might raise significant

competition issues it must refer the issue to the appropriate governmental competition authority. If ICANN determines that the registry service might raise significant stability or security issues it must refer the proposal to a standing panel of experts, which has 45 days to prepare a report for the ICANN board. After receiving the report, the ICANN board has a further 30 days to make a decision on the proposed service. If the board determines that the proposed service creates a reasonable risk of an adverse effect on stability or security, VeriSign is prohibited from offering the service.
- to pay the required registry-level fees to ICANN, which include an initial fee, fixed fees and variable fees.
- to consult twice annually with ICANN on the security and/or stability of the .com registry, the DNS and the Internet.
- to develop and deploy a centralised WHOIS service for the .com gTLD.

In addition to the above obligations, the agreement imposes specific obligations on VeriSign in relation to registrars. First, VeriSign is required to make access to registry services, including the shared registration system, available to all ICANN-accredited registrars, subject to the .com Registry-Registrar Agreement set out in Appendix 8 to the agreement. The access to registry services must be provided to accredited registrars on a non-discriminatory basis. Any revisions to the .com Registry-Registrar Agreement require prior approval by ICANN. Secondly, VeriSign is prohibited from acting as a registrar with respect to the .com gTLD. Thirdly, VeriSign is prohibited from acquiring (directly or indirectly) more than a 15 per cent ownership interest in any ICANN-accredited registrar.

The agreement also includes provisions regulating the provision of basic registry services, including price regulation. The main regulatory provisions are as follows:

- VeriSign is prohibited from tying basic registry services to any other product or service.
- The maximum price that VeriSign can charge accredited registrars for registry services relating to individual domain names is regulated, with an initial cap of US$6 per name each year, and with provision for an annual increase of up to 7 per cent for four of the next six years.
- VeriSign is prohibited from engaging in price discrimination between accredited registrars.
- VeriSign must provide registrars with no less than six months' prior notice of any increase in the price for new domain name registrations, renewals and transfers.

The main obligations agreed to by ICANN under the agreement are:

- consistent with its mission and core values, to operate in an open and transparent manner;
- not to apply standards, policies, procedures or practices arbitrarily, unjustifiably, or inequitably and not to single out VeriSign for disparate treatment unless justified by substantial and reasonable cause;
- to ensure that the authoritative root will point to the TLD zone servers designated by VeriSign for the .com TLD and that any changes to the zone server designation submitted by VeriSign will be implemented within seven days.
- to include VeriSign and its administrative and technical contacts in ICANN's publication of root-zone contact information.

The agreement establishes a presumption in favour of renewal. In this regard, the term of the agreement and each later term will be renewed unless:

- ICANN has issued VeriSign with a notice of breach of the agreement, and VeriSign has failed to cure the breach within 30 days and an arbitrator or court has determined that VeriSign has been in fundamental and material breach; and
- VeriSign has failed to comply with the decision of the arbitrator or court within 10 days of the decision, or such further time as prescribed by the arbitrator or court.

The most controversial elements of the new .com Registry Agreement were the presumption of renewal in favour of VeriSign and the removal of any obligation for VeriSign to invest in registry infrastructure. The presumption of renewal is, however, little different from the presumption of renewal in the previous .com registry agreement. Moreover, the cooperative agreement between the Department of Commerce and VeriSign provides for the Department of Commerce to take investment made in improving the security and stability of the DNS into account in determining whether to approve renewal of the registry agreement.

[2.28] Registrar Accreditation Agreement

As explained at [2.11], Amendment 11 to the Cooperative Agreement between the Department of Commerce and NSI, entered into in October 1998, required NSI to develop a shared registry system (SRS) to allow for competition among registrars in the .com, .net and .org gTLDS. As explained at [2.12], following Amendment 11 to the Cooperative Agreement the Department of Commerce and ICANN decided that ICANN would be responsible for authorising registrars operating in the gTLDs. In March 1999, the ICANN board approved a Statement of Registrar Accreditation Policy and Registrar Accreditation Agreement,[148] which applied to applications for registrar accreditation through to November 1999. As further explained at [2.12], in November 1999 ICANN and NSI entered into a Registrar Transition Agreement, which provided for NSI to be accredited as a registrar under ICANN's Registrar Accreditation Agreement.[149] At the same time, the ICANN board approved a revised Registrar Accreditation Agreement for the .com, .net and .org gTLDs, which took into account some of the concerns expressed by accredited registrars that were competing with NSI.[150]

As explained at [2.14], in May 2001 the ICANN-NSI Registry Agreement was replaced by three separate registry agreements between ICANN and VeriSign, Inc. for the operation of the .com, .net and .org gTLDS. Under the agreements, VeriSign was not required to divest itself of its registrar operations, but agreed to discontinue its role as registry for the .org gTLD. The new registry agreements coincided with approval of a revised Registrar Accreditation Agreement.[151] The May 2001 Registrar Accreditation Agreement, as amended, is the current version of the registrar accreditation agreement and applies to all ICANN-accredited registrars.

Under Section 2 of the Registrar Accreditation Agreement, ICANN has general obligations with respect to all matters that impact on the rights, obligations or role of registrars:

[148] ICANN, *Statement of Registrar Accreditation Policy* (4 Mar 1999), available at http://www.icann.org/registrars/policy_statement.html; *Registrar Accreditation Agreement* (Apr 1999), available at http://www.icann.org/registrars/ra-agreement-12may99.htm.

[149] *ICANN-NSI Registrar Transition Agreement* (approved 4 Nov 1999; signed 10 Nov 1999), available at http://www.icann.org/nsi/icann-nsi-transition-04nov99.htm.

[150] Registrar Accreditation Agreement, above n 148.

[151] *Registrar Accreditation Agreement* (17 May 2001), available at http://www.icann.org/registrars/ra-agreement-17may01.htm.

- to exercise its responsibilities in an open and transparent manner;
- not unreasonably to restrain competition and, to the extent feasible, promote and encourage robust competition;
- not to apply standards, policies, procedures or practices arbitrarily, unjustifiably, or inequitably and not to single out the registrar for disparate treatment unless justified by substantial and reasonable cause; and
- to ensure, through its reconsideration and independent review policies, adequate appeal procedures for the registrar, to the extent that it is adversely affected by ICANN standards, policies, procedures or practices.

Under Section 3 of the agreement, ICANN-accredited registrars have the following obligations:

- to submit relevant details regarding the registration of domain names in the TLD for which it is accredited to the registry operator for that TLD;
- to provide public access to a Whois service concerning all active registered names sponsored by the registrar for each TLD in which it is accredited; and
- to maintain its own electronic database, as updated, containing specified data for each active registered name sponsored by it within each TLD for which it is accredited.

The agreement requires registrars to enter into an electronic or paper registration agreement with domain name holders that includes the following provisions:

- the domain name holder must provide accurate and reliable contact details to the registrar, and must promptly correct and update them during the term of the registration;
- the wilful provision of inaccurate or unreliable information, or wilful failure promptly to update information provided to the registrar, shall constitute breach of the registration contract and be a basis for cancellation of the registration;
- a registered domain name holder that licenses the use of a domain name to a third party remains the registered domain name holder of record, and is liable for harm caused by wrongful use of the registered domain name, unless it promptly discloses the identity of the licensee to a party providing reasonable evidence of actionable harm;
- the registrar must provide notice to registered domain name holders of: the purposes for which personal data are collected; the intended recipients of the data; which data are obligatory and which are voluntary; and how a name holder can access and rectify personal data held about them;
- the registrar must agree that it will not process personal data in a way that is incompatible with the purposes and limitations of which it has provided notice, and must take reasonable precautions to protect the data from loss, misuse, unauthorised access or disclosure, alteration or destruction;
- the registered domain name holder must represent that, to the best of its knowledge and belief, neither the registration of the name nor the manner in which it is used infringes the legal rights of any third party;
- in the event of disputes arising from the use of the registered name, the registered domain name holder must submit to the jurisdiction of the courts: (1) of the registered name holder's domicile and (2) where the registrar is located;
- the registered domain name holder must agree that registration of the name is subject to suspension, cancellation or transfer pursuant to any ICANN-adopted specification or

policy: (1) to correct mistakes in registering the name; or (2) for the resolution of disputes concerning the registered name; and
- the registered domain name holder must indemnify the registry operator against any claims arising from the domain name registration.

The agreement specifically provides that the registrar must abide by ICANN consensus policies that require:

- registrars cooperatively to implement a distributed capability that provides query-based Whois search functionality across all registrars; and
- reasonable and commercially practicable (a) verification, at the time of registration, of contact information associated with a registered name; or (b) periodic reverification of such information.

An important provision of the accreditation agreement requires each registrar to have in place a policy and procedures for the resolution of disputes concerning registered domain names, specifically the Uniform Dispute Resolution Policy (UDRP).

In addition to the specific policies mentioned in the agreement, pursuant to Section 4 of the agreement, accredited registrars are required to comply with all ICANN consensus policies that deal with the following matters:

- issues for which uniform or coordinated resolution is reasonably necessary to facilitate interoperability, technical reliability and/or operational stability of registrar services, registry services, the DNS or the Internet;
- registrar policies reasonably necessary to implement ICANN policies or specifications relating to a DNS registry or to registry services;
- resolution of disputes concerning the registration of registered names, including where the policies take into account use of the domain names;
- principles for allocation of registered names;
- prohibitions on warehousing of or speculation in domain names by registries or registrars;
- maintenance of and access to accurate and up-to-date contact information regarding registered names and nameservers;
- reservation of registered names that may not be registered initially or that may not be renewed due to reasons reasonably related to (a) avoidance of confusion among or misleading of users; (b) intellectual property; or (c) the technical management of the DNS or the Internet;
- procedures to avoid disruptions of registration due to suspension or termination of operations by a registry operator or a registrar; and
- the transfer of registration data upon a change in registrar.

ccTLD Governance

[2.29] ICANN and ccTLD Governance

As explained at [1.11] above, in October 1984 Jon Postel and Joyce Reynolds issued RFC 920, which incorporated the ISO-3166 list of two-letter country codes as TLDs, known as country code TLDs or ccTLDs.[152] Postel had chosen to adopt the established ISO standard so as to avoid disputes regarding the proper designation of countries.

As explained at [2.4], responsibility for managing the domain name space was given to the Information Sciences Institute (ISI), based at the University of Southern California, by the Defense Information Systems Agency (DISA) in 1985. As explained at [2.5] above, in 1988 the ISI was internally reorganised and began referring to itself as the Internet Assigned Numbers Authority (IANA), although Jon Postel and his colleagues remained responsible for most aspects of DNS governance. From 1985 to 1993, Jon Postel was personally responsible for assigning ccTLDs to specific applicants. Delegations of responsibility for ccTLDs were customarily made quite informally on a 'first-come/first-served' basis. Mueller cites Postel himself as noting that the person assigned responsibility for a ccTLD 'is generally the first person that asks for the job (and is somehow considered a "responsible person")'.[153] As further explained by Mueller, there was no explicit policy for dealing with the delegation of responsibility for ccTLDs and disputes involving delegations were dealt with informally by Postel.

In the early 1990s more disputes arose in relation to the delegation of responsibility for ccTLDs.[154] This led to Postel drafting a more explicit policy for the delegation of TLDs, which was released as RFC 1591 in March 1994.[155] RFC 1591 stated that the IANA was responsible for the delegation of TLDs. It further indicated that delegation of management of a TLD would usually take place at the same time as a new TLD was created. RFC 1591 proceeded to set out criteria for the delegation of TLDs, but did so in very broad terms. At the most general level, it outlined the following guiding principle:

> The major concern in selecting a designated manager for a domain is that it be able to carry out the necessary responsibilities, and have the ability to do a equitable, just, honest, and competent job.

Within this principle, RFC 1591 set out the following more specific criteria:

- For each domain there must be a designated manager for supervising that domain's name space. There must also be an administrative contact and a technical contact for each domain. For ccTLDs, at least the administrative contact must reside in the relevant country.
- The designated authorities are trustees for the delegated domain, and have a duty to serve the community. The designated manager is the trustee of the TLD for both the nation, in the case of a ccTLD, and the global Internet community.
- Concerns about 'rights' and 'ownership' of domains are inappropriate. It is appropriate to be concerned about 'responsibilities' and 'service' to the community.
- The designated manager must be equitable to all groups in the domain that request domain names. This means that the same rules are applied to all requests, all requests must be processed in a non-discriminatory fashion, and academic and commercial (and other) users are treated on an equal basis.
- Significantly interested parties in the domain should agree that the designated manager is the appropriate party. The IANA tries to make any contending parties reach agreement among themselves, and generally takes no action to change things unless all the contending parties agree. Only in cases where the designated manager has substantially misbehaved would the IANA step in.

[152] J Postel and J Reynolds, 'Domain Requirements 1', RFC 920 (Oct 1984).
[153] Mueller, above n 19, 89.
[154] Mueller, above n 19, 125.
[155] J Postel, 'Domain Name System Structure and Delegation', RFC 1591 (Mar 1994).

- The designated manager must do a satisfactory job of operating the DNS service for the domain. That is, the actual management of the assigning of domain names, delegating subdomains and operating nameservers must be done with technical competence.
- For any transfer of the designated manager trusteeship from one organisation to another, the higher-level domain manager (IANA in the case of TLDs) must receive communications from both the old and the new organisations that assure the IANA that the transfer is mutually agreed.

Soon after the release of RFC 1591, however, the Internet entered its period of rapid growth. This meant that RFC 1591, including the utopian public trustee model underlying the RFC, was soon overtaken by events as control of TLDs became potentially commercially valuable.[156] Moreover, while RFC 1591 foreshadowed the establishment of an Internet DNS Names Review Board (IDNB) to review disputes regarding the delegation of TLDs, this body failed to eventuate. An especially controversial area was the role of national governments in delegations, or redelegations, of responsibility for ccTLDs. This was reflected in October 1997, when the IANA released a ccTLD news memo, which stated, in part:

> The IANA takes the desires of the government of the country very seriously, and will take them as a major consideration in any transition discussion.[157]

As explained at [2.12], in December 1998 an agreement was entered into between the University of Southern California and ICANN for the transfer of the IANA functions performed by ISI to ICANN.[158] As explained at [2.17] above, in February 2000, the US government entered into a contract with ICANN for the performance of the IANA function, which replaced the contract with ISI. The authority of ICANN over the managers of ccTLDs, however, remained somewhat ambiguous.

From the time of its establishment in March 1999, the Government Advisory Committee (GAC) campaigned for a role in the delegation and redelegation of ccTLDs. Nevertheless, in May 1999, ICANN released a policy statement on ccTLD administration and delegation, known as ICP-1.[159] ICP-1 was effectively an update of RFC 1591, reflecting the evolution of IANA delegation policy to May 1999.

In July 2004, ICANN requested the ccNSO to prepare a broad set of criteria, known as an accountability framework, to provide the basis for formal agreements between ICANN and ccTLD managers. The ccNSO Council agreed to the *Guidelines for ccTLD managers Accountability Framework discussions with ICANN* in December 2005, and the guidelines were published in January 2006.[160] To reflect the diversity of ccTLD management arrangements, ICANN developed two options for formalising the relationship between ICANN and ccTLD managers. First, an agreement can be based on ICANN's Accountability Framework document,[161] which reflects the key elements of the ccNSO's guidelines, including clauses covering the obligations of the ccTLD manager and ICANN, dispute resolution and

[156] See Mueller, above n 19, 126–7. See also J Klensin, 'Reflections on the DNS', RFC 1591, and Categories of Domains', RFC 3071 (Feb 2001).

[157] IANA, *CCTLD News Memo #1* (23 Oct 1997), available at http://www.iana.org/cctld/cctld-news1.htm.

[158] *University of Southern California/ICANN Transition Agreement* above n 79.

[159] ICANN, *Internet Domain Name System Structure and Delegation (ccTLD Administration and Delegation)*, Internet Coordination Policy 1 (ICP-1, May 1999), available at http://www.icann.org/icp/icp-1.htm.

[160] ccNSO Council, *Guidelines for ccTLD managers Accountability Framework discussions with ICANN* (6 Jan 2006), available at http://ccnso.icann.org/announcements/files/af-guidelines-14dec05.pdf.

[161] ICANN, *Accountability Framework document* (12 Feb 2006), available at http://www.icann.org/cctlds/accountability-framework-12feb06.pdf.

termination. Secondly, an agreement can be based on an exchange of letters, which includes some elements of the ccNSO's guidelines, but is less detailed. The templates developed by ICANN, both for its accountability framework and exchange of letters, represent the starting point for negotiations between ICANN and ccTLD managers.

The WSIS

[2.30] The WSIS and the IGF

World summits are international conferences convened by the United Nations to address important global issues, such as the environment, population or racism.[162] A proposal for an information society summit was initially made by the ITU in 1998.[163] In December 2001, the UN General Assembly endorsed a World Summit on the Information Society (WSIS) to be held in two phases over the period 2003–2005.[164] The first phase of the WSIS was held in Geneva from 10 to 12 December 2003 and the second phase in Tunis from 16 to 18 November 2005. The WSIS was coordinated by a High-level Summit Organizing Committee (HLSOC) established under the patronage of the UN Secretary-General. The ITU was the UN agency that played the leading role in organising the WSIS. The WSIS differed from other UN summits in two main respects: first, it was held in two phases; and, secondly, the role of civil society was more formalised than previous summits, with the creation of an official 'civil society bureau'.

A world summit essentially ratifies the work conducted in preparatory meetings—which include consultations with UN member states, experts, the private sector and NGOs—that precede the summit. World summits customarily produce two sorts of documents: a declaration of principles, which defines the normative vision of the summit; and a plan of action, which attempts to translate the principles into actions. Thus, the Geneva phase of the WSIS resulted in the Geneva Declaration of Principles[165] and the Geneva Plan of Action.[166] Similarly, the Tunis phase of the WSIS resulted in the Tunis Commitment[167] and the Tunis Agenda for the Information Society.[168]

The Geneva Declaration of Principles included high-level statements setting out a 'common vision' of the information society and a set of key principles for building an inclusive information society. The key principles included the following general statement:

> all stakeholders should work together to: improve access to information and knowledge; build capacity; increase confidence and security in the use of ICTs; create an enabling environment at all

[162] See eg H Klein, 'Understanding WSIS: An Institutional Analysis of the UN World Summit on the Information Society' (2004) 1 *Information Technologies and International Development* 3.

[163] International Telecommunication Union, World summit on the information society, Resolution 73 (Minneapolis, 1998), available at http://www.itu.int/council/wsis/R73.html.

[164] UN General Assembly Resolution 56/183 (21 Dec 2001).

[165] World Summit on the Information Society, *Declaration of Principles. Building the Information Society: a global challenge in the new Millennium*, Document WSIS-03/GENEVA/DOC/4-E (12 Dec 2003).

[166] World Summit on the Information Society, *Plan of Action*, Document WSIS-03/GENEVA/DOC/5-E (12 Dec 2003).

[167] World Summit on the Information Society, *Tunis Commitment*, Document WSIS-05/TUNIS/DOC/7-E (18 Nov 2005).

[168] World Summit on the Information Society, *Tunis Agenda for the Information Society*, Document WSIS-05/TUNIS/DOC/6(Rev 1)-E (18 Nov 2005).

> levels; develop and widen ICT applications; foster and respect cultural diversity; recognize the role of the media; address the ethical dimensions of the Information Society; and encourage international and regional cooperation.[169]

The Geneva Plan of Action included two important practical outcomes. First, the Plan of Action reinforced the call made by the Declaration of Principles for the UN Secretary-General to establish a working group on Internet governance to report before the Tunis phase of the WSIS, setting out the terms of reference for the group.[170] Secondly, the Plan of Action called for the establishment of a UN task force to review the adequacy of existing financial mechanisms for meeting the development challenges of information and communications technologies (ICT) and to report before the Tunis phase.[171]

As explained at [2.2], the Working Group on Internet Governance (WGIG), which was established pursuant to the WSIS mandate, published its report in June 2005.[172] The WGIG report adopted a working definition of Internet governance, identified public policy issues relevant to Internet governance and assessed the adequacy of existing arrangements and recommended proposals for action. The most important proposal for action was to establish a new, multi-stakeholder forum to address Internet-related public policy issues. The WGIG report also set out four possible models for Internet governance, three of which would either replace ICANN or make it accountable to a new international body that would effectively take over the role of the US Department of Commerce. As explained at [2.15], US government bodies and the US Congress responded to proposals for reforming DNS governance by issuing statements and resolutions opposing moves to transfer control of the DNS, or ICANN, to the United Nations.

Prior to the Tunis summit, divisions had emerged between nations such as China, Brazil, South Africa and Syria, that favoured UN supervision of the DNS, and nations such as the US, Canada and Australia, that supported the current governance arrangements. The European Union had initially supported the US position but, during preparatory meetings for the Tunis summit, shifted its position to support a greater role for the UN. On 15 November 2006, immediately before the Tunis summit, a compromise was reached whereby the existing DNS governance arrangements were recognised in return for an agreement to create a new UN forum to discuss Internet governance, to be known as the Internet Governance Forum. The compromise was embodied in the outcomes from the Tunis phase.

The Tunis Commitment reaffirmed the Geneva Declaration of Principles stating, for example:

> We reaffirm the commitments made in Geneva and build on them in Tunis by focusing on financial mechanisms for bridging the digital divide, on Internet governance and related issues, as well as on follow-up and implementation of the Geneva and Tunis decisions, as referenced in the Tunis Agenda for the Information Society.[173]

The Tunis Agenda recommended improvements in existing ICT financing mechanisms and approved the establishment of a Digital Solidarity Fund, designed to complement existing funding mechanisms. In relation to Internet governance, the Tunis Agenda essentially

[169] World Summit on the Information Society above n 4, para 19.
[170] World Summit on the Information Society above n 6, para 13.
[171] *Ibid*, para 27.
[172] Working Group on Internet Governance, above n 7. The report was supplemented by a more detailed background report: WGIG, *Background Report* (June 2005).
[173] World Summit on the Information Society, above n 167, para 7.

implemented the compromise reached before the Tunis summit. Thus, it included the statement that:

> We recognise that the existing arrangements for Internet governance have worked effectively to make the Internet the highly robust, dynamic and geographically diverse medium that it is today, with the private sector taking the lead in day-to-day operations, and with innovation and value creation at the edges.[174]

On the other hand, as explained at [2.2], the Tunis Agenda recognised that Internet governance was broader than DNS governance, and called on the UN Secretary-General to create a new UN forum for multi-stakeholder policy dialogue, to be known as the Internet Governance Forum (IGF).[175] The first meeting of the IGF was held in Athens from 30 October to 3 November 2006.

The terms of reference for the IGF are set out at [2.2] above.

[174] World Summit on the Information Society above n 168, para 55.
[175] *Ibid*, para 72.

3

Uniform Domain Name Dispute Resolution

Domain Name Disputes

[3.1] 'Cybersquatting'

The DNS was designed by non-lawyers and developed before the Internet began to be used commercially. The founders of the Internet were predominantly concerned with technical matters, not with potential legal conflicts. Consequently, when the DNS was established, the difficulties that might arise from conflicts between Internet identifiers and real-world identifiers—including those, such as trade marks, in which someone has legal rights—were not taken into account. Before the commercialisation of the Internet, it was impossible to predict the problems that later arose.

From the mid-1990s, however, the growing commercialisation of the Internet and the emergence of the World Wide Web were associated with a change in the status and functions of domain names. Taking advantage of the fact that domain names are designed to be easily remembered by people, businesses began to use them as a means of identifying themselves and their products. It is unsurprising that, since then, domain names have developed into an accepted and important form of identification, both on-line and off-line.

The commercial use of domain names exposed a fundamental disjunction between domain names, on the one hand, and real world identifiers in which people have rights, on the other. The disjunction between the DNS and intellectual property rights, especially trade mark rights, was explained by the World Intellectual Property Organization (WIPO), in the important report referred to at [3.4], in the following terms:

> The DNS was designed for its own internal purposes: to ensure connectivity in a technically coherent manner and to do so in a way which was simple and easy for human users to understand and use. Over the same period as the DNS has demonstrated its outstanding success in achieving its design objectives, however, it has become a victim of its own success as the applications of the Internet have expanded into all spheres of activity and as enterprises and persons have begun to include their domain names in the standard identification apparatus that they use for the purposes of business and social communication.[1]

The disjunction arises because a domain name is not merely an Internet address but, as it is an address in human-friendly language, it also communicates a meaning. Consequently, the use of a real world identifier, such as a trade mark or famous name, as a domain name may communicate some association between the domain name and the owner of the real world identifier.

[1] World Intellectual Property Organization (WIPO), *The Management of Internet Names and Addresses: Intellectual Property Issues,* Report of the WIPO Internet Domain Name Process (30 Apr 1999), available at http://www.icann.org/amc/en/processes/process1/report/, para [46].

The commercialisation of the Internet created the possibility for opportunists to attempt to profit from the business goodwill or reputation in real world identifiers, such as trade marks or famous names, by registering the identifiers as domain names. The possibility for opportunistic profiteering arose essentially because registration of domain names in the original open gTLDs—.com, .net and .org—was on a 'first-come, first-served' basis, with no vetting of applications. This meant that there was no need for an applicant to establish any association with an identifier in order for the applicant to register the identifier as a domain name.

Some of the opportunistic practices associated with the registration of real world identifiers became known by the term 'cybersquatting', which was clearly derived by analogy with the term 'squatter', meaning a person who occupies real property without lawful authority. The term 'cybersquatting', which is sometimes used interchangeably with the term 'cyberpiracy', is not capable of a precise definition. The WIPO report, referred to at [3.4], attempted to distinguish between 'cyberpiracy' and 'cybersquatting' on the basis that 'the former term relates to a violation of copyright in the content of websites, rather than to abusive domain name registrations'.[2] Because of the difficulties in defining 'cybersquatting', WIPO preferred to use what it regarded as the more precise terminology of 'abusive registration of a domain name'.

There have, nevertheless, been attempts to define 'cybersquatting'. In 1999, the United States introduced the federal Anticybersquatting Consumer Protection Act (ACPA),[3] to deal with some opportunistic practices relating to trade marks used as domain names. In general terms, the ACPA defines 'cybersquatting' as:

> the registration or use of a trade mark as a domain name in bad faith with an intention to profit from the mark.[4]

Similarly, WIPO, in its 2004 *Guide to WIPO Domain Name Dispute Resolution,* expressly adopted the term 'cybersquatting', which it defined as:

> the pre-emptive, bad faith registration of trademarks as domain names by third parties who do not possess rights in such names.[5]

These definitions, however, beg the question of what precisely is meant by 'bad faith' practices in relation to domain names. This is best explained by identifying the opportunistic practices that are commonly regarded as falling within the term 'cybersquatting'. A distinction may, in general, be drawn between what might be termed 'classical cybersquatting', on the one hand, and other practices that may be more loosely termed cybersquatting', on the other.

In relation to 'classical cybersquatting, the United States Ninth Circuit, in a case brought under the ACPA, has stated that:

> Cybersquatting is the Internet version of a land grab. Cybersquatters register well-known brand names as Internet domain names in order to force the rightful owners of the marks to pay for the right to engage in electronic commerce under their own name.[6]

[2] WIPO, above n 1, para [170].

[3] Codified As the Lanham Act § 43(c), 15 USC § 1125(c).

[4] 15 USC § 1125(d).

[5] WIPO Arbitration and Mediation Center, *Guide to WIPO Domain Name Dispute Resolution* (2004), available at http://www.wipo.int/freepublications/en/arbitration/892/wipo_pub_892.pdf.

[6] *Interstellar Starship Services, Ltd v Epix, Inc,* 304 F 3d 936, 946 (9th Cir, 2002).

In other words, 'classical cybersquatting' involves the registration of a real world identifier, such as a trade mark, followed by a demand for payment from the legal owner of the identifier in return for a transfer of the domain name. This practice was the predictable outcome of the policy of granting a domain name to the first applicant to pay, together with the low cost of registering a domain name compared with the commercial value of identifiers, such as trade marks, in which there has been a substantial investment to create business goodwill. Given the relatively high costs of defending a real world identifier, such as a trade mark, those with an interest in the identifier, such as trade mark owners, had an incentive to protect their goodwill by paying to acquire the domain name.

The reason for prohibiting 'classical cybersquatting' was explained by the United States Fourth Circuit in another ACPA case in the following terms:

> Cybersquatting is considered wrong because a person can reap windfall profits by laying claim to a domain name that he has no legitimate interest in or relationship to.[7]

This explanation, however, requires further analysis. The registration of a real world identifier, such as a trade mark, as a domain name confers a degree of leverage on the domain name registrant. This leverage arises because a domain name may be used to cause harm to the owner of the real world identifier. For example, a domain name that corresponds to a well-known trade mark might be used to dilute or tarnish the trade mark, by including material, such as pornographic or other offensive material, on a web site to which the domain name resolves. Moreover, a domain name can be used to redirect Internet users, contrary to the interests of a trade mark owner, to other on-line locations, such as the web sites of competitors of the trade mark owner.

The registration of a domain name that corresponds to a trade mark, followed by an offer to sell the domain name to the trade mark owner, can therefore be interpreted as a demand for payment coupled with an express or implied threat to harm the trade mark owner. This strategy is likely to be successful, as the registration of a real world identifier as a domain name, in and of itself, confers the potential to harm the legitimate owner of the identifier. In this sense, an offer to sell a domain name to the owner of a real world identifier, such as a trade mark, may be interpreted as a demand for money coupled with an express or implied threat, which is analogous to extortion. As the dissenting panellist in a domain name dispute resolution decision stated:

> Most of the true abusive registration cases involve someone who registers a name in the hopes of extorting money from the trademark owner, or to harass them, or to injure their business.[8]

As the extortionate demands of 'classical cybersquatters' involve social costs, including costs incurred by trade mark owners in monitoring domain name registrations for potentially harmful 'cybersquatting', and no corresponding benefits, there is a good case for providing a system that will deter these practices, and that is less costly than legal actions brought through the courts.[9]

A potential difficulty arising from this analysis is that, apart from an express threat coupled with a demand for payment, extortion may consist of an implied threat to cause harm. This suggests that, in certain circumstances, the mere registration of a real world identifier

[7] *Harrods Ltd v Sixty Internet Domain Names,* 302 F 3d 214, 238 (4th Cir, 2002).

[8] *Davidoff & Cie SA v Dario Muriel,* NAF Case No FA129124 (30 Jan 2003).

[9] For an analysis of the complex economics of threats see R Coase, 'The 1987 McCorkle Lecture: Blackmail' (1988) 74 *Virginia Law Review* 655; S Shavell, 'An Economic Analysis of Threats and their Illegality: Blackmail, Extortion and Bribery' (1993) 141 *University of Pennsylvania Law Review* 1877.

as a domain name by someone with no rights in the identifier may amount to an implied threat to use the domain name in a way that harms the owner of the real world identifier. Accordingly, the strategy of registering a real world identifier as a domain name may successfully elicit an offer to purchase the domain name from the owner of the identifier, even without an express demand for payment. This means that in practice it may be difficult to distinguish registration of an identifier, such as a trade mark, as a domain name, with an intention to elicit an offer from the trade mark owner, from registrations that are not accompanied by an implied threat to harm the trade mark owner.

Apart from 'classical cybersquatting', other forms of opportunistic practices relating to domain names may be grouped into two main categories: those that harm the interests of the owners of real world identifiers, such as trade mark owners; and those in which the domain name registrant attempts to 'free ride' on the goodwill or reputation of the owner of the real work identifier. As explained above, the extortionate demands associated with 'classical cybersquatting' depend upon the ability of the domain name registrant to inflict harm on the owner of a real world identifier. The ability to inflict harm arises because of the essential nature of domain names as Internet addresses. As an Internet address, a domain name can be used to attract Internet users to content that is placed on a web page to which the domain name resolves, or to redirect Internet users to other on-line locations. The use of a real world identifier, such as a trade mark, creates an association with the real world identifier which will attract Internet users. Once Internet users are attracted by the domain name, the interests of the owner of the real world identifier may be harmed by content that is posted on a web site to which the domain name resolves, or by redirecting Internet users to other on-line locations in a way that harms the owner of the identifier. For example, a domain name that reflects a trade mark may be used to redirect Internet users to sites operated by competitors of the trade mark owner.

In addition to practices directed mainly at harming the owners of real world identifiers, a domain name that corresponds to a real world identifier may be registered and used to 'free ride' off the goodwill or reputation of the owner of the real world identifier. For example, the goodwill in a trade mark may attract users to the domain name registrant's own web site, which may include commercial content. The content may include products or services that compete with those of the trade mark owner, or unrelated commercial products or services, such as pornographic content. Moreover, the domain name registrant may 'free ride' on the goodwill or reputation of the owner of a real world identifier by attracting Internet users, then redirecting them to other on-line locations. For example, a domain name that corresponds to a trade mark might be used to redirect Internet users to sites operated by competitors of the trade mark owners, or to commercial sites that are unrelated to the trade mark owner, including pornographic sites or sites offering fraudulent material, such as 'phishing' sites. In such circumstances, the domain name registrant may receive payments, such as 'click-through' revenue, for redirecting users to other on-line locations.

'Cybersquatting' therefore encompasses a variety of opportunistic practices relating to domain names that correspond to real world identifiers and that harm the interests of the rightful owners of real world identifiers, such as trade marks. It also encompasses opportunistic practices involving unfairly trading on the goodwill or reputation of the owners of real world identifiers. The main challenge in developing rules aimed at preventing such practices is to proscribe the diverse range of conduct that may amount to undesirable 'cybersquatting', while not inhibiting the legitimate registration and use of domain names.

This chapter explains the development of the most important mechanism for resolving disputes between domain name registrants and trade mark holders concerning undesirable 'cybersquatting' practices, known as the Uniform Domain Name Dispute Resolution Policy (UDRP). It explains the *travaux préparatoires*, and other background information that is essential for a proper understanding of the UDRP, and presents an outline and overview of the main features of the UDRP. Before doing so, it briefly explains the dispute resolution policy that pre-dated the UDRP.

[3.2] NSI Dispute Resolution Policy

The first, rudimentary domain name dispute resolution policy was developed by NSI. The process by which, in the mid-1990s, NSI became solely responsible for the registration of domain names in the open gTLDs, including both registry and registrar functions, is explained at [2.4]-[2.5]. As conflicts between domain names and intellectual property rights emerged, NSI presented an obvious target for litigation. In 1994 the owner of the registered trade mark 'knowledgenet' commenced an action against the holder of the domain name <knowledgenet.com>, which joined NSI as a defendant.[10] Although the litigation was subsequently settled, it prompted NSI to attempt to protect itself from future actions by developing its own dispute resolution policy.[11]

Under NSI's Domain Dispute Resolution Policy Statement, issued in July 1995, each domain name registrant was contractually required to represent that it had a *bona fide* intention to use the name and that the name did not infringe any third party intellectual property rights. Moreover, the policy provided that if a complainant in a dispute over a domain name could establish that it had a trade mark in the name that was registered under United States law, NSI could suspend use of the domain name provided that the domain name registrant could not establish either that it owned a trade mark in the name or that it had used the name prior to the grant of the trade mark to the complainant. Although NSI's policy was designed to protect it from litigation, it did not have this effect as a number of registrants of domain names suspended pursuant to the policy initiated actions against NSI.[12]

History of the UDRP

[3.3] Origins of the UDRP

As explained at [2.8], the current system of domain name governance, centred on ICANN, developed from recommendations made in the 1998 NTIA final report, known as the White Paper or Magaziner Report. The White Paper included the following statement of policy regarding domain name dispute resolution:

[10] See Amended Complaint, *KnowledgeNet, Inc v D L Boone & Co Network Solutions, Inc, and Digital Express Group, Inc*, US DC (ND Ill (ED), No 94-C-7195. For the history of the litigation see M Mueller, *Ruling the Root* (Cambridge, Mass, The MIT Press, 2002) 120.

[11] Network Solutions, Inc, *Domain Dispute Resolution Policy Statement* (July 1995), subsequently updated and revised three times: *Domain Dispute Resolution Policy Statement* (Revision 03, 25 Feb 1998). For a discussion of the NSI dispute resolution policy see: C Oppedahl, 'How Is a Domain Name Like a Cow?' (1997) 15 *John Marshall Journal of Computer and information Law* 437.

[12] See Mueller, above n 10, 121–2, fn 26, citing *Roadrunner Computer Systems v Network Solutions, Inc* (E D Va, 26 Mar 1996); *Giacalone v Network Solutions, Inc* (ND Cal, 14 June 1996); *Network Solutions, Inc v Clue Computing* (D Colo, 21 June 1996).

> The U.S. Government will seek international support to call upon the World Intellectual Property Organization (WIPO) to initiate a balanced and transparent process, which includes the participation of trademark holders and members of the Internet community who are not trademark holders, to (1) develop recommendations for a uniform approach to resolving trademark/domain name disputes involving cyberpiracy (as opposed to conflicts between trademark holders with legitimate competing rights), (2) recommend a process for protecting famous trademarks in the generic top level domains, and (3) evaluate the effects, based on studies conducted by independent organizations, such as the National Research Council of the National Academy of Sciences, of adding new gTLDs and related dispute resolution procedures on trademark and intellectual property holders. These findings and recommendations could be submitted to the board of the new corporation for its consideration in conjunction with its development of registry and registrar policy and the creation and introduction of new gTLDs.[13]

The White Paper further indicated that, whatever the precise nature of any dispute resolution mechanism, it should be directed at resolving disputes involving cybersquatting and cyberpiracy, and not at settling disputes between two or more parties each with legitimate interests in a name. The NTIA explained that such disputes were properly settled in an appropriate court, not by a dispute resolution mechanism.

Following from the US proposal, the World Intellectual Property Organization (WIPO) initiated a multi-staged, international consultative process to consider the future shape of a domain name dispute resolution system. WIPO is a specialised, treaty-based agency of the United Nations, which has the following objectives:

> (i) to promote the protection of intellectual property through cooperation among States and, where appropriate, in collaboration with any other international organization;
> (ii) to ensure administrative cooperation among the Unions.[14]

The traditional functions performed by WIPO consisted of activities such as developing model national laws or preparing experts' reports to support the development of international treaties. The initiating of the process to develop a domain name dispute resolution system represented a considerable departure from WIPO's traditional role. In effect, in undertaking the task, WIPO was acting in response to a request from an agency of a single government, namely the NTIA, to convene a study to be submitted to a private corporation, eventually constituted as ICANN, which could implement proposals recommended by WIPO without any further consultation with national governments, including the WIPO Member States. Nevertheless, WIPO proceeded with the first stage of its process, which consisted in consultations on the terms of reference for its process, in July 1998, prior to receiving the formal approval of its Member States.[15] WIPO did eventually receive this formal approval to conduct the international process called for in the White Paper in September 1998.[16]

[13] National Telecommunications and Information Administration (NTIA), *Management of Internet Names and Addresses* 63 Fed Reg 31,741 (1998) para [8].

[14] Convention Establishing the World Intellectual Property Organization,,signed at Stockholm on 14 July 1967 and as amended on 28 Sept 1979, Art 3. The 'Unions' referred to are defined to mean 'the Paris Union, the Special Unions and Agreements established in relation with that Union, the Berne Union, and any other international agreement designed to promote the protection of intellectual property whose administration is assumed by the Organization according to Article 4(iii)': Art 2.

[15] WIPO, above n 1, para [27].

[16] *Ibid*, para [25].

[3.4] The WIPO First Process Report

As a result of its consultative process, WIPO produced an Interim Report in December 1998.[17] The Interim Report included the following recommendations:

- That a new system of global dispute resolution be established to protect intellectual property on the Internet and that all domain name registrants be contractually required to submit to the dispute resolution system.
- That the dispute resolution system apply to all types of intellectual property disputes on the Internet, including disputes involving copyright and personality rights, as well as disputes over trade marks.
- That procedures be established to protect famous trade marks by excluding them altogether from the DNS database for new gTLDs.

After a further period of consultation following the publication of the Interim Report, WIPO issued its Final Report in April 1999. As a result of comments received in response to the Interim Report, the recommendations of WIPO's Final Report were considerably less ambitious. As the Final Report pointed out:

> The majority of commentators found . . . [the proposals of the Interim Report] . . . to be too broad insofar as the draft recommendations suggested that domain name applicants be required to submit to a mandatory administrative dispute-resolution procedure in respect of any intellectual property dispute arising out of the domain name registration.[18]

The main recommendations of the WIPO Final Report were as follows:

- The adoption of a uniform dispute resolution policy, known as an 'administrative procedure', to be made mandatory through contractual clauses imposed on all domain name registrants in open gTLDs;
- The scope of the process be limited to cases of 'deliberate, bad faith, abusive registration of trade marks as domain names', thereby effectively excluding proceedings alleging violations of trade names, geographical indications or personality rights;
- Complainants to be limited to trade mark holders;
- Remedies to be limited to cancellation or transfer of a domain name, and the allocation of costs, meaning that damages would not be available;
- Parties to retain the ability to initiate litigation in national courts, in which case the decision of the national court prevails over the administrative determination under the dispute resolution policy;
- Determinations under the dispute resolution policy to be effectively self-executing, by appropriate changes being made to the DNS database; and
- A mechanism to be established to exclude famous and well-known marks from registration as domain names in new gTLDs.

The Final Report proposed the following definition of 'abusive registration' to be applied in the administrative procedure:

[17] WIPO, *The Management of Internet Names and Addresses: Intellectual Property Issues,* Interim Report of the WIPO Internet Domain Name Process (23 Dec 1998), available at http://www.wipo.int/amc/en/processes/process/rfc/3/index.html.

[18] WIPO), above n 1, para [134].

> The registration of a domain name shall be considered to be abusive when all of the following conditions are met:
>
> (i) the domain name is identical or misleadingly similar to a trade or service mark in which the complainant has rights; and
> (ii) the holder of the domain name has no rights or legitimate interests in respect of the domain name; and
> (iii) the domain name has been registered and is used in bad faith.[19]

The report further elaborated on what would amount to bad faith registration and use by proposing that the following circumstances be taken as evidence of bad faith:

> (a) an offer to sell, rent or otherwise transfer the domain name to the owner of the trade or service mark, or to a competitor of the owner of the trade or service mark, for valuable consideration; or
> (b) an attempt to attract, for financial gain, Internet users to the domain name holder's website or other on-line location, by creating confusion with the trade or service mark of the complainant; or
> (c) the registration of the domain name in order to prevent the owner of the trade or service mark from reflecting the mark in a corresponding domain name, provided that a pattern of such conduct has been established on the part of the domain name holder; or
> (d) the registration of the domain name in order to disrupt the business of a competitor.[20]

WIPO explained the reasoning behind the three cumulative elements of abusive registration in the following terms:

> The cumulative conditions . . . make it clear that the behavior of innocent or good faith domain name registrants is not to be considered abusive. For example, a small business that had registered a domain name could show, through business plans, correspondence, reports, or other forms of evidence, that it had a bona fide intention to use the name in good faith. Domain name registrations that are justified by legitimate free speech rights or by legitimate non-commercial considerations would likewise not be considered to be abusive. And, good faith disputes between competing right holders or other competing legitimate interests over whether two names were misleadingly similar would not fall within the scope of the procedure.[21]

In explaining the reasoning behind limiting the procedure to abusive registration of trade marks or service marks, the Final Report stated:

> registrations that violate trade names, geographical indications or personality rights would not be considered to fall within the definition of abusive registration *for the purposes of the administrative procedure*. Those in favor of this form of limitation pointed out that the violation of trademarks (and service marks) was the most common form of abuse and that the law with respect to trade names, geographical indications and personality rights is less evenly harmonized throughout the world, although international norms do exist requiring the protection of trade names and geographical indications.[22]

With regard to the application of legal principles developed under national trade mark laws to the interpretation of the proposed elements of abusive registration, the Final Report proposed that the panel of decision-makers appointed to determine the matter should be

[19] WIPO), above n 1, para [171].
[20] *Ibid.*
[21] *Ibid*, para [172].
[22] *Ibid*, para [167] (emphasis in original).

responsible for determining which law, or rules of law, to apply. In particular, the Final Report stated that the panel of decision-makers should:

> to the extent necessary, make reference to the law or rules of law that it determines to be applicable in view of the circumstances of the case. Thus, for example, if the parties to the procedure were resident in one country, the domain name was registered through a registrar in that country and the evidence of the bad faith registration and use of the domain name related to activity in the same country, it would be appropriate for the decision-maker to refer to the law of the country concerned in applying the definition.[23]

[3.5] ICANN's Adoption of WIPO Recommendations

The 1998 NTIA White Paper recommended that the proposed new corporation, which was eventually constituted as ICANN, adopt policies whereby:

> Domain name registrants would agree, at the time of registration or renewal, that in cases involving cyberpiracy or cybersquatting (as opposed to conflicts between legitimate competing rights holders), they would submit to and be bound by alternative dispute resolution systems identified by the new corporation for the purpose of resolving those conflicts. Registries and Registrars should be required to abide by decisions of the ADR system.[24]

In accordance with the procedure envisaged by the White Paper, the WIPO recommendations were submitted to the newly-established ICANN on 30 April 1999. At its May 1999 board meeting the ICANN board endorsed the principle of adopting a uniform dispute resolution policy for registrars in the open gTLDs. The board passed a resolution encouraging accredited registrars to formulate a model dispute resolution policy for voluntary adoption. At the same time, the board commenced an internal process to review WIPO's proposals.[25] First, the board referred WIPO's proposals concerning a dispute resolution policy to the DNSO with a request that the board be provided with a report by 31 July 1999. Secondly, the board referred WIPO's proposals regarding the exclusion of famous and well-known names to the DNSO with a request that the board be provided with a report at the earliest practicable time after the board meeting scheduled for August 1999.

Although the organisational process for the establishment of the DNSO was not yet complete, two DNSO working groups were established to review and report on WIPO's proposals for dispute resolution and for the exclusion of famous and well-known names, respectively. DNSO Working Group-A, which was responsible for reviewing WIPO's recommendations concerning a uniform dispute resolution process, submitted its final report to the DNSO Names Council on 29 July 1999.[26] The DNSO Names Council adopted the working group report and, with minor amendments, sent it to the ICANN board as a consensus recommendation on 3 August 1999.[27] The working group report made five recommendations, including that all ICANN-accredited registrars should be required to adopt the UDRP recommended by WIPO. Other recommendations made by the working group included the

[23] *Ibid*, para [176].

[24] NTIA, above n 13, 'Revised Policy Statement'.

[25] Mueller points out that the reference to the DNSO followed opposition to ICANN adopting the WIPO FinalReport in full at the May 1999 board meeting: see Mueller, above n 10, 193.

[26] DNSO Working Group-A, *WG-A Final Report to the Names Council* (29 July 1999), available at http://www.dnso.org/dnso/notes/19990729.WGA-report.html.

[27] DNSO Working Group-A, *WG-A Final Report to the ICANN Board* (3 Aug 1999), available at http://www.icann.org/dnso/wga-final-report.htm.

need to address the situation where a domain name registrant who has been unsuccessful under the UDRP is effectively prevented from 'appealing' because of the absence of an applicable cause of action; and that WIPO be asked to consider the development of an independent set of rules for the UDRP that is not based on civil or common law in order to deal with the problem of ascertaining the applicable law in a dispute.

Meanwhile, as explained at [2.28], in March 1999 the ICANN board had approved a Statement of Registrar Accreditation Policy and Registrar Accreditation Agreement, which enabled the accreditation of new registrars to compete with NSI for registering domain names in the gTLDs. A group of registrars, including NSI, concerned with the lack of a dispute resolution policy and acting pursuant to the May 1999 board resolution, used the WIPO proposals as the basis for a draft policy and rules of procedure, which was submitted to ICANN on 20 August 1999.[28] Within a matter of days, an ICANN staff report was released with detailed proposals on a dispute resolution policy that were similar to the registrars' proposals,[29] and that was presented to a public forum held before the August 1999 ICANN board meeting.

On 26 August 1999 the ICANN board accepted the DNSO's recommendation that ICANN adopt a uniform dispute resolution policy for the open gTLDs and directed ICANN's President, with the assistance of ICANN staff and counsel, to prepare implementation documents for approval by the board to be put in place within 45 days. The board gave the following guidance on matters to be taken into account in the preparation of the implementation documents:

1. The registrar's Model Dispute Resolution Policy should be used as a starting point;
2. The president or his delegate should convene a small drafting committee including persons selected by him to express views and consider the interests of the registrar, non-commercial, individual, intellectual property, and business interests;
3. In addition to the factors mentioned in paragraph 171(2) of the WIPO report, the following should be considered in determining whether a domain name was registered in bad faith:

    (a) Whether the domain name holder is making a legitimate noncommercial or fair use of the mark, without intent to misleadingly divert consumers for commercial gain or to tarnish the mark;
    (b) Whether the domain name holder (including individuals, businesses, and other organizations) is commonly known by the domain name, even if the holder has acquired no trademark or service mark rights; and
    (c) Whether, in seeking payment for transfer of the domain name, the domain name holder has limited its request for payment to its out-of-pocket costs.

4. There should be a general parity between the appeal rights of complainants and domain name holders.
5. The dispute policy should seek to define and minimize reverse domain name hijacking.[30]

Pursuant to the board resolution a small drafting committee was formed to advise on the development of a dispute resolution policy and rules of procedure. Although members of the drafting committee provided input to the drafting process, the implementation docu-

[28] *Model Domain Name Dispute Resolution Policy for Voluntary Adoption by Registrars* (20 Aug 1999), available at http://www.icann.org/santiago/registrar-dispute-policy.htm and *Model Rules for Uniform Domain Name Dispute Resolution Policy* (20 Aug 1999).

[29] ICANN Staff Report, *Uniform Dispute Resolution Policy for gTLD Registrars* (24 Aug 1999), available at http://www.icann.org/santiago/udrp-staff-report.htm.

[30] ICANN, *Resolutions approved by the Board Santiago Meeting*, 26 Aug 1999, available at http://www.icann.org/santiago/santiago-resolutions.htm, Resolution [99.83].

ments were drafted by ICANN's counsel, Louis Touton, with ICANN's staff, and not by the drafting committee.[31] On 29 September 1999 a draft statement of dispute resolution policy and a draft set of rules of procedure, based on the draft policy prepared by the registrars and revised to reflect the guidance contained in the ICANN board resolution, were posted on the ICANN Web site for public comment. The draft documents were accompanied by an ICANN staff report that explained how the board's resolutions were implemented and seeking comment on a small number of outstanding issues.[32]

ICANN received a considerable number of comments on the draft documents, many of which recommended changes in the policy itself. One area of contention concerned the scope of the definition of abusive registration in the draft policy. In this respect, the International Trademark Association (INTA) and other trade mark owners proposed expanding the scope of the policy to include practices that would fall outside the policy as drafted. In particular, trade mark interests were concerned that the definition of bad faith required both registration *and* use of a domain name in bad faith. The trade mark owners, concerned that the policy would not apply to practices known as 'passive warehousing', in which large numbers of domain names would be registered but not used, advocated expanding the definition to apply to bad faith registration *or* use. Others argued that applying the policy to mere registration would result in an expansion of trade mark law, which conventionally requires some commercial use for there to be an infringement. The drafting committee established by ICANN was unable to agree on this issue, which was eventually resolved by ICANN's counsel, Touton, deciding to retain the requirement of bad faith registration *and* use.[33]

Following the period for public comment, minor amendments were made to the draft implementation documents and ICANN's President submitted a staff report, commonly known as the Second Staff Report, on the draft policy and on the response to the public comments to the ICANN board,[34] recommending approval of the implementation documents. The Second Staff Report is an important source of background information on the policy deliberations that eventually resulted in the final form taken by the UDRP. The report explained that, as the implementation documents were required to comply with the August 1999 board resolutions, they were based on the registrar's model policy and rules of procedure, as modified to take into account the additional factors identified in the board's resolution. In other words, the implementation documents were designed to reflect the policy decisions made by the board in August 1999 and therefore could not take into account public comments that challenged the policy itself.

In response to the Second Staff Report, on 24 October 1999 the ICANN board approved the implementation documents and authorised the President to implement the uniform dispute resolution policy using the implementation documents. As a result, ICANN is regarded as having formally adopted the following two policies on 26 August 1999:

[31] For criticism of the drafting process see AM Froomkin, 'A Catalog of Critical Process Failures; Progress on Substance; More Work Needed' (13 Oct 1999), available at http://www.icann.org/comments-mail/comment-udrp/current/msg00107.html.

[32] ICANN, *Staff Report on Implementation Documents for the Uniform Dispute Resolution Policy* (29 Sept 1999) ('*First Staff Report*'), available at http://www.icann.org/udrp/staff-report-29sept99.htm.

[33] See AM Froomkin, 'ICANN's "Uniform Dispute Resolution Policy"—Causes and (Partial) Cures' (2002) 67 *Brooklyn Law Review* 605, 655–6.

[34] ICANN, *Second Staff Report on Implementation Documents for the Uniform Dispute Resolution Policy* (24 Oct 1999) ('*Second Staff Report*'), available at http://www.icann.org/udrp/udrp-second-staff-report-24oct99.htm.

- Uniform Domain Name Dispute Resolution Policy (the 'UDRP') (Implementation Documents Approved: 24 October 1999; Policy Adopted: 26 August 1999).
- Rules for Uniform Domain Name Dispute Resolution Policy (the 'UDRP Rules') (Implementation Documents Approved: 24 October 1999; Policy Adopted: 26 August 1999).

The UDRP is set out in Appendix 2, while the UDRP Rules are set out in Appendix 3. Subsequently, on 29 November 1999, WIPO was approved as the first dispute resolution service provider.

In accordance with the May 1999 ICANN board resolutions, a second DNSO working group, Working Group-B, was established to review WIPO's recommendations for creating a mechanism to exclude famous and well-known marks. The issue of establishing additional protection for famous marks proved to be contentious, with Working Group-B eventually presenting its report to the DNSO Names Council in April 2000.[35] The report indicated that there appeared to be consensus that there was no need to create a universally famous marks list for exclusion from the DNS. The report further indicated that there appeared to be some agreement short of consensus of the need for some additional protection for trade mark owners in new gTLDs that would take the form of a 'sunrise' provision. The sunrise provision would allow all trade mark owners, not simply owners of famous marks, to pre-register trade marks during a limited 'sunrise period' prior to registrations being open to the general public. As explained at [3.13], this proposal was later implemented in policies applying to the new .biz and .info gTLDs.

[3.6] WIPO and ICANN's UDRP: Differences

ICANN's UDRP and the UDRP Rules were based on the proposals made in WIPO's Final Report. Nevertheless, in the course of ICANN's internal process for reviewing WIPO's report some significant modifications were made to WIPO's proposals. The most important difference between WIPO's recommendations and the policies adopted by ICANN was the rejection of ICANN's proposal to establish a mechanism for excluding famous and well-known marks from registration as domain names in new gTLDs. The decision to reject the proposal for excluding famous marks has had a significant influence on decisions made under the UDRP where, in general, famous or well-known marks are treated more favourably than other marks. There are, however, other important differences between WIPO's recommendations and the UDRP and UDRP Rules adopted by ICANN.

As explained at [3.5], ICANN's UDRP and UDRP Rules were based on the model policy and rules of procedure drafted by a group of registrars as modified to take into account the further matters identified by ICANN's board in August 1999. The main differences between WIPO's proposals and ICANN's UDRP and UDRP Rules are as follows:

- *Wording of first element of abusive registration*
  WIPO's Final Report had proposed that the first element of abusive registration be that the domain name be 'identical or misleadingly similar' to a trade mark or service mark in which the complainant has rights. ICANN changed the wording of this element by replacing 'misleadingly similar' with 'confusingly similar'. The modification was evidently made to reflect the wording in the US Lanham Act.[36]

[35] DNSO Working Group-B, *WG-B Report* (17 Apr 2000), available at http://www.icann.org/dnso/wgb-report-17apr00.htm.

[36] See Froomkin, above n 33, 654.

- *Clarification of circumstances in which domain name holder will have rights to or legitimate interests in the domain name*
  The second element of abusive registration requires the complainant to establish that the domain name holder has 'no rights or legitimate interests' in the domain name. The August 1999 board resolution required consideration to be given to whether the domain name holder is making a legitimate non-commercial or fair use of the mark, without intent misleadingly to divert consumers for commercial gain or to tarnish the mark; and whether the domain name holder is commonly known by the domain name. ICANN implemented this resolution by adding a new paragraph 4(c) to the UDRP, which itemises the circumstances in which a domain name holder will have rights to or legitimate interests in the domain name. Paragraph 4(c) essentially provides that a domain name holder will have rights to or legitimate interests in the domain name if:

(i) before any notice of the dispute, there is use of, or demonstrable preparations to use, the domain name in connection with a bona fide offering of goods or services; or
(ii) the domain name holder has been commonly known by the domain name, even if he or she has acquired no trade mark or service mark rights; or
(iii) the domain name holder is making a legitimate non-commercial or fair use of the domain name, without intent for commercial gain to misleadingly divert consumers or to tarnish the trade mark or service mark at issue.

- *Clarification of circumstances amounting to evidence of bad faith*
  The third element of abusive registration requires the complainant to establish that the domain name has been registered and is being used in bad faith. The WIPO Final Report set out a non-exclusive list of circumstances to be taken as evidence of bad faith. The August 1999 board resolution required consideration to be given to whether, in seeking payment for transfer of the domain name, the domain name holder has limited its request for payment to its out-of-pocket costs. Paragraph 4(b) of the UDRP sets out a non-exclusive list of circumstances which amount to evidence of bad faith. The list reflects the circumstances recommended by WIPO with relatively minor drafting changes. In particular, to take into account the board resolution, the first circumstance evidencing bad faith was reworded in paragraph 4(b)(i) to encompass:

  > circumstances indicating that you have registered or you have acquired the domain name primarily for the purpose of selling, renting, or otherwise transferring the domain name registration to the complainant who is the owner of the trade mark or service mark or to a competitor of that complainant, for valuable consideration in excess of your documented out-of-pocket costs directly related to the domain name.

- *Parity of appeal*
  The August 1999 board resolution included the principle that there 'should be a general parity between the appeal rights of complainants and domain name holders'. The drafting committee established by ICANN considered two aspects of the lack of parity between complainants and domain name holders in the policy and rules that had been drafted by the registrars. First, there was no way to ensure that a complainant would be subject to a jurisdiction in which an unsuccessful domain name holder could bring an action. Secondly, under the policy drafted by the registrars, the domain name holder could obtain an automatic stay of UDRP proceedings by presenting evidence that it had commenced an action against the complainant in any court. These aspects of the lack of parity were,

to an extent, addressed by ICANN in the UDRP and the UDRP Rules. First, ICANN introduced paragraph 3(b)(xiii) of the UDRP Rules, which requires the complainant to submit, in relation to any challenges to a UDRP decision, to the jurisdiction of courts in at least one specified mutual jurisdiction. The UDRP Rules define 'mutual jurisdiction' to mean a jurisdiction at the location of either the principal office of the registrar for the domain name or the domain name holder's address as shown in the Whois database. Secondly, the UDRP was modified to provide, in paragraph 4(k), for implementation of a UDRP decision to be stayed only if the domain name holder submits documentary evidence that an action has been commenced against the complainant in a mutual jurisdiction to which the complainant has submitted. These changes to the UDRP and UDRP Rules, however, fail to address the difficulties faced by an unsuccessful domain name holder in establishing a cause of action upon which a UDRP decision can be challenged.

- *Reverse domain name hijacking*
  The August 1999 board resolution included the principle that the UDRP 'should seek to define and minimize reverse domain name hijacking'. In accordance with the resolution, paragraph 1 of the UDRP Rules defines 'reverse domain name hijacking' to mean:

  > using the Policy in bad faith to attempt to deprive a registered domain name holder of a domain name.

  The procedural rules were modified to include provisions to deter reverse domain name hijacking. First, under paragraph 15(e) of the UDRP Rules, if a UDRP panel finds that a complaint was brought in bad faith it must state this in the decision and declare that the complaint constitutes an abuse of the administrative proceeding. Secondly, under paragraph 2(a) of the UDRP Rules, the requirements for the dispute service provider to provide notice of the complaint to the domain name holder were enhanced. Although members of the drafting committee established by ICANN suggested that a footnote be included in the UDRP Rules to refer to guidelines for identifying what would amount to reverse domain name hijacking, ICANN rejected this until experience with UDRP proceedings had been accumulated.

- *Procedures for selection of dispute-resolution service providers and panellists*
  The most important outstanding issue identified in the September 1999 ICANN staff report concerned the procedures to be used for selecting dispute resolution service providers and panellists.[37] In response to public comments, the staff report submitted to the board in October 1999 included detailed recommendations on procedures for selecting dispute resolution providers and panellists.[38] These recommendations were implemented in the UDRP and UDRP Rules as follows:

  —Under paragraph 4(d) of the UDRP, the complainant is responsible for selecting the dispute resolution service provider from among the providers approved by ICANN by submitting the complaint to that provider.
  —Under paragraph 3(b)(iv) of the UDRP Rules, the complainant may elect to have the dispute decided by a single-member panel or by a three-member panel.
  —Under paragraph 5(b)(v) of the UDRP Rules, if the complainant has elected a single-member panel, the respondent may elect to have the dispute decided by a three-member panel.

[37] ICANN, *First Staff Report,* above n 32.
[38] ICANN, *Second Staff Report,* above n 34, para [4.15].

—Under paragraph 19(a) of the UDRP Rules, the complainant pays all fees for the dispute resolution process unless the respondent elects to have a three-member panel, in which case each party must pay one half of the fees for the dispute resolution process.

- *Burden of proof*
  The WIPO Final Report and the model policy prepared by the registrars were not entirely clear on where the burden of proof for establishing the elements of abusive registration should lie.[39] ICANN addressed this issue by specifying, in paragraph 4(a) of the UDRP, that the complainant bear the onus of proving each of the three substantive elements of abusive registration. It would seem that paragraph 4(a) does no more than clarify what was already implicit in the WIPO Final Report and the registrar's model policy.[40] Nevertheless, the September 1999 ICANN staff report suggested that clarifying that the complainant bears the burden of proof would also minimise reverse domain name hijacking.[41]

The UDRP

[3.7] Abusive, Bad Faith Registration

The three elements of abusive, bad faith registration are set out in paragraph 4(a) of the UDRP. Paragraph 4(a) further provides that in proceedings under the UDRP the complainant must establish each of the three elements. The elements of abusive, bad faith registration are as follows:

- *The First Element*
  The first element that a complainant is required to establish is that the domain name 'is identical or confusingly similar to a trade mark or service mark in which the complainant has rights'.

  This element implements the recommendation of the WIPO Final Report that the procedure be limited to disputes involving trade marks or service marks and not be extended to disputes involving other intellectual property rights.[42] In this respect, the WIPO report stated:

  > There is evidence that . . . [deliberate abusive registrations of domain names] . . . extends to the abuse of intellectual property rights other than trademarks and service marks, but we consider that it is premature to extend the notion of abusive registration beyond the violation of trademarks and service marks at this stage.[43]

- *The Second Element*
  The second element that a complainant is required to establish is that the domain name holder has 'no rights or legitimate interests in respect of the domain name'.

[39] As Froomkin has observed, '[a]lthough complainants usually have the burden of proof, and in the absence of anything to the contrary, it seems plausible to assume that WIPO intended this to be the case, one might nevertheless have wished that WIPO had spelled it out': Froomkin, above n 33, 648.

[40] *Ibid*, 654.

[41] ICANN, *First Staff Report*, above n 32; see also ICANN, *Second Staff Report*, above n 34, para [4.10].

[42] WIPO, above n 1,paras [167]–[168].

[43] *Ibid*, para [168].

This element implements the recommendation of the WIPO Final Report that the scope of the procedure be confined to disputes between a complainant with trade mark rights and a respondent with no rights or interests in the domain name, and not include disputes on the merits between parties with competing rights in the domain name. In this respect, the report stated that:

> the scope of the procedure is limited so that it is available only in respect of deliberate, bad faith, abusive, domain name registrations or "cybersquatting" and is not applicable to disputes between parties with competing rights acting in good faith.[44]

The report further stated that:

> good faith disputes between competing right holders or other competing legitimate interests over whether two names were misleadingly similar would not fall within the scope of the procedure.[45]

The policy underlying the limitation established by the second element of abusive registration is clearly that complex issues arising in disputes between parties with competing rights or interests in a domain name are properly dealt with by a court, and not by the expedited UDRP procedure. The ICANN staff report submitted to the board in October 1999 explained the limitation in the following terms:

> the fact that the policy's administrative dispute-resolution procedure does not extend to cases where a registered domain name is subject to a legitimate dispute (and may ultimately be found to violate the challenger's trademark) is a feature of the policy, not a flaw. The policy relegates all "legitimate" disputes—such as those where both disputants had longstanding trademark rights in the name when it was registered as a domain name—to the courts; only cases of abusive registrations are intended to be subject to the streamlined administrative dispute-resolution procedure.[46]

- *The Third Element*

  The third element that a complainant is required to establish is that the domain name 'has been registered and is being used in bad faith'. This element represents the gist of a complaint of abusive, bad faith registration.

As explained at [3.5], in the course of ICANN's process for reviewing the WIPO recommendations, trade mark interests argued that the wording of the third element should be redrafted to apply to bad faith registration or use, so as not to require the complainant to establish bad faith registration *and* use. As further explained, however, this suggested modification was rejected by ICANN.

What will amount to bad faith registration and use of a domain name is elaborated in a list of four non-exclusive circumstances that are set out in paragraph 4(b) of the UDRP. The specific circumstances that will amount to evidence of bad faith registration and use are as follows:

> (i) circumstances indicating that the domain name holder has registered or has acquired the domain name primarily for the purpose of selling, renting, or otherwise transferring the domain name registration to the complainant who is the owner of the trademark or service mark or to a competitor of that complainant, for valuable consideration in excess of the domain name holder's documented out-of-pocket costs directly related to the domain name.

[44] WIPO, above n 1, para [135].
[45] *Ibid*, para [172].
[46] ICANN, *Second Staff Report*, above n 34, para [4.1(c)].

The first circumstance captures the clearest instances of what might be called 'classic' cybersquatting: the registration of a domain name with the intention of profiting from it by offering to sell it to the owner of rights in a trade mark or service mark that corresponds to the domain name. This provision, which applies to those activities that were of greatest concern to trade mark owners, is based on circumstance (a) proposed in the WIPO Final Report, which is set out at [3.4]. The wording of paragraph 4(b)(i), however, differs from the wording set out in WIPO's Final Report in two important respects.[47] First, WIPO's proposal was confined to offers to sell or transfer the domain name to the trade mark owner, whereas paragraph 4(b)(i) encompasses all circumstances indicating that the domain name has been registered for the purpose of selling or transferring it to the trade mark holder. Paragraph 4(b)(i) therefore catches a broader spectrum of behaviour than that caught by WIPO's proposed wording. Secondly, as explained at [3.6], the August 1999 ICANN board resolution required that consideration be given to whether the domain name holder has limited its request for payment to its out-of-pocket costs. The resolution was reflected in the drafting of paragraph 4(b)(i) by inclusion of the requirement that the domain name holder must have registered the domain name for the purpose of transferring it for an amount in excess of the 'out-of-pocket costs directly related to the domain name'. Despite the clarification that a domain name holder may recover reasonable out-of-pocket expenses, a central difficulty in the application of the first specific circumstance remains distinguishing legitimate offers to sell from abusive, bad faith registration.

(ii) the domain name holder has registered the domain name in order to prevent the owner of the trademark or service mark from reflecting the mark in a corresponding domain name, provided that the domain name holder has engaged in a pattern of such conduct.

The second circumstance faithfully reflects circumstance (c) proposed in the WIPO Final Report, which is set out at [3.4] above. This circumstance was included to deal with concerns of trade mark owners that cybersquatters would react to a prohibition on actively attempting to sell domain names by merely registering the domain names and passively waiting for trade mark owners to make offers to purchase the names.[48] In other words, some form of words had to be adopted to prevent cybersquatters from evading the prohibition on making offers to sell but effectively achieving the same result by mere registration. As with the first circumstance, the second circumstance raises difficulties in distinguishing legitimate registration of a domain name from bad faith registration. The problem with formulating the circumstance was to determine the activities which, in addition to registration of a domain name, would amount to evidence of bad faith. The distinction was drawn by requiring the domain name holder to have engaged in a 'pattern of conduct'. In comments received on ICANN's draft statement of policy released in September 1999, trade mark owners and other commentators suggested that the policy should include a more detailed explanation of the circumstances in which a pattern of bad faith registration would be found to exist. The staff report submitted to the ICANN board in October 1999, however, rejected these suggestions, concluding that 'the issue of whether a pattern is present will depend on the fact pattern presented in each case and that, until experience is accumulated with the policy, prescribing a more detailed definition is not warranted'.[49]

[47] See Froomkin, above n 33, 657.
[48] *Ibid*, 647.
[49] ICANN, *Second Staff Report*, above n 34, End Note 1.

(iii) the domain name holder has registered the domain name primarily for the purpose of disrupting the business of a competitor.

The third circumstance faithfully reflects circumstance (d) proposed in the WIPO Final Report, which is set out at [3.4]. This circumstance differs from the other three circumstances in that it is aimed at preventing a form of 'unfair competition' rather than protecting trade marks. In explaining the proposed definition of abusive registration, the WIPO Final Report referred to Article 10*bis* of the Paris Convention for the Protection of Industrial Property, which provides, in part:

(1) The countries of the Union are bound to assure to nationals of such countries effective protection against unfair competition.
(2) Any act of competition contrary to honest practices in industrial or commercial matters constitutes an act of unfair competition.

The third circumstance was evidently introduced by WIPO in response to concerns of some trade mark owners that competitors might register a domain name solely for the purpose of preventing the trade mark owner from using the domain name. Froomkin has explained the reasoning behind the introduction of the third circumstance in the following terms:

> Some mark holders feared that a firm which was first in its industry to understand the Internet's importance might warehouse its competitors' names to deny them the use of their mark as a domain. Since the name was being warehoused, rather than used, there would be no use, much less commercial confusion. And since there was no chance that the name would be offered for sale, the first WIPO category would not apply either.[50]

The third specific circumstance clearly applies to behaviour that would not be captured by the second circumstance, as it does not require evidence of a 'pattern of conduct'. The third circumstance may therefore conceivably apply to a single act of registering a domain name. The distinction between legitimate registration and bad faith registration is, however, drawn by the requirement that the registration must be 'primarily for the purpose of disrupting the business of a competitor'. This means that the complainant trade mark owner must, at least, establish that the domain name holder is a competitor.

(iv) by using the domain name, the domain name holder has intentionally attempted to attract, for commercial gain, Internet users to its web site or other on-line location, by creating a likelihood of confusion with the complainant's mark as to the source, sponsorship, affiliation, or endorsement of its web site or location or of a product or service on its web site or location.

The fourth circumstance reflects, with some changes in wording, circumstance (b) proposed in the WIPO Final Report, which is set out at [3.4]. The 'classic' case of cybersquatting, consisting of registering a trade mark as a domain name in order to offer it for sale to the trade mark owner, is captured by the first specific circumstance. The second major concern of trade mark owners was the development of practices involving the use of a trade mark in a domain name in order to attract unsuspecting Internet users to the domain name holder's Web site, or to divert users to other Web sites, including sites advertising competitive products or services and cyberporn sites. This circumstance is designed to capture these practices.

[50] Froomkin, above n 33, 647. As Froomkin explains at fn 126, the issue was referred to as the <kaplan.com> problem in the mistaken belief that the Princeton Review had registered and warehoused the <kaplan.com> domain name to prevent registration by the Kaplan Education Center.

The fourth specific circumstance differs from the other three in two important respects. First, whereas the other three circumstances refer to the registration of a domain name, the fourth applies to the use of a domain name. Secondly, the fourth circumstance requires that the use of the domain name create a 'likelihood of confusion' with the complainant's trade mark. By applying to activities involving the commercial use of a mark in a way that creates a likelihood of confusion with the complainant's mark, this circumstance most closely reflects the fundamental objective of trade mark law; namely, to prevent confusion as to the source of a product or service. In this sense, the fourth circumstance can be seen as a form of surrogate for conventional actions for trade mark infringement. On the other hand, however, the circumstance differs somewhat from an action for trade mark infringement in that it applies to the commercial use of a trade mark as a domain name specifically for the purpose of attracting users to a Web site or 'other on-line location'. The fourth circumstance therefore makes it abundantly clear that use of a trade mark as a domain name for the purpose of commercial gain by attracting unsuspecting Internet users to a Web site or on-line location is a prohibited activity.

While the four specific circumstances set out in paragraph 4(b) are designed to elaborate on the circumstances in which there will be an infringement of the third element of abusive, bad faith registration, there is a peculiar inconsistency between the wording of the third element, on the one hand, and the wording of the four circumstances, on the other hand. As explained at [3.5], ICANN rejected suggestions to change the wording of the third element from bad faith registration *and* use to bad faith registration *or* use. The wording of the third element of abusive, bad faith registration therefore suggests that it will only be established where there is both registration and use of a domain name. The problem with this interpretation is that none of the four circumstances refers to both registration and use: the first three circumstances refer only to registration of a domain name, while the fourth refers only to use of a domain name. This inconsistency in wording was significant in the reasoning in the most important panel decision interpreting the third element of abusive, bad faith registration, *Telstra Corporation v Nuclear Marshmallows,*[51] which is dealt with at [7.10].

[3.8] The Affirmative Defences

As explained at [3.5], the August 1999 ICANN board resolution required that a number of factors that had not been identified in the WIPO Final Report be considered in determining whether a domain name was registered in bad faith. Two of the additional factors were: whether the domain name holder is making a legitimate non-commercial or fair use of the mark; and whether the domain name holder is commonly known by the domain name. These factors were incorporated in paragraph 4(c) of the UDRP, which identifies three circumstances in which a domain name holder has rights or legitimate interests in a domain name.

Paragraph 4(c) operates in the following manner. The second element of abusive, bad faith registration requires a complainant to establish that the domain name holder has no rights or legitimate interests in respect of the domain name. The wording of the second element obviously begs the question of what exactly amounts to a 'right or legitimate interest'. Paragraph 4(c) deals with this issue by setting out three, non-exclusive circumstances in

[51] WIPO Case No D2000-0003 (18 Feb 2000).

which a domain name holder can establish that he or she has rights or legitimate interests in the domain name at issue. The three circumstances, which effectively act as affirmative defences, are as follows:

(i) before any notice to the domain name holder of the dispute, use of, or demonstrable preparations to use, the domain name or a name corresponding to the domain name in connection with a bona fide offering of goods or services.

Unlike the other two defences, the first affirmative defence was not one of the factors identified by the ICANN board in August 1999. Froomkin points out that trade mark interests objected to the inclusion of this defence essentially on the basis that it would enable cybersquatters to evade the policy by fabricating preparations to use a domain name without any actual intention of doing so.[52] Against this, it was contended that early registration of a domain name has become an essential part of an e-commerce business plan.

(ii) the domain name holder (as an individual, business, or other organization) has been commonly known by the domain name, even if he or she has acquired no trademark or service mark rights.

The second affirmative defence closely reflects the wording of the second factor identified in the August 1999 ICANN board resolution. Froomkin points out that this defence was sometimes known as the 'pokey' clause.[53] This terminology arose from the case in which Prima Toy Company, the owners of trade marks in the words 'Gumby' and 'Pokey', threatened to sue a 12-year-old boy who was the registered domain name holder of the domain name <pokey.org>.[54] The domain name resolved to the boy's personal Web site. The boy's father had registered the domain name to reflect his son's nickname, which was 'pokey'.

(iii) the domain name holder is making a legitimate noncommercial or fair use of the domain name, without intent for commercial gain to misleadingly divert consumers or to tarnish the trademark or service mark at issue.

The third affirmative defence closely reflects the wording of the first factor identified in the August 1999 ICANN board resolution. It is important to note that paragraph 4(c)(iii) does not create a general defence for every legitimate non-commercial or fair use of a domain name, but the defence is confined to circumstances where there is no intent misleadingly to divert users for commercial gain, or to tarnish the mark for commercial gain. As Froomkin points out, the third defence is poorly worded.[55] In particular, the wording of paragraph 4(c)(iii) leaves it open for arguments that activities which amount to a legitimate non-commercial or fair use of a domain name may, in some circumstances, infringe the policy.

The most important issue raised by this provision is probably whether or not the limitation of the defence to circumstances in which there is no tarnishment of the trade mark restricts the freedom of expression of domain name holders, including the ability to register and use domain names for criticism sites. This depends upon the interpretation of tarnishment, which was a concept imported from US trade mark law. The staff report submitted to the ICANN board was sufficiently concerned with comments that were critical of the use of the term 'tarnishment' that it included the following statement in an end-note to the report:

[52] Froomkin, above n 33, 661.
[53] *Ibid.*
[54] *Ibid*, 630.
[55] *Ibid*, 662–5.

> Several commentators indicated that the concept of 'tarnishment' in paragraph 4(c)(iii) might be misunderstood by those not familiar with United States law or might otherwise be applied inappropriately to noncommercial uses of parody names and the like. Staff is not convinced this is the case, but in any event wishes to point out that "tarnishment" in paragraph 4(c)(iii) is limited to acts done with intent to commercially gain.[56]

While these comments deal with a potential problem with the interpretation of the third affirmative defence, it does not dispose of the argument that the wording of the provision might be used to interpret the policy as prohibiting some activities that would amount to a legitimate non-commercial or fair use of a domain name. At the same time, it is important to bear in mind that circumstances in which a domain name holder may demonstrate rights or legitimate interests in a domain name are not limited to the three affirmative defences specifically identified in paragraph 4(c), which is expressly stated to be a non-exclusive list.

[3.9] Implementation of the UDRP

The ICANN UDRP establishes the substantive rules governing disputes between trade mark owners and domain name registrants. It is known as a 'mandatory administrative procedure'.[57] The UDRP Rules, which are dealt with in Chapter 4, deal with procedural matters, including notice, appointment of a dispute resolution panel and timing of decisions. As explained at [3.5], the ICANN board adopted the UDRP and UDRP Rules on 26 August 1999. As explained at [2.28], in November 1999, the ICANN board approved a revised Registrar Accreditation Agreement for the .com, .net and .org gTLDs.[58] Clause II.K of the revised agreement provided as follows:

> During the term of this Agreement, Registrar shall have in place a policy and procedure for resolution of disputes concerning SLD names. In the event that ICANN adopts a policy or procedure for resolution of disputes concerning SLD names that by its terms applies to Registrar, Registrar shall adhere to the policy or procedure.

As further explained at [2.28], a revised Registrar Accreditation Agreement was approved by ICANN in May 2001.[59] Clause 3.8 of the Registrar Accreditation Agreement provides:

> During the Term of this Agreement, Registrar shall have in place a policy and procedures for resolution of disputes concerning Registered Names. Until different policies and procedures are established by ICANN under Section 4, Registrar shall comply with the Uniform Domain Name Dispute Resolution Policy identified on ICANN's website (www.icann.org/general/consensus-policies.htm).

As a result, the UDRP applies, as a condition of accreditation, to all ICANN-accredited registrars in each of the open gTLDs:.aero, .biz, .cat, .com, .coop, .info, .jobs, .mobi, .museum, .name, .net, .org, .pro, .tel and .travel. This implements the recommendation in the WIPO Final Report to 'make available a uniform administrative dispute-resolution procedure be adopted for all open gTLDs'.[60] As the report explained, in the sense used in the UDRP, 'uniform' means:

[56] ICANN, *Second Staff Report*, above n 34, End Note 2.
[57] UDRP, para 4.
[58] *Registrar Accreditation Agreement* (approved: 4 Nov 1999; posted: 9 Nov 1999).
[59] *Registrar Accreditation Agreement* (17 May 2001), available at http://www.icann.org/registrars/ra-agreement-17may01.htm.
[60] WIPO, above n 1, para [157].

(i) The procedure should be available in all open gTLDs . . . At least in the open gTLDs, the non-availability of the procedure in any gTLD would lead to uneven protection for intellectual property rights and could cause any gTLD in which the procedure was not available to become a haven for predatory practices in respect of intellectual property rights.
(ii) The scope of the procedure and the procedural rules pursuant to which it is conducted should be the same in all open gTLDs. Again, differences in the scope of the procedure in the open gTLDs could lead to uneven protection for intellectual property rights.[61]

In addition to the open gTLDs, the UDRP has been adopted, sometimes without and sometimes with modifications, for disputes in a considerable number of ccTLDs.

The UDRP is required to be incorporated by ICANN- accredited registrars in the registration agreements for all applicable gTLDs. In this respect, paragraph 1 of the UDRP states:

> This Uniform Domain Name Dispute Resolution Policy (the "Policy") has been adopted by the Internet Corporation for Assigned Names and Numbers ('ICANN'), is incorporated by reference into your Registration Agreement, and sets forth the terms and conditions in connection with a dispute between you and any party other than us (the registrar) over the registration and use of an Internet domain name registered by you.

Domain name registrants in the relevant gTLDs are contractually bound to submit to the dispute resolution process under paragraph 4(a) of the UDRP. ICANN's ability to control the DNS means that, once a decision is made under the policy, the decision is directly enforced by changes made to the DNS. As the WIPO Final Report put it:

> Commentators in favor of an administrative policy for the resolution of domain name disputes universally supported the need for the determinations of the dispute-resolution procedure to be directly enforced. The possibility of such direct enforcement exists through the registration authorities and, indeed, constitutes one of the major reasons why an administrative procedure could be made workable and efficient in respect of domain name disputes.[62]

As explained at [3.14], the direct enforcement of decisions by means of changes to the DNS is associated with the limited remedies available under the UDRP, which are restricted to cancellation of the domain name, or transfer to the complainant.

[3.10] Dispute-resolution Service Providers

Paragraph 4(d) of the UDRP provides:

> The complainant shall select the Provider from among those approved by ICANN by submitting the complaint to that Provider.

On 29 November 1999, ICANN approved WIPO as the first dispute-resolution service provider and, on 9 December 1999, the first proceeding, *World Wrestling Federation Entertainment, Inc v Michael Bosman*,[63] was commenced. Disputes submitted to WIPO are conducted by the WIPO Arbitration and Mediation Center. Shortly following ICANN's approval of WIPO, the following three dispute-resolution service providers were approved:

- The National Arbitration Forum (NAF), an international administrator of alternative dispute resolution services that is based in Minneapolis and Los Angeles, was approved by ICANN on 23 December 1999.

[61] WIPO, above n 1, para [155].
[62] *Ibid*, para [215].
[63] WIPO Case No D1999-0001 (14 Jan 2000).

- Disputes.org/eResolution consortium (DeC), a Montreal-based dispute settlement provider that included a number of North American academics, was approved by ICANN on 1 January 2000.
- CPR Institute for Dispute Resolution (CPR), a non-profit-making dispute resolution provider that supplies mediators and arbitrators, and that is based in New York, was approved by ICANN on 22 May 2000.

On 16 October 2000, ICANN's approval of the DeC consortium was transferred to eResolution (eRes). In late 2001, however, eRes went into voluntary bankruptcy, and ceased accepting UDRP complaints from 30 November 2001. In January 2007, CPR ceased to receive UDRP complainants.

On 28 February 2002, ICANN approved a new dispute settlement provider, the Asian Domain Name Dispute Resolution Centre (ADNDRC), a joint undertaking by the China International Economic and Trade Arbitration Commission (CIETAC), the Hong Kong International Arbitration Centre (HKIAC) and the Korean Internet Address Dispute Resolution Committee (KIDRC), with offices in Beijing, Hong Kong and Seoul.

Each of the ICANN-approved dispute-resolution service providers has its own supplemental rules, which deal with procedural matters, including communications with the service provider, which the service provider applies in addition to the UDRP Rules. UDRP panellists supplied by the dispute-resolution service providers include practising lawyers, academics and retired members of the judiciary.

WIPO Second Process Report

[3.11] The WIPO Second Process Report

On 28 June 2000, a request was made to the Director General of WIPO by the Australian Minister for Communications, on behalf of 19 WIPO Member States, to develop, through a consultative process, recommendations for dealing with 'bad faith, abusive, misleading or unfair use' of certain identifiers other than trade marks as domain names.[64] In particular, the request, which was subsequently endorsed by the WIPO General Assembly, specified that the process should deal with the following identifiers:

- personal names;
- International Non-proprietary Names (INNs) for pharmaceutical substances;
- the names of international intergovernmental organisations (IGOs);
- geographical indications, geographical terms or indications of source; and
- trade names.

The common feature of the identifiers specified in the request was that the international legal protection for the identifiers is much less developed than the protection for trade marks, with considerable divergence in the protection given to these identifiers under national legal systems.

An International Non-proprietary Name (INN) for pharmaceuticals is a unique name that is used to identify pharmaceutical substances, or active pharmaceutical ingredients, that is

[64] See http://wipo2.wipo.int/process2/rfc/letter.html.

selected by the World Health Organization (WHO), which maintains a cumulative list of over 8,000 INNs. Once a name is included on the list, private rights cannot be acquired in the INN.

Names of intergovernmental organizations (IGOs) are the names of international, public sector organisations, such as the United Nations Organization (UNO), the International Monetary Fund (IMF), the International Telecommunications Union (ITU), the World Health Organization (WHO) and the World Trade Organization (WTO). The names and acronyms of IGOs are protected pursuant to Article 6*ter* of the Paris Convention,[65] which prohibits the registration and use of, inter alia, the names or abbreviations of IGOs as trade marks or elements of trade marks. WIPO administers a procedure for dealing with requests for protection under Article 6*ter*, and maintains a notification list of the names of IGOs that have been accepted for protection. Once a name is included on the list, registration and use of the name as a trade mark are prohibited unless the registration or use falls within an exception.

Geographical indications are geographical names that identify the source of goods or services, and which are subject to an international regime that prohibits the use of false indications of geographical source. Geographical terms, on the other hand, are names that identify geographical locations, such as countries, regions, cities, towns, villages, streets, rivers and mountains.

In response to the request from its Member States, WIPO conducted an international consultative process, modelled on the first WIPO domain name process and known as the Second WIPO Process, resulting in an Interim Report that was released on 12 April 2001.[66] Following consultations on the findings of the Interim Report, a final report, the WIPO Second Report, was released on 3 September 2001.[67]

The main recommendations of the WIPO Second Report were as follows:

- In relation to INNs, the report recommended the establishment of a simple notification mechanism to protect an INN against the registration of an identical domain name.
- In relation to the names and acronyms of IGOs, the report recommended that the member states of IGOs work towards the establishment of an administrative dispute-resolution procedure, similar to the UDRP, for dealing with complainants about domain names that are identical, or confusingly similar, to the names of IGOs.
- In relation to personal names, the report found that there were no existing international norms of protection, with national legal systems adopting a diversity of approaches. Given the lack of a developed international legal framework, the report suggested that the most appropriate way for the international legal protection of personal names to advance would be through the international community to further develop international norms, and not the modification of the UDRP.
- In relation to geographical identifiers, the report acknowledged that there are existing international norms that prohibit false and deceptive indications of geographical source on goods. Nevertheless, as these norms are confined to the use of terms on goods, and as there is no agreed international list of geographical indications, the report concluded that

[65] Paris Convention for the Protection of Industrial Property (Commercial No. 28 (1884), C 4043), as revised at Brussels on December 14, 1900, at Washington on June 2, 1911, at the Hague on November 6, 1925, at London on June 2, 1934, at Lisbon on October 31, 1958, at Stockholm on July 14, 1967 and as amended on September 28, 1979 ('Paris Convention'), available at http://www.wipo.int/export/sites/www/amc/en/docs/report-rfc3.pdf.

[66] WIPO, *The Recognition of Rights and the Use of Names in the Internet Domain Name System*, Second Process Interim Report (12 Apr 2001), available at http://www.wipo.int/export/sites/www/amc/en/docs/report-final2.pdf.

[67] WIPO, *The Recognition of Rights and the Use of Names in the Internet Domain Name System*, Report of the Second WIPO Internet Domain Name Process (3 Sept 2001) (the WIPO Second Report).

the international framework needs to be further developed before the use of geographical indications as domain names could be adequately addressed.

- In relation to geographical terms per se, the report concluded that, in the absence of an existing international framework, there was a need for the international community to decide whether such a framework should be developed. In particular, the report recommended that the protection in the gTLDs of country names, and the names of administratively recognised regions and municipalities, should be further considered in appropriate intergovernmental fora.
- In relation to trade names, given the significant differences between the legal protection of trade names under national laws, the report recommended that no action be taken to protect trade names from their registration and use as domain names.

While the WIPO Second Report did not recommend extending, or modifying, the UDRP to protect the identifiers dealt with in the report, as explained at [5.17]–[5.18], the report's analysis of the protection of personal names and geographical terms under the UDRP clarified the protection of those identifiers under the UDRP, and influenced subsequent UDRP decisions that have dealt with the protection of personal names and geographical identifiers. Paragraphs [5.17]–[5.18] deal with the Second Report's analysis of the protection of personal names and geographical names in more detail.

[3.12] Response to the Report

After receiving the WIPO Second Report, the WIPO member states, at a meeting held from 24 September to 3 October 2001, decided to refer the report to the WIPO Standing Committee on the Law of Trademarks, Industrial Designs and Geographical Indications (SCT). Following two Special Sessions, the SCT reached the following conclusions:[68]

- In relation to INNs, while acknowledging that many of the delegations supported the protection of INNs in the domain name system, it was decided not to recommend a particular form of protection, but that the WIPO Secretariat, in conjunction with the WHO, should continue to monitor the situation.
- In relation to the names and acronyms of IGOs, the SCT recommended that the UDRP be modified to deal with complaints by an IGO concerning the registration and use of IGOs that have been communicated under Article 6*ter* of the Paris Convention. The US delegation disassociated itself from this recommendation.
- In relation to personal names, the SCT decided to recommend that no action be taken.
- In relation to geographical indications, the SCT decided that, given the need for fundamental questions regarding the legal protection of geographical indications to be resolved, it was not timely to make any definitive decisions.
- In relation to geographical terms, the SCT noted that most delegations supported a form of protection for country names in the domain name system. In this regard, the delegations supported a system that would be based on a new list of the names of countries, and that would protect both the exact names of countries and misleading variations in all TLDs, including gTLDs and ccTLDs. The delegations of Australia, Canada and the United States disassociated themselves from this recommendation.

[68] See WIPO Standing Committee on the Law of Trademarks, Industrial Designs and Geographical Indications, Second Special Session on the Report of the Second WIPO Internet Domain Name Process, Geneva, 21–24 May 2002.

- In relation to trade names, noting that delegations were divided on the protection of trade names under the UDRP, the SCT recommended that member states keep the matter under review.

At their meeting held from 23 September to 1 October 2002, the WIPO member states essentially adopted the SCT recommendations. In doing so, they noted that the SCT was to continue its consideration of the protection of geographical indications. On 21 February 2003, WIPO communicated its decisions in favour of protecting the names and acronyms of IGOs, and country names, to ICANN. Following this, pursuant to the WIPO decisions, the SCT has continued to consider the protection of geographical indications and country names.

Subsequently, on 14 March 2002, in response to the reservation of a number of country names in the .info gTLD, the ICANN board agreed to reserve 329 country names, based on the ISO 3166-1 standard, in the .info gTLD, to be made available for registration by the governments and public authorities of the areas associated with the names.[69] The reservation of the country names was made with the approval of Afilias, the registry operator of the .info gTLD. The reservation of the country names was, however, restricted to the .info gTLD, and did not apply to domain names in other gTLDs.

On 23 March 2005, the Legal Advisers to the United Nations sent a letter to ICANN supporting WIPO's recommendations to protect the names and acronyms of IGOs in the domain name system. There has, however, been no response to the concerns expressed by the UN.

Other Dispute-resolution Systems

[3.13] Other ICANN Dispute Resolution Systems

As explained at [3.9], the UDRP applies to disputes between trade mark owners and domain name registrants in relation to domain names registered in the following gTLDs: aero, .biz, .cat, .com, .coop, .info, .jobs, .mobi, .museum, .name, .net, .org, .pro, .tel and .travel. In addition to the UDRP, ICANN has adopted a number of other dispute resolution systems that apply to domain name disputes in particular gTLDs.

There are two main additional categories of dispute resolution policies that apply to domain name disputes in the gTLDs: policies that are designed to minimise disputes between trade mark owners and domain name registrants by providing a mechanism to protect trade mark owners during an initial start-up or sunrise period for a new gTLD; and disputes which provide for registrations in gTLDs that have eligibility requirements to be challenged on the ground that the registrant does not comply with the particular eligibility requirements.

ICANN adopted the following dispute resolution policies that were designed to protect trade mark interests during an initial 'start-up period' for the .biz gTLD and an initial 'sunrise period' for the .info gTLD.

- *Start-up Trade Mark Opposition Policy (STOP).* The STOP Policy, which was designed by Neulevel, the registry operator for the .biz gTLD, was available to trade mark owners who

[69] ICANN, Regular Meeting of the Board, 14 Mar 2002, available at http://www.icann.org/minutes/minutes-14mar02.htm, Resolution [2.43].

enrolled in the IP Claim Service during the start-up period of the .biz gTLD, which ran from 25 June to 21 September 2001. Under the policy, once a trade mark owner enrolled a trade mark, the registry operator compared all applications for registration in the .biz gTLD with the trade mark, and notified domain name applicants of any competing IP claim. If the applicant decided to proceed with registration of the domain name, the domain name was placed 'on hold' for 30 days and the trade mark owner was notified of the application. The trade mark owner then had 20 days in which to file a complaint under the STOP policy. The elements of the STOP policy were similar to the elements of the UDRP, except that the third element of the STOP policy required only that the complainant establish that the disputed domain name was registered *or* used in bad faith, rather than registered *and* used in bad faith. During the start-up period, 801 STOP complaints were filed with the STOP dispute-resolution service providers, WIPO and NAF. Decisions under the STOP policy are sometimes referred to by UDRP panels.

- *Sunrise Challenge Policy (SCP).* The SCP, which was designed by Afilias, the registry operator for the .info gTLD, applied during the .info sunrise period, which ran from 25 July 2001 to 31 August 2001. During this period, owners of trade marks having a national effect that were issued before 2 October 2000 could register a domain name that was identical to the mark in the .info gTLD. Sunrise registrations could be challenged by third parties prior to 26 December 2001, under the Sunrise Dispute Resolution Policy, on the basis, for example, that the registrant did not own a relevant mark. Of the 51,764 sunrise registrations, 22,000 were either successfully challenged or cancelled without challenge.

The following dispute resolution policies essentially allow for the registration of domain names in particular gTLDs to be challenged on the basis that the registrants fail to meet the eligibility requirements.

- *Restrictions Dispute Resolution Policy (RDRP).* The RDRP applies to registrations in the .biz gTLD, which may be challenged on the ground that the domain name is not being, or will not be, used primarily for bona fide business or commercial purposes. The approved dispute-resolution service providers for the RDRP are the ADNDRC, the NAF and WIPO.
- *Charter Eligibility Dispute Resolution Policy (CEDRP).* The CEDRP applies to registrations in the sponsored gTLDs, .aero, .coop, .museum and .travel, which may be challenged on the ground that the registrant does not comply with the eligibility requirements of the particular sponsored gTLD. WIPO is the only approved dispute-resolution service providers for the CEDRP.
- *Eligibility Requirements Dispute Resolution Policy (ERDRP).* The ERDRP applies to registrations in the unsponsored, restricted .name gTLD, which may be challenged on the grounds that the registrant does not comply with the eligibility requirements for registration in the .name gTLD. The approved dispute-resolution service providers for the ERDRP are the NAF and WIPO.
- *.cat Eligibility Requirements Dispute Resolution Policy (.cat ERDRP).* The .cat ERDRP applies to registrations in the sponsored .cat gTLD, which may be challenged on the ground that they do not comply with the eligibility requirements.
- *Qualification Challenge Policy (QCP).* The QCP applies to registrations in the unsponsored, restricted .pro gTLD, which may be challenged on the ground that the registrant does not comply with the registration qualifications. The approved dispute-resolution service providers for the QCP are the ADNDRC, the NAF and WIPO.

- *Intellectual Property Defensive Registration Challenge Policy (IPDRCP).* The IPDRCP applies to intellectual property defensive registrations, which may be made by trade mark owners in relation to domain names in the .pro gTLD, and which may be challenged on the basis that the trade mark owner does not meet the qualifications for a defensive registration. WIPO is the only approved dispute-resolution service provider for the IPDRCP.
- *Transfer Dispute Resolution Policy (TDRP).* The TDRP applies to disputes between domain name registrars for the .biz, .com, .info, .name, .net, .org, and .pro gTLDs, where a domain name registrant transfers, or attempts to transfer, a domain name to another registrar. TDRP disputes can only be initiated by domain name registrars in the applicable gTLD. The ADNDRC and the NAF are the approved dispute-resolution service providers for the TDRP.
- *Eligibility Reconsideration Policy (ERP).* The ERP is not an ICANN policy, but is a policy of SITA, the sponsor of the .aero sponsored gTLD. It provides for challenges to the decisions by the sponsor regarding eligibility to register in the .aero gTLD. WIPO is the only approved dispute-resolution service provider for the ERP.

Limited Remedies

[3.14] Limited Remedies under the UDRP

The UDRP is designed to be a simple, efficient, cost-effective procedure that deals with 'cybersquatting' disputes between trade mark owners and domain name registrants. To ensure that the objectives of the policy are met, the remedies available under the UDRP are limited. The WIPO Final Report stated that, given the limited objectives of the policy:[70]

> it seems appropriate that the remedies that could be awarded by the neutral decision-maker be limited to the status of the domain name registration itself and actions in respect of that registration. In other words, monetary damages to compensate for any loss or injury incurred by the owner of an intellectual property as a result of a domain name registration should not be available under the procedure. Such a restrictive approach to remedies would underline the nature of the procedure as an administrative one, directed at the efficient administration of the DNS, which is intended to be complementary to other existing mechanisms, whether arbitration or court procedures.

Accordingly, paragraph 4(i) of the UDRP provides:

> The remedies available to a complainant pursuant to any proceeding before an Administrative Panel shall be limited to requiring the cancellation of your domain name or the transfer of your domain name registration to the complainant.

In most cases, the UDRP panel will grant the remedy requested by a successful complainant, which will usually involve a transfer of the domain name. In certain limited circumstances, however, where a complainant requests transfer of a domain name, cancellation is considered to be a more appropriate remedy. In particular, where a domain name consists of two trade marks that are owned by two different trade mark owners and a successful complaint is brought by one trade mark owner, then, in the absence of permission

[70] WIPO, above n 1, para [183].

by the other mark owner, it is appropriate to cancel the domain name rather than transfer it to the complainant. In *Dr Ing hc F Porsche AG v Automotive Parts Solutions*,[71] for example, the disputed domain name <911hyundai.com> consisted of a combination of the complainant's '911' mark and the 'HYUNDAI' mark that was owned by a third party car manufacturer. The panellist held that, in the absence of a complaint brought by both trade mark owners or a letter of support from the third party mark owner, transfer of the domain name to the complainant would violate the rights of the third party mark owner. Consequently, the panellist decided that the domain name should be cancelled rather than transferred.

Similarly, in *Lilly ICOS LLC v Tudor Burden*,[72] the disputed domain names, <cialisapcalis.com> and <cialis-viagra.info>, included the complainant's 'CIALIS' mark together with the marks 'APCALIS' and 'VIAGRA', which were owned by third parties. In the absence of a letter of support, or other authorization, from the owners of the other trade marks, the panellist held that the domain names should be cancelled rather than transferred, to avoid violating the rights of the third party trade mark owners.

Fundamental Tensions and WIPO Overview

[3.15] UDRP Fundamental Tensions

There are fundamental underlying tensions in the application of the UDRP that, essentially, arise from different interpretations of the act of registering a domain name that is identical or confusingly similar to someone else's trade mark. It is possible to identify two main views of the act of registering an identical or confusingly similar domain name:

- the registration of a domain name that reflects a trade mark by someone other than the trade mark owner is inherently suspect, as it is difficult to envisage a legitimate use of the domain name; or
- people should be free to register and use terms as domain names, including the many terms in which there are trade mark rights, unless the registration and use of a domain name that reflects a mark is abusive and in bad faith.

To understand the basis for these fundamentally different views it is first necessary to appreciate the differences between traditional trade mark law, on the one hand, and anti-cybersquatting laws, on the other. To begin with, traditional trade mark law is aimed primarily at protecting marks against source confusion. For this reason, in traditional actions for trade mark infringement, the trade mark owner must establish that the mark is used as a mark, meaning as a source identifier, in connection with the same goods or services in relation to which the mark is registered.[73] Anti-cybersquatting laws, on the other hand, are aimed at preventing extortionate and abusive conduct relating to the registration and use of a mark as a domain name, such as the registration of a mark as a domain name for the purpose of making credible threats to harm the trade mark owner.

[71] WIPO Case No D2003-0725 (17 Nov 2003).
[72] WIPO Case No D2004-0794 (20 Dec 2004).
[73] See, eg, Trade Marks Act 1994 (UK) ss 10(1), (2).

The essential difference between the two forms of law is best understood by examining what is required for a trade mark owner to bring an action against a person who does no more than register the mark as a domain name. It is difficult to bring a traditional trade mark infringement action because the mere registration of the mark as a domain name does not automatically amount to use of the mark as a source identifier. The disjunction between traditional trade mark law and the use of a mark as a domain name is illustrated by cases in which trade mark owners have attempted to bring actions under conventional principles of trade mark law against defendants that have registered a mark as a domain name.

In the principal English case, *British Telecommunications plc v One in a Million*,[74] for example, the defendants registered the marks of well-known English companies as domain names, such as <marksandspencer.com> and <britishtelecom.com>, with the intention of reselling the names to the highest bidders. The trade mark owners alleged that the mere registration of the marks as domain names amounted to passing off or a breach of section 10(3) of the Trade Marks Act 1994. Section 10(3) establishes a ground of infringement where a sign that is identical or similar to a mark that has a reputation in the United Kingdom is used in relation to goods or services that are not similar to those for which the mark is registered. In the Court of Appeal, Aldous LJ held that registration of the mark as a domain name constituted passing off as it amounted to a representation that the registrant of the domain name was connected or associated with the owner of the mark, and that a domain name that reflects a mark is, itself, an 'instrument of fraud', as any realistic use of the mark would result in a passing off. In relation to the action under section 10(3), the Court of Appeal upheld the High Court judgment that registration of the well-known marks as domain names infringed the provision. Assuming that section 10(3) requires the use of the mark to be trade mark use, Aldous LJ held that seeking to sell the domain names in connection with the service of reselling domain names was a commercial use in relation to a dissimilar service sufficient to satisfy section 10(3).

Both in relation to the action for passing off and the action under section 10(3), however, it can be seen that the court struggled to apply traditional doctrines because the mere registration of a mark as a domain name does not necessarily result in source confusion among consumers. As a result, the Court of Appeal stretched traditional principles to characterise the registration of a well-known mark as a domain name as an infringement, largely on the basis that the registration amounts to a threat to use the domain name in a way that creates source confusion, or otherwise takes unfair advantage of the goodwill in the mark.

US courts have experienced similar difficulties in applying conventional trade mark principles to defendants who merely registered trade marks as domain names. As explained at [5.24], these difficulties arise mainly because the US Lanham Act defines an infringing use as a 'use in commerce', and because it is difficult to characterise the registration of a mark as a domain name as a use in commerce in the requisite sense.

In *Avery Dennison v Sumpton*,[75] for example, the trade mark owners brought an action for trade mark dilution against a defendant who had registered trade marks as domain names in the .net gTLD for use as vanity e-mail addresses. The Ninth Circuit Court of Appeals reversed the first instance judgment, in part, on the basis that the registration of the marks as domain names for this purpose did not constitute a 'commercial use' of the marks. In this respect, the court stated:

[74] [1999] 1 WLR 903.
[75] 189 F 3d 868 (9th Cir, 1999).

Appellants do not use trademarks qua trademarks as required by the case law to establish commercial use. Rather, Appellants use words that happen to be trademarks for their non-trademark value.[76]

Given the disjunction between traditional trade mark principles and the problem presented by cybersquatting, it is unsurprising that much confusion has been engendered by attempts to apply traditional trade mark doctrines in the context of rules aimed at preventing extortionate and abusive practices in relation to the registration and use of marks as domain names. The confusion essentially arises because a domain name that reflects a mark does not need to create source confusion in order for cybersquatting to be successful; all that is necessary is for the domain name, in itself, to pose a credible threat of harm to the trade mark owner.

The problem is that a domain name that is identical or confusingly similar to a trade mark may, in itself, be seen as posing a threat because of its potential to attract Internet users that are confused by an association with the trade mark owner. A domain name that is identical or similar to a trade mark poses this threat because domain names are not simply Internet addresses but have a semantic content that conveys a meaning to users. A domain name that reflects a trade mark may therefore communicate some association or relationship with the trade mark owner. The resulting confusion is, however, not the source confusion that is traditionally the concern of trade mark law, but a lesser form of confusion, which may well be easily dispelled once a consumer visits a web site to which the domain name resolves. In this sense, the confusion is analogous to the pre-sales confusion, known as 'initial interest confusion', which is recognised under US trade mark law, and explained further at [5.23].

The differences in the treatment of the act of registering a domain name that is identical or confusingly similar to a trade mark, that are referred to above arise from differences in interpreting the potential for further confusion (or other harmful uses of the domain name) created by the initial confusion. The fundamental differences in how to treat the registration of an identical or confusingly similar domain name may be illustrated by reference to UDRP decisions in which the differences have emerged.

In *Chernow Communications, Inc v Jonathan D Kimball*,[77] for example, the complainant, which owned rights in the 'C-COM' mark, brought a complaint against the registrant of the <ccom.com> domain name. In reply, the respondent claimed that it had registered the domain for its own use, as it combined the letter 'C', as a shorthand for 'computers', with the term 'com', taken from the .com gTLD. The domain name resolved to a web site that indicated that it was 'under construction', and that the respondent could be contacted 'for info about partnership opportunities or domain availability'. The majority of the panel held that each of the three elements of the UDRP had been made out, while the dissenting panellist considered that the complainant had failed to make out the first and third elements.

The most important difference between the majority and the dissent in *Chernow* was in relation to the third element, which requires the complainant to establish that the domain name was registered, and is being used, in bad faith. While the respondent's conduct did not fall within any of the presumptive circumstances set out in paragraph 4(b) of the UDRP, the majority held that bad faith had been established based on the 'totality of circumstances':[78] the respondent had constructive notice of the complainant's mark; no active *bona fide* use of the mark had been made in the period of almost three years since the domain name was

[76] 189 F 3d 868, 880 (9th Cir, 1999).
[77] WIPO Case No D2000-0119 (18 May 2000).
[78] See para [7.4].

registered; the respondent's site suggested that the respondent be contacted regarding the domain name's availability; and the site included a counter registering the number of visitors.

The dissenting panellist in *Chernow*, on the other hand, declined to find for the complainant largely on the basis that the admissible evidence failed to establish that the respondent had registered the domain name in bad faith. In this respect, the dissenting panellist considered that the only relevant admissible evidence was that the respondent had registered the domain name nearly seven months after the complainant registered its mark. Pointing out that prior registration of a trade mark does not automatically mean that registration of the corresponding domain name is in bad faith, the dissent held for the respondent.

A similar division emerged in *Time Inc v Chip Cooper*,[79] where the respondent registered the disputed domain names <lifemagazine.com>, <lifemagazine.net> and <lifemagazine.org>, and claimed that he had registered the domain names to create a site for collectors of old issues of 'LIFE' magazine. The domain names did not resolve to active sites and, in response to a request from the complainant to transfer the domain names, the respondent inquired whether the complainant would be interested in assisting in developing a site for 'LIFE MAGAZINE' collectors.

The majority of the panel in *Time Inc* held that, although the respondent's conduct did not fall within any of the presumptive circumstances set out in paragraph 4(b), the complainant had established that the domain names had been registered, and were being used, in bad faith. In reaching this conclusion, the majority took into account the following circumstances: the 'LIFE MAGAZINE' mark is one of the best known marks in the United States and it is inconceivable that the respondent would not be aware of it; because of the strong reputation in the mark, members of the public would believe that the domain name holder was associated with the complainant trade mark owner; any realistic use of the domain name must involve a misrepresentation of an association with the respondent; after a year of sitting on the <lifemagazine.com> domain name, the respondent registered the complainant's mark in the .net and .org gTLDs; and no active use was made of the domain names.

The dissenting panellist in *Time Inc*, however, would have dismissed the complaint on the basis that there was no concrete evidence of bad faith on behalf of the respondent. In finding that there was no evidence of bad faith, the dissent rejected the suggestion that a famous mark automatically means that there can be no legitimate use of a corresponding domain name and pointed out that 'the UDRP does not intend to make mere registration actionable'. Moreover, the dissenting panellist concluded that the respondent's explanations were consistent with those of a genuine fan of the magazine, taking into account the general approach that, where there is no concrete evidence of bad faith, panellists should generally err in favour of the respondent.

These two decisions illustrate the extent to which the different interpretations of the act of registering an identical or confusingly similar domain name influence the weight to be placed on the evidence of the circumstances surrounding the registration and use of a domain name in particular disputes, and the inferences that may be legitimately drawn from the evidence. An analysis of the majorities and the dissents in the decisions also suggests that the coherence and consistency of decisions under the policy could be improved if the source of the differences were to be directly addressed.

[79] WIPO Case No D2000-1342 (13 Feb 2001).

To begin with, it is common ground that the objectives of the UDRP are limited, and that the policy does not confer presumptive rights on trade mark owners over all reflections of a mark in domain names. That said, it seems important for there to be a more rigorous analysis of the registration of a trade mark as a domain name than that which was undertaken in the original WIPO reports. For example, it would seem reasonable that, while ownership of a famous mark should not confer rights over all corresponding domain names, a person who registers a domain name that is identical to a famous mark should bear some onus of establishing that there is a legitimate reason for registering the mark as a domain name. Similarly, in relation to weaker marks, if it is established that the registrant knew, or ought to have known, of the mark, then it does not seem unreasonable to expect some explanation for the behaviour. In other words, in these circumstances, mere registration of the mark may give rise to a suspicion that the registrant may take advantage of the domain name by damaging the trade mark owner or free-riding on the goodwill in the mark, corresponding to the first view of the registration of a mark as a domain name mentioned above.

On the other hand, it is important to acknowledge that a system, such as the UDRP, necessarily has limited objectives. In particular, the policy is directed only at preventing extortionate and abusive practices and not, for example, at suppressing criticism of the trade mark owner. If the evidence does not indicate that the mark has been registered, or is being used, for an illegitimate commercial purpose that, in some way, takes advantage of the trade mark, then the conduct of a domain name registrant properly falls outside the scope of the policy.

Absent evidence that the domain has been registered, or is being used, for an illegitimate commercial purpose, then, corresponding to the second view of the registration of a mark set out above, people should be free to register terms as domain names, even if the term reflects a mark. Moreover, where the evidence gives rise to a suspicion that a mark has been registered for an illegitimate purpose, the onus must shift to the domain name registrant to make out a credible case that the mark has, in fact, been registered for a legitimate purpose.

In the end, the differences between the two views of the registration of a domain name that reflects a mark that are identified in this paragraph are important mainly in assessing the weight to be given to the evidence presented to a panel. In this respect, difficult cases will still often depend upon the credibility placed on the evidence before a panel which, of its nature, may consist of little more than statements from the parties, which may be difficult to independently substantiate. In assessing the evidence, however, it may be helpful to bear in mind the seemingly limitless creativity displayed by many of the respondents in the decisions discussed in this book in attempting to explain away behaviour that, on its face, is difficult to explain.

[3.16] WIPO Overview of UDRP Decisions

Given the large number of decisions under the UDRP and the need to promote consistency in decision making, WIPO presented the idea of developing a concise overview of the opinions of UDRP panels on particular issues to a meeting of WIPO domain name panellists, convened in Geneva in October 2004. Receiving a positive response, the WIPO Arbitration and Mediation Center produced an informal overview of panel opinions on key substantive and procedural questions that commonly arise in UDRP decisions. The overview, entitled the *WIPO Overview of WIPO Panel Views on Selected UDRP Questions* (the *WIPO Overview)* was released on 23 March 2005.[80]

[80] WIPO, *WIPO Overview of WIPO Panel Views on Selected UDRP Questions* (23 Mar 2005), available at http://www.wipo.int/amc/en/domains/search/overview/index.html.

The *WIPO Overview* takes the form of a concise statement of the views of panellists on a number of significant issues that have arisen in the application of the UDRP, coupled with the citation of a small number of decisions that support the propositions in the overview. The statement of views identifies those issues in relation to which a consensus has emerged among UDRP panellists, as well as qualifications to the consensus views. The statement also identifies those issues where there is no consensus among UDRP panellists and, where appropriate, provides a statement of the views of the majority and minority of panellists on the relevant issue. Finally, the overview identifies issues where neither a consensus nor a majority view has yet to emerge, and which continue to divide UDRP panellists.

The *WIPO Overview* is intended purely to assist UDRP panellists to identify trends and patterns in UDRP decisions, and is not binding on panellists. As the *Overview* explains:

> While predictability remains a key element of dispute resolution systems, neither this informal overview nor prior panel decisions are binding on panelists, who will make their judgments in the particular circumstances of each individual proceeding. Also, it remains the responsibility of each case party to make its own independent assessment of decisions relevant to its case. In particular, parties should note that the present overview cannot serve as a substitution for each party's obligation to argue and establish its case under the UDRP.

Since the *WIPO Overview* was released, it has proven to be an extremely useful resource for UDRP panellists, being cited by WIPO panellists, but also by panellists of other UDRP dispute-resolution service providers, including NAF panellists. The discussion of UDRP decisions in Chapters 4 to 7 extracts the relevant statement of the views of panellists from the *WIPO Overview* where appropriate.

Soon after the release of the *WIPO Overview*, in March 2005, the panellist in *Fresh Intellectual Properties, Inc v 800Network.com, Inc*,[81] explained the value, from the point of view of ensuring fairness of outcomes, of applying the Consensus Views where these are identified by the *WIPO Overview*. In this respect, the panellist stated:

> Although the WIPO Decision Overview is not precedential in nature, it does reflect a studied and considered summary of consensus positions culled from the decisions of numerous panelists during the first five years of administration of the UDRP. When such a consensus has developed, it is incumbent upon panels to follow the consensus (or the majority view) to promote consistency among UDRP decisions.[82]

Although UDRP panellists are not bound by a strict doctrine of precedent, then, the interests of fairness and consistency mean that, in practice, panellists will not generally depart from a Consensus View, even if they disagree with it. Where there is no Consensus View on an issue but it is possible to identify a majority view, it seems that the majority view will usually be applied, unless the UDRP panel finds an argument not to apply the majority view to be compelling. In *Stevland Morris v Unofficial Fan Club*,[83] for example, the panellist, citing a majority view from the *WIPO Overview*, pointed out that, although the views set out in the *Overview* do not amount to precedents, where the view is the majority view, a respondent needs to produce considerable 'evidence' for it to be disregarded.

The precedential value of UDRP panel decisions is dealt with in more detail at [4.2].

[81] WIPO Case No D2005-0061 (21 Mar 2005).

[82] This approach has been reiterated by the panellist in *Fresh Intellectual Properties* in a number of subsequent decisions, including *Wellquest International, Inc v Nicholas Clark*, WIPO Case No D2005-0552 (19 July 2005) and *Alain-Martin Pierret v Sierra Technology Group, LLC, eeParts, Inc v E E All Parts Corp*, NAF Case No FA481753 (14 July 2005).

[83] NAF Case No FA453986 (22 June 2005).

4

UDRP Procedures

The UDRP Rules

[4.1] The UDRP Rules

The procedural rules that apply to UDRP complaints are established by the Rules for Uniform Domain Name Dispute Resolution Policy ('the UDRP Rules').

The UDRP Rules are set out in Appendix 3. The UDRP Rules, which are applied by all UDRP dispute-resolution service providers, are based on the recommendations for procedural rules made in the WIPO Final Report. The report explained that the aims of the proposed procedural rules were:

(i) to ensure due process or fairness in the conduct in the procedure so that each party has an equal and adequate opportunity to present its case; and
(ii) to inform the parties how the procedure will be conducted, what they will be required to do, when they will be required to do it and what the powers of the decision-maker are.[1]

As the Final Report explained, the procedural rules were designed to be relatively simple, in order to take into account differences in national legal procedural traditions and the mandatory application of the rules to all domain name registrants in the applicable gTLDs.[2] A draft of the proposed UDRP procedural rules was included as Annex V to the WIPO Final Report.

As explained at [3.5], on 20 August 1999, a group of registrars submitted a draft dispute resolution policy and rules of procedure to ICANN.[3] As further explained, in accordance with the ICANN board resolution of 26 August 1999, the draft procedural rules were used by an ICANN drafting committee to prepare a draft set of rules of procedure, that was made available for public comment on 29 September 1999. Following the public consultation period, on 24 October 1999 the draft implementation documents were submitted to the ICANN board, together with the Second Staff Report. The Second Staff Report included comprehensive explanations of the reasons for adopting the particular procedural rules that were submitted to the ICANN board, and especially the modifications that were recommended by the drafting committee. On 24 October 1999, the ICANN board approved the implementation documents, including the modifications recommended in the Second Staff Report.

[1] World Intellectual Property Organization (WIPO), *The Management of Internet Names and Addresses: Intellectual Property Issues,* Report of the WIPO Internet Domain Name Process (30 Apr 1999), available at http://www.icann.org/amc/en/processes/process1/report/, para [180].

[2] *Ibid,* para [181].

[3] See *Model Rules for Uniform Domain Name Dispute Resolution Policy* (20 Aug 1999).

The main modifications to the procedural rules, that were accepted by ICANN, and which are explained in more detail at [3.6], were as follows:

- *Parity of appeal.* Concerns were expressed, first, that unsuccessful domain name registrants would have difficulty in obtaining judicial review of UDRP decisions cancelling or transferring a disputed domain name and, secondly, that respondents could obtain an automatic stay of UDRP proceedings by bringing an action before a court. To address the first concern, paragraph 3(b)(xiii) was inserted into the UDRP Rules, to require complainants to submit to the jurisdiction of courts in at least one specified mutual jurisdiction. To address the second concern, paragraph 4(k) was introduced into the UDRP, which provides for stays only where the domain name registrant submits documentary evidence that an action has been commenced against the complainant in a mutual jurisdiction to which the complainant has submitted under paragraph 3(b)(xiii) of the UDRP Rules.
- *Reverse domain name hijacking.* In accordance with the August 1999 board resolution, paragraph 15(e) of the UDRP Rules provides that 'reverse domain name hijacking' is an abuse of procedure. Moreover, under paragraph 2(a) of the UDRP Rules, the requirements for notifying the domain name registrant of a complaint were enhanced. Finally, the time for a domain name holder to seek court review of an adverse UDRP decision was extended.
- *Procedures for selection of dispute-resolution service providers and panellists.* As explained at [3.6], detailed procedures were introduced for selecting dispute-resolution service providers and UDRP panellists.
- *Burden of proof.* Paragraph 4(a) of the UDRP was modified to clarify that the complainant bears the burden of proof for each element of the UDRP.

As recommended by the Second Staff Report, ICANN decided that matters such as word limits and guidelines for complaints, responses and opinions be left to supplemental rules established by the UDRP dispute-resolution service provider. Each of the ICANN-approved dispute-resolution service providers identified at [3.10] has its own supplemental rules for dealing with these procedural matters.

Precedent and the UDRP

[4.2] Precedential Value of Panel Decisions

The UDRP does not implement a formal doctrine of *stare decisis*, meaning that UDRP panellists are neither strictly bound to follow previous UDRP panels, nor formally bound to treat previous panel decisions as persuasive. As the panellist pointed out in *Tata Sons Limited v Hasmukh Solanki*:[4]

> the Uniform Policy and the Uniform Rules do not provide that an Administrative Panel is bound by or required to follow precedents.

[4] WIPO Case No D2001-0974 (25 Sept 2001). See also *Société des Hotels Meridien SA v United States of Moronica*, D2000-0405 (27 June 2000) (stating that 'the principle of *stare decisis* does not apply in these proceedings').

In this respect, the WIPO Final Report stated:

> While it is desirable that the use of the administrative procedure should lead to the construction of a body of consistent principles that may provide guidance for the future, the determinations of the procedure should not have (and cannot have) the effect of binding precedent in national courts. It would be up to the courts of each country to determine what weight they wish to attach to determinations made under the procedure.[5]

The approach adopted by UDRP panellists to the precedential value of previous decisions involves balancing two main considerations. On the one hand, given the truncated, expedited nature of the procedure, in which panellists are required to reach decisions based upon limited evidence within demanding time constraints, a strict doctrine of precedent cannot be applied. On the other hand, however, basic principles of fairness, requiring that 'like cases should be decided alike', mean that a degree of consistency is highly desirable in UDRP decision-making.[6] Moreover, referring to earlier decisions dealing with similar facts necessarily improves the efficiency of decision-making by removing the need for panellists to 're-invent the wheel' for each new dispute.[7] To ensure the fairness, legitimacy and efficiency of decision-making, UDRP panellists have adopted a recognisable form of judicial reasoning, in which previous decisions are cited, and panellists distinguish the reasoning of a previous decision where there is a disagreement about the approach to be adopted. In other words, UDRP panellists display a high degree of comity with respect to the decisions of other panellists that, especially where there is a consensus view on a particular matter, approaches an informal doctrine of precedent.

The *WIPO Overview* includes the following Consensus View on the question of the degree of deference owed by UDRP panels to previous panel decisions dealing with similar factual and legal issues:

> The UDRP does not operate on a strict doctrine of precedent. However panels consider it desirable that their decisions are consistent with prior panel decisions dealing with similar fact situations. This ensures that the UDRP system operates in a fair, effective and predictable manner for all parties.[8]

The most comprehensive analysis of the role of precedent in UDRP decisions is the discussion of the issue by the panellist in *Howard Jarvis Taxpayers Association v Paul McCauley*.[9] In that decision, the panellist analysed the precedential value of UDRP decisions in the course of considering one of the most difficult and divisive issues to arise under the UDRP: whether a respondent can have a legitimate interest in a domain name that is identical to the complainant's trade mark where the domain name is used solely for bona fide, non-commercial criticism. Referring to the division among UDRP panels on this issue, the panellist stated:

> Disagreements of this type between panels creates a challenge for parties, panelists and providers. Parties in UDRP proceedings are entitled to know that, where the facts of two cases are materially indistinguishable, the complaints and responses will be evaluated in a consistent manner regardless

[5] WIPO, above n 1, para 150(v).

[6] See, eg, R Cross and JW Harris, *Precedent in English Law*, 4th edn (Oxford, Oxford University Press, 1991).

[7] See JT McCarthy, *Trademarks and Unfair Competition*, 4th edn (Eagan, Mich, Thompson/West, 2006) para § 25:74.75, note 35.

[8] WIPO, *WIPO Overview of WIPO Panel Views on Selected UDRP Questions* (23 Mar 2005) available at http://www.wipo.int/amc/en/domains/search/overview/index.html.

[9] WIPO Case No D2004-0014 (22 Apr 2004).

of the identity of the panelist; this goal is undermined when different panels can be expected to rule differently on the same types of facts. Panelists, too, are disadvantaged by these disagreements; they would be able to more efficiently evaluate cases and draft decisions if they knew that they could rely on a shared, consistent set of UDRP principles. If such consistency could be achieved, it also would assist providers, who could assign panelists to cases without any concern that panelist choice may itself inject bias into the system, and would encourage more cost effective decision making as parties could rely on a single panelist rather than having to request a three member panel in order to ensure balance. More generally, because no system of justice can long endure if its decisions are seen as random, consistency will help support the very legitimacy of the UDRP itself.

Although there was no consensus view on this particular issue, the panellist proceeded to a consider the approach to be adopted where there is a consensus. In this respect, the panellist emphatically argued that, to promote consistency, a consensus view should always be followed. In support of this approach, the panellist pointed out that, as the UDRP is not binding on the parties, an unsatisfied party is always able to relitigate the matter before a national court.

Consequently, the panellist in *Howard Jarvis* advocated a strict approach of adhering to a consensus view, even where the particular panellist disagrees with it. As the panellist explained this position:

even if the consensus supports a decision that individuals believe to be wrong or bad policy, parties would at least know how such decisions would be decided under the Policy, and could elect instead to pursue their claims in court if they believed the judicial system would provide a more just result.

Finally, the panellist considered the approach to be taken where a later decision of a national court is at variance with the UDRP consensus view. In these circumstances, the panellist maintained that panellists should take the decision of the court into account and, especially where both parties are citizens of the jurisdiction of the relevant court, may apply the decision of the court in preference to the UDRP consensus view.

The desirability of adhering to the approach adopted in previous decisions, even where the particular panellist disagrees with that approach, has been emphasised by a number of panellists. A particularly good example of the application of UDRP precedent in preference to the panellist's own views, is the decision in *PAA Laboratories GmbH v Printing Arts America.*[10] In that dispute, the panellist was required to determine whether the renewal of a disputed domain name in bad faith amounted to bad faith registration under paragraph (4)(a)(iii) of the UDRP. Although this was clearly an egregious example of renewal of a domain name in bad faith, as the panellist pointed out, the previous UDRP decisions to have considered the issue had consistently held that abusive renewals do not amount to bad faith registrations for the purpose of the policy. As explained at [7.8], the position has since been adopted as the consensus view in the *WIPO Overview.* In the interests of ensuring consistency, the panellist felt compelled to apply the approach adopted by previous panels, despite the fact that his own view was that bad faith renewal should fall within the scope of the policy. In explaining his conclusion on the question of bad faith registration the panellist observed:

In making its finding, the Panel wishes to clarify that its decision under this element is based on the need for consistency and comity in domain name dispute 'jurisprudence'. Were it not for the persuasive force of the cited decisions, this Panel would have expressed the view that paragraph 2 of

[10] WIPO Case No D2004-0338 (13 July 2004).

the Policy demonstrates that references to 'registration' in the Policy were probably intended to be references to 'registration or renewal of registration'. Absent the consistency of approach which has found favour with numerous earlier panels, this Panel would have seen no good reason for a renewal not to be considered as equivalent to 'registration' in the context of the objectives of the Policy.

Even where no consensus view has been identified in the *WIPO Overview*, there are good reasons for UDRP panellists to treat previous decisions with a high degree of comity or persuasiveness. As the majority put it in *Time Inc v Chip Cooper:*[11]

> The majority believes that the UDRP procedure should be governed by the rule of law, rather than by the individual consciences of the panelists. If a principle enunciated in a decision is well-reasoned and repeatedly adopted by other panels, the majority believes that absent compelling reasons which require a determination otherwise, the rule established should be respected. The majority believes that potential users of the UDRP are entitled to some degree of predictability. Counseling one who is considering filing a Complaint should consist of more than, 'It depends what panelist you draw.'

UDRP panellists accordingly treat previous decisions with a high degree of deference, even where there is no consensus view. In *Lilly ICOS LLC v Christina Poe*,[12] for example, the panellist maintained that panels should apply an approach to all previous decisions that is 'analogous' to the practice of adhering to the WIPO Consensus Views.

As explained at [3.16], where there is no consensus view but it is possible to identify a majority view and a minority view, then panellists will usually apply the majority view, unless there are compelling reasons not to. This appears to be the approach adopted in *Stevland Morris v Unofficial Fan Club*,[13] where the panellist indicated that considerable 'evidence' would need to be presented for a majority view to be disregarded.

UDRP Procedures

[4.3] The Complainant

Paragraph 3(a) of the UDRP Rules provides that 'any person or entity' may initiate a UDRP proceeding by submitting a complaint that complies with the UDRP and UDRP Rules to an ICANN-approved dispute-resolution service provider. Although paragraph 3(a) refers to 'any person or entity' in the singular, UDRP panels have held that multiple persons may bring one complaint. In this respect, the panellist in *Asprey & Garrard Limited and Garrard Holdings Limited v www.24carat.co.uk*,[14] stated:

> The Uniform Domain Name Dispute Resolution Policy ('the Policy') and the Rules for Uniform Domain Name Dispute Resolution Policy ('the Rules') do not expressly provide for multiple Complainants, but the Panel considers that, as in other documents, the singular may be interpreted as including the plural where appropriate.

[11] WIPO Case No D2000-1342 (13 Feb 2001).
[12] WIPO Case No D2006-1155 (29 Jan 2007).
[13] NAF Case No FA453986 (22 June 2005).
[14] WIPO Case No D2001-1501 (25 February 2002).

As the first element of the UDRP requires a complainant to establish rights in a trade mark, however, where there are multiple complainants, in order to succeed each must have rights in the mark. This will usually mean that there must be some relationship between the complainants, meaning that the complainants are, for instance, a parent company and a subsidiary or members of the same group of companies.

For example, in *Société Générale and Fimat International Banque v Lebanon Index,*[15] the complainants were a parent company and a subsidiary. In rejecting the respondent's submission that a complaint can only be filed by a single complainant, the panel stated:

> the naming of a parent and subsidiary as Complainants is permissible in a proceeding in which the Respondent is using domain names which are confusingly similar to trademarks owned by the parent company and its subsidiary. It is preferable that a dispute concerning the domain names be considered in a single proceeding than in a multiplicity of proceedings.

At the same time, it needs to be borne in mind that, as explained at [5.22], although the majority of decisions have held that that a company that is related to the trade mark owner, such as a subsidiary or parent, can have rights in the mark sufficient for paragraph 4(a)(i) of the UDRP to be satisfied, the panellist in *PartyGaming Plc v Kriss Vance*[16] held that there is an additional requirement where the complainant is a parent of a subsidiary company that is the registered mark owner. As further explained, the panellist in that decision concluded that, in these circumstances, a parent will only have rights in the mark under paragraph 4(a)(i) if it has some right to use or otherwise deal with the mark itself, or if the subsidiary holds the mark in trust for others within the group, including the parent company. Where there is more than one complainant and a panel decides to transfer the domain name, the panel may determine the transferee or may leave this decision to be determined by the complainants.

The following are examples of decisions that have held that a complaint may be brought under paragraph 3(a) of the UDRP Rules by multiple complainants.

- *Asprey & Garrard Limited and Garrard Holdings Limited v www.24carat.co.uk.*[17] The complaint was brought by two related corporations, and the panellist was prepared to assume that the complainants were in common control. The panellist ordered that the disputed domain name be transferred to the complainants and, if it was not possible to transfer the domain name to the complainants jointly, to be transferred to Asprey & Garrard.
- *Magna International Inc and Donnelly Corporation v Brian Evans.*[18] The complainants published a press release announcing an agreement to merge pursuant to which Donnelly Corporation would become a wholly-owned subsidiary of Magna International Inc. In a dispute concerning the <magnadonnelly.com> domain name, the panellist held that the involvement of two complainants did not vitiate the complaint.
- *Staples, Inc, Staples The Office Superstore, Inc, Staples Contract and Commercial, Inc v John Morgan.*[19] The complainants were related corporations and the panellist held that they had sufficient interest in the disputed domain name, <stpales.com>, to be joined in the action.

[15] WIPO Case No D2002-0760 (1 Nov 2002).
[16] WIPO Case No D2003-0456 (2 June 2006).
[17] WIPO Case No D2001-1501 (25 Feb 2002).
[18] WIPO Case No D2002-0898 (5 Nov 2002).
[19] WIPO Case No D2004-0537 (20 Sept 2004).

- *Hunter Douglas Industries, BV, and Hunter Douglas Window Fashions, Inc v Erik Little.*[20] The complainants were related corporations and the panellist held that, although not the trade mark owner, Hunter Douglas Window Fashions had rights in the mark. The panellist ordered that the disputed domain name, <hunterdouglad.com>, be transferred to 'the complainants', leaving the identification of the transferee to the complainants.
- *adidas International Marketing BV and adidas-Salomon AG v Sonia Eguia.*[21] The complainants were members of the one group of companies, the first complainant being the registered owner of the 'ADIDAS' mark and the second complainant being the registered owner of the 'TUNIT' mark. In a dispute involving the <adidastunit.com> domain name, the panel accepted the filing of the complaint by the two complainants.

[4.4] The Complaint

Paragraph 3(b) of the UDRP Rules provides that the complaint must be submitted to the dispute-resolution service provider in hard copy and in electronic form and shall:

- request that the complaint be submitted for decision in accordance with the UDRP and UDRP Rules;
- provide the name, postal and e-mail addresses, and the telephone and telefax numbers of the complainant and of any representative of the complainant;
- specify a preferred method for communications directed to the complainant;
- designate whether the complainant elects to have the dispute decided by a single-member or a three-member panel and, in the event that the complainant elects a three-member panel, provide the names and contact details of three candidates to serve as one of the panellists;
- provide the name of the respondent and all information known to the complainant regarding how to contact the respondent;
- specify the domain name or domain names that are the subject of the complaint;
- identify the registrar with whom the domain name or domain names are registered;
- specify the trade marks or service marks on which the complaint is based and, for each mark, describe the goods or services with which the mark is used;
- describe the grounds on which the complaint is made;
- specify the remedies sought;
- identify any other legal proceedings that have been commenced or terminated in connection with the disputed domain names;
- state that a copy of the complaint has been sent or transmitted to the respondent in accordance with paragraph 2(b) of the UDRP Rules;
- state that the complainant will submit to the jurisdiction of the courts in at least one specified mutual jurisdiction;
- conclude with a prescribed statement that: (a) waives any claims or rights against the dispute-resolution service provider and panellists (except for deliberate wrongdoing), the registrar, the registry administrator and ICANN; and (b) certifies that the information in the complaint is complete and accurate; and
- annex any documentary or other evidence, including any trade mark or service mark registration.

[20] WIPO Case No D2005-0460 (15 June 2005).
[21] WIPO Case No D2006-0475 (9 June 2006).

Paragraph 1 of the UDRP Rules defines 'mutual jurisdiction' to mean:

> a court jurisdiction at the location of either (a) the principal office of the Registrar (provided the domain-name holder has submitted in its Registration Agreement to that jurisdiction for court adjudication of disputes concerning or arising from the use of the domain name) or (b) the domain-name holder's address as shown for the registration of the domain name in Registrar's Whois database at the time the complaint is submitted to the Provider.

Paragraph 3(c) of the UDRP Rules provides:

> The complaint may relate to more than one domain name, provided that the domain names are registered by the same domain-name holder.

Requests for consolidating multiple disputes are dealt with by paragraph 10(e) of the rules, which states:

> A Panel shall decide a request by a Party to consolidate multiple domain name disputes in accordance with the Policy and these Rules.

Although the UDRP Rules do not expressly provide for complaints to be amended to include additional domain names, UDRP panels have decided that this is permissible in certain circumstances. After a review of the relevant decisions, the panellist in *General Electric Company v Edison Electric Corp*[22] concluded that UDRP panels will permit a complaint to be amended by the inclusion of additional domain names where the following criteria are met:

> the factual backgrounds of the additional and initial Complaints must be closely connected so that it is expedient to determine them together in order to avoid the risk of irreconcilable decisions resulting from separate proceedings; the additional Complaint must be filed against the same Respondent (as required by paragraph 3(c) of the Rules); the Parties must be treated with equality and each Party must be given a fair opportunity to present its case in accordance with paragraph 10(b) of the Rules; and the Complainant must pay any additional fees which would have been payable if the additional names had been included in the original Complaint.

In identifying the above criteria, the panellist rejected the suggestion that had been made by the panellist in *A & F Trademark, Inc v Party Night, Inc*,[23] that the additional complaint must be based on exactly the same legal and factual grounds as the original complaint. At the same time, the panellist in *General Electric* identified the following additional considerations that panels should take into account in deciding whether to permit additional domain names to be included in a complaint:

> the requirement in paragraph 11(c) of the Rules that the proceeding takes place with due expedition; the objective of the Policy of providing a cost-effective system for resolving straightforward domain name disputes; the need to secure compliance with other aspects of the Rules, such as certification, submission to a mutual jurisdiction and waiver of legal claims against the registrar, registry and dispute resolution service provider in relation to the additional domain names; and the desirability of verifying the registration details and locking of the additional domain names during the proceeding.

22 WIPO Case No D2006-0334 (13 Aug 2006).
23 WIPO Case No D2003-0172 (28 May 2003).

[4.5] Serving the Respondent

Under paragraph 4(a) of the UDRP Rules, the dispute-resolution service provider must review the complaint for compliance with the UDRP and UDRP Rules and, if the complaint complies, must forward the complaint to the respondent within three calendar days of receiving the fees payable by the complainant. Paragraph 2(a) of the rules provides that, in notifying the respondent of the complaint, the dispute-resolution service provider must employ 'reasonably available means' that are calculated to achieve actual notice. The following measures will discharge the service provider's responsibility to serve the complaint on the respondent:

- sending the complaint to all postal-mail and facsimile addresses that are shown in the domain name's registration data in the Registrar's Whois database for the registered domain name holder, the technical contact and the administrative contact and supplied by the registrar to the service provider for the registration's billing contact; and
- sending the complaint by e-mail to the e-mail addresses for the technical, administrative and billing contacts; to 'postmaster@<the contested domain name>; and to any e-mail addresses, or e-mail links, on an active web-page to which the disputed domain name resolves; and
- sending the complaint to any address the respondent has notified the provider it prefers and, to the extent practicable, to all other addresses provided to the dispute-resolution provider by the complainant.

[4.6] The Respondent

Paragraph 1 of the UDRP defines the 'respondent' to mean:

> the holder of a domain-name registration against which a complaint is registered.

This clearly refers to the entity that is recorded as the registered domain name holder in the records of the registrar for the disputed domain name. Problems have arisen, however, with certain domain privacy services or proxy registration services, under which a third party, usually in return for a fee, agrees to enter its information in the 'WHOIS' database in place of the true registrant's. As the panel observed in *WWF-World Wide Fund for Nature v Moniker Online Services LLC and Gregory Ricks:*[24]

> This is becoming an increasingly common occurrence. Under this arrangement the name of the true registrant is not displayed in the Registrar's WHOis database. It is said that proxy registrations have become popular with the rise of automatic harvesting of email and postal addresses and telephone numbers for the purpose of email spamming, direct marketing by mail and phone and also for fraud and identity theft.

In such circumstances, a complaint will still be properly filed if it identifies the registrant of record as the respondent. As the panel in *WWF* explained, the UDRP Rules seem to contemplate only the recorded holder of the registration as the respondent. Nevertheless, where it becomes apparent that the registrar of record is a proxy service provider, then the practice has developed of requiring the complainant to amend the complaint to add the party using

[24] WIPO Case No D2006-0975 (1 Nov 2006).

the privacy service to the proxy service provider or, in certain circumstances, to substitute the party using the service as the complainant. In the *WWF* case, for example, the panel required the complaint to be amended to show the user of the privacy service as well as the proxy service provider, despite the initial objections of the complainant to doing so. In that decision, the panel noted that it is for the proxy service provider to disclose the name of the privacy service user to the dispute-resolution service provider and the UDRP panel. Moreover, the panel cautioned that, once a proxy service provider becomes aware of a complaint, it should not change the 'WHOIS' information to reflect the identity of the privacy service user. Altering the registration details after a complaint has been lodged would appear to be a breach of paragraph 8(a)(i) of the UDRP, which states:

> You may not transfer your domain name registration to another holder (i) during a pending administrative proceeding brought pursuant to paragraph 4 or for a period of fifteen (15) business days . . . after such proceeding is concluded . . .

In *Ohio Savings Bank v 1&1 Internet, Inc and David Rosenbaum*,[25] however, the registrar changed the registration record to identify the party using the privacy service as the registrant after the complaint had been filed. In that decision, the panellist noted that including the party using the service as a complainant after notice has been served on the proxy service provider may well cause delays in that party being notified of the complaint, but that this is a necessary price of using a domain privacy service. In relation to the practice of adding the user of the privacy service as a complainant, the panellist observed:

> This practice is sensible and has the benefit of trying to get notice of the proceeding to the party that is most affected by the proceeding. However, it may simply be that one of the disadvantages of using a privacy service—to be weighed against the advantages of using such a service—are delays or lack of notice of proceedings that rely on the information contained in the registration record to provide notice. By this the Panel does not intend to suggest that such a delay or lack of notice is insignificant, but that they may simply be one of the consequences of the choice to use such a service, particularly where proceedings under the Policy look to and rely on information in the registration record maintained by the registrar.

In that particular dispute, as the dispute-resolution service provider had served the complaint on both the privacy service provider and the user of the service, the requirements for service under paragraph 2(a) of the UDRP Rules had clearly been satisfied.

[4.7] The Response

Paragraph 5(a) of the UDRP Rules provides:

> Within twenty (20) days of the date of commencement of the administrative proceeding the Respondent shall submit a response to the Provider.

The form of the response is prescribed by paragraph 5(b), which requires it to be submitted in hard copy and in electronic form and to:

- respond specifically to the statements and allegations contained in the complaint and include any and all bases for the respondent to retain registration of the disputed domain name;

[25] WIPO Case No D2006-0881 (15 Sept 2006).

- provide the name, postal and e-mail addresses, and the telephone and telefax numbers of the respondent and of any representative authorised to act for the respondent in the administrative proceeding;
- specify a preferred method for communications directed to the respondent for both electronic-only material and hard copy;
- if the complainant has elected a single-member panel, state whether the respondent elects to have the dispute decided by a three-member panel;
- if either the complainant or respondent elects a three-member panel, provide the names and contact details of three candidates to serve as one of the panellists;
- identify any other legal proceedings that have been commenced or terminated in connection with the disputed domain names;
- state that a copy of the response has been sent or transmitted to the complainant in accordance with paragraph 2(b) of the UDRP Rules, which generally requires communication to be made by the preferred means stated by the complainant;
- conclude with a statement certifying that the information in the response is complete and accurate; and
- annex any documentary or other evidence upon which the respondent relies.

The requirement under paragraph 5(b)(i) of the UDRP Rules that the respondent should respond specifically to the allegations in the complaint suggests that the response should identify the grounds for establishing any rights or legitimate interests in the domain name, although this is not expressly required.

While a complaint that does not comply with the formal requirements specified in paragraph 3(b) will be dismissed, the majority of UDRP panels have held that they have a discretion to consider responses, even if they do not comply with the formal requirements set out in paragraph 5(b) of the UDRP Rules. This reflects the discretion conferred on UDRP panels under paragraph 4(d) of the UDRP Rules to determine the admissibility of evidence. In *Internet Opportunity Entertainment Limited v Zubee.com Networks Inc*,[26] for example, the panellist noted that formally incorrect responses have been considered in the following circumstances:

- where the respondent wrongly submitted the response to the complainant and ICANN, but not to the dispute-resolution service provider: see *Oberoi Hotels Pvt Ltd v Aurun Jose*.[27]
- where the response was on time but was submitted in handwriting: see *Cable News Network LP, LLP v Manchester Trading*.[28]
- where the response exceeded the limit on the number of words imposed by the WIPO Supplementary Rules: see *Süd-Chemie AG v tonsil.com*.[29]
- where the response was submitted by fax: see *Veritas DGC Inc v The Collectors Source*.[30]

In the *Internet Opportunity Entertainment* decision itself, the respondent failed to meet the requirements of paragraph 5(b), by submitting an e-mail response to the complainant and by not responding specifically to the statements and allegations contained in the complaint. Although the panellist held that the respondent was formally in default of the requirements of paragraph 5(b), he indicated that, to the extent that the material submitted

[26] WIPO Case No D2006-1086 (5 Dec 2006).
[27] WIPO Case No D2000-0263 (7 June 2000).
[28] NAF Case No FA93634 (17 Mar 2000).
[29] WIPO Case No D2000-0376 (3 July 2000).
[30] NAF Case No FA94425 (8 May 2000).

informally by the respondent was relevant, it was taken into account in the interests of ensuring that the decision was based on all available information.

[4.8] Late Responses

UDRP panels have a discretion to consider late responses under paragraph 5(d) of the UDRP Rules, which provides:

> At the request of the Respondent, the Provider may, in exceptional cases, extend the period of time for the filing of the response. The period may also be extended by written stipulation between the Parties, provided the stipulation is approved by the Provider.

At the same time, paragraph 14(a) states:

> In the event that a Party, in the absence of exceptional circumstances, does not comply with any of the time periods established by these Rules or the Panel, the Panel shall proceed to a decision on the complaint.

There is a division of opinion among panellist in relation to what amounts to 'exceptional cases' in which a panel may grant a respondent's request to file a late response. Early UDRP panel decisions adopt a relatively lenient approach to applications for filing late responses on the basis that the UDRP Rules were new and relatively unknown. In *Talk City, Inc v Michael Robertson,*[31] for example, the respondent sent an e-mail to the dispute-resolution service provider two days after the expiry of the deadline for submitting a response, which coincided with the appointment of the administrative panel. The panellist appeared willing to consider the e-mail, but he did not give any weight to the allegations made in it as it failed to include a certification that the information was complete and accurate, as required by paragraph 5(b)(viii).

Panels are more likely to exercise the discretion to consider a late response where it was received before the panel was appointed. In *The Board of Governors of the University of Alberta v Michael Katz,*[32] for example, the respondent filed a response that complied with the formalities required by paragraph 5(b) one day after the expiry of the deadline for submitting the response. The panellist exercised her discretion to consider the response as it was received before the panellist was appointed and any substantive review had commenced.

Over time, UDRP panels have been less inclined to accept late responses, as it is assumed that there is more familiarity with the UDRP Rules. In *1099 Pro, Inc v Convey Compliance Systems, Inc,*[33] for example, the respondent did not file a response within the prescribed time but, on receiving a notification of default from the dispute-resolution service provider, submitted an e-mail response before the panel was constituted. The majority of the panel maintained that, in the usual course of events, panels should not consider responses received after the deadline. In this respect, the majority stated:

> it is important to apply the Rules as written, absent a good reason. Otherwise, parties will feel free to disregard deadlines and Respondents will regularly submit late responses. Thus, absent good cause, the majority of the Panel holds that late-filed responses should be disregarded.

[31] WIPO Case No D2000-0009 (29 Feb 2000).
[32] WIPO Case No D2000-0378 (22 June 2000).
[33] WIPO Case No D2003-0033 (1 Apr 2003).

In this particular dispute, although one of the panellists would have allowed the late response, the majority held that the respondent's claim that it was confused about the UDRP procedures was not a sufficient explanation for the delay in filing the response, and refused to consider it.

Since the *1099 Pro* decision, an even more rigorous approach to scrutinising late responses has been advocated by the panellist in *Mobile Communication Service Inc v WebReg, RN.*[34] In that dispute, the respondent's legal representative submitted a request to submit a late response under paragraph 5(d), claiming only that the legal representative was working on another UDRP response at the time. As, in this case, there was no indication that additional time was needed to gather evidence, the panellist decided that this was not an 'exception' to the general rule that late responses will not be accepted. In reaching this conclusion, the panellist formulated the following strict approach to assessing requests for filing a late response:

> absent exceptional circumstances supported by good cause, panels should not accept late-filed responses, even in the absence of prejudice.

At the same time, the panellist acknowledged that, in certain limited circumstances, there are legitimate reasons for accepting a late response. In this respect, the panellist noted that legitimate extensions had been granted in limited 'exceptional' circumstances such as in the following decisions:

- *Gaiam, Inc v Nielsen.*[35] A late response was accepted after service of the complaint on the respondent was delayed.
- *DK Bellevue, Inc v Landers.*[36] A late response was accepted when the response was filed on the first business day after the deadline that fell on a weekend, due to the respondent's good faith belief that the deadline would be extended in accordance with US litigation practice.

[4.9] The Panel and Panel Decisions

Paragraph 6(b) of the UDRP Rules provides that, where neither the complainant nor the respondent has opted for a three-member panel, the provider shall appoint a single-member panel from its publicly available list of approved panellists. Where there is a single-member panel, the complainant is solely responsible for the fees, and the parties have no influence on the identity of the panellist. Under paragraph 3(b)(iv) of the UDRP Rules, the complainant may elect to have the dispute determined by a three-member panel and, in so doing, submit the names of three candidates to serve as one of the panellists. If the complainant selects a three-member panel, it is solely responsible for the fees. Where a complainant does not opt for a three-member panel, paragraph 5(b)(v) provides for the respondent to elect to have a three-member panel and, in so doing, submit the names of three candidates to serve as one of the panellists. In these circumstances, the complainant must submit the names of three candidates to serve as one of the panellists within five calendar days of the communication of the response. If the respondent selects a three-member panel, the applicable fees are shared equally between the parties.

[34] WIPO Case No D2005-1304 (24 Feb 2006).
[35] NAF Case No FA112469 (2 July 2002).
[36] WIPO Case No D2003-0780 (24 Nov 2003).

Where there is a single-member panel, the dispute-resolution service provider must appoint the panellist within five calendar days of receiving a response from the respondent or, in the absence of a response, within five calendar days of the expiry of the deadline for submitting a response. Where a party has opted for a three-member panel, the dispute-resolution service provider must endeavour to appoint one panellist each from the lists of candidates submitted by the complainant and the respondent. In these circumstances, the third panellist is to be appointed from a list of five candidates submitted by the dispute-resolution service provider to the parties, in a manner that reasonably balances the preferences of both parties. Where a three-member panel has been chosen, the dispute-resolution service provider must appoint the panel with five calendar days of the provider submitting the list of potential candidates to the parties.

Paragraph 15(b) of the UDRP Rules provides:

> In the absence of exceptional circumstances, the Panel shall forward its decision on the complaint to the Provider within fourteen (14) days of its appointment . . .

In order to ensure transparency and encourage consistency of decision-making, paragraph 15(d) of the UDRP Rules requires decisions to be in writing and to provide the reasons on which the decision is based, as well as the date of the decision and the identity of the panellists.

Under paragraph 16(a) of the UDRP Rules, the dispute-resolution service provider must communicate the full text of the decision to each party, the registrar for the disputed domain name and ICANN. Moreover, under paragraph 16(b), the service provider is required to publish the full decision on a publicly accessible Web site, except for any parts of the decision that have been redacted pursuant to paragraph 4(j) of the UDRP.

Paragraph 10 of the UDRP Rules sets out the following broad powers of UDRP panels:

> (a) The Panel shall conduct the administrative proceeding in such manner as it considers appropriate in accordance with the Policy and these Rules.
> (b) In all cases, the Panel shall ensure that the Parties are treated with equality and that each Party is given a fair opportunity to present its case.
> (c) The Panel shall ensure that the administrative proceeding takes place with due expedition. It may, at the request of a Party or on its own motion, extend, in exceptional cases, a period of time fixed by these Rules or by the Panel.
> (d) The Panel shall determine the admissibility, relevance, materiality and weight of the evidence.
> (e) A Panel shall decide a request by a Party to consolidate multiple domain name disputes in accordance with the Policy and these Rules.

Choice of Law and Language

[4.10] Choice of Law

The UDRP introduced new, global rules aimed at preventing the abusive registration and use of trade marks as domain names. In devising the new rules, however, WIPO drew upon legal principles developed under national trade mark and unfair competition laws. As the WIPO Final Report pointed out in explaining the formulation of the proposed elements of the UDRP:

> The case law which has developed in the application of national laws for the protection of trademarks and service marks and for protection against unfair competition . . . supports the prohibition of the predatory and parasitical practices that would be caught under the definition of abusive registration . . .[37]

The Final Report went on to explain the approach that was envisaged for the application of national legal principles to the proposed administrative procedure in the following terms:

> In applying the definition of abusive registration given above in the administrative procedure, the panel of decision-makers appointed in the procedure shall, to the extent necessary, make reference to the law or rules of law that it determines to be applicable in view of the circumstances of the case. Thus, for example, if the parties to the procedure were resident in one country, the domain name was registered through a registrar in that country and the evidence of the bad faith registration and use of the domain name related to activity in the same country, it would be appropriate for the decision-maker to refer to the law of the country concerned in applying the definition.[38]

Consequently, the Final Report made the following recommendation in relation to the legal principles to be applied by the decision-making panel:

> in applying the definition of abusive registration, the panel of decision-makers shall, to the extent necessary, apply the law or rules of law that it determines to be appropriate in view of all the circumstances of the case.[39]

This recommendation is embodied in paragraph 15(a) of the UDRP Rules, which confers the following broad discretion on UDRP panels in determining the legal principles to apply in deciding a UDRP dispute, including whether to apply national legal principles:

> A Panel shall decide a complaint on the basis of the statements and documents submitted and in accordance with the Policy, these Rules and any rules and principles of law that it deems applicable.

To begin with, it is clear that the discretion conferred by paragraph 15(a) means that UDRP panels may take into account national legal principles where it is appropriate to do so, and there are many decisions in which panellists have referred to national legislation and judicial decisions. In particular, recourse has been had to national legal principles in determining difficult questions relating to whether a complainant is able to establish rights in an unregistered trade mark or service mark under paragraph 4(a)(i) of the UDRP. Reference is also commonly made to national legal principles in deciding whether a respondent has rights or legitimate interests in a disputed domain name under paragraph 4(a)(ii) of the policy.

Where both parties are residents of the same legal jurisdiction, there is no difficulty in applying any relevant legal principles of that jurisdiction, and such principles will be readily applied. As the panellist put it in *The Rockport Company, LLC v Gerard A Powell*:[40]

> Since all parties are domiciled in the United States, to the extent that it would assist the Panel in determining whether the Complainant has met its burden as established by Paragraph 4(a) of the Policy, the Panel will look to the principles of the law of the United States.

In *EAuto, LLC v Triple S Auto Parts*,[41] for example, the disputed domain name <eautolamps.com> incorporated the complainant's 'EAUTO' mark, which had been

[37] WIPO, above n 1, para [175}.
[38] *Ibid*, para [176].
[39] *Ibid*, para [177(ii)].
[40] WIPO Case No D2000-0064 (6 Apr 2000).
[41] WIPO Case No D2000-0047 (24 Mar 2000).

registered pursuant to § 2(f) of the US Lanham Act,[42] creating a presumption that the mark is distinctive. As both of the parties were US residents, the panellist considered that it was appropriate to consider and apply US law in determining that the complainant had rights in the mark under paragraph 4(a)(i).

National principles of trade mark and unfair competition law have been applied by UDRP panellists in many decisions involving parties that are residents of the same legal jurisdiction. The following decisions illustrate the application of national legal principles where the parties are from the same nation.

- *Westfield Corporation, Inc v Graeme Michael Hobbs.*[43] As both parties were based in Australia, the panellist referred to Australian law in determining whether the disputed domain name was confusingly similar to the complainant's Australian registered marks. In deciding to apply Australian law, the panellist stated:

  > As the dispute is one concerning the similarity between a domain name and a trade mark, it is clearly appropriate that the principles of law to be applied should be those derived from the law of trade marks and unfair competition. As the Respondent is located in Australia and as the Complainants are each subsidiaries of an Australian corporation, it is also appropriate that the Panel should look to the Australian law for guidance on the question of whether or not the name and the marks are identical or confusingly similar. That said however, the principles adopted by the Australian courts do not differ significantly from that adopted by the courts in other common law jurisdictions.

- *Sara Lee Southern Europe, SL v SL Protección de Dominios.*[44] As both parties were located in Spain, the panellist applied Spanish law.
- *Tribeca Film Center, Inc v Lorenzo Brusasco-Mackenzie.*[45] The complainant had no registered rights in the 'TRIBECA FILM CENTER' mark, but claimed common law rights. The panellist applied US law to find that the complainant had rights in the mark on the basis that it had acquired distinctiveness with a specific segment of the general public.
- *Israel Harold Asper v Communication X Inc.*[46] As both parties were situated in Canada, the panellist applied Canadian law in determining whether the complainant had trade mark rights in his personal name.
- *Myer Stores Limited v Mr David John Singh.*[47] Even though both parties were based in Australia, the respondent claimed that US law applied to the dispute as the registrar was located in the United States. The respondent further claimed that the complainant had no rights in its 'MYER' mark as it had not registered the mark in the United States. In rejecting the respondent's claims, the panellist held that the applicable law was the law of the jurisdiction in which both parties were located which, in this case, was the law of the Australian state of New South Wales, incorporating Australian federal law.
- *Autosales Incorporated v Don Terrill.*[48] The panellist followed the *EAuto* decision to hold that registration of a mark under the US Lanham Act creates a rebuttable presumption that the mark is inherently distinctive. In reaching this conclusion, the panellist explained that, as both parties were domiciled in the United States, he could 'look to rules and principles of law set out in decisions of the courts of the United States'.

[42] 15 USC § 1052(f) (US).
[43] WIPO Case No D2000-0227 (18 May 2000).
[44] WIPO Case No D2000-1690 (30 Jan 2001).
[45] WIPO Case No D2000-1772 (10 Apr 2001).
[46] WIPO Case No D2001-0540 (11 June 2001).
[47] WIPO Case No D2001-0763 (10 July 2001).
[48] WIPO Case No D2001-1341 (24 Feb 2002).

- *Cable News Network LP, LLP v Elie Khouri.*[49] The panel considered that it could apply principles of US law where the complainant was a corporation based in the United States and the domain name was registered through a US registrar, even though the respondent was not domiciled in the United States. In that dispute, however, the respondent had previously resided in the United States for approximately 15 years.
- *Robilant & Associati Srl v POWERLAB snc.*[50] As both parties were based in Italy, the panellist referred to Italian law to determine that the complainant had rights in the unregistered 'ROBILANT' mark.

As illustrated in detail in Chapter 5, an understanding of national principles of trade mark law is often essential in determining whether a complainant has rights in an unregistered trade mark or service mark pursuant to paragraph 4(a)(i) of the UDRP. At the same time, it is important to appreciate that, in relation to most aspects of a claim of abusive, bad faith registration, the principles of domain name dispute resolution are distinct from those of trade mark infringement. In this respect, the following comments of the panellist in *Israel Harold Asper v Communication X Inc*[51] are instructive:

> While the Panel agrees that in this matter references to Canadian law are applicable as both the Complainant and the Respondent are situated in Canada, it must also be borne in mind that there are essential differences between trademarks and domain names, and between the bases for proving causes of action available in various courts as contrasted with what is necessary to enable this Panel to reach its determinations under the Policy and Rules. This is also the case in regard to the remedies available through domestic courts and those obtainable under the Policy. With trademarks, there can be more than one registrant of the same mark and its effect may be limited by the category of goods to which it relates, and the geographic boundaries of its territorial protection. There are no such limitations with domain names. Due to the latter's technological limitations and borderless use, only one entity can be registered in each gTLD name throughout the world. Thus, there is justification for different legal principles to apply to the Uniform Domain Name Dispute Resolution Policy, its Rules, and the interpretation given these by Administrative Panels operating thereunder. It is these Rules with which all applicants for the registration of domain names agree to be bound.

Given the significant differences between national principles of trade mark and unfair competition law, where the parties are based in different legal jurisdictions, UDRP panellists will generally apply principles developed under the policy itself and not principles of a particular national law. In *Fox News Network, LLC v C&D International Ltd,*[52] for example, the complainant was a US-based corporation and the respondent was based in China. Given that the parties were from different legal jurisdictions, the panellist decided the dispute primarily based on the UDRP and UDRP Rules, and the statements and documents submitted. In doing so, the panellist adopted and applied the following general principle:

> With regard to the applicability of national law, the general rule is that where both parties and the registrar are from one jurisdiction, then the Panel is at liberty to consider the national law from that jurisdiction; however, where they are not, then only the Policy and Rules should apply.

Where the parties are based in different legal jurisdictions, then, UDRP panellists are hesitant about applying national legal principles, and will not usually do so.

49 NAF Case No FA117876 (16 Dec 2002).
50 WIPO Case No D2006-0991 (5 Oct 2006).
51 WIPO Case No D2001-0540 (11 June 2001).
52 WIPO Case No D2004-0108 (22 July 2004).

[4.11] Proper Language of the Proceedings

Paragraph 11(a) of the UDRP Rules deals with the proper language of UDRP proceedings, and provides:

> Unless otherwise agreed by the Parties, or specified otherwise in the Registration Agreement, the language of the administrative proceeding shall be the language of the Registration Agreement, subject to the authority of the Panel to determine otherwise, having regard to the circumstances of the administrative proceeding.

Translations of documents submitted in languages other than the proper language of the proceedings are provided for by paragraph 11(b), which states:

> The Panel may order that any documents submitted in languages other than the language of the administrative proceeding be accompanied by a translation in whole or in part into the language of the administrative proceeding.

In relation to the proper language of the proceeding, the WIPO Consensus View states the basic rule that:

> The language of the proceeding is the language of the registration agreement, unless both parties agree otherwise.[53]

As the following decisions illustrate, the language of the registration agreement will usually be applied where there has been a dispute between the parties concerning the proper language of the proceedings, and UDRP panels will rarely exercise their discretion under paragraph 11(a) to apply another language, subject to the one exception outlined below.

- *Telstra Corporation Limited v Telsra com.*[54] The complainant, an Australian corporation, brought a complaint against the respondent, which was based in Spain, in relation to the <telsra.com> domain name, which was used to redirect Internet users to a pornographic Web site. After being served with the complaint, the respondent filed a response in Spanish. The dispute-resolution service provider sent an e-mail to the respondent explaining that, as the language of the registration agreement was English, the respondent was required to file an English translation of the response. After the respondent failed to comply with the request, the appointed panellist issued a procedural order under paragraph 11(b) requiring the respondent to submit an English translation of the response. As a translation was not supplied, the respondent's submission was not taken into account, as it did not comply with paragraph 11(a).
- *Fondation Le Corbusier v Monsieur Bernard Weber.*[55] The complainant, a French foundation, filed a complaint against the respondent, who was based in Switzerland, in relation to 32 domain names that incorporated the 'LE CORBUSIER' mark. As the language of each of the registration agreements was English, the panel determined that the proper language of the proceedings was English, even though the complaint had initially been filed in French. In reaching this decision, the panel took into account the fact that the complaint had been translated into English, and that the respondents' legal representatives were English.

[53] WIPO, above n 8, para 4.3.
[54] WIPO Case No D2003-0247 (21 July 2003).
[55] WIPO Case No D2003-0251 (4 July 2003).

- *Advanced Magazine Publishers Inc v Computer Dazhong.*[56] The complainant, a US corporation, filed a complaint against the respondent, which was based in China, in relation to the <voguedress.com> domain name. Although the registrar informed the dispute-resolution service provider that the language of the registration agreement was English, the respondent's lawyers claimed that the 'true' registrar was a Chinese company, which was not an ICANN-accredited registrar. The dispute-resolution service provider decided to extend the period for filing a response and informed the respondent that it would permit a response in either English or Chinese. After the respondent filed a response in Chinese and a three-member panel was appointed, the complainant requested the panel to review the decision on the proper language of the proceedings. Given that the alleged Chinese registrar was not an ICANN-accredited registrar, the panel decided that English was the proper language of the proceedings, and issued an order under paragraph 11(b) for the respondent to file an English translation of the response.

The *WIPO Overview* includes the following exception to the Consensus View, which explains the circumstances in which UDRP panels will exercise their discretion to choose a language other than the language of the registration agreement as the proper language of the proceedings:

> In certain situations, where the respondent can clearly understand the language of the complaint, and the complainant would be disadvantaged by being forced to translate, the language of proceedings can remain the language of the complaint, even if it is different to the language of the registration agreement.[57]

The following decisions illustrate the circumstances in which a panel may decide that the proper language of the proceedings is a language other than that used in the registration agreement.

- *L'Oreal SA v MUNHYUNJA.*[58] The complainant, a French company, filed a complaint against the respondent, which was based in Korea, in respect of the <shuuemura.com> domain name, which was registered with a Canadian registrar at the time of the complaint. After the complaint was filed, the domain name was transferred to a new registrant with a new Korean registrar. The panellist held that the respondent had engaged in a practice known as 'cyberflying'—meaning 'an attempt to avoid or delay judicial or UDRP proceedings by changing domain registration details or registrars after learning of a complaint'[59] —contrary to paragraph 8(a) of the UDRP. Although, following the transfer of the domain name the language of the registration agreement was Korean, which would usually be the proper language under paragraph 11(a), the panellist held that the proper language of the proceedings was English. In reaching this conclusion, the panellist took the following considerations into account: the disadvantage to the complainant of conducting the proceeding in Korean; the possibility that a new panel might need to have been appointed; the fact that the initial registration agreement was in English; and the fact that the respondent had sufficient time to prepare a response in English.
- *Deutsche Messe AG v Kim Hyungho.*[60] The complainant, a German company, filed a complaint against the Korean respondent in relation to the <cebit.com> domain name, which

[56] WIPO Case No D2003-0668 (12 Dec 2003).
[57] WIPO, above n 8, para 4.3.
[58] WIPO Case No D2003-0585 (17 Nov 2003).
[59] See, eg, *Fifth Third Bancorp v Secure Whois Information Service,* WIPO Case No D2006-0696 (14 Sept 2006).
[60] WIPO Case No D2003-0679 (13 Nov 2003).

was registered pursuant to a Korean registration agreement. Taking into account the fact that the respondent had displayed a proficiency in English in communications with the complainant, including an offer to sell the domain name to the complainant, and that the use of Korean would unduly delay the proceeding, the panellist decided, in the interests of fairness to both parties, that the proper language of the proceedings should be English, but that the respondent was permitted to submit documents in Korean.

- *General Electric Company v Edison Electric Corp.*[61] The language of the registration agreements for the disputed domain names <edisonge.com>, <edison-ge.com> and <ee-ge.com> was English, whereas the language of the registration agreement for the disputed domain name <eege.com> was Chinese. The dispute-resolution service provider notified the complainant that the complaint in relation to the <eege.com> domain name was deficient as it was filed in English rather than Chinese. After further communications with the parties, the service provider informed the parties in both English and Chinese that it had decided to accept the complaint in English, but would accept a response in either English or Chinese, and leave it to the panel to determine the proper language of the proceedings under paragraph 11(a). Following a request from the complainant to add additional domain names, the panellist issued a procedural order permitting the inclusion of the additional domain names in the complaint and directing that the language of the proceedings would be English. The panellist determined that the language of the proceedings for the disputed domain name <eege.com> should be English, despite the Chinese language registration agreement, taking into account the English proficiency displayed in the respondent's communications, as well as the additional expense and delay that would be incurred if the proceedings were held in different languages for the different domain names. In reaching this conclusion, the panellist observed:

> The spirit of paragraph 11 is to ensure fairness in the selection of language by giving full consideration to the parties' level of comfortability with each language, the expenses to be incurred and possibility of delay in the proceeding in the event translations are required and other relevant factors

Supplemental Submissions and Refiling

[4.12] Supplemental Submissions

Paragraph 12 of the UDRP Rules states:

> In addition to the complaint and the response, the Panel may request, in its sole discretion, further statements or documents from either of the Parties.

This paragraph needs to be read together with the important provision in paragraph 10(b) of the rules, which provides:

> In all cases, the Panel shall ensure that the Parties are treated with equality and that each Party is given a fair opportunity to present its case.

In general, UDRP panels have concluded that supplemental submissions should be requested only in limited circumstances. The panellist in *The EW Scripps Company v*

[61] WIPO Case No D2006-0334 (13 Aug 2006).

Sinologic Industries,[62] after reviewing previous decisions, concluded that panels should accept additional evidence or submissions only in 'exceptional circumstances'. In that decision, the panellist gave the following example of the circumstances in which additional submissions may be accepted:

> where the party could not reasonably have known the existence or relevance of the further material when it made its primary submission; that if further material is admitted, it should be limited so as to minimise prejudice to the other party or the procedure; and that the reasons why the Panel is invited to consider the further material should, so far as practicable, be set out separately from the material itself.

In *Creo Products Inc v Website In Development,*[63] on the other hand, the panellist identified the following circumstances in which panellists will request supplemental submissions under paragraph 12:

> There may be situations in which an administrative panel considers that the complainant has not proved that the complaint satisfies all the requirements of the Uniform Policy, but that it is highly likely the complainant could do so upon the submission of further arguments and/or evidence.

The *Creo Products* approach was applied by the panellist in *Credit Suisse Group v Milanes-Espinach,*[64] who confirmed that the circumstances in which a panel will request a supplemental submission are 'unusual'. In that dispute, the panellist requested a supplemental submission from the complainant where, because of the brevity of the complaint, the panellist was not convinced that the complainant could satisfy all the elements of the UDRP, but had formed the view that it was highly likely that the complainant could establish a successful case with the submission of further arguments or evidence. Similarly, in *Auto-C, LLC v MustNeed.com,*[65] the panellist considered that the complaint did not contain adequate evidence that the complainant, as opposed to other members of the complainant's group of companies, was the owner of a mark on which the complaint was based, but formed the view that it was highly likely that the complainant could provide proof of ownership if it was permitted to submit further evidence. Consequently, the panellist requested the complainant to make a supplemental submission under paragraph 12 clarifying its rights under paragraph 4(a)(i) of the UDRP.

In *General Electric Company v Edison Electric Corp,*[66] the panellist held that the panel's discretion to request further submissions includes a discretion to accept some parts of a supplemental submission and reject other parts. Applying the approach set out in the *EW Scripps* decision to the complainant's supplemental submission, the panellist decided that some parts of the submission complied with the criteria for accepting additional submissions, whereas other parts of the submission were rejected on the basis that they did not comply with the requirement that additional material should be limited so as to minimise prejudice to the proceedings.

A strict reading of paragraph 12 might suggest that supplemental submissions can only be made in response to a request from the panel. This reading appears to have been accepted by the panellist in *EW Scripps,* who adopted the following approach to additional submissions under paragraph 12:

[62] WIPO Case No D2003-0447 (1 July 2003).
[63] WIPO Case No D2000-1490 (19 Jan 2001).
[64] WIPO Case No D2000-1376 (20 Mar 2001).
[65] WIPO Case No D2004-0025 (30 Mar 2004).
[66] WIPO Case No D2006-0334 (13 Aug 2006).

> Under the Policy and the Rules, parties have no right to submit additional arguments or evidence. However, the Panel may, in its sole discretion, request further statements or documents from the parties under paragraph 10 of the Rules; and a party's request may be regarded as an invitation to the Panel to exercise this discretion.

In general, however, the majority of UDRP panels that have considered the issue have held that, in certain circumstances, panels will entertain unsolicited supplemental submissions. In this respect, there are important differences in the way the issue is dealt with by different dispute-resolution service providers.

Among WIPO panellists there is a split between the majority of panel decisions, which have held that UDRP panels have a discretion to accept unsolicited supplementary submissions, and the minority of decisions, which consider that supplementary submissions can only be made in response to a request from a panel. The WIPO majority view is as follows:

> While panels must be mindful of the need for procedural efficiency, they have discretion to accept an unsolicited supplemental filing from either party, bearing in mind the obligation to treat each party with equality and ensure that each party has a fair opportunity to present its case. It is also helpful if the party submitting its filing can show its relevance to the case and why it was unable to provide that information in the complaint or response. Panelists who accept a supplemental filing from one side often allow the other party the opportunity to file a reply to the supplemental filing.[67]

The following are examples of decisions by WIPO panels that have accepted unsolicited supplemental submissions.

- *Delikomat Betriebsverpflegung Gesellschaft mbH v Alexander Lehner*.[68] The response was received within the specified time, but did not comply with all of the formal requirements set out in paragraph 5(b) of the UDRP Rules. Despite this, the panellist decided to consider it on the basis that it would not prejudice the complainant. After the response was filed, the complainant made two unsolicited supplemental submissions. Noting that paragraph 12 of the UDRP Rules does not provide for a party to file supplemental submissions without the leave of the panel, the panellist adopted the view that paragraph 12 should be read with paragraph 10(b). On this interpretation, paragraph 12 does not preclude acceptance of unsolicited submissions, which panels may accept in accordance with the paragraph 10(b) duty to give each party a 'fair opportunity' to present its case. In this dispute, the panellist accepted the two supplemental submissions on the basis that they addressed matters that could not have been addressed in the complaint.
- *AutoNation Holding Corp v Rabea Alawneh*.[69] After the appointment of the three-member panel, WIPO received a supplemental submission from the complainant that introduced new arguments and evidence, and a supplemental submission from the respondent arguing that panels could not accept supplemental submissions. In rejecting the respondent's contention that unsolicited submissions were impermissible, the panel emphasised its broad discretion, conferred under paragraph 10(a) of the rules, to 'conduct the administrative proceeding in such manner as it considers appropriate'. At the same time, the panel noted that 'panels generally frown on receiving additional submissions' because of the delays which such submissions may occasion. Consequently, the panel considered that the discretion to accept unsolicited submissions should be exercised on a 'case-by-case basis'. In this dispute, the panel accepted and considered the additional submissions of the parties.

[67] WIPO, above n 8, para 4.2.
[68] WIPO Case No D2001-1447 (20 Feb 2002).
[69] WIPO Case No D2002-0058 (1 May 2002).

- *De Dietrich Process Systems v Kemtron Ireland Ltd.*[70] The panellist accepted an unsolicited supplemental submission from the complainant and issued a procedural order giving the respondent an additional 10 days to respond to the supplemental filing. In reply, the respondent submitted an analysis of the UDRP Rules claiming that panels had no discretion to accept unsolicited submissions. The panellist subsequently issued a request for supplemental submissions under paragraph 12, and the respondent filed a reply claiming that the panellist was in breach of the UDRP Rules. The panellist accepted and considered both the requested and the unsolicited supplemental submissions.

A minority of decisions by WIPO panellists have held that panels cannot accept unsolicited submissions. The WIPO minority view is as follows:

> Unless the panel specifically solicits a supplemental filing, it will not consider a supplemental filing in its decision.[71]

The minority view interprets paragraph 12, and the absence of any express provision in the UDRP Rules for parties to submit supplemental filings, as precluding panels from considering unsolicited supplemental submissions. The panellist in *Viacom International Inc v Rattan Singh Mahon,*[72] expressed the minority view in the following unqualified terms:

> There can be no doubt that neither the Complainants nor the Respondents has a right to file supplementary submissions subsequent to the Complaint and the Response. Supplementary submissions can only be filed in response to a request for such from the Panel.

Panellists that adopt the WIPO minority view are also concerned that accepting unsolicited submissions would unduly prolong UDRP proceedings contrary to paragraph 10(c) of the UDRP Rules, which requires panels to 'ensure that the administrative proceeding takes place with due expedition'. In the *Viacom* decision, for example, the panellist expressed the view that panellists should request supplemental submissions under paragraph 12 only in exceptional circumstances, otherwise the UDRP process would become significantly more 'resource-consuming' for all parties, thereby undermining its role as an expeditious, cost-effective alternative to litigation.

While the WIPO Supplementary Rules do not address the issue of supplemental submissions, the NAF UDRP Supplementary Rules provide for the parties to make additional submissions within a strict time limit. In this respect, paragraph 7(a) of the NAF Supplementary Rules states:

> A party may submit additional written statements and documents to the Forum and the opposing party(s) within five (5) Calendar Days after the date the Response was received by the Forum, or, if no Response has been filed, the last date the Response was due to be received by the Forum.

Although NAF panellists would appear to have no discretion to reject supplemental submissions filed under paragraph 7(a), panellists have no obligation to take such submissions into account.[73] At the same time, there is a division among NAF panellists as to whether the panel has a discretion to accept supplemental submissions filed after the deadline set by paragraph 7(a).

[70] WIPO Case No D2003-0484 (24 Sept 2003).
[71] WIPO, above n 8, para 4.2.
[72] WIPO Case No D2000-1440 (22 Dec 2000).
[73] See, eg, *Deep Foods, Inc v Jamruke, LLC,* NAF Case No FA648190 (10 Apr 2006).

[4.13] Refiled Complaints

Paragraph 4(k) of the UDRP provides that the administrative proceedings do not prevent the parties from submitting the dispute to a court of competent jurisdiction. There is, however, nothing in the UDRP or UDRP rules that addresses the question whether UDRP panels can accept a refiled complaint. As defined by the panellist in *Creo Products Inc v Website In Development*,[74] a refiled complaint is:

> A subsequent complaint (whether or not precisely the same in substance) in relation to the same domain name, that is filed by the same complainant against the same respondent to a previous complaint.

UDRP panels agree that there is a discretion to accept a refiled complaint, but that the discretion must be exercised cautiously, and only in limited circumstances. In this respect, the WIPO Consensus View states:

> A refiled case involves the complainant submitting a second complaint involving the same domain name(s) and the same respondent(s) as in an earlier complaint that had been denied. A refiled case may only be accepted in limited circumstances. These circumstances include when the complainant establishes in the complaint that relevant new actions have occurred since the original decision, or that a breach of natural justice or of due process has occurred, or that there was other serious misconduct by the panel or the parties in the original case (such as perjured evidence). A refiled complaint will also be accepted if it includes newly presented evidence that was unavailable to the complainant during the original case.[75]

The panellist in *Parker Hannifin Corporation v East Bay Website Company*,[76] accepted the principle that complaints could be refiled but, referring to the principles of relitigation of cases in common law jurisdictions, held that panels should only accept refiled complaints in circumstances analogous to:

- serious misconduct on the part of a judge, juror, witness or lawyer;
- perjured evidence having been offered to the Court;
- the discovery of credible and material evidence which could not have been reasonably foreseen or known at trial; and
- a breach of natural justice.

In relation to the circumstances in which new evidence might justify accepting a refiled complaint, the panellist accepted the following three criteria that have been applied by common law courts:

- it must be shown that the evidence could not have been obtained with reasonable diligence for use at the trial;
- the evidence must be such that, if given, it would probably have an important influence on the result of the case, although it need not be decisive; and
- the evidence must be such as is presumably to be believed or, in other words, it must be apparently credible, although it need not be incontrovertible.

[74] WIPO Case No D2000-1490 (19 Jan 2001).
[75] WIPO, above n 8, para 4.4.
[76] WIPO Case No D2000-0703 (10 Nov 2000).

In *Creo Products Inc v Website In Development,*[77] the panellist accepted the principle that, to allow unsuccessful complainants to redress potential injustice, panels should be able to accept refiled complaints but, to prevent unmeritorious attempts to relitigate, refiled complaints should be permitted in limited circumstances only. In examining the circumstances in which panels should accept refiled complaints, the panellist drew an important distinction between:

- refiled complaints that concern the act which formed the basis of the original complaint; and
- refiled complaints that concern acts that have occurred subsequent to the original decision.

In relation to the first category of refiled complaints, the panellist accepted the view adopted in *Parker Hannifin* that the circumstances for accepting a refiled complaint must be limited, but concluded that the four circumstances identified in that dispute were not closed. In relation to refiled complaints that concern acts that have occurred after the original decision, the panellist observed that, in essence, such circumstances amount to a new dispute. With the exception that the acts relied upon in the second category of refiled complaints should not, in substance, be the same as those relied upon in the original complaint, the panellist considered that such complaints should be treated in the same way as new complaints.

In addition to identifying the circumstances in which panels may entertain refiled complaints, the panellist in *Creo Products* formulated the following general principles to apply to both categories of refiled complaints:

> First, the burden of establishing that the Refiled Complaint should be entertained under the Uniform Policy rests on the refiling complainant. Secondly, that burden is high. Thirdly, the grounds which allegedly justify entertaining the Refiled Complaint need to be clearly identified by the refiling complainant. Fourthly, the dispute resolution service provider with whom the Refiled Complaint has been filed has responsibility for determining if, *prima facie*, the refiling complainant has pleaded grounds which might justify entertaining the Refiled Complaint. Fifthly, where the dispute resolution service provider determines that, *prima facie*, the refiling complainant has pleaded grounds which might justify entertaining the Refiled Complaint (and that the other formal requirements of the Uniform Policy are satisfied), the Refiled Complaint should be submitted to an administrative panel, for determination of whether the Refiled Complaint should be entertained (and, if so, of the merits of the claim under the Uniform Policy).

In the *Creo Products* dispute itself, as the respondent had made offers to sell the disputed domain name after the original decision, the panellist accepted the refiled complaint on the basis that, insofar as it was based on acts that occurred subsequent to the original decision, it was essentially a new complaint.

In *Maruti Udyog Ltd v maruti.com,*[78] the panel applied the approach developed in *Parker Hannifin* and *Creo Products* in deciding whether the combination of new evidence and acts that occurred subsequent to the original decision justified the acceptance of a refiled complaint. The panel accepted that the new material, which included evidence of a forged birth certificate and of bad faith use of the domain name after the original decision, was sufficient for it to accept the refiled complaint. After taking into account the new evidence, the panel decided that the disputed domain name should be transferred to the complainant. Similarly, in *AB S Svenska Spel v Andrey Zacharov,*[79] the panel accepted a refiled complaint on the basis

[77] WIPO Case No D2000-1490 (19 Jan 2001).
[78] WIPO Case No D2003-0073 (5 May 2003).
[79] WIPO Case No D2003-0527 (2 Oct 2003).

of acts that had occurred following the original decision. In that dispute, the acts that had taken place—including the transfer of the disputed domain name to a new registrant within the same group of companies as the former registrant, the active use of the domain name and the fact that the complainant had acquired a registered trade mark—led the panel to conclude that the refiled complaint should be treated as a new action under the policy.

Burden of Proof

[4.14] Burden of Proof

Paragraph 4(a) of the UDRP expressly provides that the complainant bears the burden of establishing each of the three elements of the UDRP. With the exception of the second element of the UDRP, the standard of proof is the civil standard of a 'preponderance of evidence' or 'balance of probabilities'. As the panel explained in *Madonna Ciccone v Dan Parisi*:[80]

> Since these proceedings are civil, rather than criminal, in nature, we believe the appropriate standard for fact finding is the civil standard of a preponderance of the evidence (and not the higher standard of "clear and convincing evidence" or "evidence beyond a reasonable doubt"). Under the "preponderance of the evidence" standard a fact is proved for the purpose of reaching a decision when it appears more likely than not to be true based on the evidence.

Similarly, the panellist in *Grove Broadcasting Co Ltd v Telesystems Communications Limited*[81] rejected a higher standard of proof in favour of the normal civil burden of the balance of probabilities. As explained at [6.4], as the second element of the UDRP requires the complainant to establish a negative—that the respondent has no rights or legitimate interests in the disputed domain name—panels have held that, in relation to that element, the complainant need only make out a prima facie case, whereupon the burden of proof shifts to the respondent.[82]

The nature of UDRP proceedings means that the evidentiary record is often less than complete. The limited nature of the evidence, and the frequency with which UDRP respondents default, means that UDRP panels are commonly forced to rely upon inferences drawn from the available evidence. In *Madonna Ciccone v Dan Parisi*,[83] for example, the respondent registered the disputed domain name <madonna.com>, which was used for a sexually explicit Web site. The panellist inferred from the circumstances of the dispute, including the absence of any plausible explanation of the respondent's registration and use of the domain name, that the domain name had been intentionally registered to attract Internet users by trading off the reputation of the complainant, 'MADONNA'. The drawing of inferences in this way is perfectly legitimate, and often unavoidable.

The UDRP Rules confer an extremely broad discretion on the way in which panels assess allegations made by parties to UDRP disputes. Paragraph 15(a), for example, simply provides:

[80] WIPO Case No D2000-0847 (12 Oct 2000).
[81] WIPO Case No D2000-0703 (10 Nov 2000).
[82] See, eg, *Compagnie Générale des Matières Nucléaires v. Greenpeace International*, WIPO Case No D2001-0376 (14 May 2001).
[83] WIPO Case No D2000-0847 (12 Oct 2000).

> A Panel shall decide a complaint on the basis of the statements and documents submitted and in accordance with the Policy, these Rules and any rules and principles of law that it deems applicable.

Given this broad discretion, it is unsurprising that different standards have been applied by different panels in determining the evidentiary weight to be given to allegations made by the parties. Despite these differences, mere unsubstantiated allegations are generally not sufficient for the complainant to satisfy its evidentiary burden. As the panellist in *Do The Hustle, LLC v Tropic Web*[84] explained in connection with the second element of the UDRP:

> A panel is not required to blindly accept assertions offered by a respondent, any more than it is required to accept unsupported assertions offered by a complainant. It is especially important under this procedure to test respondent's assertions for evidentiary support and credibility, since normally the complainant has no opportunity to counter the respondent's assertions, while the respondent does have the opportunity to counter those of the complainant.

In that decision, the panellist pointed out that allegations must be supported by 'concrete evidence', meaning 'more than mere personal assertions'.

While much depends upon the circumstances of the particular dispute, failure to support allegations with appropriate evidence may be fatal. In *AlohaCare v PPI*,[85] for example, the respondent claimed that it had legitimate interests in the disputed domain name <alohacare.com>on the basis that its company, 'Alohacare Intelligent Monitoring', intended to use the domain name to promote e-health services. The panellist, however, rejected the respondent's claims as they were unsupported by any 'concrete evidence' of the respondent's affiliation with the company or of the alleged business plans.

Particular difficulties arise in the evaluation of factual allegations made by complainants in disputes involving defaulting respondents. These issues are discussed at [4.16].

Other Procedural Issues

[4.15] Independent Research by UDRP Panels

Given the nature of UDRP proceedings—including the electronic nature of the proceedings, the limited evidentiary record and the absence of discovery—UDRP panels may sometimes need to take a degree of independent initiative in investigating facts relating to the dispute. The flexibility required for UDRP procedures is recognised by the broad discretion conferred by paragraph 10(a) of the UDRP Rules, which provides:

> The Panel shall conduct the administrative proceeding in such manner as it considers appropriate in accordance with the Policy and these Rules.

In particular, UDRP panellists are permitted independently to investigate the way in which a disputed domain name is used by visiting a Web site to which the domain name resolves. In this respect, the *WIPO Overview* sets out the following Consensus View.

> A panel may visit the internet site linked to the disputed domain name in order to obtain more information about the respondent and the use of the domain name. The panel may also undertake

[84] WIPO Case No D2000-0624 (21 Aug 2000).
[85] WIPO Case No D2002-1160 (12 Feb 2003).

limited factual research into matters of public record if it feels that it needs that assistance in reaching a decision. If the panel feels that it requires further information to make a decision in a proceeding then it can issue a panel order to the parties.[86]

The following are examples of decisions in which the panel has conducted independent research, including by visiting Web sites linked to the disputed domain name.

- *Société des Produits Nestlé SA v Telmex Management Services.*[87] The complainant alleged that the respondent was using the disputed domain name <nestlefoods.com> to divert Internet users to its pornographic Web site. As the respondent defaulted, the panellist exercised his power under paragraph 10(a) to visit the <nestlefoods.com> Web site to investigate independently whether there was any evidence that the respondent might have rights or legitimate interests in the disputed domain name. Finding nothing to suggest that the respondent might have rights or interests, the panellist ordered the transfer of the domain name to the complainant.
- *Hasco Bastion Limited v The Trading Force Limited.*[88] The complainant claimed that the disputed domain name <hescobastian.com> redirected Internet users to the respondent's <ttforce.com> Web site, which would mislead or confuse potential customers into believing that there was an association with the complainant. On visiting the Web site, the panellist agreed with the complainant that it has the 'look and feel' of a site that a reasonable person would conclude originated from the complainant. In this dispute, the complainant also filed an unsolicited supplemental submission dealing with the beneficial ownership of the domain name which, as it did not delay the conduct of the proceeding, the panellist accepted.
- *Howard Jarvis Taxpayers Association v Paul McCauley.*[89] The respondent registered the <hjta.com> domain name for use as a site to criticise the practices and management of the complainant. In response to the complainant's claim that the domain name was being used misleadingly to divert Internet users to its Web site for commercial gain, the respondent argued that the domain name was being used for a legitimate criticism site. After visiting the respondent's site, the panellist found that it was a classic criticism site, that there was nothing misleading about the source or sponsorship of the site, and the site could not be characterised as evidencing an intent to tarnish the complainant's mark. The panellist also noted that, as the parties did not include sufficient material on the central issue of whether the use of a domain name that is identical to the complainant's mark for a criticism site can ever confer a legitimate interest on the respondent, he had undertaken 'significant independent research'.

The different approaches panels may take to exercising the discretion to visit the respondent's Web site are illustrated by two decisions concerning the one Web site, <sylviasaint.com>. In *Silvie Tomčalová aka Sylvia Saint v Juan Campos,*[90] the complainant, 'SYLVIA SAINT', claimed that the disputed domain name was being used misleadingly to redirect Internet users to a third party Web site that competed with the complainant's site, but failed to provide any evidence of the Web site to which the disputed domain name resolved. Citing the WIPO Consensus View, the panellist acknowledged that it was open to

[86] WIPO, above n 8, para 4.5.
[87] WIPO Case No D2002-0070 (2 Apr 2002).
[88] WIPO Case No D2002-1038 (31 Dec 2002).
[89] WIPO Case No D2004-0014 (22 Apr 2004).
[90] WIPO Case No D2006-0379 (5 May 2006).

him to visit the respondent's Web site but, in this case, decided that it was inappropriate to do so. In this respect, the panellist stated:

> Although the Complainant may consider this harsh, the Panel does not think it appropriate to proceed in this fashion in this case. There is the practical issue that websites change over time. The parties can not be certain what the panelist sees and what the panelist sees may not represent the true position at the relevant time. However, there is a more fundamental objection here. The Complainant must prove her case. It is not for the Panel to make the Complainant's case for her. It is one thing for a panelist to view a web site to verify a parties' assertions and quite another to embark upon an independent investigation as to what a complainant's case may be.

In concluding that the complainant had failed to discharge the evidentiary burden, the panellist was concerned that the respondent's case consisted solely of an abstract assertion, with no further explanation.

In similar circumstances, however, in a complaint involving the same complainant and domain name, the panellist in *Silvie Tomčalová, aka Sylvia Saint v Global Access*,[91] adopted a different approach. In this dispute, the complainant claimed that the disputed domain name <silviasaint.com> was being used misleadingly to divert Internet users to the respondent's Web site but, as in the *Juan Campos* dispute, failed to explain the content of the Web site or provide any print-outs. Citing the WIPO Consensus View, the panellist in this dispute decided that it was appropriate to undertake independent research by visiting the respondent's Web site. After visiting the Web site, the panellist held that, as it was being used to capitalise on the complainant's mark by offering adult entertainment DVDs, the respondent had no rights or interests in the disputed domain name, and that it was being used in bad faith.

While the respective panellists clearly adopted different approaches to the desirability of undertaking independent research, it should be borne in mind that both panellists accepted the Consensus View that panellists have a discretion to visit a respondent's Web site, and differed mainly on the evidentiary burden that a complainant must discharge to make out a prima facie case that the respondent has no rights or legitimate interests in the disputed domain name.

[4.16] Default Rules

In a considerable number of UDRP disputes the respondent fails to file a response.[92] While suggestions have been made for introducing measures to address this issue,[93] it appears that a high rate of default decisions is a permanent feature of the UDRP process. The consequences of a respondent's default are set out in paragraph 5(e) of the UDRP Rules, which provides as follows:

> If a Respondent does not submit a response, in the absence of exceptional circumstances, the Panel shall decide the dispute based upon the complaint.

The procedures for dealing with defaulting respondents are prescribed by paragraph 14 of the UDRP Rules, which provides:

[91] WIPO Case No D2006-0399 (24 June 2006).

[92] See, eg, M Geist, *Fair.com? An Analysis of the Allegations of Systematic Unfairness in the ICANN UDRP* (Aug 2001), available at http://wwwudrpinfo.com.

[93] See, eg, AM Froomkin, 'ICANN's "Uniform Dispute Resolution Policy"—Causes and (Partial) Cures' (2002) 67 *Brooklyn Law Review* 605,

(a) In the event that a Party, in the absence of exceptional circumstances, does not comply with any of the time periods established by these Rules or the Panel, the Panel shall proceed to a decision on the complaint.
(b) If a Party, in the absence of exceptional circumstances, does not comply with any provision of, or requirement under, these Rules or any request from the Panel, the Panel shall draw such inferences therefrom as it considers appropriate.

While failing to respond to a complaint certainly disadvantages the respondent, the UDRP Rules give panels considerable discretion in determining how to deal with defaulting respondents. Moreover, paragraph 14 must be read together with paragraph 10(b), which provides:

> In all cases, the Panel shall ensure that the Parties are treated with equality and that each Party is given a fair opportunity to present its case.

As paragraph 10(b) refers to 'all cases', it has been interpreted as applying to disputes in which the respondent has defaulted. In addition, the default rules do not excuse the complainant from the evidentiary burden imposed by paragraph 4(a) of the UDRP. Together, these provisions mean that, unlike in conventional legal proceedings, a respondent's default does not automatically lead to the dispute being decided in the complainant's favour. In this respect, the *WIPO Overview* sets out the following Consensus View:

> The respondent's default does not automatically result in a decision in favor of the complainant. Subject to the principles described in 2.1 above with regard to the second UDRP element, the complainant must establish each of the three elements required by paragraph 4(a) of the UDRP. While a panel may draw negative inferences from the respondent's default, paragraph 4 of the UDRP requires the complainant to support its assertions with actual evidence in order to succeed in a UDRP proceeding.[94]

As explained at [6.4], and as referred to at paragraph 2.1 of the *WIPO Overview*, the complainant will meet its evidentiary burden in relation to the second element of he UDRP once it establishes a prima facie case.

To begin with, despite the rules set out in paragraph 14 which, in the event of a default, require the panel to 'proceed to a decision' and to 'draw such inferences' from the default 'as it considers appropriate', the complainant is clearly not relieved of its evidentiary onus, under paragraph 4(a) of the UDRP, of proving each of the elements of the UDRP. As the panellist explained in *The Vanguard Group, Inc v Lorna Kang*:[95]

> The Respondent's default does not automatically result in a decision in favor of the Complainant. The Complainant must still prove each of the three elements required by Policy paragraph 4(a).

Consequently, in reaching a decision in cases involving a defaulting respondent, the panel must actively consider whether the complainant has discharged its evidentiary burden in relation to all three elements of the UDRP. Nevertheless, the UDRP Rules confer an extremely broad discretion on panels in evaluating the strength of the complainant's case. In this respect, paragraph 14(b), which permits the panel to draw 'appropriate' inferences from the respondent's default, must be read together with paragraph 15(a), which provides:

> A Panel shall decide a complaint on the basis of the statements and documents submitted and in accordance with the Policy, these Rules and any rules and principles of law that it deems applicable.

[94] WIPO, above n 8, para 4.6.
[95] WIPO Case No D2002-1064 (20 Jan 2003).

UDRP panels agree that the combination of paragraphs 14(b) and 15 means that, where a respondent defaults, the panel's decision must be 'based upon the Complainant's assertions and evidence and inferences drawn from the Respondent's failure to reply'.[96]

Given the extremely broad discretion conferred on panels in relation to how they are to treat the complainant's evidence in the event of a respondent's default, it is unsurprising that panels are split on how much weight that should be given to assertions of fact made in a complaint.

On the first view, the paragraph 14(b) invitation to draw 'appropriate' inferences from the respondent's default has led some panellists to accept all unrefuted factual allegations made by the complainant. In *Talk City, Inc v Michael Robertson*,[97] for example, the panellist stated:

> Given Respondent's failure to submit a substantive answer in a timely fashion, the Panel accepts as true all of the allegations of the complaint.

Similarly, in *ABF Freight System, Inc v American Legal*,[98] the panellist maintained:

> Because Respondent has failed to submit an answer to the Complaint in a timely fashion, and because the allegations of the Complaint, taken on their face, engender no substantial doubt, the Panel accepts as true all allegations set forth in the Complaint.

Most panellists who adopt this approach do not uncritically accept each and every allegation made in an uncontested complaint, but will accept all those allegations that are 'reasonable' or 'credible'. In *Plymouth State College v Domains, Best Domains*,[99] for example, the panellist stated:

> A respondent is not obliged to participate in a domain name dispute proceeding. But, if it fails to do so, asserted facts that are not unreasonable are taken as true and the respondent is subject to the inferences that flow naturally from the information provided by the complainant.

Similarly, panels commonly cite the following statement made by the panellist in *Vertical Solutions Management, Inc v webnet-marketing, inc*:[100]

> The Respondent submitted no response in this matter. As a result, all reasonable inferences of fact in the allegations of the Complainant will be deemed true.

In this respect, however, a distinction must be drawn between factual allegations or inferences of fact, in the one hand, and inferences that are drawn from the factual allegations, on the other. While there is disagreement among panellists concerning the weight to be attributed to factual allegations made in a complaint, UDRP panels agree that, once allegations of fact are accepted, they may draw reasonable inferences from those allegations.

As opposed to the first view, other panellists have adopted a more rigorous approach to the evidentiary burden imposed on complainants under paragraph 4(a) of the UDRP. On the second view, which is reflected in the WIPO Consensus View, it is not sufficient to rely on mere allegations, but the complainant must discharge a higher evidentiary burden. In particular, adherents of this view are likely to reject allegations that are made by a complainant where the complainant fails to support the allegations by evidence that it is in its control to produce.

96 See, eg, *Packaging World Inc v Zynpack Packaging Products Inc*, eRes Case No AF-0233 (28 July 2000).
97 WIPO Case No D2000-0009 (29 Feb 2000).
98 WIPO Case No D2000-0185 (9 May 2000).
99 WIPO Case No D2002-0939 (20 Dec 2002).
100 NAF Case No FA95095 (31 July 2000).

In *Brooke Bollea v Robert McGowan*,[101] for example, the complainant alleged that she had common law rights in her personal name, 'BROOKE HOGAN', but provided no evidence, other than an allegation that her name had acquired notoriety, that her name had become distinctive. In deciding that the complainant had failed to establish the requisite rights in her name under the first element of the UDRP, the panellist stated:

> Respondent's default does not automatically result in judgment for the Complainant and does not constitute an admission of any pleaded matter. To be more precise, Respondent's default does not mean that the Panel must accept as true factual allegations in the Complaint that are unsupported by evidence.

In rejecting the complainant's allegations, the panellist noted that evidence regarding the reputation or goodwill in the complainant's name was entirely within the knowledge and control of the complainant.

Similarly, in *TPI Holdings, Inc v LaPorte Holdings*,[102] the complainant failed to produce evidence to support factual allegations made in the complaint, including allegations relating to the popularity of the complainant's Web site. In refusing to accept those allegations that were not supported by evidence, the panellist observed:

> Statements of counsel are not evidence, and in the Panel's view the Policy requires evidence, not just factual allegations.

Moreover, referring to the split between panellists on the question of the weight to be given to allegations made by a complainant in undefended proceedings, and to the WIPO Consensus View, the panellist expressed what he took to be the emerging consensus in the following terms:

> Although this issue is not entirely settled, prevailing practice under the Policy in a default case is for the Panel not to assume Complainant's factual allegations are true simply because there has been no response, but, as in any other case, to require the complainant to prove each of the required elements under the Policy with competent evidence.

Since the publication of the *WIPO Overview*, the approach adopted in decisions such as *Brooke Bollea* and *TPI Holdings*, has been accepted by the majority of panels that have considered the issue, including the panels in *Pancil, LLC v Manila Industries*,[103] *Heidelberger Druckmaschinen AG v Wayne Graham (Trading) Limited*[104] and *Weather Shield, Mfg, Inc v Will Belak*.[105]

Whichever approach is adopted to the evaluation of allegations made by the complainant in default proceedings, the differences in the burden of proof between the second element of the UDRP and the other two elements means that uncontested allegations that a respondent has no rights or legitimate interests in the disputed domain name are more likely to be accepted than allegations about the first or third elements of the UDRP. This follows from the fact that a complainant only needs to establish a prima facie case in relation to the second element before the burden shifts to the respondent. If a respondent fails to reply to a complaint it is obviously difficult for a panel to discount a complainant's claims that the

[101] WIPO Case No D2004-0383 (29 June 2004).
[102] WIPO Case No D2006-0235 (7 Apr 2006).
[103] WIPO Case No D2006-0970 (8 Oct 2006).
[104] WIPO Case No D2006-1131 (31 Oct 2006).
[105] WIPO Case No D2007-0053 (27 Feb 2007).

respondent has no rights or interests. As the panellist stated in *The Vanguard Group Inc v Alvaro Collazo*:[106]

> Respondent has failed to submit a Response in this proceeding. Therefore, Complainant's submission has gone unopposed and the arguments undisputed. In the absence of a Response, the Panel accepts as true all reasonable allegations contained in the Complaint unless clearly contradicted by the evidence. Further, because Respondent has failed to submit a Response, Respondent has failed to propose any set of circumstances that could substantiate its rights or legitimate interests in the disputed domain name.

The complainant's evidentiary burden in relation to the second element of the UDRP is discussed at [6.4]–[6.5].

Reverse Domain Name Hijacking

[4.17] Reverse Domain Name Hijacking

As explained at [3.5], on 26 August 1999 the ICANN board, in giving guidance about the matters to be taken into account in the preparation of UDRP implementation documents, directed that the policy should seek to define and minimise reverse domain name hijacking. As further explained, the ICANN Second Staff Report of 24 October 1999 explained that the following measures had been included in UDRP and UDRP Rules to deal with the potential problem of complainants abusing the process in an attempt to obtain a disputed domain name:

- Paragraph 15(e) was introduced to the UDRP Rules, providing for panels to reach a finding that a complaint has been brought in bad faith.
- Enhanced requirements were introduced in paragraph 2(a) of the UDRP Rules to ensure, as much as practicable, that respondents receive actual notice of the complaint.
- Clarification that the complainant bears the burden of proof under paragraph 4(a) of the UDRP and the lengthening of the time for a domain name holder to seek court review of an adverse decision under paragraph 4(k) of the UDRP.

While some comments on the draft implementation documents suggested that more punitive measures should be introduced to discourage reverse domain name hijacking, the Second Staff Report indicated that this was considered to be outside the scope of the envisaged procedure.

'Reverse domain name hijacking' is defined by paragraph 1 of the UDRP Rules to mean:

> using the Policy in bad faith to attempt to deprive a registered holder of a domain name.

Paragraph 15(e) of the UDRP Rules, which deals with reverse domain name hijacking provides, in relevant part:

> If after considering the submissions the Panel finds that the complaint was brought in bad faith, for example in an attempt at Reverse Domain Name Hijacking or was brought primarily to harass the domain-name holder, the Panel shall declare in its decision that the complaint was brought in bad faith and constitutes an abuse of the administrative proceeding.

106 NAF Case No FA349074 (1 Dec 2004).

A finding of reverse domain name hijacking does not automatically follow from a failure of the respondent to make out each element of the UDRP, but must be affirmatively established by the respondent. As the majority pointed out in *Jazeera Space Channel TV Station v AJ Publishing aka Aljazeera Publishing:*[107]

> The onus of proving bad faith is on the Respondent, and mere lack of success of the Complaint is not of itself sufficient to constitute reverse domain name hijacking.

At the same time, as is clear from the wording of paragraph 15(e), a finding of reverse domain name hijacking can be made even if the respondent does not request such a finding. As the panellist in *Goway Travel Limited v Tourism Australia*[108] put it:

> Under the Rules, the Panel must also consider whether Complainant is guilty of Reverse Domain Name Hijacking ('RDNH'). That is true notwithstanding that Respondent has not requested such a finding, and even in the absence of a Response. The Rules specifically put the burden on the Panel to determine whether a complainant has tried to use 'the Policy in bad faith to attempt to deprive a registered domain name holder of a domain name.'

This means that, as in the *Goway* decision, the fact that the respondent fails to file a response does not prevent a panel from finding reverse domain name hijacking. In any case, it needs to be borne in mind that, given the limited nature of the remedies under the UDRP, there are no significant consequences that follow from a finding of reverse domain name hijacking, in and of itself.

There are two possible interpretations of what is meant by the 'complaint' in paragraph 15(e). On one interpretation, paragraph 15(e) applies only where the complaint as a whole was brought in bad faith, meaning that there could be no finding of reverse domain name hijacking in a complaint involving multiple domain names where, for example, the complaint against only one of the domain names was brought in bad faith. On the other view, however, paragraph 15(e) permits panels to make a finding of reverse domain name hijacking in relation to each separate domain name that is included in a complaint.

In *Fondation Le Corbusier v Monsieur Bernard Weber*[109] the panel applied a common sense approach by adopting the second interpretation of paragraph 15(e). In that dispute, the respondent registered a number of domain names that included the 'LE CORBUSIER' mark, including some domain names that included variations of the term 'LE CORBUSIER CENTER'. While the complainants successfully established rights in the 'LE CORBUSIER' mark, the respondents had operated a museum in Zurich known as 'CENTRE LE CORBUSIER' since 1967. The panel held that, while the complainant had successfully established each of the elements of the UDRP in relation to the domain names that incorporated the 'LE CORBUSIER' mark, it failed to establish that the respondent had no rights or legitimate interests in those domain names that were variations of the term 'CENTRE LE CORBUSIER'. On the basis that the complainant had known that the respondents had rights or interests in the term 'CENTRE LE CORBUSIER' since the museum of that name was established in 1967, the panel held that the complaint had been brought in bad faith in relation to those domain names that included that term.

[107] WIPO Case No D2005-0309 (19 July 2005).

[108] WIPO Case No D2006-0344 (6 June 2006). See also *Rodale, Inc v Cambridge,* WIPO Case No DBIZ2002-0153 (28 June 2003); *Howard Jarvis Taxpayers Association v Paul McCauley,* WIPO Case No D2004-0014 (22 Apr 2004).

[109] WIPO Case No D2003-0251 (4 July 2003).

In reaching its finding of reverse domain name hijacking in *Le Corbusier*, the panel held that, to make out a case that a complaint has been brought in bad faith under paragraph 15(e), the respondent must establish one of the following circumstances:

- the complainant knew of the respondent's unassailable right or legitimate interest in the disputed domain name or the clear lack of bad faith registration and use, and nevertheless brought the complaint in bad faith; or
- that the complaint was brought in knowing disregard of the likelihood that the respondent possessed legitimate interests; or
- that the complainant knew it had no rights in the trade mark or service mark upon which it relied and nevertheless brought the complaint in bad faith.

The majority in the *Jazeera Space Channel* decision pointed out that, in determining whether a complaint has been brought in bad faith, 'whether a complainant should have appreciated at the outset that its complaint could not succeed, will often be an important consideration'.

The following are examples of decisions that have found that a complaint was brought in bad faith under paragraph 15(e) in each of the categories identified by the panel in *Le Corbusier*. At the same time, it should be recognised that these circumstances are illustrative only, and the categories of reverse domain name hijacking are neither closed nor mutually exclusive.

The complainant knew of respondent's rights or legitimate interests in the disputed domain name or clear lack of bad faith registration and use.

- *Deutsche Welle v DiamondWare Limited.*[110] The complainant brought a complaint in relation to the <dw.com> domain name, which corresponded to the initials of the respondent's trade name. The majority of the panel held that the complaint had been brought in bad faith, as there was no evidence on which to base a claim that the domain name was registered and used in bad faith. The dissenting panellist, however, rejected the finding of reverse domain name hijacking on the basis that the respondent had failed to inform the complainant of its bona fide use of the domain name in communications prior to the complaint being filed.
- *Goldline International, Inc v Gold Line.*[111] The complainant brought a complaint in relation to the disputed domain name <goldline.com>, while before the complaint was filed the respondent informed the complainant of its business plans for the 'GOLD LINE' name. The panel found that the complaint had been brought in bad faith as the complainant must have known that the respondent's registration and use of the mark could not, on any fair interpretation, amount to bad faith.
- *Futureworld Consultancy (Pty) Limited v Online advice.*[112] The complainant brought a complaint in relation to the <futureworld.com> domain name, which the respondent registered for use in connection with her vision for the future. Concluding that the complainant knew or should have known, at the time it filed the complaint, that it could not prove that the domain name was registered in bad faith, the panel declined to find reverse domain name hijacking.

[110] WIPO Case No D2000-1202 (2 Jan 2001).
[111] WIPO Case No D2000-1151 (4 Jan 2001).
[112] WIPO Case No D2003-0297 (18 July 2003).

- *Kur- und Verkehrsverein St Moritz v Domain Finance Ltd.*[113] The complainant brought a complaint in relation to the <stmoritz.com> domain name, which the respondent used for a Web site that contained information about 'ST MORITZ'. The majority of the panel held that the complaint was brought in bad faith, as there was no reasonable basis for the complainant's claim that the domain name was being used in bad faith.
- *Jazeera Space Channel TV Station v AJ Publishing aka Aljazeera Publishing.*[114] The complainant brought a complaint in relation to the <aljazeera.com> domain name, while the respondent had since 1992 published a magazine under the name 'AL JAZEEERA', which is an Arabic term that refers to the Arabian peninsular. The panel held that, as the domain name was registered before the complainant had acquired rights in the 'AL JAZEERA' mark, the complainant had not established that the mark was registered in bad faith. The majority of the panel held that, as the complainant should have known that its claim of bad faith registration could not succeed, the complaint had been brought in bad faith. The dissenting panellist, however, held that the complaint had not been brought in bad faith as it had been brought by mistake, and was not motivated by malice.
- *ZZounds Music, LLC v Zounds.*[115] The complainant brought a complaint in relation to the <zounds.com> domain name, which the respondent planned to use in connection with his business activities. The panel found that this was a case of reverse domain name hijacking on the basis that the complainant should have known that the domain name was not confusingly similar to the respondent's mark, that the respondent had rights or interests in the disputed domain name or that the complainant would be unable to prove bad faith.

The complaint was brought in knowing disregard of the likelihood that the respondent possessed legitimate interests.

- *Smart Design LLC v Carolyn Hughes.*[116] The complainant brought a complaint in relation to the <smartdesign.com> domain name apparently knowing that the respondent had a bona fide intention to use the domain name in connection with an e-commerce business. The panellist held that, as the complainant knew of the respondent's plans and as it should have known that its claims could not conceivably succeed, the complaint had been brought in bad faith.
- *Goway Travel Limited v Tourism Australia.*[117] The complainant brought a complaint in relation to the <downunder.travel> domain name and the panellist held that the attempt to obtain a transfer of the generic domain name was indicative of reverse domain name hijacking.
- *The Exit Light Co, Inc v N/A c/o Gary Glass.*[118] The complainant brought a complaint in relation to the <theexitstore.com> domain name, which the respondent used for selling emergency lighting and exit signs. Given the generic nature of the domain name, the panellist held that the complainant had acted recklessly, or with knowing disregard, of the likelihood that the respondent had legitimate interests in the domain name and had brought the complaint in bad faith.

[113] WIPO Case No D2004-0158 (14 June 2004).
[114] WIPO Case No D2005-0309 (19 July 2005).
[115] NAF Case No FA817093 (24 Nov 2006).
[116] WIPO Case No D2000-0993 (18 Oct 2000).
[117] WIPO Case No D2006-0344 (6 June 2006).
[118] NAF Case No FA849010 (1 Feb 2007).

The complainant knew it had no rights in the trademark or service mark upon which it relied and nevertheless brought the complaint in bad faith.

- *Koninklijke KPN NV v Telepathy Inc.*[119] The complainant brought a complaint in relation to the <moneyplanet.com> domain name, even though the respondent's registration of the domain name preceded the complainant's application to register the 'MONEY PLANET' mark. The majority found that, as the complaint had no possibility of success, this was an example of reverse domain name hijacking.
- *Aspen Grove, Inc v Aspen Grove.*[120] The complainant brought a complaint in relation to the <aspengrove.com> domain name, even though the only evidence of rights in the 'ASPEN GROVE' mark was an application for trade mark registration. The panel found that this was an instance of reverse domain name hijacking on the basis that the complainant had no rights in the mark, the respondent had demonstrated rights in the domain name and the fact that the respondent had registered the domain name two years before the complainant came into existence.
- *Her Majesty The Queen, in right of her Government in New Zealand v Virtual Countries, Inc.*[121] The complainant brought a complaint in relation to the <newzealand.com> domain name but, given evidence of the complainant's response to a WIPO survey, the panel found that, when the complaint was launched, the complainant was well aware that it had no trade mark rights in the geographical term 'NEW ZEALAND'. On this basis the panel upheld the respondent's allegation of reverse domain name hijacking.
- *Dan Zuckerman v Vincent Peeris.*[122] The complainant brought a complaint in relation to the <shoes.biz> domain name, knowing that the word 'SHOES' is generic and that it had no trade mark rights in the mark. On this basis, the panel found that the complaint had been brought in bad faith under paragraph 15(e).
- *Mess Enterprises v Scott Enterprises, Ltd.*[123] The complainant brought a complaint in relation to the <mess.com> domain name on the basis of an application to register the 'MESS.COM' mark that was made two and a half years after the respondent registered the disputed domain name. As the complainant knew that it had no relevant rights in the mark, the panel found that 'this was one of the most egregious examples of reverse domain name hijacking that any of the Panelists has thus far ever seen'.

Panellists have pointed out that there is often a fine line between the aggressive enforcement of trade mark rights and reverse domain name hijacking. In *Lockheed Martin Corporation v The Skunkworx Custom Cycle*,[124] for example, the panellist made the following observations:

> In this Panel's mind, a boundary, though quite elusive, amorphous and factually-dependent, exists under UDRP jurisprudence between, on the one hand, a complainant's conduct which constitutes, in any given situation, permissible aggressive enforcement of its mark and, on the other, that which constitutes harassment and abuse of the administrative process, i.e., reverse domain name hijacking. The Policy does not define the metes and bounds of that boundary. As such, it remains for each

[119] WIPO Case No D2001-0217 (7 May 2001).
[120] WIPO Case No D2001-0798 (5 Oct 2001).
[121] WIPO Case No D2002-0754 (27 Nov 2002).
[122] WIPO Case No DBIZ2002-0245 (12 Aug 2002).
[123] WIPO Case No D2004-0964 (25 Jan 2005).
[124] WIPO Case No D2004-0824 (18 Jan 2005).

> administrative panel to assess the conduct of a corresponding complainant and determine whether, under the specific facts then presented in the record, that conduct has crossed the boundary or not.

The following decisions illustrate the sometimes difficult distinction between an aggressive attempt to enforce trade mark rights, which is perfectly legitimate, and a bad faith abuse of the process.

- *Sustainable Forestry Management Limited v SFM.com.*[125] The complainant filed a complaint in relation to the <sfm.com> domain name, but only commenced using its 'SFM' mark six months after the respondent registered the domain name. The majority held that, in these circumstances, the complaint was completely devoid of merit and had been brought in bad faith. The dissenting panellist, however, held that, as there was no reason to conclude that the complainant appreciated the difficulties that prior registration of the domain name created, this was not a case of reverse domain name hijacking.
- *Lockheed Martin Corporation v The Skunkworx Custom Cycle.*[126] The complainant filed a complaint in relation to the <skunkworxcc.com> domain name, which the respondent used in connection with its motorcycle customisation business. Although the panellist held that the respondent had rights or legitimate interests in the domain name, he held that, while the complainant had aggressively attempted to enforce its 'SKUNK WORKS' mark, its conduct was not sufficiently egregious to amount to reverse domain name hijacking.
- *Howard Jarvis Taxpayers Association v Paul McCauley.*[127] The complainant filed a complaint in relation to the <hjta.com> domain name, which was used for a criticism site. Although the panellist found that the complaint had been brought in an effort to suppress the criticism, he held that, given the split among UDRP panellists about whether use of a domain name constitutes a legitimate interest, a finding of reverse domain name hijacking was not warranted.
- *Rudy Rojas v Gary Davis.*[128] The complainant filed a complaint in relation to the <nativestyles.net> and <nativestyles.org> domain names, claiming that he had superior rights in the 'NATIVE STYLES' mark, as he had commenced using it before the respondent. The complainant's evidence of common law rights in the mark was, nevertheless, scanty. Moreover, as the respondent had used the term 'NATIVE STYLES' in connection with its clothing since 2002, the panel held that the respondent had legitimate interests in the domain name. In considering whether this was a case of reverse domain name hijacking, the panel noted that the complainant had advanced a questionable legal theory and presented a misleading picture of the case. While this might have led to a conclusion that the complaint was brought in bad faith, as the respondent also presented misleading evidence, the panel declined to find reverse domain name hijacking in this case.
- *Deutsche Post AG v NJDomains.*[129] The complainant filed a complaint in relation to the <post.com> domain name, which the respondent used for a personalised e-mail service. The majority held that this was not a case of reverse domain name hijacking, as the complainant might have legitimately misunderstood the scope of its registered rights in marks that included the term 'POST'. The dissenting panellist on this point, however, believed

[125] WIPO Case No D2002-0535 (13 Sept 2002).
[126] WIPO Case No D2004-0824 (18 Jan 2005).
[127] WIPO Case No D2004-0014 (22 Apr 2004).
[128] WIPO Case No D2004-1081 (18 Apr 2005).
[129] WIPO Case No D2006-0001 (1 Mar 2006).

that the complaint had been brought in bad faith as the complainant could not reasonably have believed, at that time that it filed the complaint, that it could prove that the domain name was registered in bad faith.

- *CPFilms, Inc v Solar Lunar Performance Film.*[130] The complainant filed a complaint in relation to the <lunarsolarfilm.com> domain name, which the respondent used for its window tinting business. The panellist held that the disputed domain name was not confusingly similar to the complainant's 'LLUMAR' mark. In finding, nevertheless, that the complaint had not been brought in bad faith, the panellist stated:

> For a panel to find that a complainant has engaged in reverse domain name hijacking (RDNH), much more is required on the part of a complainant than just tenacious enforcement of its marks - even if its claim under the Policy is ultimately proven to be erroneous.

Equitable Defences

[4.18] Equitable Doctrines and Defences

UDRP panels have a considerable amount of flexibility in the rules and principles to apply in settling domain name disputes under paragraph 15(a) of the UDRP Rules, which specifies:

> A Panel shall decide a complaint on the basis of the statements and documents submitted and in accordance with the Policy, these Rules and any rules and principles of law that it deems applicable.

Despite the broad discretion conferred by paragraph 15(a), the majority of panels have rejected the applicability of equitable defences, such as laches, estoppel, forfeiture, waiver or unjust enrichment. Similarly, the majority of panels that have considered the issue have held that other equitable doctrines, such as 'unclean hands', do not apply to the policy.

A number of decisions have held that the equitable defence of laches has no application under the UDRP. In *The Hebrew University of Jerusalem v Alberta Hot Rods,*[131] for example, the complainant, which owned the registered French trade mark 'ALBERT EINSTEIN', brought a complaint against the registrant of the <alberteinstein.com> domain name. In reply, the respondent successfully argued that it had rights or legitimate interests in the domain name, but also that the complaint was barred by laches, as the complaint was not brought until more than five years after the disputed domain name was registered. In rejecting the availability of the defence under the UDRP, the panel stated:

> The Policy is part of the domain name registration agreement. The Administrative Proceeding is brought pursuant to that agreement, the issue for determination being whether the grounds set out in the Policy for transfer or cancellation have been established. There is no limitation period in the Policy. The remedy available in an Administrative Proceeding under the Policy is not equitable. Accordingly, the defence of laches has no application.

The application of the defence of laches was similarly rejected by the panellist in *The EW Scripps Company v Sinologic Industries.*[132] In that dispute, the respondent claimed that it had

[130] WIPO Case No D2002-0616 (7 Oct 2002).
[131] WIPO Case No D2002-0616 (7 Oct 2002).
[132] NAF Case No FA220007 (21 Feb 2004).

rights or legitimate interests in the disputed domain name <commercialappeal.com>, as it had used the domain name for a web site for almost four years without the complainant taking action. In relation to the argument that the complainant was debarred as a result of the delay, the panellist concluded as follows:

> If the requirements of a valid complaint under the Policy are established, the Policy does not provide any defence of laches. This accords with the basic objective of the Policy of providing an expeditious and relatively inexpensive procedure for the determination of disputes relating to egregious misuse of domain names. The availability of defences such as laches could result in significant delay and expense. Although a Panel is required by paragraph 15(a) of the Rules to decide the dispute in accordance with any rules and principles of law that it deems applicable, this would not justify the adoption of a substantive defence not provided in the Policy and running counter to its objectives. If a respondent considers that the bringing of the complaint was inequitable because of delay, it can submit the matter to a court of competent jurisdiction.

In *Anna Nicole Smith v DNS Research, Inc,*[133] on the other hand, the respondent successfully argued that it had rights or legitimate interests in the <annanicolesmith.com> domain name on the basis of use of the domain name for a fan site without objection from the complainant for a period of two years. In deciding that the respondent had successfully established rights in the domain name, the panellist observed:

> failing to act within a reasonable period of time from the date of registration and use, Complainant has acquiesced to Respondent's registration and use of the subject domain and has waived any rights, even if they were tenuous, to request that the domain name be transferred.

Despite this statement, the application of a doctrine of waiver under the policy has been rejected by subsequent panel decisions. In *Tom Cruise v Network Operations Center,*[134] for example, the respondent referred to the *Anna Nicole Smith* decision as support for the proposition that the complainant's 10-year delay in bringing a complaint amounted to a waiver of rights. In rejecting the application of a doctrine of waiver under the policy, the panel pointed out that the decisions cited by the panellist in *Anna Nicole Smith* did not support the proposition that failing to act within a reasonable time constitutes a waiver. Consequently, the panel agreed with the majority of panels in rejecting outright the possibility of an equitable defence, concluding simply:

> This Panel does not accept that there is meaningful precedent under the Policy for refusing to enforce trademark rights on the basis of a delay in bringing a claim following use of a disputed domain name.

While laches is not a substantive defence under the policy, a delay in bringing a complaint may be taken into account in determining whether the respondent has acted in bad faith. As the panellist put it in *Mattel, Inc v Eastwind Groups, Ltd:*[135]

> The defence of laches is not available under the Policy . . . The fact that Complainant did not bring this Complaint until the disputed domain name had been registered for over 5 years requires the Panel to consider very carefully whether Complainant has discharged its burden of proving its case on the balance of probabilities. However, the elapsed time is not in itself a ground for dismissal.

[133] WIPO Case No D2006-0560 (5 July 2006).
[134] NAF Case No FA849009 (12 Jan 2007).
[135] NAF Case No FA209572 (29 Sept 2003).

It may be more difficult for a complainant to establish bad faith where the respondent has relied upon a lack of action by the complainant to build up a business based on the disputed domain name. In *Square Peg Interactive Inc v Naim Interactive Inc*,[136] for example, the respondent argued that it used the disputed domain name <gototrafficschool.com> to operate a bona fide business in traffic school services, and claimed that the complainant was barred from bringing the complaint by laches. The panellist held that, while laches was not a defence that was recognised by the policy, a complainant's delay in seeking relief is relevant to determining whether the respondent has built up legitimate rights in the domain name and whether it is using the domain name in bad faith. In that dispute, the panellist considered that the respondent had built up a bona fide business by the use of the domain name, and that there was insufficient evidence to establish that the respondent had no rights or interests in the domain name or that it had acted in bad faith.

The *Square Peg* decision has been referred to by a number of subsequent panel decisions that have found that a complainant's delay in enforcing its rights is one factor to be taken into account in determining whether a respondent has built up rights in the disputed domain name or acted in bad faith. In *First Look Studios, Inc v New.net Incorporated*,[137] for example, the respondent claimed that the complainant's nine-year delay in bringing a complaint had permitted it to build up its rights in the domain name through use and foreign registrations. Citing the *Square Peg* decision, the panel found that the nine-year delay in bringing the complaint had allowed the respondent to acquire rights or interests in the <firstlook.com> domain name, which were reinforced by evidence that the respondent had acquired foreign registrations for the 'FIRST LOOK' mark.

[136] WIPO Case No D2003-0447 (1 July 2003).
[137] NAF Case No FA874264 (21 Feb 2007).

5

Identical or Confusingly Similar Domain Names

Overview

[5.1] Overview

The first element of abusive, bad faith registration, set out in paragraph 4(a)(i) of the UDRP, requires the complainant to prove that:

> the domain name is identical or confusingly similar to a trademark or service mark in which the complainant has rights.

This element has two components. First, the complainant must establish that he or she has rights in a trade mark or service mark. Secondly, the complainant must establish that the domain name is identical or confusingly similar to the trade mark or service mark in which he or she has rights. In the words of one panellist:

> The Policy requires the Panel to consider, first, whether the Complainant has rights in a relevant trade mark; and secondly, whether the Domain Name is the same as, or is confusingly similar to, that mark.[1]

Trade Marks

[5.2] Trade Mark

The wording of the first element of abusive, bad faith registration raises the question of what is meant by a 'trade mark' for the purpose of the UDRP. The answer to this question effectively establishes the scope of the UDRP.

As explained at [3.4], the WIPO Final Report recommended limiting the dispute-resolution procedure to abusive registration of trade marks or service marks and suggested that registrations which violate trade names, geographical indications or personality rights would not fall within the procedure. As further explained at [3.5], this recommendation was adopted by ICANN and embodied in the first element of abusive, bad faith registration.

The UDRP does not define the terms 'trade mark' or 'service mark'. Recourse to principles established under national trade mark laws is therefore necessary to determine whether

[1] See *Julie Brown v Julie Brown Club*, WIPO Case No D2000-1628 (13 Feb 2001); *Celine Dion and Sony Music Entertainment (Canada) Inc v Jeff Burgar*, WIPO Case No D2000-1838 (13 Feb 2001).

indicia are, or are not, 'trade marks' or 'service marks'. As explained at [3.4], WIPO's Final Report proposed that the decision-making panel be responsible for determining the law to apply to particular disputes and, in doing so, should be able to refer to the law that it determines is applicable to the circumstances of the case. As further explained at [4.10], this recommendation was embodied in paragraph 15(a) of the UDRP Rules and, in practice, UDRP panels often refer to national legal principles where both parties reside in the same legal jurisdiction. The reference to national legal principles is especially important in determining the meaning of a 'trade mark' in paragraph 4(a)(i). As the panellist said in *Evolution USA, Inc v Alexei Doicev*:[2]

> What is and is not a trademark for the purposes of the Policy is something that can only be determined by reference to local law

The need to refer to national laws to determine whether indicia are trade marks reflects the established position under international trade mark law. The principal international treaty dealing with trade marks, the Paris Convention,[3] has never attempted to define the term 'trade mark'. The failure of the Paris Convention to define what is meant by a 'trade mark' is consistent with one of the most fundamental principles of international trade mark protection, the principle of territoriality. Although the principle has been said to be ambiguous, it has been defined by Ladas to mean that:

> the protection of a trademark in a certain country depends exclusively on the law of that country, and that the effects of a trademark ownership by use or registration in a country do not reach beyond the borders of that country.[4]

The principle of territoriality is reflected in the principle of independence of protection, established under Article 6(3), which states:

> A mark duly registered in a country of the Union shall be regarded as independent of marks registered in the other countries of the Union, including the country of origin.[5]

Despite the principle of territoriality, Ladas recounts that there has been some discussion of the desirability of including a definition of a trade mark in the Convention, including the adoption of a resolution at the Berlin Congress of 1963, which provided that:

> A mark is a sign capable of distinguishing the products or services of a person or of a group of persons. The distinctive character of a mark, in respect of the goods or services designated by it, arises from the nature of the sign or of the use which has been made of it.[6]

The TRIPS Agreement[7] establishes a mechanism for enforcing the Paris Convention, by requiring that member states comply with Articles 1–12 and 19 of that Convention.[8] Article 15.1 of the TRIPS Agreement includes the following statement on what is capable of amounting to a trade mark:

[2] WIPO Case No D2006-0086 (24 Mar 2006).

[3] Paris Convention for the Protection of Industrial Property (Commercial No 28 (1884), C 4043), as revised at Brussels on December 14, 1900, at Washington on June 2, 1911, at the Hague on November 6, 1925, at London on June 2, 1934, at Lisbon on October 31, 1958, at Stockholm on July 14, 1967 and as amended on September 28, 1979.

[4] SP Ladas, *Patents, Trademarks, and Related Rights: National and International Protection, Vol. II* (Cambridge, Mass., Harvard University Press, 1975) §732.

[5] See JT McCarthy, *Trademarks and Unfair Competition*, 4th edn (Eagan, Minn, Thompson/West 2006) § 29:1.

[6] See Ladas, above n 4, §569.

[7] Agreement on Trade-Related Aspects of Intellectual Property Rights, Annex 1C of the Marrakesh Agreement Establishing the World Trade Organization (concluded on 15 Apr 1994; entered into force on 1 Jan 1995).

[8] Art 2(1).

> Any sign, or any combination of signs, capable of distinguishing the goods or services of one undertaking from those of other undertakings, shall be capable of constituting a trademark. Such signs, in particular words including personal names, letters, numerals, figurative elements and combinations of colours as well as any combination of such signs, shall be eligible for registration as trademarks. Where signs are not inherently capable of distinguishing the relevant goods or services, Members may make registrability depend on distinctiveness acquired through use. Members may require, as a condition of registration, that signs be visually perceptible.

Although this is helpful, it remains the case that the meaning of the term 'trade mark' in the first element of the UDRP is to be determined by reference to national trade mark laws. The meaning of the term is often far from straightforward. An understanding of the early history of trade mark law assists to explain the meaning.

[5.3] Early History of Trade Mark Law

In the last half of the nineteenth century inadequacies with the common law protection of marks resulted in movements to introduce a system of registration of trade marks in both the United States and the United Kingdom. The first US federal trade mark law was introduced in 1870.[9] In 1879, however, the 1870 Act was declared unconstitutional on the basis that the patent and copyright clause of the US Congress did not confer power on Congress to legislate in relation to trade marks.[10] US federal trade mark law is therefore enacted pursuant to the Commerce power, which means that it cannot regulate intrastate use of trade marks. It was not until 1946 that a comprehensive federal trade mark Act, known as the Lanham Act,[11] was introduced.

In the United Kingdom, the Trade Marks Registration Act 1875 first introduced a system of registration of trade marks and established the first United Kingdom Register of Trade Marks. The legal effect of the 1875 Act was considered by the English Court of Appeal in *British Telecommunications Plc v One in a Million Ltd*,[12] the principal English case dealing with whether registration of a domain name can amount to trade mark infringement. The court maintained that, prior to the 1875 Act, there were two common law causes of action, one for infringement of a trade mark and the other for passing off.[13] The difference between the two actions was that the action for trade mark infringement depended upon use of the mark, whereas the action for passing off depended upon reputation. As explained at [5.7] below, however, this is an over-systematisation of nineteenth century English law.

The 1875 United Kingdom Act did not establish a cause of action for infringement of a registered trade mark, evidently on the basis that an action could be brought under the general law.[14] The Act did, however, make registration a prerequisite to the institution of proceedings for infringement of trade marks as defined under that Act. This provision was apparently introduced with the intention of removing the right to bring a common law

[9] An Act to Revise, Consolidate and Amend the Statutes Relating to Patents and Copyrights (US), Act of 8 July 1870, 16 Stat at L 198, ss 77–84.

[10] *Trade-mark Cases* (1879) 100 U.S. 82, 25 L Ed 550. The patent and copyright clause of the US Constitution grants Congress the power 'to promote the progress of science and the useful arts, by securing for limited times to authors and inventors the exclusive right to their respective writings and discoveries': US Constitution, Art 1 s 8 cl 8.

[11] Federal Trademark Act of 1946 ('Lanham Act'), 15 USC §§ 1051–1141n (1946) (US).

[12] [1999] FSR 1.

[13] [1999] FSR 1, 8.

[14] C Wadlow, *The Law of Passing Off*, 3rd edn (London, Sweet & Maxwell, 2004) para [1.38].

action for trade mark infringement. Under an amendment introduced in 1876,[15] the prohibition against bringing proceedings for infringement of unregistered trade marks was relaxed for trade marks in use before 13 August 1876 that had been refused registration. The English courts subsequently interpreted the legislative restriction on bringing a common law action for trade mark infringement narrowly so as to preserve the common law right to bring an action in passing off.[16] Later trade mark Acts specifically preserved the right to bring an action for passing off.[17]

The most important consequence of the complex history of English trade mark law is that, while it is clear that common law rights may arise independently of registration, it is important to be precise about the legal requirements for common law trade mark rights. For example, common law trade mark rights are not co-extensive with the English action for passing off but are, nevertheless, protected by means of the action for passing off. The requirements for common law rights under English law are dealt with at [5.7], whereas common law rights under US law are dealt with at [5.8]. The requirements for legal rights to arise in unregistered marks in civil law systems are dealt with at [5.9].

[5.4] Legislative Definitions of 'Trade Mark'

As explained at [5.2], the meaning of the term 'trade mark' in the first element of the UDRP is to be derived from national trade mark laws. The process of referring to national laws to determine whether or not a complainant has trade mark rights was explained by one UDRP panellist in the following terms:

> Unregistered trademark rights do not exist in some nebulous way across the breadth of the countries in which a complainant proves it has a reputation. These rights derive from national laws and do not exist divorced from such laws. They therefore depend upon the extent to which the laws of a particular country recognise or do not recognise that the activities of the complainant provide unregistered rights in that country. A personal name is capable of qualifying as an unregistered trademark recognised by the Policy, but in determining whether it does in a particular case a panel will look to the jurisdictions in which those rights are claimed. The panel will then determine whether the complainant can show that the name provides an unregistered right in at least one of these jurisdictions.[18]

In further considering what is meant by 'trade mark' it is helpful briefly to survey the definitions in some regional and national legislation.

Article 2 of the 1988 European Trade Mark Directive provides:

> A trade mark may consist of any sign capable of being represented graphically, particularly words, including personal names, designs, letters, numerals, the shape of goods or their packaging, provided that such signs are capable of distinguishing the goods or services of one undertaking from those of other undertakings.[19]

[15] Trade Marks Registration (Amendment) Act 1876 (UK).

[16] *Great Tower Street Co v Langford & Co* (1888) 5 RPC 66; *Faulder & Co Ltd v O and G Rushton Ltd* (1903) 20 RPC 477.

[17] Trade Marks Act 1905 s 45; Trade Marks Act 1938 s 2; Trade Marks Act 1994 s 2(2). See also Trade Marks Act 1905 (Cth) s 53B; Trade Marks Act 1995 (Cth) s 230.

[18] *Antonio de Felipe v Registerfly.com*, WIPO Case No D2005-0969 (19 Dec 2005).

[19] First Council Dir 89/104 of 21 Dec 1988 to approximate the laws of the Member States relating to trade marks (Trade Mark Directive) [1989] OJ L40/1.

Although the United Kingdom implementation of the Trade Mark Directive adopts Article 2 as a definition, it has been argued that Article 2 does not define a trade mark, but merely outlines what a trade mark may consist of.[20] Nevertheless, the current United Kingdom trade mark law, the Trade Marks Act 1994, defines a 'trade mark' to mean:

> any sign capable of being represented graphically which is capable of distinguishing goods or services of one undertaking from those of other undertakings.
>
> A trade mark may, in particular, consist of words (including personal names), designs, letters, numerals or the shape of goods or their packaging.[21]

The current federal US trade mark law is the 1946 Lanham Act.[22] The Lanham Act defines a trade mark as including:

> any word, name, symbol, or device or any combination thereof adopted and used by a manufacturer or merchant to identify his goods and distinguish them from those manufactured or sold by others.[23]

The legislative definition adopted in the Lanham Act is similar to a definition that had much earlier been proposed by the US Supreme Court, which in 1878 had stated:

> a trade-mark may consist of a name, symbol, figure, letter, form or device, if adopted and used by a manufacturer or merchant in order to distinguish the goods he manufactures or sells, to distinguish the same from those manufactured or sold by another . . .[24]

The current Australian trade mark law, the Trade Marks Act 1995, defines a 'trade mark' as:

> a sign used, or intended to be used, to distinguish goods or services dealt with or provided in the course of trade by a person from goods or services so dealt with or provided by any other person.[25]

From this admittedly incomplete survey, together with statements such as Article 15.1 of the TRIPS Agreement, it is suggested that a trade mark consists of the following uncontroversial elements:

- A sign (or other indicia);
- That is capable of distinguishing the products of a trader from those of another;
- Because it is either inherently distinctive, or because it has become distinctive through use.

These elements reflect the functions of trade marks, which Ladas identifies as:

> protection of the public against confusion and deception by the identification of source of origin of particular products as distinguished from similar products of another source, and the protection of the trademark owner's trade and business and his advantageous relations with the public symbolized by his trademark.[26]

[20] D Kitchin *et al*, *Kerly's Law of Trade Marks and Trade Names*, 14th edn (London, Sweet & Maxwell, 2005), paras [2.012]–[2-017].

[21] Trade Marks Act 1994 (UK) s 1(1).

[22] Lanham Act, 15 USC §§ 1051–1141n (1946).

[23] Lanham Act § 45, 15 USC § 1127.

[24] *McLean v Fleming* (1878) 96 US 245, 24 L Ed 828.

[25] Trade Marks Act 1995 (Cth) s 17.

[26] See Ladas, above n 4, § 564.

[5.5] 'Trade Mark' in the *Union Label* Case

As the term 'trade mark' is conventionally defined by national legislation establishing systems of registration of trade marks, it has been rare for common law courts to consider the meaning of the term in the abstract. In one prominent case, however, the Australian High Court was required to address the issue. As explained above, in 1879 the US Supreme Court held that the patent and copyright clause of the US Constitution did not confer power on Congress to make laws in relation to trade marks. Unlike the US Constitution, however, under the Australian Constitution the federal Parliament has the power 'to make laws for the peace, order, and good government of the Commonwealth with respect to . . . copyrights, patents of inventions and designs, and trade marks'.[27] In the *Union Label Case*,[28] the High Court was required to decide whether a mark which could be registered to indicate that goods were the product of an individual worker or an association of workers was a trade mark for the purposes of the Constitution. In the course of deciding that the 'workers' mark' was not a trade mark, the High Court held that trade mark (in relation to goods):

> meant a mark which is the visible symbol of a particular kind of incorporeal or industrial property consisting in the right of a person engaged in trade to distinguish by a special mark goods in which he deals, or with which he has dealt, from the goods of other persons.[29]

Griffiths CJ went on to explain that the concept of a trade mark includes the following five 'distinct elements':

> (1) A right which is in the nature of property;
> (2) The owner of the right must be a person, natural or artificial, engaged in trade;
> (3) The right is appurtenant or incident to the dealing with goods in the course of trade;
> (4) The owner has such an independent dominion over the goods to which the mark is to be affixed as to entitle him to affix it to them; . . .
> (5) The mark distinguishes the goods as having been dealt with by some particular person or persons engaged in trade; . . .[30]

Although the social and economic functions of trade marks have changed since the early twentieth century, there is no reason to assume that the comments of Griffiths CJ are any less relevant today than they were when the decision was handed down.

[5.6] Service Mark

The expansion of the functions of trade marks in the twentieth century led to the introduction of the concept of a service mark. A service mark is a sign (or other indicia) that is capable of distinguishing the services of a trader from those of other traders. Ladas attempted to distinguish service marks from trade marks in the following terms:

> Service marks are quite different from trademarks. Goods are subject to many transactions and they may be more or less durable. They may change hands many times and may remain in the ultimate purchaser's possession for an indefinite time. The mark shown thereon testifies to their producer at

[27] Australian Constitution, s. 51 (xviii).
[28] *Attorney-General (NSW) v The Brewery Employees' Union of New South Wales* (1908) 6 CLR 469.
[29] (1908) 6 CLR 469, 512–3 per Griffiths CJ.
[30] (1908) 6 CLR 469, 513 per Griffiths CJ.

all times. Service marks identify services as between the grantor of the service and the customer, which service ceases to exist after the performance of the service. The function of the service mark is to bring the service of the business of the user of such mark to the attention of the public. Thus its function is essentially an advertising function.[31]

The importance of services in modern economies has, perhaps, made this distinction less relevant.

The United States was the first jurisdiction to protect service marks in the Trademark Act of 1946. The service mark concept was recognised at the international level by the 1963 Lisbon Revision of the Paris Convention, which adopted Article 6*sexies*. Article 6*sexies* provides:

The countries of the Union undertake to protect service marks. They shall not be required to provide for the registration of such marks.

'Unregistered' Trade Marks

[5.7] Common Law Trade Marks and Passing Off

As explained at [5.9], the UDRP protects unregistered trade marks to the same extent as it protects registered marks. It is therefore important to understand the basis on which national legal systems may protect unregistered marks. Unregistered marks are protected under English-oriented legal systems principally by the action for passing off.

Before the introduction of statutory systems of registration of trade marks, the common law developed to protect trade marks as an incident of a more general protection afforded to traders against competitors who falsely represented that their goods or business were those of the trader. The origins, and early juridical basis, of the common law protection of trade marks are uncertain.[32] Nevertheless, it appears that early cases were regarded as a variety of, or analogous to, the common law action for deceit, thereby requiring an intention to deceive.[33] The requirement of actual deceit might have impeded the development of the action. In 1838, however, in *Millington v Fox*,[34] Lord Cottenham LC decided that an injunction could be awarded against a defendant who had acted innocently. This decision paved the way for the development of the action for passing off in equity.

By the middle of the nineteenth century the law was gradually being systematised, with the courts beginning to use the terms 'pass off' or 'palm off'. In 1842, for example, Lord Langdale MR stated the English law in the following terms:

I think that the principle on which both the courts of Law and of Equity proceed, in granting relief and protection in cases of this sort, is very well understood. A man is not to sell his own goods under the pretence that they are the goods of another man . . .[35]

[31] Ladas, above n 4, §587.

[32] See WL Morison, 'Unfair Competition and Passing Off' (1956) 3 *Sydney Law Review* 50; Wadlow, above n 6, paras [1.24]–[1.39].

[33] See, eg, *Crawshay v Thompson* (1842) 4 Man & G 357; 134 ER 146; *Rodgers v Nowill* (1847) 6 Hare 325; 67 ER 1191; *Leather Cloth Co Ltd v American Leather Cloth Co Ltd* (1865) 11 HLC 523; 11 ER 1435 (HL).

[34] (1838) 3 My & Cr 338; 40 ER 956.

[35] *Perry v Truefitt* (1842) 6 Beav 66, 73; 49 ER 749, 752.

The precise basis for equitable intervention was, nevertheless, unclear. In a series of cases in the 1860s, Lord Westbury LC invoked equity's traditional concern with the protection of property to conclude that there was a property right in a trade mark that could be enforced in equity.[36] For example, in *Leather Cloth Co v American Leather Cloth Co Ltd*, Lord Westbury stated:

> the mistake of buyers in the market under which they in fact take the defendant's goods as the goods of the plaintiff, that is to say, imposition on the public, becomes the test of the property in the trade mark having been invaded and injured, and not the ground on which the court rests its jurisdiction.[37]

As explained at [5.3], the first statutory system of registration of trade marks in the United Kingdom was introduced in 1875. Wadlow suggests that, by this time, there were three 'interconnected bodies of law': the common law action for passing off, derived from the action for deceit, which required proof of actual fraud; an equitable action for passing off, evidently based on equitable fraud; and an equitable doctrine of property in trade marks, perhaps extending to property in other indicia.[38] These imperfectly differentiated causes of action appear to be the source of the over-simplified notion that, prior to the 1875 Act, there were two separate actions, a tort of passing off and a tort of trade mark infringement.[39]

As further explained at [5.3], the 1875 Act made registration a prerequisite for proceedings for trade mark infringement. While the precise legal effect of this provision, and of subsequent provisions that expressly preserved the action for passing off, is unclear, the practical effect was that the courts abandoned Lord Westbury's attempt to ground common law protection in a right of property in trade marks. As Wadlow explains:

> The final result was that the law of passing-off emerged into the last quarter of the nineteenth century with its scope unaffected but one possible theoretical basis destroyed. It could not be said that passing-off protected a right of property in a trade mark, because if it did so it would have been no more than the abolished common law infringement action under another name. By the same token, to have admitted that the action for passing-off had sometimes protected a right of property in a trade name would have confined it for the future to cases not involving the use of marks on goods, which the courts were reluctant to do.[40]

The view that the action for passing off protected a property right in trade marks was conclusively rejected by Lord Parker in *Spalding v Gamage*,[41] which established that the tort protects business goodwill against misrepresentation. In this respect, Lord Parker stated:

> There appears to be considerable diversity of opinion as to the nature of the right, the invasion of which is the subject of what are known as passing-off actions. The more general opinion appears to be that the right is a right of property. This view naturally demands an answer to the question property in what? Some authorities say property in the mark, name, or get-up improperly used by the defendant. Others say, property in the business or goodwill likely to be injured by the misrepresentation. Lord Herschell in *Reddaway v Banham*[42] expressly dissents from the former view; and if the right invaded is a right of property at all, there are, I think, strong reasons for preferring the

[36] See, eg, *Edelsten v Edelsten* (1863) 1 De GJ & S 185; 46 ER 72; *Hall v Barrows* (1863) 4 De G J & S 150; *Leather Cloth Co v American Leather Cloth Co Ltd* (1863) 4 De G J & S 137; 46 ER 868.
[37] (1863) 4 De GJ and Sm 1347; 46 ER 868, 870.
[38] Wadlow, above n 14, para [1.38].
[39] *Ibid*, paras [1.38]–[1.39].
[40] *Ibid*, para [1.39].
[41] (1915) 32 RPC 273.
[42] [1896] AC 199.

latter view. In the first place, cases of misrepresentation by the use of a mark, name, or get-up do not exhaust all possible cases of misrepresentation.[43]

During the twentieth century, the English action for passing off was extended by the courts to apply to such a diversity of circumstances that it is difficult to define the limits of the action with precision.[44] Nevertheless, English and Australian courts have generally adopted two forms of the tort, known as the 'classic trinity' and the 'extended form'.[45] The 'classic trinity' requires a claimant to establish three things: reputation or goodwill deserving protection; misrepresentation by the defendant likely to confuse or mislead the public; and damage suffered, or likely to be suffered, as a result of the misrepresentation. The three elements of the 'classic trinity' were authoritatively reiterated by the House of Lords in 1990 in *Reckitt & Colman Products Ltd v Borden*.[46]

The House of Lords had earlier formulated the 'extended form' of the action—in the famous 1979 *Advocaat* case[47]—in an attempt to deal with circumstances that fail to fit neatly within the 'classic trinity'. In that case, Lord Diplock identified the following five characteristics that are required to establish an action for passing off:

> (1) misrepresentation (2) made by a trader in the course of trade, (3) to prospective customers of his or ultimate consumers of goods or services supplied by him, (4) which is calculated to injure the business or goodwill of another trader (in the sense that this is a reasonably foreseeable consequence) and (5) which causes actual damage to a business or goodwill of the trader by whom the action is brought or (in a *quia timet* action) will probably do so.[48]

Whichever formulation is adopted—the 'classic form' or the 'extended form'—it remains the case that the modern action for passing off is designed to prevent misrepresentations that damage the goodwill of a trader, not to protect a property right in trade marks. At the same time, however, the action for passing off[49] remains the principal means, under English-oriented common law systems, of protecting indicia that are not registered under a statutory system of trade mark registration.

The clear implication of this analysis of the history and development of the action for passing off is that, in applying the first element of the UDRP, it is important to distinguish between the extent to which the action for passing off may protect a trade mark and its wider operation in protecting against misrepresentations that damage goodwill. In other words, the action for passing off is relevant to the first element of the UDRP only insofar as it protects common law marks, while its wider operation of protecting all manifestations of goodwill is irrelevant.

[5.8] The US Unfair Competition Tort

English and US law dealing with unregistered trade marks have common roots in the early development of the action for passing off by the English courts, but diverged in the early

[43] (1915) 32 RPC 273, 284.

[44] As Gummow J stated in *Conagra Inc v McCain Foods (Australia) Pty Ltd* (1992) 106 ALR 465, 518: 'the law of passing off contains sufficient nooks and crannies to make it difficult to formulate any satisfactory definition in short form'.

[45] See Kitchin *et al*, above n 20, paras [15-004]–[15-014].

[46] [1990] RPC 341.

[47] *Erven Warnink v Townend* [1980] RPC 31.

[48] [1980] RPC 31, 93.

[49] Together with overlapping statutory causes of action, such as s 52 of the Trade Practices Act 1974 (Cth), which prohibit misleading or deceptive conduct.

twentieth century with the emergence of the broad American tort of 'unfair competition'. Until then, unfair competition law had been equated with 'passing off' or 'palming off'.

The most significant development in the emergence of an expanded American doctrine of unfair competition was the 1918 US Supreme Court decision in *International News Service v Associated Press*,[50] in which the Court held that the scope of the equitable action for unfair competition extended beyond passing off. It is now well established that 'passing off' or 'palming off', as well as trade mark infringement, is merely a species of a broader American doctrine of unfair competition law.[51] Unlike US law, English and Australian courts have consistently refused to recognise a general tort of unfair competition.[52]

US law relating to unfair competition and trade mark infringement underwent considerable change during the twentieth century. The common law initially drew a distinction between 'technical trademarks' and 'trade names'.[53] A 'technical trademark' was an arbitrary symbol or word that was automatically distinctive, and so inherently capable of functioning as a trade mark. 'Trade names', on the other hand, were descriptive terms that were not inherently distinctive, including geographical terms and personal names. A 'trade name' was not protected by trade mark law but, upon proof of a secondary meaning, could be protected by unfair competition law.

This traditional distinction was abandoned by the 1946 Lanham Act, which combined the two forms of mark in a new definition of a 'trade mark', effectively conferring the same level of protection on 'trade names' as that afforded 'technical trademarks'. At the same time, the principles underlying the traditional distinction were preserved in the distinction drawn between inherently distinctive marks and non-inherently distinctive marks. Unlike English-oriented trade mark law, where rights in a trade mark may arise from registration independently of use of a mark, under US trade mark law rights and priority over others arise mainly from actual use.[54] In other words, apart from constructive use, a trade mark can only be registered under the Lanham Act if it has actually been used in connection with goods or services.[55] As McCarthy puts it:

> Apart from the concept of a constructive use, it is not registration, but only actual use of a designation as a mark that creates rights and priority over others. The United States, unlike many civil law nations, has a rule of priority that is based on first-to-use, not first-to-register.[56]

As explained at [5.15] ff, under US trade mark law it is important to distinguish between distinctive marks, descriptive marks and generic terms.

[5.9] Unregistered Marks in Civil Law

The Paris Convention does not require registration as a trade mark as a condition for protection. That rights may exist independently of registration is confirmed by Article 6*bis* of

[50] 248 US 215, 63 L Ed 211 (1918).

[51] See, eg, *Flexitized, Inc v National Felexitized Corp* 335 F 2d 774 (1964).

[52] *Victoria Park v Taylor* (1937) 58 CLR 479; *Moorgate Tobacco Co Ltd v Philip Morris Ltd* (1984) 156 CLR 414; *Mogul Steamship Co v McGregor Gow* (1889) 23 QBD 598; *Lonrho v Shell Petroleum* [1981] AC 173.

[53] See McCarthy, above n 5, §§ 4:3–4:9.

[54] *Ibid*, § 16:1. Rights can arise otherwise than by actual use in circumstances where there is a constructive use: Lanham Act § 7(c), 15 USC § 1057(c).

[55] McCarthy, above n 5, § 16:1.

[56] *Ibid*, § 16:18.

the Convention, which protects the rights of a prior user of a well-known mark. Article 6*bis*(1) provides:

> The countries of the Union undertake, ex officio if their legislation so permits, or at the request of an interested party, to refuse or to cancel the registration, and to prohibit the use, of a trademark which constitutes a reproduction, an imitation, or a translation, liable to create confusion, of a mark considered by the competent authority of the country of registration or use to be well known in that country as being already the mark of a person entitled to the benefits of this Convention and used for identical or similar goods. These provisions shall also apply when the essential part of the mark constitutes a reproduction of any such well-known mark or an imitation liable to create confusion therewith.

As national laws in civil law jurisdictions differ in relation to the basis for the acquisition of rights, the position of unregistered marks must be examined against the background of the relevant national law.

Section 4 of the current German Trade Mark Law, for example, states:[56a]

> Trade mark protection shall accrue
>
> 1. by registration of a sign as a trade mark in the Register kept at the Patent Office;
> 2. through the use of a sign in the course of trade insofar as the sign has acquired a secondary meaning as a trade mark within the affected trade circles; or,
> 3. by notoriety as a trade mark within the meaning of Article 6bis of the Paris Convention for the Protection of Industrial Property.

Registered Trade Marks

[5.10] Registered Trade Marks under the UDRP

A complainant has rights in a mark under paragraph 4(a)(i) if the trade mark or service mark has been registered. The WIPO Consensus View is that:

> If the complainant owns a registered trademark then it satisfies the threshold requirement of having trademark rights.[57]

In *Thomas B Hudson v Red Machi*,[58] for example, the respondent contended that the complainant could have no rights in the registered trade mark 'HOUSELAW' because it comprised two generic terms. The panel, however, confirmed that the complainant had rights by virtue of registration of the trade mark with the US Patent and Trademark Office (USPTO). Similarly, in *General Machine Products Company, Inc v Prime Domains*,[59] the panel decided that the complainant had rights in the registered trade mark, 'CRAFTWORK', even though the registration was for stylised versions of the mark, which consisted of two generic words. At the same time, the panel suggested that the fact that the name consisted of generic terms might make it easier for the respondent to establish that it had a legitimate interest in the name under paragraph 4(a)(ii) of the UDRP.

[56a] Trademark Law (as amended 16 July 1998), English translation at http://www.jpo.go.jp/shiryon_e/s_sonata_e/fips_e/germany/tl/mokuji.htm.

[57] WIPO, *WIPO Overview of WIPO Panel Views on Selected UDRP Questions* (23 Mar 2005) available at http://www.wipo.int/amc/en/domains/search/overview/index.html, para 1.1.

[58] NAF Case No FA133759 (21 Jan 2003).

[59] NAF Case No FA92531 (16 Mar 2000).

Registration of a mark on the Principal Register of the USPTO gives rise to a presumption of validity under US law. As McCarthy points out:

> A registration on the Principal Register is prima facie evidence of the validity of the registered mark, of the registration of the mark, of the registrant's ownership of the mark, and of the registrant's exclusive right to use the registered mark.[60]

Evidence of registration on the Principal Register therefore establishes rights in the mark sufficient for the UDRP. Although respondents have questioned the validity of trade mark registrations, UDRP panels are not equipped to determine the complex issues involved in deciding whether or not a registration has been validly made. UDRP panels therefore properly refuse to question the validity of a trade mark registration. The response of the panellist to a challenge to a service mark registered on the USPTO Principal Register in *Argyle Diamond Sales Limited v Van Daaz*[61] explains why UDRP panels should refuse to examine the validity of registered marks:

> The Respondent in its Response has gone to considerable lengths in attempting to persuade the Panel that the Service Mark consists of purely descriptive or generic words, in which the Complainant can have no legitimate rights . . . [T]he Panel is not prepared to entertain that sort of inquiry. The only relevant matter which the Panel believes it can consider under paragraph 4(a)(i) of the Policy is whether or not the Complainant has a legal right in the Service Mark, and quite obviously it does. Indeed, the Respondent does not suggest otherwise. Whether or not the Complainant *should have been given* the particular right which forms the basis for its Complaint under paragraph 4(a)(i) of the Policy, is not a matter with which the Panel can be concerned in this administrative proceeding. Still less is it for a Panel dealing with an administrative proceeding such as this to inquire into procedural matters relating to the obtaining of the trademark or service mark on which a complainant relies. Those are matters for the Courts in the relevant jurisdiction or jurisdictions where the marks are registered.

[5.11] Collective and Certification Marks

Association marks distinguish the business of members of an association from other businesses, and are used to indicate geographical origins, mode of manufacture or quality of goods. There are two kinds of association marks: collective marks and certification marks. A collective mark distinguishes the goods or services of members of an association from those of other businesses, whereas a certification mark indicates the origin, material, mode of manufacture or quality of goods or services. Article 7*bis*(1) of the Paris Convention provides:

> The countries of the Union undertake to accept for filing and to protect collective marks belonging to associations the existence of which is not contrary to the law of the country of origin, even if such associations do not possess an industrial or commercial establishment.

Registration of a collective mark or certification mark under a national law is sufficient to establish rights in the mark for the purpose of paragraph 4(a)(i) of the UDRP. In *Consorzio del Formaggio Parmigiano Reggiano v La casa del Latte di Bibulic Adriano,*[62] for example, the complainant was a non-profit-making organisation that owned registered collective marks under Italian law, including a number of registrations for 'PARMIGIANO

[60] McCarthy, above n 5, §19:9.
[61] WIPO Case No D2001-1323 (31 Dec 2001).
[62] WIPO Case No D2003-0661 (21 Oct 2003).

REGGIANO' and 'PARMESAN'. The panellist concluded that proof of registration of the collective marks satisfied paragraph 4(a)(i).

[5.12] Location of Jurisdiction of Registration

Provided that a mark is registered, the jurisdiction in which the mark is registered is irrelevant for the purpose of paragraph 4(a)(i). The WIPO Consensus View is that:

> The location of the registered trademark and the goods and/or services it is registered for are irrelevant when finding rights in a mark.[63]

This means, for example, that if the disputed domain name is registered in a particular geographical jurisdiction, there is no requirement for the complainant to establish that he or she has rights in a registered mark in that jurisdiction.

Thus, in *Bennett Coleman & Co v Lalwani*,[64] the complainant had registered trade marks in India. The respondent submitted that the complainant had no rights in the trade marks because it had no trade marks registered in the United States, which was where the domain names were registered. The panel, however, decided that, given the global nature of the Internet, the place in which rights in a trade mark are created is of no significance. In this respect, the panellist observed:

> Since the essence of the Internet is its world-wide access, consideration of the propriety of domain name registrations cannot be confined to comparisons with trade mark registrations or other rights in the country where the site is hosted. It is therefore relevant to consider the effects of a domain name registration and use in relation to the registration and use of identical and confusingly similar trade marks in other countries. Equally, it is not conclusive against a complaint that the Complainant's mark could not have been registered in a particular country.

Similar comments were made by the panellist in *UEFA v Funzi Furniture*.[65] In that dispute, the complainant, football's European governing body, had obtained an International Trademark Registration for the mark 'CHAMPIONS LEAGUE'. The respondent attacked the mark on the basis that it lacked distinctiveness and that the term 'Champions League' was used by many other sporting competitions, including other football competitions. The panel decided that, even though the registration had not created exclusive rights in the mark, this was irrelevant for the purpose of determining whether the complainant had rights in the mark. In so deciding, the panellist stated:

> If the onus on the Complainant was to establish exclusive rights across all categories of goods and services and across all territorial boundaries, the object of the Policy (namely to combat bad faith registration and use of domain names) would be defeated. The purpose of paragraph 4(a)(i) of the Policy is simply to ensure that the Complainant has a bona fide basis for making the Complaint.

A clear statement of the principle that the jurisdiction in which the trade mark rights arise is not relevant under the UDRP was made by the panel in *Koninklijke KPN NV v Telepathy Inc.*[66] In that dispute, the complainant, a Dutch corporation, had registered

[63] WIPO, above n 11, para 1.1.

[64] WIPO Case No D2000-0014 (11 Mar 2000). See also *Bennett Coleman & Co Ltd v Long Distance Telephone Company*, WIPO Case No D2000-0015 (11 Mar 2000); *Thaigem Global Marketing Limited v Sanchai Aree*, WIPO Case No D2002-0358 (16 July 2002).

[65] WIPO Case No D2000-0710 (22 Oct 2000).

[66] WIPO Case No D2001-0217 (7 May 2001).

Benelux marks, and applied for further Benelux registrations. In deciding that the complainant had satisfied the first element of the UDRP, the panel stated:

> The Policy does not require that the mark be registered in the country in which the Respondent operates. It is sufficient that a complainant can demonstrate a mark in some jurisdiction.

In that case, the complainant did not present any evidence of the registrations, or of the applications for registration. While lack of evidence of registration will normally be fatal to a complaint,[67] if there is no response to the complaint a panel may accept allegations of registration as correct without further documentation.[68]

[5.13] Registration where Full Rights not Granted

Registration of a trade mark under a national system of registration that does not create full legal rights in the mark, or where registration is automatic or unexamined, is usually insufficient to establish rights in the mark for the purpose of paragraph 4(a)(i). If a mark is registered under a system that does not confer full rights, the best way to establish rights in the mark is by evidence that there are unregistered rights in the mark.

The WIPO Consensus View is as follows:

> In certain, highly limited circumstances, a panel may examine the circumstances of trademark registration to determine whether the registration satisfies UDRP requirements. Trademark registrations that are automatic or unexamined (such as US state registrations as opposed to US federal registrations) may not always be owed the same deference as examined registrations.[69]

An example of the cautious approach of panellists to registrations that do not confer full rights is *PwC Business Trust v Ultimate Search*,[70] where the complainant, a trustee for Price Waterhouse Coopers, had registered the mark 'PWC' in a number of countries including Bahrain, Brunei, Cambodia, China, Costa Rica, Hong Kong, Malawi, Mexico, New Zealand, Saudi Arabia, Switzerland, Thailand and Turkey. The respondent contended that some of these registrations were 'bare registrations', meaning that they did not provide a presumption of validity in the jurisdictions where they were issued. The panel found that there were 'serious doubts as to the value of some of these registrations in terms of proving that the Complainant has valid and valuable registered trade marks of "PWC" '. It concluded, however, that it was sufficient that some of the registrations, such as the New Zealand registration, created rights in the mark.

One form of registration that has consistently been found to create no rights for the purpose of paragraph 4(a)(i) is registration on the USPTO's Supplemental Register. The Supplemental Register was established by the Lanham Act to enable US registrants to obtain registration of marks in foreign countries, especially Latin American countries.[71] A mark registered under the Supplemental Register creates no substantive rights in the mark beyond any common law rights and, in particular, does not create a presumption of validity.[72]

[67] See, eg, *Rodale, Inc v Christianne Schelling*, WIPO Case No DBIZ2002-00130 (30 June 2002).

[68] See, eg, *Transpark LLC v Network Administrator*, NAF Case No FA135602 (21 Jan 2003).

[69] WIPO, above n 57, para 1.1.

[70] WIPO Case No D2002-0087 (22 May 2002).

[71] McCarthy, above n 5, § 19:33.

[72] *Clairol, Inc v Gillette Co* 389 F 2d 264 (2d Cir 1968); *American Personnel, Inc v Management Recruiters* 165 USPQ 173 (1970).

Moreover, registration on the Supplemental Register is only available for marks that are not registrable on the Principal Register, meaning that the mark is not distinctive of the applicant's goods or services.[73] Thus, registration of a mark on the Supplemental Register does not confer the same rights in a mark as registration in the Principal Register, but is prima facie evidence that the mark is descriptive and lacks distinctiveness.

In *America's Community Bankers Corporation v Charles R. Wing*,[74] the complainant had registered the marks 'AMERICA'S COMMUNITY BANKER' and 'AMERICA's COMMUNITY BANKERS' on the Supplemental Register. The panellist concluded that, as the complainant relied solely on registration of the marks and presented no evidence that the marks had acquired distinctiveness, the marks were descriptive. As a result, the claimant had not established any rights in the marks.

A similar result was reached in *CyberTrader, Inc v Gregory Bushell*.[75] In that case, the complainant had registered the marks 'CYBERTRADER' and 'CYBERBROKER' on the Supplemental Register. The Panellist decided that a failure to produce evidence that the marks had become distinctive was fatal to the complainant's case, stating:

> In my opinion both CYBERTRADER and CYBERBROKER are merely descriptive of a business that provides some type of trading or brokering transactions over the Internet and are, therefore, not eligible for federal trademark protection absent a showing of secondary meaning. The USPTO's refusal to place these marks on the Principal Register supports my opinion. Complainant may very well have evidence establishing secondary meaning in the marks CYBERTRADER and CYBERBROKER, but none is presented in the record here.

In *John Gard v Francesco Spina*,[76] the complainant applied to register the mark 'SCUBA.BIZ' in the Supplemental Register after the respondent had registered the corresponding domain name. The panellist decided that the complainant had no rights in the mark, concluding that:

> not only does the mere application to register give rise to no rights, the fact that registration is sought in the Supplemental Register is evidence that there were no rights at common law at the time of application.

The same conclusion was reached in *Rodale, Inc v Cambridge*.[77] In that dispute, the complainant had registered the mark 'SCUBADIVING.COM' on the Supplemental Register. The panel concluded that registration on the Supplemental Register was evidence that the mark was descriptive and lacked secondary meaning. In reaching this conclusion, the panellist stated that registration on the Supplemental Register 'is powerful evidence that the PTO believes that the mark "scuba diving" is at least descriptive, and that Complainant lacks persuasive evidence to the contrary (or else it would have submitted it to the PTO rather than agree to registration on the supplementary registry)'.

Another form of registration that has been considered by panels is registration of a mark as an official mark under section 9(1)(n)(iii) of the Canadian Trade-marks Act, which prohibits the use of a registered official mark of a public authority as a trade mark, otherwise than with the consent of the public authority. UDRP panels have concluded that

[73] See *Re Consolidated Foods Corp*, 200 USPQ 477 (1978); *Application of Clorox Co*, 578 F 2d 305 (US Court of Customs and Patent Appeals, 1978).

[74] WIPO Case No D2000-1780 (5 Mar 2001).

[75] WIPO Case No D2001-1019 (30 Oct 2001).

[76] WIPO Case No DBIZ2002-00167 (27 July 2002). For a similar case decided under the UDRP see *Robert Chiappetta dba Discount Hydroponics v CJ Morales*, WIPO Case No D2002-1103 (20 Jan 2003).

[77] WIPO Case No DBIZ2002-00153 (28 June 2002).

registration as an official mark will confer rights for the purpose of paragraph 4(a)(i). In *Her Majesty the Queen in right of Ontario, Canada v Venture Labour.com Inc.*,[78] for example, the complainant's mark 'SERVICEONTARIO' was published as an official mark, and had been used in association with services, such as the renewal of drivers' licences. The panel decided that the complainant had rights in the mark on the basis of common law rights acquired from its use, as well as statutory rights arising from its registration as an official mark. Similarly, in *Canadian Hockey Association v Mrs Jello, LLC*,[79] the complainant's mark 'HOCKEY CANADA' had been registered as an official mark. The panel concluded that, given that registration as an official mark conferred more rights than registration as a trade mark, such registration conferred rights for the purpose of paragraph 4(a)(i).

[5.14] Applications for Registration

The majority view of UDRP panels is that an application for registration of a trade mark is, in itself, insufficient to create rights in the mark for the purpose of paragraph 4(a)(i). If an application for registration is pending, a complainant may assert rights in the mark by establishing that he or she has unregistered rights in the mark. A minority of decisions have concluded, often with little analysis, that an application for registration can give rise to rights in a mark.[80] The correct view, however, is that whether or not an application for registration creates rights depends upon whether the application confers any rights under the relevant national law.[81] It may also be that rights may be held to arise prior to registration of a mark where the trade mark opposition period has expired.

A good illustration of the principle that an application for trade mark registration does not confer rights in a mark is the decision in *Aspen Grove, Inc v Aspen Grove*.[82] In that dispute, the complainant presented evidence showing that an application for registration of the mark 'ASPENGROVE' had been published for public comment. In deciding that this was insufficient to establish rights in the mark, the panel stated that it:

> is not aware of any rights conferred by a trademark application either under the Policy or at US law. Certainly, under the Policy, while Panels have been willing to recognize that proof of a valid and subsisting trademark registration is prima facie evidence of trademark rights, no such presumption arises from a pending application to register a mark.

An explanation of the reasoning underlying the principle was given by the panellist in *First Tuesday Limited v The Startup Generator*.[83] In that dispute, the complainant had filed applications for registration of the mark 'WIRELESS WEDNESDAY' in more than 10 countries around the world. In deciding that the applications did not create rights in the marks, the panellist summarised the position under US law in the following terms:

[78] WIPO Case No D2004-0408 (23 July 2004).

[79] WIPO Case No D2005-1050 (20 Dec 2005).

[80] See, eg, *Phone-N-Phone Services (Bermuda) Ltd v Shlomi (Salomon) Levi*, WIPO Case No D2000-0040 (23 Mar 2000); *Michael Aston v Dena Pierrets*, NAF Case No 117322 (27 Sept 2002); *La Française des Jeux v L Welsr*, WIPO Case No D2002-0305 (16 June 2002); *Bart van den Bergh Merk-Echt BV v Kentech Company Ltd*, WIPO Case No D2005-0127 (24 Mar 2005).

[81] *Fashiontv.com GmbH v Mr Chris Olic*, Case No D2005-0994 (8 Dec 2005).

[82] WIPO Case No D2001-0798 (5 Oct 2001).

[83] WIPO Case No D2000-1732 (12 Feb 2000).

> Trademark rights are acquired by virtue of use of the trademark, or its registration based upon use or a priority document in another country. Accordingly, neither the United States application nor the community trademark application, nor the specific applications in particular countries, provide evidence in and of themselves of trademark rights having been acquired by Complainant in WIRELESS WEDNESDAY.

A more complete summary of the position was given in *Mk-Net-Work v IVE Technologies*,[84] where the panel stated:

> a trademark application is not in itself sufficient evidence of a trademark or service mark in which the Complainant has rights, since applications are often opposed or may be refused for lack of distinctiveness. The only types of right that amount to trademark rights are, in addition to registered marks, unregistered or common law trademark rights.
>
> Pending trademark application is not in itself sufficient evidence that the Complainant has generated goodwill in a mark which it wishes to protect. Combined with other information about the use and renown of the alleged mark an application can be relevant to a panel finding common law trademark rights.

Although some decisions have suggested that rights may arise from applications for registration, as the panellist correctly pointed out in *Fashiontv.com GmbH v Mr Chris Olic*,[85] the status of an application for registration is properly determined by reference to the position under the relevant national trade mark law. As the panellist observed, the reasoning in decisions that have held that an application can give rise to rights in a mark has been limited, with the decisions often being based upon unregistered rights arising from use, rather than the application. In *Fashiontv.com* itself, the panellist was required to decide whether an application for registration conferred rights where, under the Trade Marks Act 1994 (UK), registration confers rights from the date of filing. Referring to comments made in *HQ UK Limited v Head Quarters*,[86] the panellist pointed out that this creates difficulties insofar as the validity of a mark is determined at registration, not application. While the UK Act deals with this by providing that proceedings cannot be commenced until the date of registration, there is no equivalent provision in the UDRP. This led the panellist to conclude that, for the purpose of paragraph 4(a)(i), rights did not arise until registration. At the same time, the panellist did not rule out the possibility that, in certain circumstances, rights may arise prior to registration. Without deciding the matter, the panellist offered the following as an example of where rights might arise prior to registration:

> where the . . . opposition period has passed, an opposition or oppositions have been filed but the nature of these oppositions are such that it is clear that ultimately the complainant will be awarded registration in at least one trademark class *regardless* of the outcome of those oppositions. In such a case it may be argued that if a complainant is able to put forward uncontested evidence to the effect that this is the position, it is neither necessary nor fair to insist that the complainant awaits formal registration before commencing UDRP proceedings.

The *Fashiontv.com* decision clearly illustrates the competing considerations involved in deciding whether or not there are rights in a mark prior to registration. On the one hand, UDRP panels are hardly equipped to determine the merits of an application for registration. On the other hand, however, publication of an application for registration may create the

[84] WIPO Case No D2004-0302 (5 July 2004).
[85] WIPO Case No D2005-0994 (8 Dec 2005).
[86] WIPO Case No D2003-0942 (5 Feb 2004).

opportunity for a potential cybersquatter to register a corresponding domain name.[87] This latter reasoning is evidently behind the suggestion by one panellist that a complainant would have rights in an 'immediately impending mark known to Respondent' at the time the domain name at issue was registered.[88] The residual uncertainty regarding the position of marks prior to registration suggests that, where an application for registration has been made but registration has yet to be granted, a complainant should generally either provide evidence of unregistered rights in the mark or wait until the mark has been registered before lodging a complaint. As explained at [5.12], the time at which rights must arise in a mark for a complaint to be made under the UDRP is the time of lodging the complaint, not the time of registration of the corresponding domain name.

[5.15] Time at Which Rights Arise under the UDRP

Historically, there has been disagreement among panellists about the time at which rights must arise in a mark under paragraph 4(a)(i). The WIPO Consensus View, however, is that:

> Registration of a domain name before a complainant acquires trademark rights in a name does not prevent a finding of identity or confusing similarity. The UDRP makes no specific reference to the date of which the owner of the trade or service mark acquired rights. However it can be difficult to prove that the domain name was registered in bad faith as it is difficult to show that the domain name was registered with a future trademark in mind.[89]

After considerable controversy, the consensus is that the time for determining whether the complainant has rights in the mark is the time the complaint is made, and not the time that the domain name is registered. In other words, a complainant can establish rights in the mark at any time prior to the complaint being made. The fact that the complainant has no rights at the time the domain name is registered is therefore irrelevant to the first element of the UDRP, and will be dealt with in determining whether the registration has been in bad faith. Where the domain name is registered before rights have arisen in a trade mark, the domain name will usually not be registered in bad faith, as the registrant will not be aware of any rights the complainant may have. The WIPO Consensus View on this point is that:

> Normally speaking, when a domain name is registered before a trademark right is established, the registration of the domain name was not in bad faith because the registrant could not have contemplated the complainant's non-existent right.[90]

If, however, the domain name is registered in expectation of rights that may arise, such as where a respondent is aware of the potential rights of a complainant and registers the domain name to take advantage of those potential rights, the domain name may be registered in bad faith. The circumstances in which a complainant has no rights at the time of registration, but nevertheless registration may be in bad faith, are explained in more detail at [7.13].

It might initially be thought that a domain name registrant should not be liable for registering a name that includes, or is confusingly similar to, a mark at a time when there are no rights in the mark. This view was supported in a considerable number of panel decisions.

[87] See, eg, *Boehringer Ingelheim Corporation v Kumar Bhatt*, NAF Case No FA95011 (11 Aug 2000).
[88] *Ezcommerce Global Solutions, Inc v AlphaBase Interactive*, WIPO Case No D2002-0943 (22 Nov 2002).
[89] WIPO, above n 57, para 1.4.
[90] *Ibid*, para 3.1.

For example, in *Phoenix Mortgage Corporation v Tom Toggas,*[91] the complainant had registered the mark 'E-MORTGAGE', but the domain name <e-mortgage.com> was registered before the effective date of the trade mark registration. The panellist decided that the complainant had failed to prove that it had prior rights in the mark, and so did not satisfy the first element of the UDRP. In coming to this conclusion, the panellist stated that paragraph 4(a)(i) of the UDRP:

> necessarily implies that the Complainant's rights predate the Respondent's registration and use of the Domain Name. Any other interpretation would allow a junior trademark user to challenge a prior domain name registration, a possibility that is obviously contrary to the intent of the Policy and to trademark law generally.

This interpretation has now been conclusively rejected by panellists, the consensus being that the acquisition of rights in a mark after the domain name has been registered will not prevent the first element of the UDRP from being satisfied. For example, in *Countrywide Financial Corporation, Inc v Aziz Popal,*[92] the complainant applied to register the 'COUNTRYWIDE FINANCIAL' mark in August 2002. The mark was first used in November 2002, and was finally registered in November 2004. The respondent registered the domain name <countrywidefinancial.com> before any of these events. The panel, citing previous decisions,[93] held that the acquisition of rights in a mark after the respondent has registered the domain name does not prevent a finding of identity or confusing similarity under paragraph 4(a)(i) of the UDRP, but may be relevant in determining whether registration was in bad faith under paragraph 4(a)(iii).

The two factors supporting the consensus view were probably best explained by the dissenting panellist in *Firstgate Internet AG v David Soung.*[94] The panellist first pointed out that the wording of paragraph 4(a)(i)—requiring that the domain name 'is identical or confusingly similar to a trademark or service mark in which the complainant has rights'—is expressed in the present tense. This consideration is, however, hardly determinative as the policy takes the form of a representation made by the domain name registrant at the time the domain name is registered and, consequently, cannot be construed to refer to the situation at the time the complaint is lodged. Secondly, the panellist explained what has become the accepted rationale, namely that disbarring a complaint merely because a complainant had no rights at the time the domain name was registered would excuse some bad faith registrations. In this respect, the panellist offered the following example of the sort of circumstance that should be covered by the UDRP, but would fall outside the policy if the complainant were required to establish rights in the mark at, or before, the domain name was registered:

> A first competitor illegally learns that a second competitor is about to adopt a particular term for use in its business, as a trademark or otherwise. The first competitor in bad faith registers that term as its own domain name. The second competitor then does develop the term as its trademark. However, when the second competitor tries to register the trademark as a domain name, it learns that the first competitor has already registered it.

[91] WIPO Case No D2001-0101 (30 Mar 2001).

[92] WIPO Case No D2005-0295 (15 June 2005).

[93] The most commonly cited decision is *Digital Vision, Ltd v Advanced Chemill Systems,* WIPO Case No D2001-0827 (23 Sept 2001).

[94] WIPO Case No D2000-1311 (29 Jan 2001) (Thomas L Creel dissenting).

Unregistered Marks

[5.16] Unregistered Marks under the UDRP

Establishing rights in an unregistered mark, such as common law trade mark rights, is sufficient to make out rights in the mark for the purpose of paragraph 4(a)(i). It is clear that the UDRP is not confined to protecting rights that arise from registration. Whether or not rights arise in an unregistered mark must be determined by reference to national legal systems.

Although it is universally acknowledged that unregistered trade mark rights are sufficient under the UDRP, the WIPO Final Report did not consider the issue in any detail. The report did, however, identify one of the main shortcomings with previous dispute-resolution policies, specifically the NSI Dispute Resolution Policy, in the following terms:

> One of their important deficiencies results from their reliance on the ability of the parties to produce certain trademark certificates, without any review of the question of use of the domain name and alleged infringement. These policies are seen as not sufficiently allowing for the consideration of all legitimate rights and interests of the parties (which are not necessarily reflected in a trademark certificate), opening the door to unjust results, including for those who are not trademark owners.[95]

These comments were obviously persuasive rather than definitive. Froomkin, however, maintains that participants in the drafting of the UDRP understood that the policy would apply to common law marks.[96] Moreover, as he further points out, denying protection to common law marks could be seen as favouring legal systems in which rights are obtained exclusively by registration over systems in which rights can arise in an unregistered mark.[97]

In any case, the fact that the terms 'trademark' and 'service mark' in the first element of the UDRP are unqualified leads to the implication that registration of a mark is not a prerequisite for protection. As one panellist pointed out in a series of decisions involving the personal names of well-known English authors that were registered as domain names:[98]

> On the face of paragraph 4(a)(i) of the Policy there is no limit to registered trade marks or service marks. It simply refers to 'a trade mark or service mark in which the Complainant has rights'.[99]

Despite the absence of clear guidance in the original policy documents, from the outset panels have regarded unregistered rights as sufficient. Early decisions to reach this conclusion,

[95] World Intellectual Property Organization (WIPO), *The Management of Internet Names and Addresses: Intellectual Property Issues,* Report of the WIPO Internet Domain Name Process (30 Apr 1999) para [149]. See also para [330].

[96] See AM Froomkin, 'ICANN's "Uniform Dispute Resolution Policy"—Causes and (Partial) Cures' (2002) 67 *Brooklyn Law Review* 605, 638.

[97] *Ibid,* fn 107.

[98] See *Julian Barnes v Old Barn Studios Limited* ,WIPO Case No D2001-0121 (26 Mar 2001); *Louis De Bernieres v Old Barn Studios Limited* ,WIPO Case No D2001-0122 (26 Mar 2001); *Antony Beevor v Old Barn Studios Limited* ,WIPO Case No D2001-0123 (26 Mar 2001); *Margaret Drabble v Old Barn Studios Limited* ,WIPO Case No D2001-0209 (26 Mar 2001) (Tony Willoughby, panellist).

[99] Similarly, in *Passion Group Inc v Usearch, Inc,* eRes Case No AF-0250 (10 Aug 2000), the panellist stated that '[h]ad ICANN intended the Policy to be restricted to cases involving only registered trademarks or service marks, it could easily have used narrower language than "a trademark or service mark in which the complainant has rights"'.

and which are commonly referred to, include *Cedar Trade Associates, Inc v Gregg Ricks,* [100] *Bennett Coleman & Co Ltd v Steven S Lalwani,*[101] *The British Broadcasting Corporation v Jaime Rentaria,*[102] *SeekAmerica Networks Inc v Tariq Masood*[103] and *BroadcastAmerica.com, Inc v Cynthia Quo.*[104] Following the early decisions, a considerable number of panels have referred to a recent edition of *McCarthy on Trademarks and Unfair Competition,* in the following terms:

> The ICANN dispute resolution policy is 'broad in scope' in that 'the reference to a trademark or service mark "in which the complainant has rights" means that ownership of a registered mark is not required—unregistered or common law trademark or service mark rights will suffice' to support a domain name Complaint under the Policy. *See McCarthy on Trademarks and Unfair Competition,* § 25:74.2, Vol. 4 (2000).[105]

The weight of authority was so overwhelming that the 2001 WIPO Second Report simply concluded that:

> The UDRP ... does not require that a complainant must hold rights specifically in a *registered* trademark or service mark. Instead, it provides only that there must be 'a trademark or service mark in which the complainant has *rights,*' without specifying how these rights are acquired. With this distinction in mind, many decisions under the UDRP have therefore determined that common law or unregistered trademark rights may be asserted by a complainant and will satisfy the first condition of the UDRP.[106]

Many panel decisions have expanded upon what the complainant must establish in order to have rights in an unregistered mark, which necessarily depends upon national legal principles. As the panellist put it in *America Online, Inc v John Deep*:[107]

> Nowhere does the Policy contain a restriction that specifically limits its reach to only registered marks. Therefore, given the apparent inclusive nature of the Policy, whether the Complainant has rights to an unregistered mark and hence protectable interest under the Policy, lies with national trademark law that governs the Respondent's actions that are the subject of the Complaint.

There are, nevertheless, important differences in the principles that apply to determine whether there are rights in unregistered marks under different national laws. The WIPO Consensus View on what needs to be shown for a complainant to successfully assert common law or unregistered trade mark rights is that:

[100] NAF Case No FA93633 (25 Feb 2000).

[101] WIPO Case No D2000-0014 (11 Mar 2000). See also *Bennett Coleman & Co Ltd v Long Distance Telephone Company* WIPO Case No D2000-0015 (11 Mar 2000).

[102] WIPO Case No D2000-0050 (23 Mar 2000).

[103] WIPO Case No D2000-0131 (13 Apr 2000).

[104] WIPO Case No DTV2000-0001 (4 Oct 2000).

[105] See *Bridal Rings Company v Abert Yemenian/Albert Yemenian,* NAF Case No FA95608 (26 Oct 2000); *Brooklyn Institute of Arts and Sciences v Fantastic Sites, Inc,* NAF Case No FA95560 (2 Nov 2000); *Home Properties v SMSOnline,* NAF Case No FA95639 (2 Nov, 2000); *CMG Worldwide Inc. v Naughtya Page,* NAF Case No FA95641 (8 Nov 2000); *Mike Warner 2001 v Mike Larsen,* NAF Case No FA95746 (15 Nov 2000); *Missing Children Minnesota v Run Yell Tell, Ltd,* NAF Case No FA95825 (20 Nov 2000); *Air Tahiti Nui v International Domains, Inc.,* NAF Case No FA114332 (8 July 2002); *Michael Aston v Dena Pierrets,* NAF Case No FA117322 (27 Sept 2002); *USA Prescriptions, Inc v usa-prescription-drugs-online.com,* NAF Case No FA128679 (23 Dec 2002); *Peter Capotosto d/b/a Armor Crafts v Steebar Corp,* NAF Case No FA135597 (16 Jan 2003); *America International Group, Inc v Rod Dobson,* NAF Case No FA146568 (8 Apr 2003).

[106] WIPO, *The Recognition of Rights and the Use of Names in the Internet Domain Name System,* Report of the Second WIPO Internet Domain Name Process (3 Sept 2001) (the WIPO *Second Report,* available at http://www.wipo.int/export/sites/www/amc/en/docs/report-final2.pdf, para [182] (emphasis in original).

[107] NAF Case No FA96795 (14 May 2001).

> The complainant must show that the name has become a distinctive identifier associated with the complainant or its goods and services. Relevant evidence of such 'secondary meaning' includes length and amount of sales under the mark, the nature and extent of advertising, consumer surveys and media recognition. The fact that the secondary meaning may only exist in a small geographic area does not limit complainant's rights in a common law trademark. Unregistered rights can arise even when the complainant is based in a civil law jurisdiction.[108]

Although this statement is accurate as far as it goes, it represents a considerable simplification of national legal principles. In order better to reflect national legal principles, a more precise analysis is required. For example, a more detailed analysis is required adequately to distinguish unregistered trade mark rights from other forms of legal right that may protect a name. Accordingly, the applicable national legal principles are discussed in more detail in the paragraphs immediately below.

[5.17] Unregistered Marks: Civil Law Jurisdictions

Rights may arise in unregistered marks in civil law jurisdictions where the law of the relevant jurisdiction provides for such rights. It is of the utmost importance, however, that unregistered rights in a mark be sufficiently substantiated under the relevant national law. As one panellist has put it:

> A panel deciding a case under the Policy will not always have knowledge of the laws of the jurisdiction in which unregistered trademarks are claimed. A complainant bears the burden of showing that it has rights recognised by Paragraph 4(a)(i) of the Policy. Therefore any complainant claiming unregistered trademark rights would be wise not only to provide evidence of the activities that are alleged to found those rights but also to provide evidence that as a matter of law those activities are recognised as giving unregistered trademarks rights in that jurisdiction. This is particularly so if relevant unregistered rights in that jurisdiction have not been unambiguously recognised in previous UDRP decisions.[109]

A good illustration of the principles that may apply in determining whether there are unregistered rights in a mark in civil law jurisdictions is *Skattedirektoratet v Eivind Nag.*[110] In that decision, the central tax office for Norway brought a complaint against the registered owner of the <skatteetaten.com> domain name. As the term 'skatteetaten' (meaning 'tax authority') was not registered as a trade mark, the tax authority relied upon sections of the Norwegian Trade Mark Law that recognise trade mark rights through use and recognition over time. The panel held that there were unregistered rights in the term 'skatteetaten' as it was not a generic term, but was a descriptive term that had become distinctive through use.

The *Skateedirektoratet* decision may be compared with a decision on similar facts in *The Republic of Turkey v Haval Kurdistan,*[111] which concerned the domain name <genel-kurmay.com>. The complainant in that dispute, the official representative of the Turkish Armed Forces, claimed unregistered trade mark rights in the term 'Genelkurmay', which is the abbreviated Turkish expression for 'Turkish General Staff'. The Turkish General Staff had exclusive rights to use the term under Turkish law, but the term was not registered as a trade mark. Under the Turkish Trade Mark Law, trade mark rights are confined to rights

[108] WIPO, above n 57, para 1.7.
[109] *Antonio de Felipe v Registerfly.com,* WIPO Case No D2005-0969 (19 Dec 2005).
[110] WIPO Case No D2000-1314 (18 Dec 2000).
[111] *The Republic of Turkey v Haval Kurdistan,* WIPO Case No D2006-0184 (17 May 2006).

that arise from registration, with the exception that the law provides for unregistered rights in well known marks under Article 6*bis* of the Paris Convention. The panel held that the complainant had no trade mark rights in the term, as the exclusive rights to use the term were not trade mark rights and there was insufficient evidence that the term was a well known mark.[112] The decision illustrates the importance of a careful application of national trade mark laws in determining whether unregistered rights arise in a civil law jurisdiction.

[5.18] Common Law Rights under US Law

As explained at [5.3], under US trade mark law rights in a mark arise mainly from actual use of the mark, not registration. In determining the requirements for establishing common law rights in a mark under US law it is essential to distinguish between inherently distinctive marks, descriptive marks and generic terms, as common law rights can only arise in a mark that either is distinctive or has become distinctive through use. As the Supreme Court explained in *Two Pesos, Inc v Taco Cabana, Inc:*

> The general rule regarding distinctiveness is clear: an identifying mark is distinctive and capable of being protected if it *either* (1) is inherently distinctive *or* (2) has acquired distinctiveness through secondary meaning.[113]

Although the principles for determining whether there are rights in an unregistered mark are similar in jurisdictions that recognise such rights, the importance of US law in UDRP decisions merits a more complete description of US legal principles.

[5.18.1] Inherently Distinctive Marks

Inherently distinctive marks are marks that are fanciful, arbitrary or suggestive.[114] A fanciful mark is a term that has been invented to operate as a trade mark. An arbitrary mark is a word or other indicia that is not invented but that, although in common use, does not suggest or describe the product in relation to which the mark is used. In other words, the relationship between the mark and the product is arbitrary. A suggestive mark is a mark that suggests, but does not describe, the product in relation to which it is used. As the panellist in *Media West-GSI, Inc v EARTHCARS.COM, INC.*[115] explained:

> Only marks that are arbitrary or fanciful in the sense of bearing no logical relationship to the product or service for which they are used, or suggestive in the sense of requiring imagination to be associated with a product or service, are deemed inherently distinctive.[116]

Rights in inherently distinctive marks are acquired upon adoption and use of the mark in commerce, whereupon there is an irrebutable presumption that the designation is a trade mark. Protection of inherently distinctive marks does not require proof of secondary meaning or consumer recognition. As one US court has put it:

[112] See also *Genelkurmay Baflkanligi v Genelkurmay Inc*, WIPO Case No D2001-1279 (18 Dec 2001).

[113] 505 US 763, 120 L Ed 2d 615, 23 USPQ 2d 1081, 1084 (1992) (emphasis in original).

[114] McCarthy, above n 5 § 11:4, citing *Two Pesos, Inc v Taco Cabana, Inc* 505 US 763, 120 L Ed 2d 615, 23 USPQ 2d 1081 (1992).

[115] WIPO Case No D2000-0463 (28 July 2000).

[116] Citing *Wal-Mart Stores, Inc v Samara Bros., Inc*, 120 S Ct 1339, 1343 (2000).

> The presumption that a fanciful word or mark becomes distinctive and identifies the source of goods on which it is used immediately after adoption and bona fide first use is basic in trademark law.[117]

Although even a quantitatively small amount of sales may establish rights in an inherently distinctive mark, it is necessary to establish that the mark has been continuously in use until the present in order to establish ownership of an inherently distinctive mark.[118] In general, use of a designation in preparing to establish a business will not be sufficient for rights to arise in the mark.[119] On the other hand, a number of US cases have held that prominent use of a mark in pre-sales publicity aimed at consumers may be sufficient for rights to arise.[120] In any case, the Lanham Act defines 'use in commerce' to mean 'the bona fide use of a mark in the ordinary course of trade, and not made merely to reserve a right in a mark'.[121]

A large number of UDRP decisions have concluded that a complainant has established common law rights in an inherently distinctive mark on the basis of continuous use. For example, in *Tuxedos By Rose v Hector Nunez,*[122] the mark 'TUXEDOS BY ROSE' had been used in connection with a tuxedo shop since 1957. In January 1998, the complainant purchased the business, including any rights in the mark. The respondent registered the domain names <rosetuxedo.com> and <tuxedobyrose.com> in February 2000. The domain names resolved to a Web site that contained promotional material for the respondent's tuxedo shop. The panellist concluded that the complainant's rights to the mark are 'ongoing and continuous in that it has been used for approximately 43 years'. On this basis, the panellist decided that the complainant had common law rights in the mark. Similarly, in *Sammy's Management Company v Keith Wimbley,*[123] the complainant established use of the mark 'SAMMY'S' in relation to a strip club or men's club since the 1970s. The respondent registered the domain name <sammysgirls.com> in mid-2002. The panellist concluded that the complainant had common law rights in the mark based upon continuous use of the 'SAMMY'S' mark since at least as early as the 1970s.

The requirement of 'use in commerce' was further explained by the panellist in *Geniebooks.com Corporation v William E Merritt.*[124] In that dispute, the complainant developed a preliminary strategy for a new electronic publishing and book distribution business known as 'GENIEBOOKS.COM'. After learning of the complainant's plans, the respondent, who was involved in an ongoing dispute with the complainant, registered the domain name <geniebooks.com>. Although books had not been sold under the name, the complainant was able to establish that it had incorporated under the name, had secured financing for the company using the name, had worked with Web site developers using the name and had a web hosting agreement using the name. The respondent argued that the complainant merely had rights in a business name, not trade mark rights. In the course of deciding that the complainant had rights in the mark, the panellist referred to the Lanham Act definition of 'use' in relation to inherently distinctive marks in the following terms:

117 *Blisscraft of Hollywood v United Plastics Co* 294 F 2d 694 (1961), cited by McCarthy, above n 5, § 11.4.
118 *Ibid,* § 16.9, citing *DM & Antique Import Corp v Royal Saxe Corp* 311 F Supp 1261 (1970).
119 *Ibid,* § 16.12.
120 *Ibid,* § 16.13.
121 Lanham Act § 45, 15 USC § 1127 (definition of 'use in commerce').
122 NAF Case No FA95248 (17 Aug 2000).
123 WIPO Case No D2002-0912 (9 Dec 2002).
124 WIPO Case No D2000-0266 (26 July 2000).

> Actual use in commerce, without more, establishes protectable trademark rights under United States common law. The 'use in commerce' standard requires (a) a bona fide use of a trademark or service mark in the ordinary course of trade, and (b) not made merely to reserve a right in the mark.

Applying the definition, the panellist concluded:

> There is no need for evidence of an actual sale to establish trademark rights, as long as the company is not merely trying to reserve a right in a mark. Here it is clear that geniebooks.com is not merely trying to reserve a mark, but has used and intends to continuously use the mark in the ordinary course of its trade as an electronic book publisher.

Given that the requirements for establishing rights in an inherently distinctive mark are less onerous than the requirements for establishing rights in a descriptive mark, determining whether a mark is inherently distinctive or descriptive is of considerable importance. The distinctions between fanciful, arbitrary, suggestive and descriptive marks are, however, sometimes difficult to make. As one US court has put it:

> Although meant as pigeon-holes, these useful labels are instead central tones in a spectrum; they tend to merge at their edges and are frequently difficult to apply.[125]

The difficulties involved in distinguishing between the categories were illustrated in *Cepheid Corporation v Healthexpert LLC.*[126] In that decision, the complainant had applied for registration of the mark 'GENEXPERT' in connection with a DNA bioanalysis device but, as the mark was not yet registered, the complainant had to rely on common law rights. The complainant submitted that the mark was fanciful, whereas the respondent contended that it was descriptive when used in conjunction with a DNA bioanalysis system. The panellist, however, disagreed with both parties, concluding that the mark was suggestive. In reaching this conclusion, the panellist adopted the most popular test used by the US courts to distinguish between suggestive and descriptive marks, known as the 'degree of imagination' test. As formulated by Judge Weinfield in *Stix Products, Inc v United Merchants Manufacturers, Inc,*[127] the test is as follows:

> A term is suggestive if it requires imagination, thought and perception to reach a conclusion as to the nature of the goods. A term is descriptive if it forthwith conveys an immediate idea of the ingredients qualities or characteristics of the goods.

In applying the test, the panellist took the view that 'imagination, thought and perception' would be required to conclude that 'GENEXPERT' was a DNA bioanalysis system. As the mark was suggestive, not descriptive, it was entitled to protection without proof of secondary meaning.

Given the problems of categorising marks, and especially distinguishing between suggestive and descriptive marks, where there is some doubt a complainant would be well advised to include evidence that an unregistered mark has acquired a secondary meaning through use.

[125] *Soweco, Inc v Shell Oil Co,* 617 F 2d 1178 (5th Cir, 1980).
[126] WIPO Case No D2001-1272 (10 Dec 2001).
[127] 295 F Supp 479, 160 USPQ 777 (1968).

[5.18.2] Descriptive Marks

Common law rights only arise in marks that are categorised as 'descriptive' if a secondary meaning in the mark is established.[128] In other words, once a descriptive mark has acquired a secondary meaning it becomes distinctive, thereby giving rise to common law rights.

As McCarthy explains, a mark is 'descriptive' if it is descriptive of:

- the intended purpose, function or use of the goods.
- the size of the goods.
- the provider of the goods or services.
- the class of users of the goods or services . . .
- a desirable characteristic of the goods or services . . .
- the nature of the goods or services . . .
- the end effect upon the user.[129]

McCarthy further explains that marks that are 'laudatory', in the sense of extolling the merit of a product, are also categorised as 'descriptive'.[130]

A descriptive mark will acquire a secondary meaning when consumers come to associate the mark with a single source. The Supreme Court of North Carolina has explained secondary meaning as follows:

> When a particular business has used words publici juris for so long or so exclusively or when it has promoted its product to such an extent that the words do not register their literal meaning on the public mind but are instantly associated with one enterprise, such words have attained a secondary meaning. That is to say, a secondary meaning exists when in addition to their literal, or dictionary meaning, words connote to the public a product *from a unique source.*[131]

In order for a descriptive mark to acquire a secondary meaning, it is not necessary that consumers be able to recognise the actual identity of the producer or seller. As McCarthy points out:

> All that is necessary to establish a secondary meaning is that the ordinary buyer associates the mark with a single, albeit anonymous, source. The buyer need not know the corporate or personal name of the source.[132]

Many UDRP decisions have considered whether or not a complainant has established common law rights in a descriptive mark through the acquisition of a secondary meaning. A number of decisions have adopted the following statement from a Californian court in *Visser v Macres*[133] to explain what amounts to secondary meaning:

> Secondary meaning is the consumers' association of the mark with a particular source or sponsor and is established out of long association of the name with the business, whereby the name and the business have become synonymous in the mind of the public, submerging the primary meaning of the name in favor of its meaning as a word identifying that business.[134]

[128] McCarthy, above n 5, § 11.15, citing *Two Pesos, Inc v Taco Cabana, Inc* 505 US 763, 120 L Ed 2d 615, 23 USPQ 2d 1081 (1992). See also *Armstrong Paint & Varnish Works v Nu-Enamel Corp* 305 US 315, 83 L Ed 195 (1938); *Wal-Mart stores, Inc v Samara Bros, Inc,* 120 S Ct 1339 (2000).

[129] *Ibid,* § 11.16.

[130] *Ibid,* § 11.17.

[131] *Charcoal Steak House, Inc v Staley* 263 NC 199, 139 SE 2d 185 (1964) (emphasis in original).

[132] McCarthy, above n 5, § 15.8.

[133] 214 Cal App 2d 249, 137 USPQ 492 (1963).

[134] See, eg, *Marty Rodriguez Real Estate, Inc v Lancaster Industries,* WIPO Case No D2000-1468 (24 Dec 2000); *Bill Withers v Robert Dominico,* WIPO Case No D2000-1621 (28 Jan 2001); *United Artists Theatre Circuit, Inc v Domains for Sale Inc* WIPO Case No D2002-0005 (27 Mar 2002).

Other panel decisions have adopted the following definition of secondary meaning from *Inwood Laboratories, Inc v Ives Laboratories, Inc:*[135]

> To establish secondary meaning, a manufacturer must show that, in the minds of the public, the primary significance of a product feature or term is to identify the source of the product rather than the product itself.[136]

Whether a mark has a secondary meaning is a question fact, meaning that it rests on the evidence presented by the complainant. In determining this issue, a number of panel decisions have referred to the following six, non-exclusive factors that were identified by the Second Circuit in *Centaur Communications Ltd v A\S\M\ Communications, Inc:*[137]

> (1) advertising expenditures, (2) consumer surveys linking the mark to a source, (3) unsolicited media coverage of the service, (4) sales success, (5) attempts to plagiarize the mark, and (6) length and exclusivity of the mark's use.[138]

Similarly, in *Candy Direct, Inc v italldirect2u.com,*[139] the panellist stated that the relevant 'evidence of secondary meaning includes length and amount of sales under the mark, the nature and extent of advertising, consumer surveys and media recognition'.

Under US law there are three possible sources of evidence of secondary meaning:

- Direct or circumstantial evidence tending to prove that the relevant buyer class associates the mark with the alleged owner;[140]
- Proof of substantially exclusive and continuous use as a mark in commerce for five years, which gives rise to a rebuttable presumption of secondary meaning;[141] and
- Evidence of ownership of prior registration of the same mark for closely related goods.[142]

A considerable number of UDRP decisions have dealt with the acquisition of a secondary meaning by a descriptive mark. The decisions illustrate the importance of the evidence presented to establish secondary meaning. At the same time, it needs to be borne in mind that the UDRP Rules contemplate only limited evidentiary presentations by the parties, and that the UDRP is not designed to determine conclusively whether or not a complainant has trade mark rights.

The following decisions are examples of the circumstances in which the evidence has been sufficient to establish secondary meaning:

- *Passion Group Inc v Usearch, Inc*[143]—use of the name 'JOB*POSTINGS*' in connection with a careers magazine for students since 1998, together with the revenue generated from advertising.

[135] 456 US 844, 851, n 11; 102 S Ct 2182, 2187, n 11 (1982).

[136] See, eg, *2E Corporation v Imagisys, Inc,* eRes Case No AF-0162 (15 May 2000); *Lowestfare.com LLC v US Tours & Travel, Inc,* eRes Case No AF-0284 (9 Sept 2000); *Media West-GSI, Inc v EARTHCARS.COM, INC,* WIPO Case No D2000-0463 (28 July 2000).

[137] 830 F 2d 1217, 1221 (1987). This list was formulated in *Thompson Medical Co v Pfizer, Inc,* 753 F 2d 208 (2d Cir 1985). For a similar list see *Echo Travel, Inc v Travel Associates, Inc,* 870 F 2d 1264 (7th Cir 1989).

[138] See, eg, *Media West-GSI, Inc v EARTHCARS.COM, INC,* WIPO Case No D2000-0463 (28 July2000); *Nintendo of America, Inc v Garrett N. Holland,* WIPO Case No D2000-1483 (11 Jan 2001); *United Artists Theatre Circuit, Inc v Domains for Sale Inc,* WIPO Case No D2002-0005 (27 Mar 2002).

[139] NAF Case No FA514784 (22 Aug 2005).

[140] McCarthy, above n 5, §§ 15:61; 15:69–15:73.

[141] Lanham Act § 2(f), 15 USC § 1052(f); McCarthy, above n 5, § 15:62.

[142] McCarthy, above n 5, § 15:63.

[143] eRes Case No AF-0250 (10 Aug 2000).

- *Smart Design LLC v Carolyn Hughes*[144]—use of the name 'SMART DESIGN' in connection with designing and developing consumer products, predominantly for Fortune 500 companies, including that the complainant company had received many industry awards, and was known as one of the top design consultancies in the US.
- *Great Plains Metromall, LLC v Gene Creach*[145]—use of the name 'THE GREAT MALL OF THE GREAT PLAINS' in connection with developing and operating a mall since 1996, including the expenditure of substantial sums on advertising and promotion, and registration and extensive use of the domain name <greatmallgreatplans.com>.
- *Universal City Studios, Inc v David Burns*[146]—use of the name 'ADAM-12' in connection with a well-known television programme, and related merchandising, for over 25 years.
- *The EW Scripps Company v Sinologic Industries*[147]—use of the name 'THE COMMERCIAL APPEAL' in connection with a significant local newspaper in Memphis, Tennessee, since 1894.
- *Hachette Filipacchi Media US v URTHERE Productions, Inc*[148]—use of the name 'CHARTER FLEET INTERNATIONAL' on a magazine about charter yachts on the basis of a single copy of the complainant's magazine and production of evidence of five years of continuous use.

The following are examples of decisions in which the evidence has been insufficient to establish a secondary meaning in a descriptive mark:

- *Lowestfare.com LLC v US Tours & Travel, Inc*[149]—use of the name 'LOWESTFARE.COM' (a descriptive, laudatory designation) on a Web site under the domain name <lowestfare.com>, from which travel services were offered. The panellist refused to find that the mark was so well known that 'judicial notice' could be taken of a secondary meaning.[150]
- *Media West-GSI, Inc v EARTHCARS.COM, INC.*[151]—use of the name 'FREEPRESSCLASSIFIED.COM' on a Web site under the domain name <freepressclassified.com> in order to operate a business for classified advertisements, on the basis that the complainant had presented no evidence of consumer surveys, or even anecdotal evidence that consumers had come to associate the domain name with the complainant's business, and that the 27-month period during which the domain name had been used was not long enough for it to have acquired a secondary meaning.
- *Weatherford International, Inc. v Paul Wells*[152]—use of the name 'WELLSERV' on promotional and advertising material, the panellist concluding that the complainant had failed to present evidence of secondary meaning in the relevant public.
- *Keystone Publishing, Inc v UtahBrides.com*[153]—use of the name 'UTAHWEDDINGS.COM' since 2001 to advertise wedding services in the State of Utah via the Internet, evidenced by records of billboards, television advertising and Internet hits and affidavits from business

144 WIPO Case No D2000-0993 (18 Oct 2000)
145 NAF Case No FA97044 (18 May 2001).
146 WIPO Case No D2001-0784 (1 Oct 2001).
147 WIPO Case No D2003-0447 (1 July 2003).
148 WIPO Case No D2002-0143 (30 May 2002).
149 eRes Case No AF-0284 (9 Sept 2000).
150 For circumstances in which a court will find that a mark is so well known that it will take judicial notice that the mark has become distinctive see McCarthy, above n 5, § 15:31.
151 WIPO Case No D2000-0463 (28 July 2000).
152 NAF Case No FA153626 (19 May 2003).
153 WIPO Case No D2004-0725 (17 Nov 2004).

clients, largely on the basis that the name had been used for too short a time to have acquired a secondary meaning.

- *Lilian Vachovsky v A+Hosting Inc*[154]—use of the names 'APLUS' and 'APLUS.NET' in connection with the Web site associated with the <aplus.net> domain name, on the basis that the complainant failed to produce any evidence, apart from mere assertions, of secondary meaning.

The panellist's comments in *Weatherford International* on the importance of submitting sufficient evidence to establish secondary meaning must be borne in mind where a claim is made for common law rights in a descriptive mark under US law:

> Although Complainant asserts common law rights in the WELLSERV mark, it failed to submit any evidence indicating extensive use or that its claimed mark has achieved secondary source identity. More specifically, the Complaint references Exhibit 2 when stating, 'Weatherford has used the WELLSERV mark since 1988, and has prominently and extensively used, promoted and advertised its mark.' However, Exhibit 2 is a letter sent from the Vice President of Complainant's Intervention Services division to Respondent, which includes minimal references to the 'WellServ Ltd.' corporate entity. Although Complainant's WELLSERV product and related services may well be well-known among relevant consumers, that is a finding that must be supported by evidence and not self-serving assertions. Precedent does not support merely protecting Complainant's assertions in lieu of any supporting evidence, statements of proof (eg, business sales figures, revenues, advertising expenditures, number of consumers served, trademark applications or intent-to-use applications). Simply operating under a business name does not, per se, constitute secondary source identification sufficient to warrant a finding of common law rights in the WELLSERV mark. Complainant's burden requires proffering supportive evidence or information that indicates relevant consumers identify its alleged WELLSERV mark as representative of its services and identity.

Under US law, mere registration of a domain name is not sufficient evidence that the name has a secondary meaning, even if the associated web site is popular.[155] The position under the UDRP is illustrated by the decision in *Mark Jenkins v DB Enterprise Attn: Dan Black.*[156] In that decision, the complainant was the registered holder of the domain name <link-o-rama.com>, which consisted of links to adult sites and sites containing pornographic material. The evidence submitted by the complainant included statistics compiled by the SexTracker counter service that indicated that the site was very popular. The panellist, however, concluded that the complainant had not submitted sufficient evidence to establish that the relevant consuming public associated the domain name with the complainant and that the mere record of visits to a web site was insufficient to establish common law rights in the mark.

[5.18.3] Generic Terms

A generic term is a name of a product or service itself. Famous examples of generic terms include the product names, 'ASPIRIN'[157] and 'CELLOPHANE'.[158] A generic term cannot operate to indicate the origin or source of a product. It follows that a generic term cannot be distinctive and cannot function as a trade mark.[159]

154 WIPO Case No D2006-0703 (26 Oct 2006).
155 *In re Candy Bouquet International, Inc*, 73 USPQ 2d 1883 (2004).
156 WIPO Case No D2001-1106 (27 Nov 2001).
157 *Bayer Co v United Drug Co*, 272 F 505 (DCNY, 1921).
158 *Du Pont Cellophane Co v Waxed Products Co*, 85 F 2d 75, 81 L Ed 443 (2d Cir, 1936).
159 See McCarthy, above n 5, § 12.1.

This fundamental point was made in the following terms by the Seventh Circuit in *Liquid Controls Corp v Liquid Control Corp:*

> A generic term is one that is commonly used as the name of a kind of goods . . . Unlike a trademark, which identifies the source of a product, a generic term merely specifies the genus of which the particular product is a species.[160]

The test for determining whether a term is (or has become) generic was expressed by Learned Hand J as follows:

> The single question, as I view it, in all these cases, is merely one of fact: What do the buyers understand by the word for whose use the parties are contending?[161]

The standard for determining whether a term is generic is not whether the term has some generic significance, but whether the principal meaning of the term is generic.[162]

If a term is determined to be generic, no amount of evidence of secondary meaning can make the term distinctive, meaning that common law rights cannot arise in the term.[163] The following are examples of UDRP decisions that have found that there are no trade mark rights in a generic term:

- *Geneva Watch, Inc v Rafique Budhwani.*[164] The complainant, an importer and seller of quality, quartz watches registered the domain name <geneva-watch.com> and claimed common law rights in the term 'GENEVA WATCH' by virtue of use of the term for more than 25 years. The panellist held that the complainant, having disclaimed rights in the generic terms 'GENEVA' and 'WATCH' in registering a trade mark that included the terms, could not establish common law rights on the basis that it is impossible to establish secondary meaning in a combination of generic words.
- *Shopping.com v Internet Action Consulting.*[165] The complainant had registered the domain name <shopping.com> and offered electronic commerce services through its Web site. Although the complainant presented evidence that it had invested more than US $8 million in advertising and marketing, the panellist held that there could be no trade mark rights in the name 'SHOPPING.COM', as no amount of secondary meaning can convert a generic term into a trade mark.
- *Tough Traveler, Ltd v Kelty Pack, Inc.*[166] The complainant, a manufacturer of backpacks and child carriers, used the name 'KID CARRIER' on child carriers for 15 years and invested substantial sums in advertising, catalogues and sales material. The panellist held that the term was generic for 'child carriers', even though there are other generic terms for the product, and therefore that there were no trade mark rights in the term.
- *Successful Money Management Seminars Inc v Direct Mail Express.*[167] The complainant claimed that it used the term 'SEMINAR SUCCESS' in connection with its business of creating, selling and supporting educational seminar systems, but the panel held that the terms 'SEMINAR' and 'SUCCESS' were generic and could not acquire a secondary meaning.

[160] 802 F 2d 934, 936; 231 USPQ 579 (1986).

[161] *Bayer Co v United Drug Co* 272 F 505, 509 (DCNY 1921).

[162] McCarthy, above n 5, § 12:6.

[163] See, eg, *SS Kresge Co v United Factory Outlet, Inc* 634 F 2d 1, 208 USPQ 313 (1st Cir, 1980); *Miller Brewing Co v Falstaff Brewing Co* 655 F 2d 5, 211 USPQ 665 (1st Cir, 1981). See also McCarthy, above n 5, § 12:46.

[164] NAF Case No FA97323 (19 July 2000).

[165] WIPO Case No D2000-0439 (28 July 2000).

[166] WIPO Case No D2000-0783 (28 Sept 2000).

[167] NAF Case No FA96457 (7 Mar 2001).

[5.18.4] Composite Marks

A composite mark consists of some combination of descriptive or generic terms. The status of a composite mark depends upon the status of the composite as a whole, rather than the status of the individual terms that comprise the composite. For example, the combination of two generic words may be generic or non-generic, depending upon an assessment of the combination as a whole.

First, a combination of descriptive terms may amount to a non-descriptive mark, meaning that there is no need to establish a secondary meaning. For example, the combination mark 'SUGAR & SPICE' used in connection with bakery products was held to be non-descriptive, and therefore inherently distinctive.[168] Secondly, a combination of a descriptive and a generic term may amount to a descriptive mark, meaning that common law rights may arise upon the acquisition of a secondary meaning. Thirdly, a composite mark that contains a generic term together with a non-generic component may give rise to common law rights in the composite. For example, US courts have found that common law rights exist in the famous mark 'COCA-COLA', which was held to contain the generic term 'COLA' and the non-generic component, 'COCA'.[169] Fourthly, common law rights may arise in a composite mark consisting of two generic terms taken together, provided the composite is distinctive, for example because it has acquired a secondary meaning.[170] In such cases, however, there is a high burden on the party claiming that the term has acquired a secondary meaning, as the courts are reluctant to confer protection on words with a commonly understood meaning.

The following are examples of UDRP decisions that have applied US trade mark principles relating to composite marks to domain name disputes:

- *Pet Warehouse v Pets.Com, Inc.*[171] The complainant claimed common law rights in the composite term 'PET WAREHOUSE', but the panel rejected the claim on the basis that the composite was generic. Moreover, the panel pointed out that even if the composite could be protected the complainant had not established the high level of secondary meaning required to prove distinctiveness in such cases.
- *Tall Oaks Publishing, Inc v National Trade Publications, Inc.*[172] The complainant was the owner of the mark 'ULTRAPURE WATER' registered on the US Supplemental Register in connection with use on technical journals, and therefore had to rely on common law rights. The panellist held that, although the composite term 'ULTRAPURE WATER' was generic for a type of treated water, it was descriptive when used in connection with a technical journal. Despite limited evidence, the panellist found that the composite mark had acquired sufficient secondary meaning for rights to arise.
- *State of Florida Department of the Lottery v John Zuccarini.*[173] The complainant claimed rights in the composite marks 'FLORIDA LOTTERY' and 'FLA LOTTERY', which the panellist found were descriptive of lotteries conducted in Florida. The panellist held that the composite terms had acquired secondary meaning by virtue of extensive use in connection with lottery services.

[168] *In re Colonial Stores, Inc* 394 F 2d 549 (CCPA, 1968).
[169] See, eg, *Coca-Cola Co v Koke Co of America* 254 US 143, 65 L Ed 189 (1920); *Dixi-Cola Laboratories, Inc v Coca-Cola Co* 117 F 2d 352, 48 USPQ 164 (4th Cir, 1941).
[170] See, eg, *Mil-Mar Shoe v Shonac* 75 F 3d 1153 (1996).
[171] WIPO Case No D2000-0105 (13 Apr 2000).
[172] NAF Case No FA94346 (5 May 2000).
[173] WIPO Case No D2002-0307 (25 May 2002).

- *The Men's Wearhouse Inc v Brian Wick.*[174] The complainant, having disclaimed the exclusive right to use the term 'WAREHOUSE' in its registered mark 'THE SUIT WAREHOUSE', claimed common law rights in the composite term 'SUIT WAREHOUSE'. Assuming that the composite was not generic, the panellist concluded that the complainant had failed to establish sufficient secondary meaning in the composite term.
- *SA Bendheim Co, Inc v Hollander Glass.*[175] The complainant claimed rights in the composite mark, 'RESTORATION GLASS', which was registered on the US Supplemental Register. The panellist held that the composite mark was descriptive and had acquired a secondary meaning as a source indicator of the complainant's antique sheet glass product as a result of long use and promotion, and despite the fact that third parties made use of the descriptive term.

[5.19] 'Common Law Rights' under English Law

The history of English common law protection of trade marks is dealt with at [5.3] above. As explained at [5.3], the tort of passing off survived the introduction of the English system of registration of trade marks and, in *Spalding v Gamage,*[176] was given its modern form as a tort that protects business goodwill against misrepresentation, and not a property right in trade marks.

As further explained, it has become customary to refer to two forms of the action for passing off: the 'classic form' and the 'extended form'. The 'classic form', as restated by the House of Lords in *Reckitt & Colman Products Ltd v Borden,*[177] requires the claimant to establish three elements, known as the 'classic trinity': reputation or goodwill deserving protection; misrepresentation by the defendant likely to confuse or mislead the public; and damage suffered, or likely to be suffered, as a result of the misrepresentation. The extended form of passing off, as formulated by the House of Lords in the *Advocaat* case,[178] applies to a broader spectrum of circumstances, such as to the protection of names of products derived from a geographical area. Nevertheless, it remains important to note that passing off is neither a general tort of unfair competition, nor a general tort that applies to all misrepresentations that injure other traders.[179]

The essence of the tort of passing off is an actionable misrepresentation. As Lord Parker expressed it in *Spalding v Gammage:*

> the basis of a passing-off action being a false representation by the defendant, it must be proved in each case as a fact that the false representation was made. It may, of course, have been made in express words, but cases of express misrepresentation of this sort are rare. The more common case is, where the representation is implied in the use or imitation of a mark, trade name, or get-up with which the goods of another are associated in the minds of the public.[180]

[174] NAF Case No FA117861 (16 Sept 2002).
[175] NAF Case No FA142318 (13 Mar 2003).
[176] (1915) 32 RPC 273.
[177] [1990] RPC 341.
[178] *Erven Warnink v Townend* [1980] RPC 31.
[179] See Kitchin *et al*, above n 20, paras [15-037]–[15-041]. For example, exaggerated claims or 'puffing' about the quality of a trader's goods or false statements that they are better than a rival's will not be a passing off.
[180] (1915) 32 RPC 273, 284.

Consequently, the action for passing off does not directly confer rights on marks or other indicia,[181] but protects marks as an incident of the protection of business goodwill against misrepresentation. As Parker J explained the distinction in *Burberrys v JC Cording & Co Ltd:*

> On the one hand, apart from the laws as to trade marks, no one can claim monopoly rights in the use of a word or name. On the other hand, no one is entitled by the use of any word or name, or indeed in any other way, to represent his goods as being the goods of another to that other's injury. If an injunction be granted restraining the use of a word or name, it is no doubt granted to protect property, but the property, to protect which it is granted, is not property in the word or name, but property in the trade or goodwill which will be injured by its use.[182]

The action for passing off differs from an action for trade mark infringement in important respects. First, the claimant in an action for passing off must establish both a reputation worth protecting and a false representation which damages, or is likely to damage, the goodwill of the trader, neither of which needs to be established in an action for trade mark infringement. Secondly, although actions in passing off commonly involve the misuse of some mark or other indicia of the claimant, the action does not protect goodwill in the mark as such, but is based on protecting goodwill in the claimant's business as a whole. This means that the action for passing off is more flexible than an action for trade mark infringement, applying to a greater variety of circumstances, over and above the infringing use of a trade mark. For example, an action for passing off may lie in respect of a false representation made in relation to indicia that cannot be registered.[183] Moreover, an action in passing off may even be available against a registered trade mark owner.[184]

Although the action for passing off is not an action to protect rights in a trader's mark, as such, where the action is brought to protect business goodwill that is embodied in an unregistered mark, the mark is often referred to as a 'common law mark'.[185] Despite this terminology, the protection of the mark under the action for passing off remains parasitical on the protection of goodwill. Consequently, while there are similarities between the protection of common law marks under US trade mark law and the English action for passing off, it is important to bear in mind the extent to which different legal principles apply. While under US common law use of an unregistered mark establishes rights in the mark, the action for passing off is generally only available in relation to an unregistered mark if use of the mark by the defendant amounts to a misrepresentation. Given that the English action for passing off is directed at preventing misrepresentation, and not with establishing rights in an unregistered mark, difficulties arise in applying paragraph 4(a)(i) of the UDRP, which requires that a complainant establish rights in a mark. To the extent that the action for passing off is not, strictly speaking, an action for trade mark infringement, there is inevitably a degree of tension between the protection of common law marks under English law and under the UDRP.

[181] See, eg, *Payton & Co v Snelling, Lampard & Co* (1899) 17 RPC 48; *Burberrys v JC Cording & Co Ltd* (1909) 26 RPC 693.

[182] (1909) 26 RPC 693, 701.

[183] See, eg, *Reddaway v Banham* [1896] AC 199.

[184] See, eg, *Van Zeller v Mason, Cattley* (1907) 25 RPC 37; *Thorne v Pimms* (1907) 26 RPC 221.

[185] See Kitchin *et al*, above n 20, para [15-034].

[5.19.1] Requirements for Common Law Rights under English Law: Distinctiveness and Secondary Meaning

As explained at [5.15], under US common law, rights will arise in an unregistered mark that is inherently distinctive or a descriptive mark that has become distinctive. In bringing an action for passing off in relation to the misuse of an unregistered mark, the degree of distinctiveness of the mark is also critical to the success of the action. Under English common law, however, the extent to which an unregistered mark is distinctive is a factor to be taken into account in determining whether or not there has been a misrepresentation and not in determining whether common law rights arise in the mark. As Wadlow explains the general position:

> the question of whether a particular mark is distinctive of a certain trader is only one aspect of the wider question of whether there has been a misrepresentation. If the claimant in a typical case is successful it may be assumed that the mark relied on was distinctive of him, that the defendant's corresponding mark was confusingly similar, and that there were no other factors to prevent deception.[186]

As with US trade mark law, distinctiveness under English-oriented law is a term of art that refers to the extent to which a mark has come to be identified with the goods or services of a trader. Unlike US courts, however, courts assessing the distinctiveness of a mark in an action for passing off do not draw a hard and fast legal distinction between inherently distinctive marks and descriptive marks, as the critical issue in an action for passing off is whether the mark is distinctive in fact, and not whether it has a capacity to be distinctive.

The difference in approach follows from the different role performed by the concept of distinctiveness in the respective legal systems: under US law distinctiveness must be established in order for common law rights to arise in a mark, whereas under English-oriented law distinctiveness is a factor to be taken into account in determining whether there has been a misrepresentation. Wadlow outlines the approach taken to distinctiveness in passing off actions in the following terms:

> Passing-off is relatively unconcerned with the distinction drawn in trade mark law between inherent capacity to distinguish and distinctiveness in fact. If factual distinctiveness exists, then it does not matter whether it was achieved with ease for a mark well adapted to distinguish or with difficulty for a mark of the opposite kind. If factual distinctiveness does not exist a traditional passing-off case must fail.[187]

Nevertheless, the practical differences between the approaches taken to distinctiveness under the American and English legal systems should not be over-drawn. Although the approach taken under the law of passing off is more flexible, it is clear that less is required to establish distinctiveness in a mark with an inherent capacity to distinguish and that, correspondingly, a claimant faces a significantly higher burden in establishing that a descriptive mark has become distinctive. In other words, while a mark only becomes distinctive through use,[188] less use is required to establish distinctiveness in an inherently distinctive mark than is required for a descriptive term. As Parker J pointed out in *Burberrys v JC Cording & Co Ltd:*

[186] Wadlow, above n 14, para [8-3].
[187] *Ibid*, para [8-4].
[188] *Ibid*, para [8-9].

> the degree of readiness with which the court will infer the probability of deception must depend on the circumstances of each particular case, including the nature of the word or name, the use of which is sought to be restrained. It is important for this purpose to consider whether the word or name is prima facie in the nature of a fancy word or name, or whether it is prima facie descriptive of the article in respect of which it is used. It is also important for the same purpose to consider its history, the nature of its use by the person who seeks the injunction, and the extent to which it is or has been used by others. If the word is prima facie descriptive or be in general use, the difficulty of establishing the degree of deception is greatly increased.[189]

In determining whether a mark is distinctive, it is not necessary to show that the mark has become associated with the trader in the minds of the majority of consumers; merely that a substantial proportion of the relevant public is likely to be deceived.[190] In general terms, if a mark is inherently distinctive or arbitrary, then less is required to show that a proportion of the public will be deceived by use of the mark than if the mark is descriptive, where the mark needs to be recognised as distinctive by a much greater proportion of the relevant public.[191]

Although courts in English-oriented legal systems dealing with actions in passing off do not make the precise distinctions between arbitrary, fanciful, suggestive and descriptive marks drawn by US courts, a distinction is sometimes made between 'fancy' and descriptive words.[192] As a 'fancy' word has no obvious association with the product of a trader, it is easier to establish that the term is distinctive. A mark that is, to some extent, descriptive of a trader's product, however, requires more to show that it has become associated with the trader's product to the degree that it has become distinctive. As in US trade mark law, descriptive words will become distinctive for the purpose of passing off if they acquire a secondary meaning that denotes the trader's goods, rather than merely describing them. As Lord Hershell explained, in rejecting the contention that a defendant could not be restrained from using a descriptive term in *Reddaway v Banham*:

> the fallacy lies in overlooking the fact that a word may acquire in a trade a secondary signification differing from its primary one, and that if it is used, to persons in the trade who will understand it . . . in its secondary sense, it will none the less be a falsehood that in its primary sense it may be true.[193]

In that case, the House of Lords held that the term 'camel hair', when used in connection with the claimant's belts, had acquired a secondary meaning distinctive of the claimant's products, in addition to describing the product. Similarly, in *BM Auto Sales v Budget Rent-A-Car,*[194] the Australian High Court held that the words 'budget' and 'rent-a-car' had become so identified with the trader's business that they had developed a secondary meaning, so that their use as a mark would be protected.

As is the case under US trade mark law, however, a descriptive word will receive less protection under the law of passing off than an inherently distinctive mark. In other words, it is less likely that the use of a mark that differs, in minor respects, from a descriptive

[189] (1909) 26 RPC 693, 701.

[190] See, eg, *Saville Perfumery Ltd v June Perfect* (1941) 58 RPC 147; *Kark (Norman) Publications Ltd v Odhams Press Ltd* [1962] RPC 163.

[191] Wadlow, above n 14, para [8-40].

[192] See, eg, *Parsons v Gillespie* [1898] AC 239; *Cellular Clothing Co v Maxton & Murray* [1899] AC 326; *McCain International v Country Fair Foods* [1981] RPC 69; *10th Cantanae Pty Ltd v Shoshana Pty Ltd* (1987) 10 IPR 289; *Telmak Teleproducts (Aust) Pty Ltd v Coles Myer Pty Ltd* (1988) 12 IPR 297.

[193] [1896] AC 199, 212–3; (1896) 13 RPC 218, 229.

[194] (1977) 51 ALJR 254.

term will amount to a misrepresentation than would be the case if the mark were inherently distinctive. The principles underlying this approach were stated by Stephen J of the Australian High Court in a passage approved by the Privy Council in *Cadbury Schweppes v Pub Squash Co:*

> There is a price to be paid for the advantages flowing from the possession of an eloquently descriptive trade name. Because it is descriptive it is equally applicable to any business of a like kind, its very descriptiveness ensures that it is not distinctive of any particular business and hence its application to other like businesses will not ordinarily mislead the public. In cases of passing-off, where it is the wrongful appropriation of the reputation of another or that of his goods that is in question, a plaintiff which uses descriptive words in its trade name will find comparatively small differences in a competitor's trade name will render the latter immune from action . . .[195]

As in US trade mark law, the use of a generic term that applies to a product will not be protected by the law of passing off against the use of that term in connection with the product. Unlike US courts, however, English-oriented courts have been more likely to use the term 'descriptive' in a sense that also embraces 'generic' terms.[196] As a result, it is generally more difficult to determine under English law whether a claimant has failed in an action for passing off because the term was generic, or because it was a descriptive term that has failed to acquire a secondary meaning. Although the reluctance of English courts precisely to distinguish generic from descriptive terms has created a degree of conceptual confusion in some cases, it remains the case that an action in passing off will not lie to protect a descriptive or generic term, unless the term has become distinctive of a trader's product by virtue of use.[197]

[5.19.2] The Action for Passing Off and the UDRP: Some Problems in Practice

The first important decision on the existence of common law rights under English-oriented law for the purpose of the UDRP was *Jeanette Winterson v Mark Hogarth.*[198] The decision established the basic principle, since confirmed by many subsequent decisions, that a complainant can establish common law rights in an unregistered mark if the complainant can bring an action in passing off. As explained below, however, the right to bring an action in passing off is a necessary, but not sufficient, condition for establishing unregistered trade mark rights under the UDRP.

In *Jeanette Winterson* the complainant, the well-known English author Jeanette Winterson, brought a complaint against the registration of domain names that consisted of her personal name, namely <jeanettewinterson.com>, <jeanettewinterson.net> and <jeanettewinterson.org>. The respondent argued that a personal name was not a trade mark, directly raising the issue of the relationship between the action for passing off and the need to establish rights in a mark under paragraph 4(a)(i). The central problem the panellist had to confront was that paragraph 4(a)(i) is concerned with whether there are rights in a mark, and not whether registration of a domain name amounts to a passing off. In this respect, the panellist observed that the:

195 [1981] RPC 429, citing Stephen J in *Hornsby Building Information Centre Pty Ltd v Sydney Building Information Centre Ltd* (1978) 140 CLR 216, 229.

196 See Wadlow, above n 14, para [8-55].

197 See, eg, *Daimler Motor Co (1904) Ltd v London Daimler Co Ltd (1907)*(1907) 24 RPC 379.

198 WIPO Case No D2000-0235 (22 May 2000).

> issue is not whether the Respondent has committed passing-off by registering the three domain names in issue, it is merely whether under English common law unauthorized use of a mark can be restrained other than by an action for infringement of a trade mark. The mere fact that the right to sue for infringement of an unregistered trade mark is not available under English law does not affect a person's rights of action against another for passing-off . . . The case for decision here does not concern whether or not passing-off has occurred but whether the Complainant (Jeanette Winterson) has rights in her name sufficient to constitute a trade mark for the purposes of para. 4a of the Policy.

One difficulty facing panellists dealing with the status of trade mark rights in a famous personal name under English law is the decision of the English Court of Appeal in *Elvis Presley Trade Marks.*[199] In that case, the Court of Appeal held that the name 'ELVIS PRESLEY' was not sufficiently distinctive to be registered as a trade mark. In upholding the first instance decision of Laddie J, Robert Walker LJ stated:

> That conclusion was reached by a number of intermediate steps, one of which was the judge's finding that members of the public purchase Elvis Presley merchandise not because it comes from a particular source, but because it carries the name or image of Elvis Presley.[200]

The panellist in *Jeanette Winterson* dealt with this difficulty by pointing out that *Elvis Presley* was concerned with distinctiveness for the purpose of trade mark registration and not for determining whether there were common law rights in the name. In this respect, the panellist stated:

> *trademarks* where used in para. 4a of the Policy is not to be construed by reference to the criteria of registrability under English law [the ELVIS PRESLEY case] but more broadly in terms of the distinctive features of a person's activities. In other words, akin to the common law right to prevent unauthorized use of a name.

Although the panellist in *Jeanette Winterson* clearly distinguished between the ability to bring an action for passing off and the establishment of common law trade mark rights, the decision did not precisely identify the requirements for common law rights to arise under English law. In part, this was because the decision was concerned more with distinguishing the right to register a mark from common law rights in a mark.

The differences between registered trade mark rights and rights in a mark arising from the action for passing off were further explored by the panellist in *Julian Barnes v Old Barn Studios Limited*[201] and a series of related decisions dealing with the names of well-known English authors that were registered as domain names.[202] In each case, the respondent relied on the decision in *Elvis Presley* to support the contention that the personal name of the author lacked the requisite level of distinctiveness. The panellist, however, followed *Jeanette Winterson* in deciding that an unregistered mark did not need to satisfy the requirements for trade mark registration in order to be sufficiently distinctive to bring an action in passing off.

In reaching this conclusion, the panellist first explained the meaning of a 'common law trade mark' under English law:

[199] [1999] RPC 567.

[200] [1999] RPC 567, 587.

[201] WIPO Case No D2001-0121 (26 Mar 2001).

[202] See, eg, *Louis De Bernieres v Old Barn Studios Limited*, WIPO Case No D2001-0122 (26 Mar 2001); *Antony Beevor v Old Barn Studios Limited*, WIPO Case No D2001-0123 (26 Mar 2001); *Margaret Drabble v Old Barn Studios Limited*, WIPO Case No D2001-0209 (26 Mar 2001).

> This involves understanding what is meant by 'passing off', because the easiest way of defining a common law trade mark is to say that it is an unregistered mark used by its proprietor in the course of trade, the unauthorized use (or imitation) of which by another trader will lead to passing off.

The panellist proceeded to explain how, given that an action for passing off could be available in relation to marks that are incapable of registration, the *Elvis Presley* decision is irrelevant in determining whether there are common law rights in a mark. As the panellist pointed out, *Elvis Presley* decided that the singer's name was not distinctive of the merchandise in relation to which registration was sought, and was not concerned with the broader question of whether an action could be brought in passing off to restrain the unauthorised use of the name as a mark.

Finally, the panellist addressed the central question whether the action in passing off, which is directed against misrepresentations arising from the unauthorised use of a mark, could be said to give rise to rights in a mark for the purpose of paragraph 4(a)(i). As the panellist explained this issue:

> The question still remains as to whether the Policy extends to common law trade marks. On the face of paragraph 4(a)(i) of the Policy there is no limit to registered trade marks or service marks. It simply refers to 'a trade mark or service mark in which the Complainant has rights'. The problem (if there be a problem) stems from the fact that common law rights in passing off do not, strictly speaking, confer rights in anything. They constitute rights (inter alia) to restrain misrepresentations, which are likely to damage the claimant's goodwill. As indicated above, those misrepresentations are commonly made by use of the claimant's name or something similar to it, the claimant's name being a symbol of the claimant's goodwill. It is for this reason that, loosely, they are often referred to as 'rights in' the relevant common law trade mark. Certainly it is fair to describe them as 'rights in respect of' the relevant common law trade mark, to borrow the terminology from paragraph 4(a)(ii) of the Policy.

Unfortunately, like the panellist in *Jeanette Winterson,* the panellist in *Julian Barnes* was not precise in distinguishing between the right to bring an action in passing off and common law rights in a mark. As pointed out in subsequent decisions dealing with the protection of personal names, this can easily lead to the false assumption that establishing a right to bring an action for passing off is sufficient for there to be common law trade mark rights. As these decisions indicate, however, in addition to being able to bring an action for passing off to protect a name, it is necessary for the name to have acquired a secondary meaning, so that it operates as a mark. For example, in a carefully reasoned decision in *R E 'Ted' Turner and Ted Turner Film Properties, LLC v Mazen Fahmi,*[203] the majority of the panel decided that the complainant, 'TED TURNER', had no common law rights in his personal name as there was insufficient evidence that he had 'ever offered goods or services under that name, or that that name has acquired any secondary meaning in relation to any goods or services'. The complexities involved in establishing common law trade mark rights in a personal name under English law are dealt with further at [5.20.3].

As a general rule, then, in order for there to be common law trade mark rights under English law, in addition to being able to bring an action for passing off the complainant must establish that the term has become distinctive. An additional complexity that arises under English law concerns the difference between the level of distinctiveness required for registration of a trade mark and the level of distinctiveness required to establish passing off.

[203] WIPO Case No D2002-0251 (4 July 2002).

As pointed out in *Kerly's Law of Trade Marks and Trade Names*, the distinctiveness required for registration of a mark is greater than the distinctiveness required for passing off.[204] The reasons for this difference are explained as follows:

> The principles developed in passing off cases allow a flexible approach to the facts of a particular case. If the case involves a sign with a primary descriptive meaning, the court can effectively alter the extent to which it requires the plaintiff to demonstrate a secondary meaning by balancing the interests of the parties and the public. For example, the plaintiff may have to endure some confusion and small differences may suffice, and a different result may obtain if the court considers that people really are being misled. By contrast, registration of a trade mark confers an exclusive right—a monopoly right . . . In order to justify the exclusive right which covers the wide range of situations which fall within the infringement provisions, it is suggested that the applicant must demonstrate that a secondary distinctive meaning has displaced the primary descriptive meaning to a greater extent than may be required for success in passing off.[205]

The following principles arise from the above discussion of the problems of establishing common law trade mark rights by means of the English action for passing off:

- A common law trade mark in an English-oriented legal system is an unregistered mark the unauthorised use of which will give rise to an action for passing off.
- It is not every action in passing off that will amount to common law trade mark rights, but only the sub-set of the action for passing off that involves misrepresentations arising from the unauthorised use of a mark.
- Therefore, in order for an action for passing off to arise for unauthorised use of the term, the term must be a mark. This means that the term must have become distinctive (acquired a secondary meaning) by virtue of use.
- The distinctiveness required to bring an action for passing off is different from, and may be less than, the level of distinctiveness required for a mark to be registered.

[5.19.3] The Action for Passing Off and the UDRP: Some Examples

The inquiry into whether there are common law rights in a mark under English-oriented legal systems so as to satisfy the UDRP often resolves to the question whether or not the mark is distinctive. This is still something of a simplification of the law but, especially since the publication of the *WIPO Overview* in March 2005, most panels have concentrated on whether or not the mark has acquired a secondary meaning. The following are examples of UDRP decisions that have found there to be common law trade mark rights under English-oriented legal systems:

- *Australian Trade Commission v Matthew Reader*.[206] The complainant, the Australian federal government's export agency, claimed common law trade mark rights in the term 'AUSTRADE'. The panel found that the name had acquired a secondary meaning through use, including the investment of over AUS$10 million to raise awareness of its services and use on publicity material. In reaching this conclusion, the panel held that the combination of the generic terms 'AUS' and 'TRADE' was not itself generic, given the long association of the term with the services of the complainant.

[204] Kitchin *et al*, above n 20, para [8-133].
[205] *Ibid*, para [8-136].
[206] WIPO Case No D2002-0786 (12 Nov 2002).

- *ITIPS Ltd v Saeed Ahmed.*[207] The complainant claimed common law rights in the mark 'ACTUALTESTS.COM' on the basis that it had over 50,000 paying customers and had engaged in extensive advertising. The panel held that although the term was descriptive it had become distinctive through use. In reaching this conclusion, the panel pointed out that distinctiveness is based solely on use in English-oriented legal systems.
- *Harbord Real Estate v Austin Robinson.*[208] The complainant claimed common law trade mark rights in the name 'HARBORD REAL ESTATE'. Citing the WIPO Consensus View, the panellist held that the name had become distinctive by virtue of trade conducted under it since 1981, evidence of substantial sales, extensive advertising and industry awards.
- *Control Techniques Limited v Lektronix Ltd.*[209] The complainant claimed common law trade mark rights in the name 'CONTROL TECHNIQUES'. Citing the WIPO Consensus View, the panellist held that the name, although generic, had become distinctive as 'the Complainant has been selling its products worldwide under the name CONTROL TECHNIQUES since 1985, it has advertised extensively under that name, its products are recognized by consumers and the media, it has adopted the name as its company and business name, its domain names have adopted the name around the world, its publications use the name and it continues to spend large sums of money on advertising and promoting the name at exhibitions and conferences'.

Personal Names

[5.20] Personal Names

One of the most controversial issues in the application of the UDRP has been the extent to which the policy is available to protect personal names, especially famous personal names. As explained at [3.4], the WIPO Final Report recommended limiting the dispute-resolution procedure to abusive registration of trade marks or service marks and suggested that registrations that violated trade names, geographical indications or personality rights would not fall within the procedure.

As explained at [3.11], the Second WIPO Process expressly addressed issues relating to the abusive, bad faith registration of personal names. In the course of examining these issues, the WIPO Second Report outlined the main trends in panel decisions concerning the application of the UDRP to personal names. The WIPO Second Report pointed out that:

> the clear weight of authority of many decisions is in favor of the application of the UDRP to the protection of personal names when they constitute trademarks.[210]

In outlining panel decisions that concluded that complainants had established common law rights in personal names, the Second Report explained that panels had identified a number of relevant factors, including:

> (i) the distinctive character or notoriety of the name and the requirement that the domain name must be "identical or confusingly similar" to it, (ii) the relationship between this distinctive character and

[207] WIPO Case No D2005-1346 (17 Mar 2006).
[208] WIPO Case No D2006-0418 (15 May 2006).
[209] WIPO Case No D2006-1052 (11 Oct 2006).
[210] WIPO, above n 106, para [179].

use of the name in connection with goods or services in commerce, and (iii) the location of the parties and the bearing that this may have on the acquisition of unregistered trademark rights.[211]

As explained at [3.11], the WIPO Second Report recommended that, at this stage, the UDRP should not be modified to accommodate broader protection for personal names than that which is provided under the UDRP.[212] As further explained, the basis for this recommendation was the absence of an 'international norm protecting personal names' and the diversity of legal approaches adopted to protect personal names under national legal systems.[213] As the Second Report explained:

> The most cogent of the arguments against modification of the UDRP is, we believe, the lack of an international norm protecting personal names and the consequent diversity of legal approaches deployed to protect personal names at the national level. We consider that this diversity would place parties and panellists in an international procedure in an untenable position and would jeopardize the credibility and efficiency of the UDRP.[214]

The Second Report therefore suggested that the 'most appropriate way' for the international legal protection of personal names to advance would be through the 'development of international norms', and not the modification of the UDRP.[215]

The result of the analysis of the application of the UDRP to personal names conducted by WIPO's Second Report, together with the recommendation against further extending the protection of personal names, is that the UDRP protects personal names that function as trade marks but does not protect personality rights. The implications of the distinction between protecting names as an incident of protecting trade marks and protecting names as part of a more general doctrine of personality rights was outlined by the Second Report in the following terms:

> Persons who have gained eminence and respect, but who have not profited from their reputation in commerce, may not avail themselves of the UDRP to protect their personal names against parasitical registrations. The UDRP is thus perceived by some as implementing an excessively materialistic conception of contribution to society.[216]

The WIPO Consensus View on the protection of personal names is as follows:

> While the UDRP does not specifically protect personal names, in situations where an unregistered personal name is being used for trade or commerce, the complainant can establish common law trademark rights in the name. Reference can be made to the test required for the common law action of passing off. Personal names that have been trademarked are protected under the UDRP.[217]

Although accurate, it is extremely important that this general statement be read with the qualification that immediately follows it in the *WIPO Overview:*

> The name in question should be actually used in trade or commerce to establish unregistered trademark rights. Merely having a famous name (such as a businessman, or religious leader) is not necessarily sufficient to show unregistered trademark rights.[218]

[211] *Ibid,* para [182].
[212] *Ibid,* para [202].
[213] *Ibid,* para [201].
[214] *Ibid.*
[215] *Ibid,* para [203].
[216] *Ibid,* para [199].
[217] WIPO, above n 57, para 1.6.
[218] *Ibid.*

The meaning of the above statements in the *WIPO Overview* can be best explained through examples drawn from UDRP decisions.

[5.20.1] Rights in Personal Names that are Registered as Trade Marks

Registration of a personal name (including a pseudonym) as a trade mark or service mark is sufficient to establish rights in the mark for the purpose of paragraph 4(a)(i) of the UDRP. In this respect, there is no difference in the treatment of personal names under the UDRP from that of any other registered marks.

A considerable number of decisions have concluded that registration of a personal name as a mark satisfies the requirements of paragraph 4(a)(i). The following are examples of decisions in which rights have been found in personal names that have been registered as marks under national trade mark laws:

- *Harrods Limited v Robert Boyd.*[219] The complainant was the owner of the mark 'DODI FAYED', which was registered as a European Community Trade Mark. The respondent subsequently registered the domain name <dodialfayed.com>, and offered it for sale at an auction Web site. The panellist had no difficulty in concluding that the complainant had rights in the mark.
- *Madonna Ciccone v Dan Parisi.*[220] The well-known entertainer was the owner of rights in the mark 'MADONNA', which was registered with the USPTO in connection with entertainment services and related goods. The respondent, who was in the business of developing Web sites, purchased the disputed domain name <madonna.com> and registered 'MADONNA' as a trade mark in Tunisia. The panellist held that the complainant had rights in the mark as a result of registration under US law, as well as common law rights arising from continuous use of the mark for entertainment services since 1979.
- *Sheryl Crow v LOVEARTH.net.*[221] The complainant, a well-known entertainer and musician, had registered the mark 'SHERYL CROW' with the USPTO in connection with entertainment and musical sound recordings. The complainant had also registered the domain names <sherylcrow.com> and <sherylcrowworld.com>. The respondent, who had registered a number of celebrity names, registered the domain name <sheryl-crow.com>, which it linked to its own Web site <celebrity-websites.com>. The panellist had no difficulty in concluding that the complainant had rights in the mark through registration with the USPTO, as well as common law rights acquired through continuous use of the mark since 1986.

[5.20.2] Rights in Personal Names that are not Registered as Trade Marks: US Trade Mark Law

The precise requirements for trade mark rights to arise in a personal name that has not been registered as a mark depend upon the application of national trade mark law principles. As the panellist pointed out in *Antonio de Felipe v Registryfly.com*:[222]

> Regardless of the exact extent of possible unregistered trademark rights for the purposes of the Policy, one thing is reasonably clear. Unregistered trademark rights do not exist in some nebulous

[219] WIPO Case No D2000-0060 (16 Mar 2000).
[220] WIPO Case No D2000-0847 (12 Oct 2000).
[221] NAF Case No FA203208 (28 Nov 2003).
[222] WIPO Case No D2005-0969 (19 Dec 2005).

> way across the breadth of the countries in which a complainant proves it has a reputation. These rights derive from national laws and do not exist divorced from such laws. They therefore depend upon the extent to which the laws of a particular country recognise or do not recognise that the activities of the complainant provide unregistered rights in that country. A personal name is capable of qualifying as an unregistered trademark recognised by the Policy, but in determining whether it does in a particular case a panel will look to the jurisdictions in which those rights are claimed. The panel will then determine whether the complainant can show that the name provides an unregistered right in at least one of these jurisdictions.[223]

Under US trade mark law personal names, including surnames and first names, are generally regarded as descriptive terms. As non-inherently distinctive terms, personal names therefore require proof of secondary meaning in order to be protected as common law marks.[224] If a personal name has become generic through use, however, it is impossible for there to be common law rights in the name.[225]

Many UDRP panel decisions have been perfunctory in the application of US trade mark law in finding rights in personal names under the UDRP. This relatively loose approach is illustrated by *Julia Fiona Roberts v Russell Boyd*,[226] the first panel decision to conclude that a complainant had established common law rights in a personal name under US trade mark law. In that decision, 'JULIA ROBERTS', the well-known film actress, brought a complaint against the registrant of the domain name <juliaroberts.com>. With very little discussion, the panel decided that:

> the name "Julia Roberts" has sufficient secondary association with Complainant that common law trademark rights do exist under United States trademark law.

A straightforward, orthodox approach to determining whether common law rights arise in a personal name under US law has been followed in a number of decisions, including *Marty Rodriguez Real Estate, Inc v Lancaster Industries*.[227] In that decision, the complainant, 'MARTY RODRIGUEZ', was known as a top estate agent, but had no registered trade mark rights in his name. The respondent, a competitor of the complainant, registered the disputed domain name <martyrodriguez.com>. After stating that a complainant must establish sufficient secondary meaning for there to be common law rights in a personal name, the panellist referred to the oft-quoted statement of what amounts to secondary meaning from *Visser v Macres* and to the factors identified in *Centaur Communications Ltd v A\S\M\ Communications, Inc*, both of which are dealt with at [5.18.2] above. In the instant dispute, the complainant was easily able to establish that the name had been used extensively in association with estate agency services since 1979 and, as a result, had acquired a secondary meaning.

One of the clearest, most accurate statements of the need for a secondary meaning under US law was provided by the concurring panellist in *Planned Parenthood Federation of America, Inc and Gloria Feldt v Chris Hoffman*,[228] who explained:

[223] See also *Arthur Golden v Galileo Asesores SL*, WIPO Case No D2006-1215 (15 Dec 2006), dealing with unregistered rights under Spanish law.

[224] McCarthy, above n 5, §13:2, citing *Herring-Hall-Marvin Safe Co v Hall's Safe Co* 208 US 554 (1908); *L.E. Waterman Co v Modern Pen Co*, 235 US 88 (1914).

[225] *Ibid*, §13:24.

[226] WIPO Case No D2000-0210 (29 May 2000).

[227] WIPO Case No D2000-1468 (24 Dec 2000).

[228] WIPO Case No D2002-1073 (21 Feb 2003) (Frederick M Abbott Panellist concurring).

> From the standpoint of countries such as the United States that accord protection to "common law" marks, a personal name may be protected under the UDRP if it has taken on the characteristics of a trademark or service mark: that is, if it has become well known among the relevant public (ie, acquired sufficient secondary meaning) to distinguish the goods or services of one enterprise from those of others . . . Whether a personal name has acquired sufficient secondary meaning so as to take on the legal characteristics of a trademark or service mark will depend on a variety of factors, including the distinctiveness of the name, the extent of its usage in connection with goods or services, and the degree to which the name is recognized by the relevant public in connection with goods or services. Because a personal name might be common among the general public, the burden to demonstrate secondary meaning might be high in a particular case.

Nevertheless, as the panellist acknowledged out in *Sibyl Avery Jackson v Jan Teluch*:[229]

> the question of the scope of protection under the Policy of personal names that may qualify for protection as an unregistered trademark . . . is not without difficulty.

These difficulties have not been fully appreciated by some panellists who have concluded that there are relevant rights merely because a complainant is famous.[230] As the *WIPO Overview* states, however, merely because a personal name is famous does not necessarily mean that there are unregistered trade mark rights in the name. In this respect, unnecessary complexities were introduced by panellists who, in some early UDRP decisions, referred to 'famous'marks under US anti-dilution law[231] and under the Anti-cybersquatting Consumer Protection Act.[232] The question whether or not a mark is 'famous' is something of a distraction as, despite a degree of uncertainty regarding the issue in US trade mark law, a famous mark is simply a mark with a high degree of distinctiveness. As McCarthy puts it:

> A trademark cannot be 'famous' unless it is 'distinctive', but it can be 'distinctive' without being 'famous'. In fact, a designation cannot be a trademark at all unless it is 'distinctive.' By definition, all 'trademarks' are 'distinctive'—very few are 'famous'.[233]

Accordingly, the key issue in determining whether there are unregistered rights in a personal name under US law is not whether the name is famous, but whether the name has acquired distinctiveness through use in connection with goods or services. Furthermore, the issue is not merely whether the personal name is used in trade or commerce, but whether the name has become distinctive, in the sense that it operates as a mark by identifying the origin or source of goods or services. In this respect, the comments of WIPO's Second Report, that persons 'who have not profited from their reputation in commerce, may not avail themselves of the UDRP' places a misleading emphasis on profit-making, rather than use as a trade mark. As the concurring panellist in *Planned Parenthood Federation of America, Inc and Gloria Feldt v Chris Hoffman*,[234] which concerned the personal name, 'GLORIA FELDT', pointed out:

[229] WIPO Case No D2002-1180 (4 Mar 2003).

[230] See, eg, *Mick Jagger v Denny Hammerton*, NAF Case No FA95261 (11 Sept 2000); *Nik Carter v The Afternoon Fiasco*, WIPO Case No D2000-0658 (17 Oct 2000); *Kevin Spacey v John Zuccarini*, NAF Case No FA96937 (8 May 2001); *Pierce Brosnan v Network Operations Center*, WIPO Case No D2003-0519 (27 Aug 2003); *Johnnie L Cochran, Jr v Brian Wick*, NAF Case No FA198946 (9 Dec 2003); *Carmen Electra v Network Operations Center*, WIPO Case No D2003-0852 (23 Dec 2003); *Tommy Lee v Netico, Inc*, WIPO Case No D2005-0915 (28 Dec 2005).

[231] Federal Trademark Dilution Act, 15 USC §§ 1125(c)–1127, as amended by Trademark Dilution Revision Act of 2006, Pub L No 109–312, 120 Stat 1730. UDRP decisions raising this issue include: *Monty and Pat Roberts, Inc v Bill Keith*, WIPO Case No D2000-0299 (9 June 2000); *Monty and Pat Roberts, Inc v J Bartell*, WIPO Case No D2000-0300 (13 June 2000); *Steven Rattner v BuyThisDomainName*, WIPO Case No D2000-0402 (3 July 2000); *Ricky Skaggs v W Ignacio Vincente*, WIPO Case No D2000-1134 (18 Dec 2000).

[232] 15 USC §§ 1125(d), 1129.

[233] McCarthy, above n 5, §24:91:25.

[234] WIPO Case No D2002-1073 (21 Feb 2003) (Frederick M Abbott Panellist concurring).

> The Panel majority asks whether 'Gloria Feldt' has been used 'commercially' in connection with a service. Because the organization with which Ms. Feldt is associated is a non-profit provider of services, I would clarify that use of the term 'commercial' should not be understood to suggest that only for-profit enterprises, or individuals, may acquire rights in common law service marks . . . To use a nation-state-neutral point of reference, the definition of a trademark in the WTO TRIPS Agreement, Article 15, does not place a condition of 'commerciality' in the sense of profit-seeking on prospective trademark holders . . . If Ms. Feldt is providing reproductive rights "services", we should not hold her to the requirement of doing so for a profit.

As explained at [5.20], WIPO's Second Report was careful to distinguish unregistered trade mark rights in personal names from personality (or publicity) rights in personal names. Panel decisions delivered since the publication of the Second Report have tended to be more careful in finding common law trade mark rights in personal names under US law than the earlier panel decisions. The following are examples of decisions delivered after WIPO's Second Report, that have dealt with the existence of common law trade mark rights in personal names, and which have been chosen to illustrate the operation of the relevant principles:

- *Kathleen Kennedy Townsend v BG Birt.*[235] The complainant, 'KATHLEEN KENNEDY TOWNSEND', was a politician, the Lieutenant Governor of the US State of Maryland and the eldest daughter of Robert F Kennedy. The panel rejected a complaint relating to the registration of domain names that included the complainant's personal name on the basis that there were no common law trade mark rights in the name. In reaching this conclusion, the panel pointed out that even though the complainant was a famous politician, protection of her personal name was 'outside the scope of the Policy since it is not connected with commercial exploitation'.
- *The Reverend Dr Jerry Falwell v Gary Cohn.*[236] The complainant, 'JERRY FALWELL', was the well-known religious figure who, amongst other activities, hosted a weekly television programme. The panel concluded that, although the complainant's name was extremely well-known, he had failed to establish that it 'has been used in a trademark sense as a label of particular goods or services'. In particular, the panel pointed out that the complainant had not provided marketing brochures, trade advertisements, or other evidence of use of his name as a trade mark.
- *Kevin Spacey v Alberta Hot Rods.*[237] The complainant, the famous actor and celebrity, brought a complaint in relation to the registered domain name <kevinspacey.com>, which was used by the respondent to redirect users to a celebrity Web site. The panel concluded that there were common law rights in the name as 'a celebrity's name can serve as a trademark when used to identify the celebrity's performance services'. In other words, the panel decided that the name, 'KEVIN SPACEY', had acquired secondary meaning in relation to performance services.
- *Dr Michael Crichton v Alberta Hot Rods.*[238] The complainant was the well-known author of books, including *The Andromeda Strain* and *Jurassic Park*. The panel pointed to the

[235] WIPO Case No D2002-0030 (11 Apr 2002).

[236] WIPO Case No D2002-0184 (3 June 2002). This decision may be compared with the more superficial treatment of the issue in *The Reverend Dr Jerry Falwell v Lamparello International,* NAF Case No FA198936 (20 Nov 2003).

[237] NAF Case No FA114437 (1 Aug 2002).

[238] WIPO Case No D2002-0872 (25 Nov 2002). See also *Dr Michael Crichton v In Stealth Mode,* WIPO Case No D2002-0874 (25 Nov 2002).

need for distinctiveness, concluding that this was satisfied by the complainant's claims to be the author of a number of well-known works.

- *Planned Parenthood Federation of America, Inc and Gloria Feldt v Chris Hoffman.*[239] The complainant, 'GLORIA FELDT', was the President of Planned Parenthood Federation of America, Inc and had spoken publicly on women's reproduction rights for over 30 years. The panel held that the complainant did not have common law rights in her name as, although she had acquired a reputation as a spokesperson, her name had not acquired a secondary meaning as the source of goods or services. The concurring panellist pointed out that, while Planned Parenthood provided services including counselling services, the complainant was not known in her individual capacity because she provides a service.
- *Sibyl Avery Jackson v Jan Teluch.*[240] The complainant was the self-published author of a best-selling novel. The panellist concluded that the complainant had common law trade mark rights on the basis that she 'uses her name in the creation, promotion, and commercial exploitation of her work and that name identifies the work to customers'.
- *Allee Willis v NetHollywood.*[241] The complainant, 'ALLEE WILLIS', was a well-known song writer, who had written songs performed by recording artists such as Bonnie Rait, Ray Charles and Aretha Franklin, and who had received numerous music awards, including a Grammy Award. The panellist determined that the complainant held common law rights in her personal name as it had acquired a secondary meaning in conjunction with her well-known artistic services. In addition, the panellist took into account the unusual nature of the personal name, on the basis that it is easier to establish a secondary meaning in a name that is uncommon.

[5.20.3] Rights in Personal Names that are not Registered as Trade Marks: the English Action for Passing Off

The importance of a precise analysis of the legal basis for determining the existence of unregistered rights in a personal name under national trade mark law was emphasised by the panellist in *Sibyl Avery Jackson v Jan Teluch.*[242] In that decision, in response to the complainant citing decisions that had been decided in accordance with English legal principles, the panellist stated that:

> These are indeed authorities for the proposition that insofar as a Complainant has unregistered trademark rights they can be used to found a claim under the Policy. However, both cases involved Complainants and Respondents based in the United Kingdom and involved a detailed analysis of the extent to which English law, through the law of passing off, recognizes unregistered trademark rights. These cases are therefore not authority for the proposition that outside of England and Wales a Complainant in the position of the Complainants in these two cases necessarily has unregistered trademark rights sufficient to avail him of the UDRP procedure.

As explained at [5.19], unregistered marks are protected in English-oriented legal systems by means of the action for passing off. An extensive case law has established that personal

[239] WIPO Case No D2002-1073 (21 Feb 2003) (Frederick M Abbott Panellist concurring).
[240] WIPO Case No D2002-1180 (4 Mar 2003).
[241] WIPO Case No D2004-1030 (17 Feb 2004).
[242] WIPO Case No D2002-1180 (4 Mar 2003).

names—including surnames, first names and pseudonyms—can acquire sufficient distinctiveness to be protected by the action for passing off.[243] As Wadlow points out:

> Personal names have often acquired distinctiveness and been protected by the action for passing-off. This was one of the most common forms the action took in the nineteenth century, when most businesses traded under the name of the proprietor or founder. So far as the existence of a cause of action is concerned, there is no distinction between personal names used as business names and those used as trade marks on goods. The name may be the real name of the plaintiff or a predecessor in business, or an assumed name, or a quite arbitrary choice. Most of the cases concern surnames (with or without forenames or initials), but forenames alone can also be distinctive . . . It may be harder to prove distinctiveness for a common name, but there is no rule of law that certain names are incapable ever of being distinctive. The question is purely one of fact.[244]

Establishing common law trade mark rights in a name is not, however, simply a matter of establishing a right to bring an action for passing off. As explained at [5.19]ff, in addition to being able to bring an action for passing off, it is necessary for a name to have acquired a secondary meaning in order for there to be common law trade mark rights in the mark under English law. This has not always been apparent from UDRP decisions dealing with personal names.

Early panel decisions dealing with the status of personal names under English-oriented legal systems tended to blur the distinction between the right to bring an action for passing off and common law trade mark rights. In *Jeanette Winterson,*[245] for example, the panellist simply concluded that 'applying English law the Complainant clearly would have a cause of action to prevent unauthorized use of the Mark JEANETTE WINTERSON in passing-off'. Similarly, the panellist in *Julian Barnes*[246] stated:

> The clearest example of a common law trade mark is a person's name, which as the Respondent has pointed out may sometimes be difficult to register as a trade mark. So, if the Panel writes a novel and falsely represents that the author is Julian Barnes, the Complainant would sue in passing off and would undoubtedly succeed. In that example 'Julian Barnes' is the common law trade mark.

The lack of attention given to the separate requirement for a secondary meaning to be established in the early decisions gave rise to considerable subsequent controversy. The Second WIPO Process, which dealt with the protection of personal names, precipitated a more sophisticated legal analysis. In this respect, the important decision in *Israel Harold Asper v Communication X Inc*[247] (*'Izzy' Asper*), which was delivered after the publication of the WIPO Second Process Interim Report but before publication of the Second Report, was a turning point.

In that dispute, the complainant, who was commonly known by his nickname 'IZZY ASPER', was the founder of the prominent Canadian multinational media company, CanWest. In a complaint about registration of the domain names <izzyasper.com>, <izzyasper.net> and <izzyasper.org>, the central issue was whether the well-known business identity could establish common law rights in his personal name. In analysing this

[243] See, eg, *Jaeger v Jaegar & Co Ltd* (1927) 44 RPC 437; *Teofani & Co v Teofani* [1913] 2 Ch 545; (1913) 30 RPC 446; *Parker & Son (Reading) Ltd v Parker* [1965] RPC 323; *Wright, Layman & Umney Ltd v Wright* (1949) 66 RPC 149; *Joseph Rodgers & Sons Ltd v WN Rodgers & Co* (1924) 41 RPC 277; *Roberts Numbering Machine Co v Davis* (1935) 53 RPC 79.

[244] Wadlow, above n 14, para [8.85].

[245] WIPO Case No D2000-0235 (22 May 2000).

[246] WIPO Case No D2001-0121 (26 Mar 2001).

[247] WIPO Case No D2001-0540 (11 June 2001).

issue, the panellist first referred to the following distinction drawn between trade mark rights and personality rights in the WIPO Second Process Interim Report:

> under trademark law, the mark itself for which protection is sought must be *used in commerce* as a precondition to relief, whereas the personality right may protect a person who does not commercially exploit his or her own identity, but nevertheless desires to prevent others from doing so.

After an analysis of UDRP decisions which had found common law trade mark rights in personal names, the panellist held that, in each case, the complainant 'either used the personal name in question as a marketable commodity, allowing his or her name or image to be used for a fee, to promote someone else's goods or services, or for direct commercial purposes in the marketing of his or her own goods and services'. Referring to the lack of precision in previous UDRP decisions dealing with the status of personal names, the panellist stated:

> It is the view of this Panel that it would be helpful in these personal name cases if the Complaint, when dealing with the criteria in 4(a)(i) identified what the nature of the commercial connection would be in the Complainant's use of the Domain Name in issue. In many cases, this will be so obvious as not to need any further embellishment. It would only be in such cases, as here, where this is not apparent that such evidence would be needed.

In the instant dispute, the panellist concluded that there was insufficient evidence that the complainant's name had been used for the purpose of merchandising or other commercial promotion of goods or services, and that it therefore had not acquired sufficient secondary meaning.

The differences between panellists who have emphasised the importance of establishing a secondary meaning in a personal name and those who have been more prepared to assume that a famous name is protected under the UDRP are illustrated by the differences between the majority and the minority in *'Ted' Turner.*[248] In that dispute, the prominent business person and philanthropist, "TED' TURNER', brought a complaint in relation to the domain name, <tedturner.com>. Referring to the *'Izzy' Asper* decision, the majority were careful to distinguish the protection of names from the protection of trade marks. In this instance, the complainant had essentially relied on his name being famous. The majority held that this was insufficient evidence of common law rights as it failed to show that 'Ted' Turner had ever offered goods or services under his name, or that his name had acquired a secondary meaning. The minority, however, rejected this conclusion, maintaining that:

> The majority has wrongly decided that the name 'Ted Turner,' although known throughout the world as a symbol of business acumen and commercial success, has not been used commercially to promote goods and services. The name 'Turner' has been used in the names of several media companies, in addition to its inclusion in the name of the Complainant Ted Turner Film Properties. The majority needs numerous pages of explanation to dispel the obvious and lose the forest for the trees. This evidence is sufficient to find the existence of a common law trademark or service mark in the name 'Ted Turner.'

From these comments it seems that the minority panellist agreed with the need for a personal name to acquire a secondary meaning for there to be common law rights in the name, but considered that this could be satisfied if a well-known name is used in a commercial context.

[248] WIPO Case No D2002-0251 (4 July 2002).

In *Chinmoy Kumar Ghose v ICDSoft.com ('Sri Chinmoy')*,[249] the panel identified two competing views on what is required for a famous personal name to be protected as a common law mark under English-oriented law: the view that all that is needed is for a famous person to establish that his or her name would be protected by an action for passing off, which the panel claimed was supported by the panellist in *Julian Barnes*, and the view that a secondary meaning needs to be established, supported by the panels in *'Izzy' Asper* and *'Ted' Turner*. The panel in *Sri Chinmoy* did not consider it necessary to resolve the conflict, but clearly preferred the view that a secondary meaning is required. The complainant in that dispute, Chinmoy Kumar Ghose—a well-known Indian yogi and peace activist, who provided meditation and charity services, and had a long list of publications—claimed common law rights in the marks 'SRI CHINMOY' and 'GURU SRI CHINMOY'. The expression 'Sri' is a title of respect in Hindi. The panel held that there was sufficient evidence of secondary meaning in the first potential mark as a result of sales of publications under the name 'SRI CHINMOY', as well as limited evidence of the name being used for a radio station and to identify a meditation centre. As the name 'GURU SRI CHINMOY' did not appear to have been used in connection with goods or services, however, the panel did not find that it had acquired a secondary meaning, although it was not necessary to reach a conclusion on this issue for the purpose of the decision.

The upshot of the analysis undertaken in these decisions is that the preferred view is that, to be protected as a common law mark, a famous personal name must have acquired a secondary meaning so as to function as an identifier of goods or services. Where a famous personal name has been used to market the complainant's own goods or services, or is used for a fee to promote someone else's goods or services, there will usually be little difficulty in establishing a secondary meaning. Thus, common law trade mark rights will readily be found in a famous author's name that is used in connection with the author's books, or in a celebrity's name that is used in connection with promoting products. As the panellist pointed out in *'Izzy' Asper*, in such cases the acquisition of secondary meaning will usually be obvious. Merely having a famous name, however, is not sufficient for there to be common law rights, even if the fame results from commercial activities.

The distinction is especially important in relation to the personal names of famous business people. The difference between a personal name that is famous because of commercial activities and a personal name that is a mark because it has become distinctive in relation to goods or services was emphasised in *'Izzy' Asper*, where the panellist point out that:

> it is less likely that a business person will use his or her name to market their own goods or services, and very unlikely that they will do so to market someone else's. Take, by way of examples, Henry Luce, Kenneth Thomson, Conrad Black and Rupert Murdoch. Are their publications marketed as a product connected to the person in question? Not likely. It is also not likely that someone would pay a fee to use these names in the promotion of a good or service with which they have no connection. Yet, each of these gentlemen is or was a famous businessman with a very high profile in the publishing business and held in high regard by many.

As the panellist further pointed out, where it is not obvious that a famous personal name has become distinctive, the onus is on the complainant to show, on the preponderance of evidence, that the name has acquired a secondary meaning.

It might be thought that any continuing controversy surrounding the status of personal names has been resolved by the publication of the *WIPO Overview* which, as set out at

[249] WIPO Case No D2003-0248 (22 May 2003).

[5.20], presents the WIPO Consensus View on this issue. It remains the case, however, that there are ambiguities in the Consensus View that still require interpretation. In particular, a careful reading of the qualification added to the Consensus View is required. The qualification added to the Consensus View, which seems to have been adapted from the distinction drawn between trade mark rights and personality rights in the WIPO Second Process Interim Report, states that:

> The name in question should be actually used in trade or commerce to establish unregistered trademark rights. Merely having a famous name (such as a businessman, or religious leader) is not necessarily sufficient to show unregistered trademark rights.[250]

In reading the first sentence of this statement, it is important to bear in mind that it is not sufficient for the personal name to be used in trade or commerce, but it must be used in trade or commerce *as an unregistered mark*. In other words, it must always be remembered that the UDRP requires the complainant to establish rights in a trade mark or service mark which, in relation to an unregistered personal name, means that the name must have acquired a secondary meaning.

Some potential problems with the wording of the WIPO Consensus View are illustrated by the decision in *Chung, Mong Koo v Individual.*[251] The complainant in that decision, 'CHUNG MONG KOO', the well-known Chairman and Chief Executive Officer of the Hyundai automotive group, needed to establish unregistered trade mark rights in his name. After extracting the WIPO Consensus View, the panellist in the decision expressed the issue in the following terms:

> the consensus view among UDRP panellists is that a complainant may show that his or her personal name has taken on such a cachet that it has become a trademark, but that to succeed in doing so the complainant will have to show that the name has actually been used in trade or commerce. It will instantly be seen . . . that the reason why some cases have failed to establish trademark rights in a personal name is that the evidence has shown only that the name is famous and not that it has been used in trade or commerce.

The panellist went on to identify the following factors to be taken into account in determining whether there is a sufficient nexus between the personal name and use of the name in trade and commerce:

> the extent to which the commercial community identifies the individual with the company, the extent to which the individual is seen by relevant media and sections of the public as the *alter ego* and driving force behind the company, the extent of the personal ownership of the company by the individual, the degree of personal control that the individual exercises over the enterprise, the extent to which the individual is identified with any major achievements of the enterprise and whether . . . it can be said that the individual and/or the company has a demonstrable interest in protecting the individual's name for commercial use.

Taking these factors into account, the panellist essentially found that the complainant was so closely identified with the Hyundai business that he had unregistered trade mark rights in his personal name. The problem with the analysis in the decision, however, is that it concentrates on use of the name in trade and commerce at the expense of any consideration of whether or not the name has acquired distinctiveness as an unregistered mark. By adopting this approach, the decision glosses over the careful distinction drawn between the personal

[250] WIPO, above n 57, para 1.6.
[251] WIPO Case No D2005-1068 (21 Dec 2005).

names of famous business people and famous names that are common law marks drawn by decisions such as *'Izzy' Asper* and *'Ted' Turner*. Moreover, the source of the problem is clearly to be found in the wording of the WIPO Consensus View, which emphasises use of a personal name in trade and commerce at the expense of the key issue, which is whether or not the name has acquired a secondary meaning.

[5.20.4] Status of Unregistered Personal Names: Summary and Examples

After the above discussion, it is useful to provide a summary of the general position with respect to personal names under the UDRP and to give examples of the categories of personal names that are likely to be protected.

The status of unregistered personal names under the UDRP depends upon the application of principles of national trade mark law. In general terms, however, under both US trade mark law and the English action for passing off, to be protected as a common law mark a personal name must have become distinctive as a source of goods or services. The panellist in *Russell Coutts v Massimo Gallotta*[252] provided a good summary of previous panel decisions dealing with when a secondary meaning can be established in a personal name in the following terms:

> a complainant may be successful if he either used his personal name as a marketable commodity, allowing his name or image to be used for a fee, to promote someone else's goods or services, or for direct commercial purposes in the marketing of his own goods or services . . .

As pointed out by the panellist in *Steven Rattner v BuyThisDomainName*,[253] 'protection at common law is determined on a case by case basis and depends greatly on the factual basis of each individual case'. Nevertheless, the panellist in *'Izzy' Asper* usefully set out a number of categories of personal names, with an indication of the likelihood of unregistered trade mark rights being found in the name, which provides the basis for the following list:

- *Entertainers (living and deceased)*—UDRP panels have been almost unanimous in finding unregistered trade mark rights in the personal names of famous entertainers which may be used as marks in connection with performance services or in promoting the goods or services of others: *Julia Fiona Roberts v Russell Boyd*[254] (famous actor); *Cho Yong Pil v ImageLand, Inc*[255] (famous Korean pop star); *Experience Hendrix, LLC v Denny Hammerton* [256] (estate of deceased musician and rock star); *Nik Carter v The Afternoon Fiasco*[257] (radio broadcaster); *Helen Folsade Adu known as SADE v Quantum Computer Services Inc*[258] (popular singer, song writer and recording artist); *Madonna Ciccone v Dan Parisi*[259] (popular singer, pop star and celebrity); *Isabelle Adjani v Second Orbit*[260] (popular actor); *Julie Brown v Julie Brown Club*[261] (well-known actor, scriptwriter, and producer of music videos); *Celine Dion v Jeff Burgar*[262] (well-known pop music artist); *Mick Jagger v Denny Hammerton* (musician, rock star and celebrity); *MPL Communications Ltd v*

[252] WIPO Case No D2006-0008 (23 Mar 2006).
[253] WIPO Case No D2000-0402 (3 July 2000).
[254] WIPO Case No D2000-0210 (30 May 2000).
[255] WIPO Case No D2000-0229 (14 May 2000).
[256] WIPO Case No D2000-0364 (4 Aug 2000).
[257] WIPO Case No D2000-0658 (23 Oct 2000).
[258] WIPO Case No D2000-0794 (7 Oct 2000).
[259] WIPO Case No D2000-0847 (12 Oct 2000).
[260] WIPO Case No D2000-0867 (7 Oct 2000).
[261] WIPO Case No D2000-1628 (13 Feb 2001.
[262] WIPO Case No D2000-1838 (13 Feb 2001).

Denny Hammerton (<paulmccartney.com> and <lindamccartney.com>; musicians, pop/rock stars and celebrities); *Estate of Frank Gorshin v [no name provided in the Whois] aka Terry Martin*[263] (estate of television and film actor); *Tommy Lee v Netico, Inc*[264] (musician, rock star and celebrity); *Silvie Tomčalová, aka Sylvia Saint v Global Access*[265] ('Sylvia Saint', an adult entertainment star); *James Christian Kimmel v jimmykimmel*[266] ('Jimmy Kimmel', an American comedian).

- *Authors*—UDRP panels have been almost unanimous in finding unregistered trade mark rights in the personal names of authors, which may be used as marks in connection with their publications: *Jeanette Winterson v Mark Hogarth*[267] (Jeanette Winterson); *Monty and Pat Roberts, Inc v Bill Keith;*[268] *Monty and Pat Roberts, Inc v J Bartell* (Monty Roberts, the 'horse whisperer');[269] *Margaret Drabble v Old Barn Studios Limited*[270] (Margaret Drabble); *Louis De Bernieres v Old Barn Studios Limited*[271] (Louis de Bernieres); *Antony Beevor v Old Barn Studios Limited*[272] (Antony Beevor); *Dr Michael Crichton v Alberta Hot Rods*[273] (Michael Crichton); *Sibyl Avery Jackson v Jan Teluch*[274] (Sibyl Avery Jackson); *Arthur Golden v Galileo Asesores SL*[275] (Arthur Golden); *Jeffrey Archer v Alberta Hotrods*[276] (Jeffrey Archer).
- *Professional athletes*—UDRP panels have been almost unanimous in finding unregistered trade mark rights in the personal names of professional athletes, clearly reflecting the marketability of sports stars: *Daniel C Marino, Jr v Video Images Productions*[277] (athlete, film actor and sports commentator); *Jules I Kendall v Donald Mayer*[278] (professional golfer); *Jaap Stam v Oliver Cohen*[279] (famous soccer player); *Pierre van Hooijdonk v SBTait*[280] (famous soccer player).
- *Business people*—Decisions regarding the names of famous business people are not consistent but, in general terms, panels distinguish between a name that is famous as a result of the commercial activities of the business person, which will not be protected, and a personal name that is a mark because it has become distinctive: *Harrods Ltd v Robert Boyd*[281] (Dodi al-Fayed); *Monty and Pat Roberts, Inc v Bill Keith*[282] (Monty Roberts, who was in the business of selling and training horses); *Monty and Pat Roberts, Inc v J Bartell*[283] (Monty Roberts); *Steven Rattner v BuyThisDomainName*[284] (Steven Rattner, a high pro-

[263] WIPO Case No D2005-0803 (31 Oct 2005).
[264] WIPO Case No D2005-0915 (28 Dec 2005).
[265] WIPO Case No D2006-0399 (24 June 2006).
[266] WIPO Case No D2006-0402 (22 May 2006).
[267] WIPO Case No D2000-0235 (22 May 2000).
[268] WIPO Case No D2000-0299 (9 June 2000)
[269] WIPO Case No D2000-0300 (13 June 2000).
[270] WIPO Case No D2001-0209 (26 Mar 2001).
[271] WIPO Case No D2001-0122 (26 Mar 2001).
[272] WIPO Case No D2001-0123 (26 Mar 2001).
[273] WIPO Case No D2002-0872 (25 Nov 2002). See also *Dr Michael Crichton v In Stealth Mode,* WIPO Case No D2002-0874 (25 Nov 2002).
[274] WIPO Case No D2002-1180 (4 Mar 2003).
[275] WIPO Case No D2006-1215 (15 Dec 2006).
[276] WIPO Case No D2006-0431 (1 June 2006).
[277] WIPO Case No D2000-0598 (2 Aug 2000).
[278] WIPO Case No D2000-0868 (26 Oct 2000).
[279] WIPO Case No D2000-1061 (4 Nov 2000).
[280] WIPO Case No D2000-1068 (4 Nov 2000).
[281] WIPO Case No D2000-0060 (16 Mar 2000).
[282] WIPO Case No D2000-0299 (9 June 2000).
[283] WIPO Case No D2000-0300 (13 June 2000).
[284] WIPO Case No D2000-0402 (3 July 2000).

file investment banker and financial adviser); *Leonard Asper v Communication X Inc*[285] (Leonard Asper, son of a famous Canadian business person); *Israel Harold Asper v Communication X Inc*[286] ('Izzy' Asper, famous Canadian business person); *R E 'Ted' Turner and Ted Turner Film Properties, LLC v Mazen Fahmi*[287] ('Ted' Turner); *Chung, Mong Koo v Individual*[288] (Chairman and CEO of Hyundai).

- *Royalty*—There are very few disputes involving royalty, but in a number of disputes involving the Estate of Diana, Princess of Wales, panels held that the Princess of Wales had common law trade mark rights in her name, but not in terms such as 'ENGLAND's ROSE' or 'THE PEOPLE'S PRINCESS': *CMG Worldwide, Inc v Naughtya Page*[289] (representative of estate of Diana, Princess of Wales); *CMG Worldwide, Inc v Steve Gregory*[290] (representative of Estate of Diana, Princess of Wales); *CMG Worldwide, Inc v Bonnie Masterson*[291] (representative of Estate of Diana, Princess of Wales); *Albert Fürst von Thurn und Taxis v Doris Eckert*[292] (representative of the European royal family Thurn und Taxis and owner of the registered 'THURN UND TAXIS' mark).
- *Politicians*—There have been very few cases involving politicians, and panels have split on the question whether a politician can establish common law rights in a personal name as a result of his or her political activities. While the decisions in *Anne McLellan* and *Hilary Clinton* held that the politicians in those disputes had established common law trade mark rights, the panels in *Kathleen Kennedy Townsend* and *Artur Mas* held that the politicians had not established unregistered trade mark rights in their personal names: *Anne McLellan v Smartcanuck.com*[293] (Anne McLellan, well-known Canadian politician); *Friends of Kathleen Kennedy Townsend v B G Birt*[294] (political committee for the Lieutenant Governor of Maryland and member of the Kennedy family); *Convergència Democràtica de Catalunya v ar mas*[295] (Arur Mas, a famous Spanish politician); *Hillary Rodham Clinton v Michele Dinoia*[296] (US Senator and former First Lady); *Jeffrey Archer v Alberta Hotrods*[297] (Jeffrey Archer, politician and author, but treated as an author for the purpose of the policy).

Geographical Terms

[5.21] Geographical Terms

As explained at [3.4], the WIPO Final Report recommended limiting the dispute-resolution procedure to abusive registration of trade marks or service marks and suggested that

285 WIPO Case No D2001-0539 (21 June 2001).
286 WIPO Case No D2001-0540 (11 June 2001).
287 WIPO Case No D2002-0251 (4 July 2002).
288 WIPO Case No D2005-1068 (21 Dec 2005).
289 NAF Case No FA95641 (8 Nov 2000).
290 NAF Case No FA95645 (7 Nov 2000).
291 NAF Case No FA97061 (13 June 2001).
292 WIPO Case No D2004-0817 (14 Dec 2004).
293 eRes Case No AF-0303a-b (25 Sept 2000).
294 WIPO Case No D2002-0451 (31 July 2002).
295 WIPO Case No DTV2003-0005 (19 Dec 2003).
296 NAF Case No FA414641 (18 Mar 2005).
297 WIPO Case No D2006-0431 (1 June 2006).

registrations that violated trade names, geographical indications or personality rights would not fall within the procedure. The decision to limit the scope of the UDRP to trade marks and service marks, and not to protect geographical names as such, has created problems for complainants interested in protecting geographical terms, including names of countries, regions and cities.

As further explained at [3.11], the Second WIPO Process expressly addressed issues relating to the abusive, bad faith, misleading or unfair use of geographical indications, geographical terms or indications of source. The WIPO Second Report first pointed out that the international legal framework for the protection of geographical identifiers, including the Paris Convention and the TRIPS Agreement, consists of two elements: a prohibition of false descriptions of the geographical source of goods and rules prohibiting the misuse of a class of geographical source indicators, known as geographical indications. The Report went on to identify two main problems in applying the international legal framework to the registration of geographical identifiers as domain names. First, the Report pointed out that, as the legal framework was developed to protect geographical identifiers of goods, it did not readily apply to the use of geographical terms as domain names, as there is often no relationship between a domain name and goods. Secondly, the Report highlighted the significant differences in the protection of geographical identifiers under national laws. Given these problems, the Report recommended against extending the UDRP to encompass complaints about the misuse of geographical identifiers, on the basis that extension of the international legal framework would be properly advanced through multilateral discussions.[298]

After dealing with the existing legal framework, the WIPO Second Report turned to an examination of policy issues relating to the protection of geographical terms per se, including country names and names of places within countries. While acknowledging the arguments in favour of protecting country names, the Report set out the following two conclusions arising from the Second Process debate:

(i) The question of the appropriateness of the registration of country names in the gTLDs is inextricably linked by some governments to what they perceive to be their national sovereign interest.
(ii) Protecting country names in the gTLDs would require or amount to the creation of new law, a function traditionally reserved for States.[299]

On the basis of these conclusions, the Second Report made the following recommendation:

> that the question of the protection in the gTLDs of country names and the names of administratively recognized regions and municipalities be further considered in the appropriate intergovernmental fora, in particular with a view to a discussion on the need for new international rules for the protection of country names.[300]

In the absence of any subsequent developments in establishing an international regime for the protection of geographical terms, those concerned with the unauthorised use of such terms as domain names must rely on the UDRP or existing national laws. The UDRP, of course, protects geographical terms only insofar as trade mark rights can be established in such terms.

The WIPO Consensus View on the protection of geographical terms or identifiers under the UDRP is as follows:

[298] WIPO, above n 106, paras [244]–[245].
[299] *Ibid*, para [286].
[300] *Ibid*, para [288].

> The report of the Second WIPO Internet Domain Name Process declined to specifically extend protection to geographical terms under the UDRP. Some geographical terms however, can be protected under the UDRP, if the complainant has shown that it has rights in the term and that the term is being used as a trademark. Normally this would require the registration of the geographical term as a trademark.[301]

This general statement must be read with the immediately following qualification, which deals with the protection of geographical terms where the complainant has no registered trade mark rights:

> It is very difficult for the legal authority of a geographical area to show unregistered trademark rights in that geographical term on the basis of secondary meaning.[302]

The extent to which the UDRP applies to geographical terms is best illustrated by examples drawn from UDRP decisions.

[5.21.1] Rights in Geographical Terms that are Included in Registered Trade Marks

The inclusion of a geographical term in a registered trade mark always gives rise to particular difficulties. The difficulties arise because, although it may be desirable to use a geographical term to indicate the source of goods or services, registration would prevent the legitimate use of the geographical term in a descriptive sense by other traders. This policy tension is explained in *Kerly's Law of Trade Marks and Trade Names* in the following terms:

> The use by traders of marks with geographical significance is both normal and problematic: it is normal to wish to indicate a connection with a particular geographic location, especially if that location gives to the product a cachet or characteristic it would not otherwise have; it is problematic if, in so doing, the trader seeks to fence off part of the commons which should be free to any other trader who does not mislead by using the geographical name.[303]

Concerns about the extent to which registration of a mark that includes a geographical term would confer a monopoly on the use of that term have resulted in specific rules relating to geographical terms under national trade mark laws.

Under US federal trade mark law, terms that describe a geographical location are not regarded as being inherently distinctive, and therefore require proof of secondary meaning in order to be registered.[304] If a geographical term is protected as a result of the acquisition of a secondary meaning, other traders retain rights to use the term in a descriptive sense under the US fair use defence.[305] In this respect, the Lanham Act provides a defence in the following terms:

> That the use of the name, term or device charged to be an infringement is a use, otherwise than as a mark . . . of a term or device which is descriptive of and used fairly and in good faith only to describe the goods or services of such party, or their geographic origin.[306]

In the United Kingdom, section 3(1)(c) of the Trade Marks Act 1994 prohibits the registration of trade marks which consist exclusively of signs or designations which indicate

[301] WIPO, above n 57, para 1.5.

[302] *Ibid.*

[303] Kitchin *et al*, above n 20, para [11-001]. See also the comments of Gummow J in *Chancellor, Masters & Scholars of the University of Oxford (t/a Oxford University Press) v Registrar of Trade Marks* (1990) 24 FCR 1, 23.

[304] Lanham Act § 2(e)(2), 15 USC § 1052(e)(2); McCarthy, above n 5, §14:1.

[305] See McCarthy, above n 5, §14:12.

[306] Lanham Act § 33(b)(4), 15 USC § 1115(b)(4).

geographical origin unless the mark has become distinctive through use. In assessing the acquisition of distinctiveness by a geographical name, the European Court of Justice has cautioned that:

> where a geographical name is very well-known, it can acquire distinctive character . . . only if there has been long-standing and intensive use of the mark by the undertaking applying for registration.[307]

Under Australian trade mark law, signs that consist wholly of a sign that is ordinarily used to indicate geographical origin are classified as marks that are not inherently adapted to distinguish goods or services.[308] As such, the mark can only be registered if it has become distinctive through use before the filing date.[309] Moreover, if a mark suggests a geographical origin, the Registrar of Trade Marks may restrict the effect of registration to that particular area.[310]

Despite the special treatment given to geographical terms under national trade mark laws, the inclusion of a geographical term in a trade mark or service mark that is registered under a national trade mark law is usually sufficient for rights in the mark to be established for the purpose of paragraph 4(a)(i) of the UDRP. Nevertheless, in applying the UDRP it remains important to preserve the policy balance between protecting the rights in registered marks that include geographical terms while ensuring that others remain free legitimately to use such terms, including using the terms as domain names. As the panellist put it in *Neusiedler Aktiengesellschaft v Vinayak Kulkami:*[311]

> Geographic names can not be monopolized by registering a trademark or company name. The use of geographic terms as such in domain names or otherwise by third parties is generally possible despite a trade-mark registration.

The preferred approach to preserving the policy balance is to deal with the scope of protection of a registered mark that includes a geographical term in determining whether the disputed domain name is confusingly similar to the mark, whether the respondent has rights or legitimate interests in the domain name or whether the domain name has been registered and is being used in bad faith. The preferred approach, which has been adopted by most panels, is illustrated by the following decisions:

- *Excelentisimo Ayuntamiento de Barcelona v Barcelona.com Inc.*[312] The complainant, the City Government of Barcelona, had a large number of trade mark registrations that included the geographical term 'BARCELONA'. In each of the relevant registrations, the term 'BARCELONA' was accompanied by the complainant's name, 'Excelentísimo Ayuntamiento de Barcelona'. Moreover, each certificate included the statement, 'This mark consists of the expression "BARCELONA", bearing written below the name of the title holder'. The panellist held that the name 'BARCELONA' was the distinctive part of the registered marks and, on this basis, that the <barcelona.com> domain name was confusingly similar to the registered mark.

[307] Case C–108/97 *Windsurfing Chiemsee Produktions- und Vertriebs GmbH v Boots- und Segel-zubehor Walter Huber & Franz Attenberger* [1999] ECR I–2779, para [50].

[308] Trade Marks Act 1995 (Cth), s 41(6), Note 1.

[309] *Ibid*, s 41(6)(a).

[310] See M Davison, K Johnstone and P Kennedy, *Shanahan's Australian Law of Trade Marks & Passing Off*, 3rd edn (Sydney, Lawbook Co, 2003) para [10.45].

[311] WIPO Case No D2000-1769 (5 Feb 2001).

[312] WIPO Case No D2000-0505 (4 Aug 2000).

- *Kur- und Verkehrsverein St Moritz v StMoritz.com.*[313] The complainant, the official organisation for the community of St. Moritz, had registered trade marks for 'ST. MORITZ' in Switzerland and 26 other countries. In a dispute involving the <stmoritz.com> domain name, the panellist held that the complainant had rights in the geographical name 'ST. MORITZ', but concluded that the respondent, who used the domain name for providing information about St. Moritz, had rights or legitimate interests in the domain name, and that there was no bad faith use and registration of the domain name.
- *Skipton Building Society v Peter Colman.*[314] The complainant had registered the mark 'SKIPTON' under United Kingdom law. The complainant had used the mark since 1853, but the registered mark was geographically limited to services outside a 50-mile radius from the town of Skipton. The panel held that the complainant had established trade mark rights in the geographical term 'SKIPTON' and, despite the geographical limitation on the registration of the mark, concluded that the respondent had no rights or legitimate interests in the disputed domain name <skipton.com>, and that there was bad faith registration and use of the domain name.
- *Brisbane City Council v Warren Bolton Consulting Pty Ltd.*[315] The complainant, the local government authority for the Australian city of Brisbane, had registered a trade mark that included the words 'BRISBANE CITY WORKS'. The registered mark consisted of an image of a diamond indented by three triangles and the words 'BRISBANE CITY WORKS'. The panellist confirmed that the complainant had rights in the registered mark, but concluded that the disputed domain name, <brisbanecity.com>, was not visually confusingly similar to the mark. In addition, the panellist held that even if the mark consisted only of the words 'BRISBANE CITY WORKS' there would be no confusing similarity, as the distinctive feature of the mark was the conjunction of the word 'WORKS' with the words 'BRISBANE CITY'.
- *FC Bayern München AG v Peoples Net Services Ltd.*[316] The complainant operated the soccer activities of the famous Bayern München football club, and had registered the mark 'FC BAYERN MÜNCHEN E.V.' in about 30 countries, including the United Kingdom and a Community Trade Mark registration. The English translation of the 'BAYERN MÜNCHEN' part of the mark is the geographical term, 'BAVARIA MUNICH'. In a dispute involving the domain names <bayernmuenchen.net> and <bayernmunchen.net>, the panellist held that the complainant had trade mark rights in the registered mark, but discounted the 'FC' and 'E.V.' parts of the registered marks as descriptive elements. The panellist went on to find that the term 'BAYERN MÜNCHEN' had acquired a secondary meaning in connection with the soccer club, but dealt with this in concluding that the respondent had no rights or legitimate interests in the domain names. The panellist also concluded that the domain names were registered and used in bad faith.
- *BAA plc, Aberdeen Airport Limited v Mr H Hashimi.*[317] The complainant, BAA plc, was the largest airport operator in the United Kingdom, and had registered the service mark, 'BAA ABERDEEN' in the United Kingdom. The registered mark consisted of the name and a device that included a small dark square with a diagonal white stripe and a dark triangle, the effect of which was to give the impression of an aircraft that has just taken off from a

[313] WIPO Case No D2000-0617 (17 Aug 2000).
[314] WIPO Case No D2000-1217 (1 Dec 2000).
[315] WIPO Case No D2001-0047 (7 May 2001).
[316] WIPO Case No D2003-0464 (15 July 2003).
[317] WIPO Case No D2004-0717 (21 Oct 2004).

runway. In a dispute concerning the domain name <aberdeenairport.com>, the panellist acknowledged that the complainant had registered rights in the mark, but concluded that the domain name was not confusingly similar to the registered mark. In reaching this conclusion, the panellist held that there was insufficient visual, aural or phonetic similarity between the domain name and the device mark to constitute 'confusing similarity'.

- *Deer Valley Resort Company v Intermountain Lodging.*[318] The complainant had registered trade mark rights in the term 'DEER VALLEY' under US law. In a dispute over the <deervalleylodging.info> domain name, the panellist held that the complainant had relevant rights in the 'DEER VALLEY' mark. Pointing out that 'there is a conflict between the rights Complainant has by way of trademark registration and the rights of the community at large to use the geographic indicator outside the scope of the trademark registrations', however, the panellist held that the respondent, who managed real properties in the Deer Valley area, had rights or legitimate interests in the domain name.
- *Angus Sholto-Douglas v Sergey Fedorov.*[319] The complainant operated Kwandwe Private Game Reserve, which covers an area of 62,000 acres in South Africa. The term 'KWANDWE' means 'Place of the Blue Cranes' in Xhosa. The complainant had registered trade marks for the term 'KWANDE' in the Republic of South Africa. The panellist held that the complainant had rights in the trade mark arising from the South African registrations as well as common law rights resulting from the mark having become distinctive of the complainant's services.

The preferred approach has not been universally applied by UDRP panels, with some panels holding that registration of a mark that includes a geographical term is not sufficient to confer trade mark rights in the geographical term. The decision by the panel in *Brisbane City Council v Joyce Russ Advertising Pty Limited*[320] illustrates the reluctance of some panels to find trade mark rights in a registered mark that includes a geographical term. In that dispute, the local government authority for Brisbane relied on registered marks for 'E-BRISBANE', 'WATER SAVE IT!—BRISBANE CITY' and 'BRISBANE CITY WORKS' in a complaint against the <brisbane.com> domain name. Unlike the panellist in *Brisbane City Council v Warren Bolton,* the panel in this dispute decided that the complainant had no trade mark rights in the words 'BRISBANE CITY' or 'BRISBANE'. The panel gave three main reasons for reaching this conclusion. First, the panel held that the trade mark registrations gave no exclusive rights to the word 'BRISBANE' because the geographical name was not the distinguishing feature of the marks. Secondly, the panel referred to the policy objections to conferring exclusive rights on a geographical term. Thirdly, the panel referred to evidence of a significant number of registrations of Australian trade names and trade marks that featured the word 'BRISBANE' to support the conclusion that the complainant had no trade mark rights in 'BRISBANE' or 'BRISBANE CITY'.

The reasoning of the panel on this point is, however, problematic. The preferable approach, as adopted by the panellist in *Brisbane City Council v Warren Bolton,* is to find that registration confers rights in a mark that includes a geographical term, then to consider whether the domain name is confusingly similar to the registered mark. Otherwise, the interests of a respondent who has legitimately registered a domain name that includes a geographical term may be protected under the other elements of the UDRP. Other decisions

[318] NAF Case No FA474344 (27 June 2005).
[319] WIPO Case No D2006-1184 (27 Nov 2006).
[320] WIPO Case No D2001-0069 (14 May 2001).

that have refused to find trade mark rights in a geographical term that is included in a registered mark include *Neusiedler Aktiengesellschaft v Vinayak Kulkami*[321] and *Chambre de Commerce et d'Industrie de Rouen v Marcel Stenzel*.[322]

In the absence of registered trade mark rights, complainants seeking to protect a geographical term must be able to establish unregistered rights in the term.

[5.21.2] Unregistered Rights in Geographical Terms

The requirements for trade mark rights to arise in a geographical term that has not been registered as a mark are determined by national laws . In general, for unregistered rights to arise in a geographical term, it is necessary to establish that the term has acquired a secondary meaning. Given that geographical terms are highly descriptive, it is more difficult for a complainant to establish a secondary meaning in geographical terms than it is in other terms.

Under US trade mark law a geographically descriptive term is not inherently distinctive, and therefore must acquire a secondary meaning in order to be protected as an unregistered mark. Geographically descriptive terms are terms that indicate geographical locations, and include continents, countries, regions, states, cities, streets and rivers.[323] To acquire a secondary meaning, a geographically descriptive term must be associated with the source of goods or services, and not simply indicate the location in which products are produced. At the same time, if a geographic term is used in an arbitrary rather than a descriptive manner, the mark will be regarded as inherently distinctive and no secondary meaning is required.[324] An example of a descriptive term that is used in an arbitrary manner is use of the term 'AMAZON' to refer to the on-line bookshop, as no one would conceive that the bookshop was connected to the Amazon River. If a geographical term is combined with a non-geographical term as part of a composite mark, the status of the composite mark is to be determined by an assessment of the mark as a whole.[325]

In order to be protected as a common law mark under the English action for passing off, a geographical term must have become distinctive of the goods or services of a trader.[326] If, however, the term retains its descriptive meaning, for example, as the locality in which a particular product is produced, then it is not a passing off for other traders to use the term in a descriptive sense. Cases in which an action for passing off is available in relation to a geographical term that has become distinctive should be distinguished from cases in which an action arises from a misrepresentation that goods are associated with a particular geographical area.[327] In the latter case, the plaintiff does not have a common law trade mark right in the geographical term.

The general position in relation to unregistered rights in geographical terms was summarised by the panellist in *The Paris Pages v Woohoo T&C Ltd*[328] in the following terms:

> Though common law marks are sufficient to trigger the Policy, nevertheless for a unregistered term to acquire common law trademark status, that term, as used in commerce by its owner, must have

[321] WIPO Case No D2000-1769 (5 Feb 2001).
[322] WIPO Case No D2000-0348 (18 June 2001).
[323] McCarthy, above n 5, §14:3.
[324] *Ibid*, §14:7.
[325] *Ibid*, §14:11.
[326] See, eg, *Brestian v Try* [1958] RPC 161; *Chelsea Man Menswear v Chelsea Girl* [1987] RPC 189; *Australian Home Loans v Phillips* (1998) 40 IPR 392.
[327] See *Bollinger v Costa Brava Wine Co Ltd* [1960] Ch 262; *Vine Products v Mackenzie* [1968] FSR 625.
[328] NAF Case No FA110763 (10 July 2002).

acquired, through its use, an appropriate level of distinctiveness such that the term has acquired sufficient secondary meaning in the mind of the relevant consuming public. Geographically descriptive terms are rarely, if ever, able to acquire secondary meaning sufficient to overcome their public recognition as a common name of a place or region.

At the same time, if a geographical term is part of a composite mark that includes other terms, including generic terms, it is easier for a complainant to establish that the mark has acquired a secondary meaning. The following are examples of disputes involving unregistered rights in names that include geographical terms:

- *Manchester Airport PLC v Club Club Limited.*[329] The complainant, the operator of 'MANCHESTER AIRPORT', claimed unregistered trade mark rights in the name, largely on the basis of use in commerce since 1986. The presiding panellist considered that the complainant had unregistered rights in the mark, but the majority held that the complainant had provided insufficient arguments or evidence to establish common law rights in the mark.
- *Wembley National Stadium Limited v Bob Thomson.*[330] The complainant, the owner of Wembley Stadium, was held to have common law rights, as well as registered trade mark rights, in the mark 'WEMBLEY STADIUM', which consisted of a geographical name followed by a generic term.
- *Sydney Markets Limited v Nick Rakis.*[331] The complainant, who operated the Sydney Markets, asserted unregistered rights in the term 'SYDNEY MARKETS'. On the basis of evidence that the complainant had used the term in business since 1997, the panellist held that the term was distinctive and therefore that the complainant had common law rights. The dispute differed from others involving geographical terms in that the geographical area 'SYDNEY MARKETS' had been named after the business, the panellist evidently concluding that the composite term had not lost its distinctiveness by virtue of its use as a geographical term.
- *The Paris Pages v Woohoo T&C Ltd.*[332] The complainant, who was based in the United States, alleged common law rights in the geographical term 'PARIS', on the basis of use in connection with the <paris.org> web site to promote business and tourism in the area of Paris, France. Unsurprisingly, the panellist concluded that the name had not acquired secondary meaning, pointing out that:

  It is absolutely inconceivable that anyone, at first instance, hearing this term would think of anything other than that particular city. Accordingly, not only does this term not have any common law significance as a trademark, it strains reason to even contemplate that this term could ever obtain such significance.

- *BAA plc, Aberdeen Airport Limited v Mr H Hashimi.*[333] In addition to a registered device mark, the complainant relied on common law rights in the mark 'ABERDEEN AIRPORT'. The evidence presented by the complainant consisted of a handbook distributed to local businesses, archived web pages and flight guide information. The panellist concluded that there was sufficient evidence of the name being used in connection with airport services to establish a common law service mark right in 'ABERDEEN AIRPORT'.

[329] WIPO Case No D2000-0638 (22 Aug 2000).
[330] WIPO Case No D2000-1233 (16 Nov 2000).
[331] WIPO Case No D2001-0932 (8 Oct 2001).
[332] NAF Case No FA110763 (10 July 2002).
[333] WIPO Case No D2004-0717 (21 Oct 2004).

- *Keystone Publishing, Inc v UtahBrides.com.*[334] The complainant asserted unregistered trade mark rights in the descriptive term 'UTAHWEDDINGS.COM', arguing that the name had acquired a secondary meaning as a result of extensive advertising and widespread name recognition in Utah. The panellist, however, concluded that the evidence was insufficient to establish a secondary meaning in the term, observing that:

  > The Panel has difficulty imagining how the public would come to identify this mark with just the Complainant in the relevant market since the end of 2001 because, as the Respondent tellingly points out, numerous other websites with highly similar or nearly identical names are contemporaneously offering similar wedding services in the State of Utah. (Response Exhibit B lists <utahweddings.info>, <weddingsutah.com>, <utahweddings.net> and many others).

- *Sydney Airport Corporation v John Crilly.*[335] The complainant alleged rights arising from a registered mark and common law rights in the name 'SYDNEY AIRPORT'. Taking into account evidence of the history of the airport, the extent of the complainant's business activities and use of the name in trade promotions, the panellist held that the complainant had acquired a common law trade mark in the name.
- *Traditional Tuscany v Maria Concetta de Concini.*[336] The British complainants, who offered holiday villas and apartments in Italy, claimed unregistered trade mark rights in the term, 'TRADITIONAL TUSCANY', but presented no evidence other than an unsigned, unsworn statement. The panellist held that there was no acceptable evidence that the complainant had acquired rights in the name.

[5.21.3] Rights of Legal Authorities for Geographical Areas

A number of UDRP complaints have been brought by legal authorities for particular geographical areas, such as cities, against registration of the geographical name as a domain name. It is absolutely clear that the legal authority for a geographical area does not automatically have relevant rights in the name of the area, even in jurisdictions where local authorities have statutory rights in the name of a locality, but must establish that the name has a secondary meaning. Local authorities have experienced considerable difficulties in establishing unregistered trade mark rights in a geographical term, such as the name of the city, on the basis that it has acquired a secondary meaning. The following are the major decisions to have considered this issue:

- *Port of Helsinki v Paragon International Projects Ltd.*[337] The complainant, the port authority for the Port of Helsinki, asserted unregistered trade mark rights in the geographical term 'PORT OF HELSINKI'. The panellist pointed out that unregistered trade mark rights can arise under Article 2 of the Finnish Trade Mark Act, which provides that 'a trade symbol shall be regarded as established if it has become generally known in the appropriate business or consumer circles in Finland as a symbol specific to its proprietor's goods'. In other words, unregistered rights in a geographical term arise under Finnish law upon proof of distinctiveness. In the absence of specific evidence of distinctiveness, such as market research, the panellist held that the port authority had failed to establish unregistered rights in the name.

[334] WIPO Case No D2004-0725 (17 Nov 2004).
[335] WIPO Case No D2005-0989 (22 Nov 2005).
[336] WIPO Case No D2006-0997 (21 Nov 2006).
[337] WIPO Case No D2001-0002 (12 Feb 2001).

- *City of Hamina v Paragon International Projects Ltd.*[338] The complainant, the legal representative of the Finnish city of Hamina, claimed unregistered trade mark rights in the geographical term 'PORT OF HAMINA'. To begin with, the panellist rejected the complainant's contention that rights to a unique geographical name should automatically belong to the owner of the geographical area. Confirming that unregistered rights can arise in a geographical term under Finnish law, the panellist differed from the panellist in *Port of Helsinki* by concluding that the name 'PORT OF HAMINA' had acquired distinctiveness in connection with services rendered in the port. The panellist reached this conclusion despite the limited nature of the evidence. In determining the extent of protection of the geographical term, however, the panellist referred to Article 6.1(b) of the 1988 European Trade Mark Directive, which provides that a trade mark shall not entitle the proprietor to prohibit a third party from using an indication of geographical origin in the course of trade, provided it is used in accordance with honest practices in industrial or commercial matters.[339] On this basis, the panellist held that the respondent, who had created a network of sites providing information about ports, had acquired a legitimate interest in the domain name <portofhamina.com>.
- *Brisbane City Council v Warren Bolton Consulting Pty Ltd.*[340] In addition to relying upon registered trade mark rights, as referred to at [5.21.1], the complainant in this dispute asserted common law rights in the geographical term, 'BRISBANE CITY'. The panellist held that, on the basis of the evidence submitted, it was impossible to conclude that the name was distinctive of the goods or services of the complainant, but remained descriptive of the geographical location. In concluding that the name was not distinctive, the panellist noted:

  > there are likely to be many traders providing goods or services in or relating to the City of Brisbane who would have a bona fide interest in using the unregistered mark BRISBANE CITY in relation to those goods or services. In these circumstances, the unregistered mark BRISBANE CITY is unable to and does not perform the function of a trademark, and so is not a trademark for the purposes of the Uniform Policy.

  The panellist distinguished the *City of Hamina* decision on the basis that the panellist in that dispute had determined that the geographical term was distinctive. The panel in *Brisbane City Council v Joyce Russ Advertising*[341] agreed that the local government authority had no common law rights in the term 'BRISBANE', referring to the difficulties in establishing the distinctiveness of geographical terms, and to the desirability of others being able to use such terms.
- *Stadt Heidelberg v Media Factory.*[342] The complainant, the German city of Heidelberg, asserted rights in the city's name, claiming that the name was legally protected under German law. The panellist, citing previous decisions (including the *Brisbane City Council* decisions) held that the complainant had no unregistered rights in the name as it failed to provide evidence that the name had acquired distinctiveness as a trade mark.
- *EThekwini Municipality v Domain Strategy Inc.*[343] The complainant, the local government authority of the city of Durban in the Republic of South Africa, asserted unregistered

338 WIPO Case No D2001-0001 (12 Mar 2001).
339 Trade Mark Directive, above n 19, Art 6(1)(b).
340 WIPO Case No D2001-0047 (7 May 2001).
341 WIPO Case No D2001-0069 (14 May 2001).
342 WIPO Case No D2001-1500 (6 Mar 2002).
343 WIPO Case No D2002-0204 (23 Apr 2002).

rights in the mark 'DURBAN EXPERIENCE'. The panellist held that the complainant had no trade mark rights in the term on the basis that the evidence fell short of establishing that the name had been used as a trade mark. Although the dispute differed from similar disputes in that the alleged mark was more than just the city name, the panellist found that the term, 'DURBAN EXPERIENCE', had been simply used to refer to and promote the city as a particular location.

- *Land Sachsen-Anhalt v Skander Bouhaouaia.*[344] The complainant, *Land* Sachsen-Anhalt, was the federal German state known in English as Saxony-Anhalt. In a dispute concerning the <sachsen-anhalt.com> domain name, the panellist held that there was insufficient evidence of unregistered trade mark rights in the name. In particular, the panellist pointed out that there was no evidence that the complainant offered or sold goods or services under the name 'SACHSEN-ANHALT'.
- *Generalitat de Catalunya v Thomas Wolf.*[345] The complainant, the Government of the Autonomous Community of Catalonia, claimed unregistered trade mark rights in the name 'GENERALITAT DE CATALUNYA' in a dispute involving the <generalitatdecatalunya.info> domain name. After pointing out that the Spanish Trade Mark Law does not provide for unregistered rights in marks as such, the panellist referred to Articles 34–35 of the Spanish Law, which provide for unregistered rights in well-known trade marks in accordance with Article 6*bis* of the Paris Convention. The panellist held that there was sufficient evidence that the name, which refers to the 'Autonomous Community of Catalonia', had become well-known in the requisite sense through the complainant's participation in advertising campaigns, book fairs, festivities, exhibitions and the mass media. On this basis, the panellist held that the complainant had unregistered trade mark rights in the term.
- *City of Lake Worth v John C Becker, Inc.*[346] The complainant, the City of Lake Worth, claimed common law trade mark rights in the geographical term 'LAKE WORTH'. The panellist pointed out that geographic names do not automatically have trade mark rights under the UDRP. Citing *Brisbane City Council v Warren Bolton,* the panellist held that the name of the city was not distinctive to the complainant.

The preponderance of UDRP decisions therefore illustrates the obstacles facing a local authority that seeks to bring a complaint under the UDRP in relation to the name of a city or region. Moreover, it is not clear that the two decisions that have found that local authorities have unregistered trade mark rights—*City of Hamina* and *Generalitat de Catalunya*—correctly applied the policy. In general, then, local authorities may have more success in pursuing actions under national laws than in bringing complaints under the UDRP.

Non-exclusive Rights

[5.22] Licensees and Other Non-exclusive Rights

As explained at [5.10], if the complainant is the registered trade mark owner, then it will have rights in the mark for the purpose of the first element of the UDRP. It is not, however,

344 WIPO Case No D2002-0273 (8 July 2002).
345 WIPO Case No D2002-1124 (14 Apr 2003).
346 WIPO Case No D2003-0576 (18 Sept 2003).

necessary for the complainant to be the registered owner of the mark, as most UDRP panels have held that something less than exclusive rights, such as the rights of a trade mark licensee, will amount to 'rights' under paragraph 4(a)(i). On the other hand, a minority of panellists have held that a complainant must have exclusive rights in the mark, and that non-exclusive rights, such as the rights conferred by a non-exclusive licence, are not sufficient to satisfy paragraph 4(a)(i).

As there is no consensus on this issue, the *WIPO Overview* sets out both a majority view and a minority view. The WIPO majority view is that:

> In most circumstances a licensee of a trademark or a related company such as a subsidiary or parent to the registered holder of a mark is considered to have rights in a trademark under the UDRP.[347]

In a statement that is commonly referred to, this view was supported by the panellist in *Smart Design LLC v Carolyn Hughes*[348] in the following terms:

> In this Panel's view the test under paragraph 4(a)(i) of the Policy, which makes no mention of 'exclusive rights' is or ought to be a relatively easy test for a Complainant to satisfy, its purpose simply being to ensure that the Complainant has a bona fide basis for making the Complaint in the first place.

The WIPO minority view, on the other hand, is that:

> A complainant that holds a non-exclusive trademark license does not have rights in a trademark under the UDRP.[349]

The panellist in *NBA Properties, Inc v Adirondack Software Corporation*[350] explained the rationale behind the minority view in the following terms:

> The Policy is believed by the Panel to envision a transfer of a disputed domain name to a complainant/trademark owner as a route to unification of control over the uses of the domain name and the trademark. However, Complainant's request for an order transferring the disputed name to Complainant in this case would place ownership of the domain name in an entity other than the trademark owner without consent from the trademark owner.

When one is analysing UDRP decisions dealing with the rights of complainants that have lesser interests than full ownership of a registered mark, the position of complainants who are trade mark licensees can be distinguished from the position of complainants who are parent companies or members of the same group of companies.

In relation to trade mark licensees, a comprehensive examination of previous decisions was conducted by the panellist in *Evolution USA, Inc v Alexei Doicev*.[351] Although the panellist referred to the two views set out in the *WIPO Overview*, after taking into account decisions handed down following the publication of the *Overview*, the panellist concluded that there were two relevant views on whether a trade mark licensee can establish rights in the mark for the purpose of paragraph 4(a)(i):

> (a) That a licence provides contractual rights in relation to a trademark and that these are sufficient to constitute rights in a trademark for the purposes of the Policy; and

[347] WIPO, above n 57, para 1.8.
[348] WIPO Case No D2000-0993 (18 Oct 2000).
[349] WIPO, above n 57, para 1.8.
[350] WIPO Case No D2000-1211 (8 Dec 2000).
[351] WIPO Case No D2006-0086 (24 Mar 2006).

(b) A licence is sufficient but only if as a matter of the law of the territory of the relevant mark that licence provides rights in the trademark over and above the contractual rights granted in that licence.[352]

It would seem that these two views have replaced the majority and minority views set out by WIPO. The differences in the two views can be illustrated by an examination of the decisions that support each of the views. As the panellist pointed out in *Evolution USA*, the 'situations addressed in these cases have been diverse and it is difficult to fully reconcile these decisions'.

The first view was summarised by the panellist in *American Civil Liberties Union of New Mexico v Vilma Morales/e:bOOm, SA*[353] who, in explaining what he considered would amount to rights under paragraph 4(a)(i), stated:

> These rights need not be rights of ownership and need not be exclusive. A licensee having the right to use a mark under its license is clearly a party having a right in the service or trademark.

Under this view, any licensee of a mark, no matter how minor, has rights in the mark for the purpose of the policy. The following decisions may be regarded as lending some support to the first view identified in *Evolution USA*.

- *Lycos Europe NV v RegionCo.*[354] Under a sub-licence with the exclusive licensee of the 'LYCOS' mark, the complainant had rights to use the mark in Europe in its own corporate name in the combined form 'Lycos Europe'. The panellist held that the UDRP is not limited to trade mark owners and that, as the complainant had rights in the mark, the fact that it could not demonstrate trade mark rights of its own did not prevent it from bringing a complaint under the policy.
- *Sunlane Media v Lucien de Cassan.*[355] The complainant claimed that it was the sole licensee of the 'LITEROTICA' mark, but provided little evidence of this. Nevertheless, the panellist was prepared to infer rights in the mark on the basis of an unchallenged signed statement from the trade mark owner indicating that she owned and controlled the complainant.
- *Telcel, CA v jerm.*[356] The complainant, who brought a complaint in relation to the <telcelbellsouth.com> domain name, was the owner of the 'TELCEL' mark and the Venezuelan licensee of the 'BELL SOUTH' mark. The panellist held that it was unnecessary for the owner of the 'BELL SOUTH' mark to be joined in the proceedings as the complainant was a licensee of the mark and, in any case, the domain name was confusingly similar to the complainant's 'TELCEL' mark.
- *FDNY Fire Safety Education Fund, Inc v Roger Miller.*[357] The complainant was a non profit-making organisation that was a licensee of the 'FDNY' marks owned by the City of New York. The panellist held that the complainant had established trade mark rights, as a licensee, in the 'FDNY' marks, including the trade mark consisting solely of the initials 'FDNY'.
- *Savin Corporation and Ricoh Corporation v Kent S Schisler.*[358] The complainant, Ricoh Corporation was the US licensee of the 'RICOH' mark and a subsidiary of the trade mark

352 WIPO Case No D2006-0086 (24 Mar 2006) para 6.11.
353 WIPO Case No D2004-0473 (23Aug 2004).
354 WIPO Case No D2000-1102 (28 Aug 2001).
355 WIPO Case No D2002-0093 (27 Mar 2002).
356 WIPO Case No D2002-0309 (5 June 2002).
357 NAF Case No FA145235 (26 Mar 2003).
358 NAF Case No FA294206 (25 Aug 2004).

owner. The panellist held that the complainant had rights in the mark under paragraph 4(a)(i).

- *DigiPoll Ltd v Raj Kumar.*[359] The complainant presented letters indicating that, although it was not the registered owner of the 'DIGIPOLL' marks, it was licensed to use the marks. The panellist held that a licence to use a trade mark is a 'right' for the purposes of the policy and that 'such a right is effective for that purpose even when another entity also has rights in the trademark'. Although the panellist considered that decisions should be made on a case-by-case basis, he concluded that the evidence in this case indicated that the complainant had rights in the mark.
- *Hunter Douglas Inc v Domaincar.*[360] The complainant was a trade mark licensee of the Dutch company Hunter Douglas Industries BV. The panellist held that, as a licensee of the 'HUNTER DOUGLAS' mark from the trade mark owner, the complainant had rights in the mark under paragraph 4(a)(i).

The second view identified in *Evolution USA* has been influenced by the decision in *NBA Properties*, in which the panellist held that a non-exclusive licensee does not have a sufficient interest to amount to 'rights' under paragraph 4(a)(i). In that dispute, the complainant was the merchandising agent for the National Basketball Association and had contractual rights to use the 'KNICKS' mark. The panellist concluded that these rights were not sufficient for the first element of the UDRP as they were rights *with respect to*, not rights *in*, the trade mark. In reaching this conclusion, the panellist added:

> There may well be circumstances in which the contract rights possessed by an exclusive licensee vest in him substantially all the powers of an owner of the licensed property. However, such circumstances have not been shown to exist here.

Accordingly, the second view, which is much stricter than the first, requires panellists to examine the rights of a trade mark licensee under the relevant national trade mark law on a case-by-case basis to determine, for example, whether the licensee is entitled to bring a legal action against trade mark infringers without joining the trade mark owner. On this view, the standing of a licensee to bring a complaint under the UDRP should be determined by principles analogous to those that are used to determine whether a licensee can bring an action for trade mark infringement under national laws. The following decisions may be taken to support the second view.

- *Toyota Motor Sales USA Inc v J Alexis Productions.*[361] The complainant, a wholly-owned subsidiary of Toyota Motor Corporation, a Japanese corporation that was the registered owner of the 'LEXUS' mark, was a non-exclusive licensee of the mark. Applying US law, the panellist held that, as a non-exclusive licensee has a right to assert rights in the mark and to bring an action to protect the mark, the complainant had satisfied the first element of the policy.
- *Intermedia Film Equities USA Inc v Varietydomains.com.*[362] The complainant, AGV, claimed to have an exclusive licence in the 'TERMINATOR' mark, while the respondent argued that the complainant had only a non-exclusive licence. After examining the licence agreement, the panellist concluded that the complainant was a non-exclusive licensee and,

[359] WIPO Case No D2004-0939 (3 Feb 2005).
[360] NAF Case No FA638982 (28 Mar 2006).
[361] WIPO Case No D2003-0624 (16 Oct 2003).
[362] WIPO Case No D2004-0065 (23 July 2004).

as such, did not have sufficient rights in the mark to satisfy paragraph 4(a)(i). The panellist gave three reasons for reaching this conclusion. First, the panellist found that, in *NBA Properties,* the only previous decision to deal with a non-exclusive licensee that was unrelated to the trade mark owner, the panel had decided that the licensee lacked standing to bring a complaint. Secondly, the panellist held that, under US law, only an exclusive licensee has standing to bring an action for trade mark infringement in its own right, and a non exclusive licensee must join the trade mark owner. Thirdly, the panellist agreed with the panellist in *NBA Properties* that the object of the policy is to unify control of an infringing domain name with the complainant's trade mark.

After explaining the two competing views, the panellist in *Evolution USA* pointed out that there were arguments in favour of each view. Moreover, given that recourse to national trade mark laws is required to determine the meaning of a trade mark under paragraph 4(a)(i), it could be argued that national laws should also be referred to in determining whether a licensee has sufficient 'rights' in the mark to bring a complaint under the UDRP. While acknowledging these arguments, however, the panellist preferred the first view, largely on the ground that requiring panels to engage in a detailed analysis of the rights of a licensee under national laws runs counter to the objectives of the UDRP as an expeditious, cost-effective form of dispute resolution. In rejecting the second, strict approach, the panellist observed that it was:

> likely to add a significant complication for licensee complainants under the Policy. If correct, a complainant may feel compelled to explain why under the law of the relevant mark the nature of his rights was sufficient for the purposes of the Policy. Is the addition of this technical requirement really in keeping with the aims of the Policy which are intended to provide a cost efficient dispute resolution mechanism for domain name disputes with a minimum of legal formalities? It certainly would appear to run counter to those cases that treat Paragraph 4(a)(i) of the Policy as 'a relatively easy test for a Complainant to satisfy'. Also it may result in different treatment of different licensees under the Policy dependant upon the geographical accident of where the licensee has rights . . . Lastly, consistency between Policy decisions is to be encouraged and the contractual approach at present appears to represent the majority view.

Despite the views of the panellist in *Evolution USA,* UDRP panellists continue to regard the status of a non-exclusive licensee under the policy as unsettled. In *9 Squared, Inc v Mass Management Limited,*[363] for example, the panellist observed:

> The consensus is that in most circumstances a licensee of a trademark or a related company such as a subsidiary or parent to the registered holder of a mark is considered to have rights in the trademark. Nonetheless, the question in this case remains unresolved since the Panel is reminded that in *NBA Properties, Inc v Adirondack Software Corp,* . . . it was held that a complainant that holds a non-exclusive trade mark license does not have rights in a trade mark under the Policy.

In that dispute, the panellist was not required to address the issue directly, as it was unable to establish that there were any trade mark rights in the unregistered mark.

Apart from the split in decisions concerning the treatment of non-exclusive licensees, most panels have agreed that a company that is related to the actual trade mark owner, such as a subsidiary or parent of the trade mark owner, can have rights in the mark sufficient for paragraph 4(a)(i) to be satisfied. The following decisions have held that companies that are related to the trade mark owner have rights in the mark for the purpose of the UDRP.

[363] NAF Case No FA811932 (3 Nov 2006).

- *Miele, Inc v Absolute Air Cleaners and Purifiers.*[364] The complainant was the American subsidiary of a Swiss corporation that was owned by a German company that was the US registered owner of the 'MIELE' mark. The panellist held that the complainant, through its relationship with its grandparent corporation, had sufficient rights in the mark for the purpose of the policy.
- *Grupo Televisa, SA v Party Night Inc.*[365] The complainant was the parent company of the registered owners of the trade marks. Supporting the view that the UDRP does not require the complainant to have exclusive rights in the mark, the panellist held that a company that is related as a subsidiary or parent to the registered owner of a mark may be considered to have rights in the mark
- *HQUK Limited v Head Quarters.*[366] The complainant was a related company of companies that had applied to register Community marks. The panellist observed that it was 'not unreasonable' to assume that, given the relationship with the trade mark applicants, it could be assumed that they would have consented to the UDRP proceedings. As the complainant was able to establish common law rights in the mark, however, it was not necessary for the panellist to decide this question.
- *Spherion Corporation v Peter Carrington.*[367] The complainant was the parent company of wholly owned subsidiaries that were the registered owners of various 'SPHERION' marks. After requesting further evidence, the panellist found that the trade mark owners were subsidiaries of the complainant and, accordingly, that it had rights in the mark for the purpose of paragraph 4(a)(i).

Although the majority of panels have decided that a parent company only needs to provide evidence of its relationship to a subsidiary that is a registered owner of the mark to establish rights in the mark for the purpose of the policy, the reasoning in decisions such as *Grupo Televisa* was criticised by the panellist in *PartyGaming Plc v Kriss Vance.*[368] In that dispute, the complainant claimed rights in various 'PARTYPOKER' marks on the basis that the registered owners of the marks were subsidiaries of the complainant. After reviewing previous decisions, however, the panellist concluded that a parent company does not automatically have rights in a mark that is owned by a subsidiary, but must independently establish rights in the mark. In this respect, the panellist maintained that:

> It is by no means obvious to this Panel that having an ability to exercise general control over some *other entity* which enjoys rights in a mark, is always going to be the same thing as having rights in that mark itself.

On this basis, the panellist preferred the view that 'the parent company must have some right to use or otherwise deal with the mark itself, whether under an express or implied licence from its subsidiary, or perhaps because the subsidiary holds the mark in trust for the parent company and/or others within the group'. In the particular dispute, the panellist took into account statements that the corporate group had an avowed strategy of integrating its marks to infer that the marks were held by the subsidiaries for the benefit of all the companies in the group, including the complainant. As either the complainant was a licensee or the subsidiaries held the mark on trust for all members of the group, the panellist concluded that the complainant

[364] WIPO Case No D2000-0756 (11 Sept 2000).
[365] WIPO Case No D2003-0796 (2 Dec 2003).
[366] WIPO Case No D2003-0942 (5 Feb 2004).
[367] WIPO Case No D2003-1027 (10 Mar 2004).
[368] WIPO Case No D2003-0456 (2 June 2006).

had rights in the mark under paragraph 4(a)(i). In a subsequent decision involving the same complainant, *PartyGaming Plc v WHOis Privacy Protection Service, Inc*,[369] the panellist in the first *PartyGaming* dispute affirmed the reasoning in that decision, and proposed the following minor amendment to the test for determining what a parent company must establish in order to have rights in a registered mark that is owned by a subsidiary:

> the parent company must have some right to use or otherwise deal with the mark itself, whether under an express or implied licence from its subsidiary, or perhaps because the subsidiary holds the mark 'in trust for others within the group including the parent company'.

Consequently, where a complaint is brought by a member of a corporate group that is not the registered owner of the mark, the complainant should be advised to produce evidence concerning both the relationship of the companies within the group, and how that relationship gives rise to specific rights in the particular complainant in relation to the mark.

Comparing Domain Names and Marks

[5.23] Comparing Marks under National Laws

The second component of paragraph 4(a)(i) of the UDRP requires the complainant to prove that the disputed domain name is identical or confusingly similar to the mark in which the complainant has rights. The wording of the test for comparing the disputed domain name with the mark draws from national trade mark laws. As trade mark law is designed to protect marks as indications of the source of goods or services, trade mark laws prohibit registration of marks that would create confusion as to source, and prohibit uses of marks that would create confusion with a registered mark. National trade mark laws therefore incorporate tests for comparing a mark that is the subject of an application for registration with an earlier registered mark, and for comparing a registered mark with an allegedly infringing mark. This paragraph reviews tests for comparing marks under some natonal laws.

Under US federal trade mark law, the Lanham Act provides that no distinctive mark shall be refused registration on the principal register unless, inter alia, it consists of a mark which so resembles a registered mark 'as to be likely, when used on or in connection with the goods of the applicant, to cause confusion, or to cause mistake, or to deceive'.[370] This test, known as the 'likelihood of confusion' test, is the same test that is used for determining infringement of both registered and common law trade marks under the Lanham Act.[371]

There are a number of different forms of confusion that are recognised by US law. While confusion as to source is the most common form of confusion, the 'likelihood of confusion' test also encompasses confusion about affiliation, connection or sponsorship.[372] In line with the fundamental objective of trade mark protection, the confusion that is relevant is confusion relating to use of a mark in connection with goods or services.[373]

[369] WIPO Case No D2006-0508 (26 June 2006).
[370] Lanham Act § 2(d), 15 USC § 1052(d).
[371] Lanham Act, § 32, 15 USC § 1114(1) (infringement of a registered mark); § 43(a), 15 USC § 1125(a) (infringement of an unregistered mark).
[372] McCarthy, above n 5, §23:8.
[373] *Ibid*, §23:11.50.

One form of confusion that is specifically recognised under US trade mark law but not under the trade mark laws of other jurisdictions is a form of pre-sales confusion, known as 'initial interest confusion'. In this form of confusion, consumers are confused about an association with the source of a registered mark because of the similarity of an allegedly infringing mark, even though they may not be confused at the time of purchase. As one court has put it:

> Initial interest confusion, which is actionable under the Lanham Act, occurs when a customer is lured to a product by the similarity of the mark, even if the customer realizes the true source of the goods before the sale is consummated.[373a]

This form of confusion is particularly relevant to use of designations on the Internet, such as the registration of a domain name that is similar to a registered mark, as actionable confusion may arise from mere use of the designation, despite the fact that the confusion is later dissipated, for example, by a user visiting a web site. In *Brookfield Communications, Inc v West Coast Entertainment Corp,*[374] for example, the use of a confusingly similar name as a metatag on a web site was held to create initial interest confusion by analogy with the effect of a misleading advertising hording on a highway. In other cases, US courts have applied initial interest confusion to the registration of a similar domain name on the basis that the domain name will attract users to a web site, even though the confusion may be dissipated once the web site is accessed.[375]

Under the 1988 European Trade Mark Directive, Article 4(1) provides that a trade mark is not to be registered:

> (a) if it is identical with an earlier trade mark, and the goods or services for which the trade mark is applied for or is registered are identical with the goods or services for which the earlier trade mark is protected;
> (b) if because of its identity with, or similarity to, the earlier trade mark and the identity or similarity of the goods or services covered by the trade marks, there exists a likelihood of confusion on the part of the public, which includes the likelihood of association with the earlier trade mark.[376]

Trade mark infringement is dealt with under Article 5(1) of the Directive, which provides that the trade mark owner is entitled to protect unconsented to use in the course of trade:

> (a) any sign which is identical with the trade mark in relation to goods or services which are identical with those for which the trade mark is registered;
> (b) any sign where, because of its identity with, or similarity to, the trade mark and the identity or similarity of the goods or services covered by the trade mark and the sign, there exists a likelihood of confusion on the part of the public, which includes the likelihood of association between the sign and the trade mark.

The provisions of the Directive dealing with the tests of identicality and confusing similarity are incorporated into the trade mark law of the United Kingdom by provisions of the Trade Marks Act 1994.[377]

[374a] *Dorr-Oliver, Inc v Fluid-Quip, Inc,* 94 F 3d 376, 382 (7th Cir, 1996).

[374] 174 F 3d 1036 (9th Cir, 1999).

[375] See *Interstellar Starship Services, Ltd v Epix, Inc,* 304 F 3d 936 (9th Cir, 2002); *Planned Parenthood Federation of America, Inc v Bucci,* 152 F 3d 920 (2d Cir, 1998).

[376] Trade Mark Directive, above n 19, Art 4(1).

[377] Trade Marks Act 1994 (UK), ss 5(1) (registration of trade mark that is identical with an earlier mark); 5(2) (registration of trade mark where likelihood of confusion with earlier trade mark); 10(1) (infringement by use of identical mark); 10(2) (infringement where likelihood of confusion with registered trade mark).

The main European authority on the 'likelihood of confusion' test, the decision of the European Court of Justice in *Sabel v Puma*,[378] stated:

> The likelihood of confusion must . . . be appreciated globally, taking into account all factors relevant to the circumstances of the case.
>
> That global appreciation of the visual, aural or conceptual similarity of the marks in question, must be based on the overall impression given by the marks, bearing in mind, in particular, their distinctive and dominant components.

The confusion that is required by the European test is confusion in relation to the origin of goods or services. As *Kerly's Law of Trade Marks and Trade Names* explains:

> It is not enough that on seeing the mark the subject of the application, the earlier mark is "called to mind" if there is no possibility of the customer being under any misapprehension as to the origin of the goods or services. The mere association which the public might make between two trade marks as a result of their analogous semantic content is not in itself a sufficient ground for concluding that there is a likelihood of confusion . . .[379]

Section 44(1) of the current Australian trade mark law, the Trade Marks Act 1995, provides that an application for registration of a mark must be rejected if the applicant's trade mark is 'substantially identical with, or deceptively similar to' a registered mark. Section 120 of the Australian Act defines trade mark infringement. Section 120(1) provides that a registered mark is infringed if a person uses as a trade mark a sign that is substantially identical or deceptively similar to the registered mark in relation to goods or services in respect of which the mark is registered. Section 120(2) provides that a registered mark is infringed if a person uses as a trade mark a sign that is substantially identical or deceptively similar to the registered mark in relation to goods or services that are of the same description as, or closely related to, the registered goods or services. The term 'deceptively similar' is defined by section 10 of the Act, which provides that a trade mark is taken to be deceptively similar to another mark if it so nearly resembles the other mark that it is 'likely to deceive or cause confusion'. The distinction between 'likely to deceive' and 'likely to cause confusion' was explained by a New Zealand court in the following terms:

> Where the deception or confusion alleged is as to the source of the goods, deceived is equivalent to being misled into thinking that the goods bearing the applicant's mark come from some other source and confused to being caused to wonder whether that might not be the case.[380]

Each of the above trade mark laws has developed more detailed rules of comparison for determining when there will be a 'likelihood of confusion' between marks.

[5.24] The Requirement of Use as a Trade Mark

In determining whether there is a likelihood of confusion between a registered mark and an allegedly infringing mark under trade mark laws, the confusion that is required relates to use as a trade mark in connection with goods or services. The reason for this was explained in relation to Article 5(1) of the Trade Mark Directive by the European Court of Justice in the *Anheuser-Busch* case:

378 Case C–251/95, [1997] ECR I–6191, [1998] RPC 199 at paras 22–23.
379 Kitchen *et al*, above n 20, para [9-045].
380 *Pioneer Hi-Bred Corn Co v Hyline Chicks Pty Ltd* [1979] RPC 410, 423 per Richardson J.

> the exclusive right conferred by a trade mark was intended to enable the trade mark proprietor to protect his specific interests as proprietor, that is, to ensure that the trade mark can fulfil its functions and that, therefore, the exercise of that right must be reserved to cases in which a third party's use of the sign affects or is liable to affect the functions of the trade mark, in particular its essential function of guaranteeing to consumers the origin of the goods . . .[381]

As explained at [3.15], the requirement that an allegedly infringing trade mark be used as a mark means that it is difficult to apply traditional trade mark infringement actions against 'cybersquatters' who register a domain name with the intention of profiting from it, but do not use the domain name as a mark. This paragraph examines these difficulties.

In the United States, the federal Lanham Act defines an infringing use as a 'use in commerce' that is likely to cause confusion, or to cause mistakes, or to deceive.[382] 'Use in commerce' means the 'bona fide use of a mark in the ordinary course of trade'.[383] Merely registering a domain name does not amount to 'use in commerce'. As one US court has put it:

> When a domain name is used only to indicate an address on the Internet, the domain name is not functioning as a trademark . . . something more than the registration of the name is required before the use of a domain name is infringing.[384]

In addition to traditional trade mark infringement, US federal trade mark law includes a developed doctrine of trade mark dilution.[385] Actions for infringement of dilution laws are not directed at protecting against confusion as to source, but at protecting the value of a 'famous' mark against actions that detract from the value of the mark as a source identifier. US anti-dilution law incorporates two main forms of dilution: 'blurring' and 'tarnishment'. 'Blurring' involves weakening an established mark by undermining its distinctiveness through use on different goods or services, even if there is no confusion as to source, while 'tarnishment' involves an essentially negative use of a mark that creates inferior or degrading associations. To amount to an unlawful dilution under the Lanham Act, there must be a use of the mark 'in commerce', which involves use of a designation as a mark or trade name in a commercial setting.

The emergence of the problem of cybersquatting led to US courts developing a third category of trade mark dilution to apply specifically to the registration of a domain name for the purpose of selling it to the owner of a corresponding famous mark. For example, in *Panavision International LP v Toeppen*,[386] the defendant registered domain names that were identical to the plaintiff's famous marks and requested payment for transfer of the domain names. The main problem confronting the plaintiff was the difficulty of establishing that there was a commercial use of the domain names, given that the defendant made no use of the domain names whatsoever. In recognising cybersquatting as a new form of unlawful dilution, however, the US courts held that Toeppen's actions amounted to a 'commercial use' of the marks as he was in the business of registering famous marks in order to sell them to the owners. This represented a considerable extension of US anti-dilution law, as a cybersquatter clearly makes no 'commercial use' of the name as a mark or trade name in the conventional understanding of that term.

The difficulties in fitting actions against cybersquatters within the framework of trade mark infringement or anti-dilution actions led the United States to adopt a *sui generis*

[381] Case C–245/02 *Anheuser-Busch v Budejovicky Budvar NP* [2004] ECR I–10989, [2005] ETMR 27, para 59.
[382] Lanham Act § 32, 15 USC § 1114(1).
[383] Lanham Act § 45, 15 USC § 1127.
[384] *Lockheed Martin Corp v Network Solutions, Inc*, 985 F Supp 949 (CD Cal, 1997).
[385] Lanham Act, 15 USC §§ 1125(c)–1127.
[386] 945 F Supp 1296 (CD Cal 1996); 141 F 3d 1316 (9th Cir, 1998).

law directed specifically against cybersquatting, in the form of the federal Anti-cybersquatting Consumer Protection Act (the ACPA).[387] The enactment of the ACPA may mean that actions against cybersquatters for trade mark dilution are either redundant or impermissible.[388]

The requirement that the relevant confusion relates to commercial use of an allegedly infringing mark as a source identifier has also created difficulties in dealing with cybersquatting under trade mark laws in English-oriented legal systems. These difficulties were central to the problems encountered in bringing an action for trade mark infringement against cybersquatting in the main English authority, *British Telecommunications Plc v One in a Million Ltd* which is also dealt with at [3.25].[389] In that case, the defendants registered a number of domain names, including <marksandspencer.co.uk> and <britishtelecom. co.uk> and attempted to sell them to the owners of the corresponding famous trade marks. Actions were brought against the defendants for passing off and for trade mark infringement under section 10(3) of the Trade Marks Act 1994 (UK). Section 10(3) provides that a registered mark is infringed by a person who uses in the course of trade a sign that is identical or similar to the registered mark, and is used in relation to goods or services that are not similar to those for which the trade mark is registered, in a way that takes unfair advantage of, or is detrimental to the distinctiveness of, the registered mark. The problem confronting the claimants in relation to the infringement action was establishing that the domain names were used in the course of trade in connection with goods or services, when the defendants had merely registered the domain names and offered them for sale. At first instance, Jonathan Sumption QC held that the domain names were used in the course of trade on the basis that:

> the use of a trade mark in the course of the business of a professional dealer for the purpose of making domain names more valuable and extracting money from the trade mark owner amounted to 'use in the course of trade'.[390]

This conclusion was upheld by the English Court of Appeal, which stated:

> The appellants seek to sell the domain names which are confusingly similar to registered trade marks. The domain names indicate origin. That is the purpose for which they were registered. Further they will be used in relation to the services provided by the registrant who trades in domain names.[390a]

Despite this analysis, it is clear that the reasoning of the English courts in extending section 10(3) to encompass cybersquatting is as strained as that of the US courts that have extended US anti-dilution law. The difficulties arise because, as explained at [3.15], the problem posed by cybersquatting is different from the problems posed by trade mark infringement or trade mark dilution. Actions to prevent cybersquatting do not fit easily within the framework of conventional trade mark law because a cybersquatter does not need to use a domain name as a trade mark. In order to succeed, the most that a cybersquatter needs to do is impliedly to threaten that the domain name may be used in a way that will harm the value of a trade mark that corresponds to the domain name. Therefore, the requirement that a domain name be used in commerce as a trade mark is an obstacle to the use of trade mark laws against cybersquatters, and is irrelevant to any laws directed specifically against cybersquatting.

[387] Lanham Act 15 USC §§ 1125(d), 1129.
[388] See McCarthy, above n 5, §25:77.
[389] (1998) 42 IPR 289.
[390] (1998) 42 IPR 289, 293.
[390a] (1998) 42 IPR 289, 307.

[5.25] Comparing Marks under the UDRP

The UDRP is aimed at preventing abusive, bad faith registration of trade marks as domain names. As explained at [3.15], it is not directed at preventing conventional trade mark infringement. The comparison required for the purpose of paragraph 4(a)(i) of the UDRP involves a comparison between the disputed domain name and the mark in which the complainant has rights. It does not involve a comparison between two marks.

The precise nature of the comparison required to be undertaken by paragraph 4(a)(i) properly depends upon identifying the purpose of the paragraph, and of the UDRP more generally. The preferred view is that the purpose of paragraph 4(a)(i) is simply to establish a threshold to determine whether the domain name is sufficiently similar to the complainant's mark for there to be a finding of abusive, bad faith registration. It is not to determine whether use of the domain name infringes the complainant's mark. Whether or not a domain name is used as a trade mark is therefore irrelevant to the comparison required by the UDRP. Moreover, UDRP panels are not equipped to decide the complex issues involved in an action for trade mark infringement. As the panellist pointed out in *Bridgestone Firestone, Inc v Jack Myers:*[391]

> the Panel does not have jurisdiction to decide claims of trademark infringement, dilution, unfair competition or other statutory or common law causes of action . . . Further, such causes of action require extensive factual development and analysis which these domain name administrative proceedings are not designed to accommodate.

In comparing marks under national trade mark laws it is essential to take into account the use of an allegedly infringing mark in connection with goods or services. In the English decision, *Pianotist Co's Application,* for example, Parker J explained the nature of the comparison as follows:

> You must take the two words. You must judge of them, both by their look and by their sound. You must consider the goods to which they are to be applied. You must consider the nature and kind of customer who would be likely to buy those goods. In fact, you must consider all the surrounding circumstances; and you must further consider what is likely to happen if each of those trade marks is used in a normal way as a trade mark for the goods of the respective owners of the marks.[392]

As the UDRP is not a trade mark infringement law, principles of trade mark law and associated rules for comparing marks should not be incorporated wholesale into the UDRP for the purpose of determining whether or not a disputed domain name is identical or confusingly similar to the complainant's mark. National trade mark laws necessarily take into account the use of an allegedly infringing mark in making a comparison. Just as whether or not a domain name is used in commerce as a mark is irrelevant to actions directed against cybersquatting, however, the use to which an allegedly infringing domain name is put should not be taken into account in comparing the domain name with the mark under paragraph 4(a)(i).

Although the differences between the comparisons required under trade mark laws and under the UDRP should be clear, some panellists have improperly incorporated trade mark principles into the comparison required by paragraph 4(a)(i). A good example of this

[391] WIPO Case No D2000-0190 (6 July 2000).
[392] (1906) 23 RPC 774, 777.

is the decision in *Wal-Mart Stores, Inc v Walsucks and Walmarket Puerto Rico,*[393] which concerned the disputed domain names <walmartcanadasucks.com>, <walmartcanada sucks.com>, <walmartuksucks.com>, <walmartpuertorico.com> and <walmartpuertoricosucks.com>. The panellist in that dispute expressly applied the eight factor test applied by the US Ninth Circuit for determining the likelihood of confusion for the purpose of trade mark law, known as the '*Sleekcraft* factors'.[394] The eight factors are: (1) strength of the mark; (2) proximity of the goods; (3) similarity of the marks; (4) evidence of actual confusion; (5) marketing channels used; (6) type of goods and the degree of care likely to be exercised by the purchaser; (7) defendant's intent in selecting the mark; and (8) likelihood of expansion of the product lines. In determining that the domain names were confusingly similar to the 'WALMART' mark, the panellist gave particular weight to the strength of the complainant's mark and the intent of the respondent in selecting the disputed domain names. The problem with the decision is its overt and unqualified reliance on principles of comparison developed under US trade mark law.

Given the differences between the comparisons under the UDRP and under trade mark laws, it is preferable for guidance on the factors to be taken into account in making the comparison under paragraph 4(a)(i) to be drawn from the developing UDRP principles than from trade mark laws. The preferred approach was explained in the following terms by the panellist in *Wachovia Corporation v Alton Flanders:*[395]

> Proof of trademark infringement is not the same as proof of the first element under the UDRP Policy. This Panel finds it more appropriate to adopt the UDRP Policy and precedent developed specifically to address cybersquatting.

This analysis does not necessarily mean that the factors taken into account for the purposes of comparison under trade mark laws are all necessarily irrelevant to the comparison under the UDRP. As explained at [5.26], the extent to which trade mark principles can be taken into account depends upon the view taken as to the precise nature of the comparison under the UDRP, including whether it requires a literal comparison or a comparison directed at avoiding source confusion.

Regardless of differences about the nature of the comparison to be undertaken between the disputed domain name and the complainant's mark, there is general agreement that considerations relating to the use of the domain name are irrelevant to the comparison. As pointed out by the panel in *Future Media Architects, Inc v Oxado SARL:*[396]

> It is well settled under paragraph 4(a)(i) of the Policy that the question of identity or confusingly similarity is evaluated solely based on a comparison of the complainant's mark and the alphanumeric string constituting the domain name at issue. *Magnum Piering, Inc v The Mudjackers and Garwood S Wilson*, WIPO Case No D2000-1525. Thus, the Panel compares the mark and the domain name alone, independent of the use factors usually considered in a traditional trademark infringement action. See *Banconsumer Service, Inc v Mary Langthorne, Financial Advisor*, WIPO Case No D2001-1367; *InfoSpace.com, Inc v Delighters, Inc d/b/a Cyber Joe's Internet Café*, WIPO Case No. D2000-0068.

393 WIPO Case No D2000-0477 (20 July 2000). See also *Diageo plc v John Zuccarini*, WIPO Case No D2000-0996 (22 Oct 2000).

394 *AMF, Inc v Sleekcraft Boats*, 599 F 2d 341 (9th Cir, 1979). Similar 8 factor tests have been applied by the US Second Circuit in *Polaroid Corp v Polarad Electronics Corp*, 287 F 2d 492 (2d Cir, 1961) (the '*Polaroid* test'); and the US *Third Restatement of Unfair Competition*, §20–3 (1995): see McCarthy, above n 5, §23:19.

395 WIPO Case No D2003-0596 (19 Sept 2003).

396 WIPO Case No D2006-0729 (31 July 2006).

A similar point has been made by the panellist in *Lyonnaise de Banque v Richard J*:[397]

> The test of confusing similarity under the Policy is confined to a comparison of the disputed domain name and the trademark alone, independent of the products for which the domain name is used or other marketing and use factors, such as the "Sleekcraft factors"—*AMF Inc v Sleekcraft Boats*, 599 F 2d 341, 346 (9th Cir. 1979), usually considered in trademark infringement (See *Arthur Guinness Son & Co (Dublin) Limited v Dejan Macesic*, WIPO Case No. D2000-1698 <guiness.com>; *Ansell Healthcare Products Inc v Australian Therapeutics Supplies Pty, Ltd*, WIPO Case No D2001-0110, <ansellcondoms.com>; *Dixons Group Plc v Mr Abu Abdullaah*, WIPO Case No D2001-0843, <dixons-online.net>; *AT&T Corp v Amjad Kausar*, WIPO Case No. D2003-0327, <attinternet.com>, <attuniversal.com>; *BWT Brands, Inc and British Am Tobacco (Brands), Inc v NABR*, WIPO Case No D2001-1480, <kool.com>; *Britannia Building Society v Britannia Fraud Prevention*, WIPO Case No D2001-0505, <britanniabuildingsociety.org>).[398]

Furthermore, as explained at [5.27], the content of a web site associated with the disputed domain name is irrelevant to the comparison under paragraph 4(a)(i).

[5.26] Literal Comparison or Source Confusion

There is agreement among panellists that the comparison required by the UDRP must focus primarily on the disputed domain name and the complainant's mark. There is, however, disagreement about the extent to which some factors that are considered relevant under trade mark laws, such as the strength of the complainant's mark or evidence of actual confusion, may be taken into account in the comparison.

To begin with, it is clear that the comparison must focus on the domain name and the complainant's mark, and is completely independent of use and marketing factors. As pointed out by the panel in *Disney Enterprises, Inc v John Zuccarini*:[399]

> The test of confusing similarity under the Policy, unlike trademark infringement or unfair competition cases, is confined to a consideration of the Disputed Domain Name and the trademark: *AltaVista Company v SMA, Inc*, (WIPO Case No. D2000-0927); *Gateway, Inc v Pixelera.com, Inc (formerly Gateway Media Productions, Inc)* (WIPO Case No D2000-0109).

There is, nevertheless, a degree of conflict among panellists about whether the comparison required for the purpose of the UDRP should be strictly confined to a literal comparison of the disputed domain name with the complainant's mark or whether issues relevant to source confusion may be taken into account. The policy itself is not explicit about which approach is to be preferred. If the main purpose of the comparison is to protect the complainant's mark against source confusion, then some (but not all) of the factors that are taken into account for the purpose of comparing marks under trade mark laws may be taken into account, as the extent of possible confusion as to source is relevant. If, however, the main purpose of the comparision is something other than protecting against source confusion, these factors may not be relevant.

[397] WIPO Case No D2006-0142 (30 Mar 2006).

[398] See also comments to like effect in *Kirkbi AG v Dinoia*, WIPO Case No D2003-0038 (9 Mar 2003); *Aquascape Designs, Inc v Vertical Axis, Inc*, NAF Case No FA629222 (7 Mar 2006); *The Neiman Marcus Group, Inc and NM Nevada Trust v Faith McGary*, NAF Case No FA757475 (22 Aug 2006); *Seiko Epson Corporation and Epson America, Inc v AOS Web Com, Inc*, NAF Case No FA823033 (27 Nov 2006).

[399] WIPO Case No D2001-0489 (19 June 2001).

The view that a strict literal comparison is required was first clearly expressed in *Magnum Piering, Inc v The Mudjackers*.[400] In that decision, the panellist formulated the 'literal comparison' approach as follows:

> Under the Policy, the question of identity and confusing similarity is evaluated based solely on a comparison between a complainant's word mark and the alphanumeric string constituting the domain name at issue. Other considerations taken into account by U.S. trademark law may properly be addressed at other stages of the analysis (such as factors that bear on a registrant's legitimate interest or bad faith), but they are not dispositive with respect to the analysis of the first UDRP factor.

In a series of decisions, one panellist has consistently justified the adoption of the 'literal comparison' approach, giving the following reasons:

> A number of considerations support the view that the appropriate test for confusing similarity under the Policy is a literal comparison of the challenged domain name and conflicting mark:
>
> (1) The language of Policy paragraph 4(a) should be interpreted in a purposive manner consistent with the Policy's remedial nature and objective—to prevent the extortionate behavior known as 'cybersquatting.'
> (2) The language of Policy paragraph 4(a) does not invoke a likelihood of confusion test, as contrasted with Policy paragraph 4(b)(iv), which expressly refers to a 'likelihood of confusion with the complainant's mark as to the source, sponsorship, affiliation, or endorsement of [the registrant's] website or location or of a product or service on [the registrant's] website or location' as an indication of bad faith domain name registration and use.
> (3) The Policy is international and, in the absence of express language, it is inappropriate to import into Policy paragraph 4(a) a test that may not exist under all national trademark laws.
> (4) The procedure contemplated by the Policy is not well suited to the kinds of factual issues presented by a conventional source confusion analysis. In particular, the Policy is intended to be a relatively quick and inexpensive process, panels must make factual determinations based upon written submissions and documentary evidence, complainants have no right of reply without the panel's permission, and parties are often not represented by counsel.[401]

These comments were specifically endorsed by another panellist in *A & F Trademark, Inc v Party Night, Inc*.[402]

A variation on the explanation for the 'literal comparison' approach was given by the panellist in *Lebanon Valley College v BDC Partners, Inc*,[403] who maintained that:

> While the class of good of each mark is a factor in analyzing whether or not two marks are confusingly similar in a trademark infringement action under the Lanham Act, it does not make sense to consider the class of goods or services under a Policy ¶ 4(a)(i) analysis. In many UDRP proceedings there is no class of goods at all connected with the at-issue domain names; for example when the domain name is registered for resale or is dormant. Furthermore, a plain reading of paragraph 4(a)(i) simply requires a panel to consider whether or not the domain name is 'identical or confusingly similar.' There is nothing in the Policy that suggests that 'confusingly similar' be given any definition other than its plain meaning. There is no basis for a panel to read a legal definition apparently excised from United States trademark law as the definition for 'identical or confusingly

[400] WIPO Case No D2000-1525 (29 Jan 2001).

[401] See *Bradford & Bingley Plc v Registrant info@fashionID.com 987654321*, WIPO Case No D2002-0499 (16 Aug 2002); *Bayerische Motoren Werke AG v (This Domain is For Sale) Joshuathan Investments, Inc*, WIPO Case No D2002-0787 (8 Oct 2002); *The Evening Store.com, Inc v Henry Chan*, WIPO Case No D2004-0305 (26 June 2004); *Bellsouth Intellectual Property Corporation v Alvaro Collazo*, WIPO Case No D2004-0572 (22 Sept 2004); *Las Vegas Sands, LLC v Full Spiral–NSW*, WIPO Case No D2006-0981 (19 Oct 2006) (Bradley J Freedman, panellist).

[402] WIPO Case No D2003-0172 (28 May 2003).

[403] NAF Case No FA703608 (3 July 2006).

similar' under paragraph 4(a)(i) of the Policy. In fact there is no reason to look beyond the domain name at all, when determining whether or not a domain name is 'identical or confusingly similar' to a respondent's domain name.

As the panellist pointed out in *A & F Trademark, Inc v Party Night, Inc,*[404] literal comparison is quite straightforward, involving an analysis of both visual and aural similarity. Moreover, as pointed out by other panels, adopting a 'literal comparison' approach means that considerations relating to the scope of the complainant's trade mark rights are relevant to the other elements of the UDRP, namely whether the respondent has legitimate interests and whether there is bad faith registration.[405]

If, however, source confusion is relevant to the comparison under paragraph 4(a)(i), a different approach to the comparison would be required. The following test for comparing the disputed domain name with the complainant's mark to avoid source confusion was formulated by the panellist in *America Online, Inc v Johuathan Investments, Inc:*[406]

> In the trade mark context the term 'confusingly similar' refers to confusion as to trade origin. Is it likely therefore that, because of the similarity between the Domain Name on the one hand and the Complainant's trade mark on the other hand, people will believe that the Domain name is associated in some way with the Complainant?

This test was essentially adopted by the panellist in *Wal-Mart Stores, Inc v Traffic Yoon*[407](<boycottwalmart.com>), who added:

> Not only must a Complainant establish similarity between the disputed domain name and its trade marks, but it must also establish that people seeing the disputed domain name will be confused by that similarity as to the trade origin of the products, services or proprietorship of the site to which the disputed domain resolves.

The difference between the two approaches is illustrated by the way in which the panellist in *Traffic Yoon* rejected the application of the *Sleekcraft* factors, while indicating that some of the factors remain relevant to the comparison under the UDRP:

> this Panel does not agree that the *Sleekcraft* test of 'likelihood of confusion' of goods bearing one trade mark with another trade mark is or should be the identical test for 'confusing similarity' under the Policy, although some of the factors enumerated in *Sleekcraft* are useful for determining confusing similarity between a trademark and a domain name.

Other panellists have also found some of the factors developed under trade mark laws to be relevant, although not determinative.[408] At the same time, however, the panellist in the <boycottwalmart.com> dispute was concerned with the evidential burden required for a source confusion analysis, concluding that:

> unless the Policy is to be asked to perform a task for which it and the administrative procedures that it involves are poorly equipped, in terms of receiving and dealing with evidence, the question needs to be more confined to one of confusing similarity when the disputed domain name and the complainant's trademarks are compared side by side through the eyes of the likely viewers of the two.

404 WIPO Case No D2003-0172 (28 May 2003).

405 See, eg, *HRB Royalty, Inc v Phayze Inc,* WIPO Case No D2003-1043 (16 Mar 2004); *Pfizer Inc v Blank Canvas Projects Limited,* WIPO Case No D2005-0125 (29 Mar 2005); *International Organization for Standardization ISO v Capaccio Environmental Engineering, Inc,* WIPO Case No D2006-1190 (28 Nov 2006).

406 WIPO Case No D2001-0918 (14 Sept 2001).

407 WIPO Case No D2006-0812 (20 Sept 2006). See also: *Wal-Mart Stores, Inc v xc2,* WIPO Case No D2006-0811 (29 Aug 2006). For an argument in favour of the 'source confusion' approach see: *Asda Group Limited v Mr Paul Kilgour,* WIPO Case No D2002-0857 (11 Nov 2002).

408 See, eg, *Dixons Group Plc v Mr Abu Abdullaah,* WIPO Case No D2000-1406 (18 Jan 2001).

The position under the UDRP should, in theory, be the same as that which applies under the US ACPA. The ACPA specifically provides that liability is to be determined 'without regard to the goods or services of the parties'. This means that the comparison of the domain name and the claimant's mark under the ACPA is, like the UDRP, different from the comparison under trade mark laws. US courts have, moreover, specifically held that the 'likelihood of confusion' test applied under US trade mark law does not apply in comparing the domain name and the claimant's mark under the ACPA. In this respect, one court has stated that:

> the 'likelihood of confusion' test of trademark infringement is 'more comprehensive' than the 'identical or confusingly similar' requirement of ACPA, as it requires considering factors beyond the facial similarity of the two marks.[409]

The preferred view of the comparison required by the UDRP should obviously depend upon the reasons for undertaking the comparison. The comparison is not to determine whether there is source confusion between the domain name and the complainant's mark, but to determine whether there is sufficient similarity for the actions of the respondent to amount to abusive, bad faith registration. In determining whether there is sufficient similarity, the potential for confusion between the domain name and the complainant's mark is relevant. But the confusion is not necessarily that the domain name is acting as a mark; some lesser form of association than source confusion may be sufficient. For example, a degree of similarity that would attract users to a web site in the belief that it is somehow associated with the complainant may be sufficient for a respondent to make an abusive use of the domain name. In this sense, analogies can be drawn with 'initial interest confusion' recognised under US law, and explained at [5.23].

These considerations all suggest that the comparison between the disputed domain name and the complainant's mark should mainly be a literal side-by-side comparison, taking into account both visual and aural similarity. This conclusion is reinforced by the fact that panels are poorly equipped to undertake the complex exercise of source comparison, that it is consistent with the approach of US courts under the ACPA and that the majority of panels appear to prefer the 'literal comparison' approach. Despite this analysis, fundamental questions remain concerning the precise nature of the association required to be established between a disputed domain name and the complainant's mark for there to be confusing similarity. These complex issues are best addressed in the context of the ongoing controversy concerning the status of 'sucks'-type domain names, which is dealt with at [5.35]ff, and especially at [5.41].

'Identical or Confusingly Similar'

[5.27] Web Site Content Irrelevant in Confusion

The content of a web site associated with a disputed domain name is irrelevant to the comparison between the domain name and the complainant's mark under the UDRP. In this respect, the WIPO Consensus View states that:

[409] *Northern Light Technology, Inc v Northern Lights Club*, 236 F 3d 57, n 14 (1st Cir, 2001); see also *Hartog & Co AS v SWIX.com*, 136 F Supp 2d 531 (ED Va, 2001).

> The content of a website (whether it is similar or different to the business of a trademark owner) is irrelevant in the finding of confusing similarity. This is because trademark holders often suffer from 'initial interest confusion', where a potential visitor does not immediately reach their site after typing in a confusingly similar domain name, and is then exposed to offensive or commercial content. The test for confusing similarity should be a comparison between the trademark and the domain name to determine the likelihood of confusion.[410]

The view that the content of a web site is not to be taken into account in the comparison under paragraph 4(a)(i) follows from the more general proposition that any use of the disputed domain name is irrelevant to the comparison (see [5.26]). The confusion that is important under the UDRP is confusion between the domain name and the complainant's mark that may, for example, attract users to the respondent's web site. The fact that the confusion may be dispelled by the content of the web site, such as by a disclaimer, is therefore of no consequence. The following are examples of decisions that have held that the content of a web site is irrelevant to confusing similarity under the policy.

- *Arthur Guinness Son & Co (Dublin) Limited v Dejan Macesic.*[411] The disputed domain name <guiness.com> was associated with a web site that discussed home brewing and sports, and included disclaimers. The respondent claimed that there was no confusing similarity, as the site did not sell similar goods to those sold by the complainant in the same geographical area. In concluding that the domain name was confusingly similar, the panellist stated that:

> there is an important distinction between the domain name on the one hand and the Website on the other. The use to which the site is put has no bearing upon the issue whether the domain name is confusingly similar to the trademark, because by the time Internet users arrive at the Website, they have already been confused by the similarity between the domain name and the Complainant's mark into thinking they are on their way to the Complainant's Website.

- *Dixons Group Plc v Mr Abu Abdullaah.*[412] The disputed domain names, <dixons-online.net> and <dixons-online.org>, resolved to web sites ostensibly for the purpose of promoting discussion among the complainant's customers. The web sites included the words 'DIXONS CUSTOMERS' and disclaimers. In concluding that the domain names were confusingly similar, the panellist disregarded the content of the web sites.
- *AT&T Corp v Amjad Kausar.*[413] The disputed domain name, <attinternet.com> and <attuniversal.com>, diverted users to a dialogue box containing the text 'CONGRATULATIONS!!! YOU'VE WON VIRTUAL REALITY CASINO! CLICK on OK and Get up to +100$ Bonus on your first deposit!'. Users who clicked on the dialogue box were diverted to a gambling Web site. In undefended proceedings, the panellist had no difficulty in finding confusing similarity, despite the contents of the web site.
- *Asset Loan Co Pty Ltd v Gregory Rogers.*[414] The respondent claimed that the disputed web site, <assetloanco.net>, was not confusingly similar, in part, because the web site associated with the domain name did not sell the same goods and services as the complainant, and that it did not use the complainant's logo. After citing the WIPO Consensus View, the panel rejected this assertion and found confusing similarity.

[410] WIPO, above n 59, para 1.2.
[411] WIPO Case No D2000-1698 (25 January 2001).
[412] WIPO Case No D2001-0843 (23 August 2001).
[413] WIPO Case No D2003-0327 (29 July 2003).
[414] WIPO Case No D2006-0300 (2 May 2006).

- *SENSIS Pty Ltd v Kevin Goodall.*[415] The respondent argued that there was no confusing similarity as the web site associated with the disputed domain name, <tradingpost11.com>, did not look like the complainant's web site and included a disclaimer. After citing the WIPO Consensus View, the panellist held that the domain name was confusingly similar.

It should be noted that, as explained at [6.9.11], the content of a web site may be referred to in determining whether a respondent has rights or legitimate interests in the disputed domain name under paragraph 4(a)(ii).

[5.28] Elements to be Ignored in Comparison

It is absolutely clear that gTLD suffixes, such as '.com' and .net', are irrelevant to the comparison between the disputed domain name and the complainant's mark required by paragraph 4(a)(i). Although, strictly speaking, the 'www' prefix that often accompanies a domain name is not part of the domain name this, too, is to be ignored for the purpose of the comparison.

Two main reasons have been given to justify ignoring the gTLD: that a gTLD is necessary, from a purely functional perspective, for any domain name;[416] and that a gTLD is generic or descriptive, and therefore cannot serve to distinguish the domain name from the complainant's mark.[417]

From a practical perspective, the precise reason for ignoring the gTLD is of no significance. As every domain name requires a gTLD, it would make no sense for the gTLD to be taken into account in any comparison between a domain name and a trade mark. As the panellist pointed out in *SENSIS Pty Ltd, Research Resources Pty Ltd v Kevin Goodall:*[418]

> The reason is that the Policy requires a true comparison to be made between the domain name and the trademark. No proper comparison could ever be made if the suffix of a domain name could itself be a point of distinction between the two expressions, as no domain name can be effective without a suffix which enables the browser to find the destination sought. Thus, to make a true comparison between a trademark and a domain name, the suffix of the domain name is ignored, enabling the real substance of the two expressions to be compared . . .

[5.29] Graphical or Design Elements

A trade mark often includes graphical or design elements in addition to text. Such marks are known as device marks. In *Minibar North America Inc v Ian Musk,*[419] the panellist identified two approaches that have been taken by panels in comparing a domain name with a device mark.

[415] WIPO Case No D2006-0793 (22 Aug 2006).

[416] See, eg, *Pomellato SpA v Richard Tonetti,* WIPO Case No D2000-0493 (7 July 2000).

[417] See, eg, *Ticketmaster Corporation v DiscoverNet Inc,* WIPO Case No D2001-0252 (9 Apr 2001); *Williams-Sonoma v John Zuccarini,* WIPO Case No D2002-0582 (9 Aug 2002); *The Vanguard Group, Inc v Lorna Kang,* WIPO Case No D2002-1064 (20 Jan 2003).

[418] WIPO Case No D2006-0793 (22 Aug 2006).

[419] WIPO Case No D2005-0035 (2 Mar 2005). See also *Evolution USA, Inc v Alexei Doicev,* WIPO Case No D2006-0086 (24 Mar 2006).

Under the first approach, the graphical elements of the mark should be ignored, as graphical elements clearly cannot be reproduced in a domain name. This approach was explained by the panellist in *Pomellato SpA v Richard Tonetti:*[420]

> Due to the specific nature of Internet domain names, however, whose input takes place by means of a keyboard, the aural and conceptual aspects of a trademark far outweigh the visual aspect.

Under the second approach, a mark that includes graphical elements cannot be identical to a domain name, but may be confusingly similar, provided that the word elements of a device mark are sufficiently prominent.[421]

The difference between the two approaches was explained by the panellist in *Schneider Electronics GmbH v Schneider UK Ltd*,[422] in the following terms:

> Since a domain name can not include a design element, the word component of a design mark is the only element that can be searched by an Internet user. Some panels have therefore concluded that the word element of a trademark is the only element that is assessed when determining confusing similarity between a trademark and domain name (see *SWATCH AG v Stefano Manfroi*, WIPO Case No D2003-0802).
>
> The Panel notes that there have also been a number of prior UDRP cases involving trademarks containing geographic terms and design elements in which the design element has been found to be an essential part of the distinctiveness of the mark in question (see for example *Spreewaldverein eV v RCS Richter Computer Systemhaus GmbH*, WIPO Case No. D2003-0614; *Brisbane City Council v Warren Bolton Consulting Pty Ltd*, WIPO Case No D2001-0047).

Although there is no consensus view on this issue, it would seem clear that a domain name cannot be identical to a device mark. In the absence of a clear view on the issue, much necessarily depends upon the details of the device mark in each case. For most purposes, however, graphic or design elements of a device mark can be safely ignored.

[5.30] Identicality

Clearly enough, a literal word-for-word reproduction of the complainant's mark in the domain name will mean that the domain name and the mark are identical, thereby satisfying the requirements for paragraph 4(a)(i). As the first element of the policy will be satisfied if the domain name is 'identical *or* confusingly similar', there is rarely a need for panels to draw a hard and fast distinction between identicality and confusing similarity.

This differs from the position under some national trade mark laws, where a distinction is drawn between the 'identicality' and 'confusing similarity' standards.[423] One reason for this distinction is that 'identicality' involves a literal comparison, whereas 'confusing similarity' raises issues of source confusion. By analogy, under the UDRP, some panellists have held that, where the disputed domain name is identical to the complainant's mark, this obviates the need to establish confusion. In *Shirmax Retail Ltd v CES Marketing, Inc*,[424] for example, the panellist stated:

[420] WIPO Case No D2000-0493 (7 July 2000). See also *EFG Bank European Financial Group SA v Jacob Foundation*, WIPO Case No D2000-0036 (22 Mar 2000).

[421] See, eg, *Mentor ADI Recruitment Ltd (trading as Mentor Group) v Teaching Driving Ltd*, WIPO Case No D2003-0654 (29 Sept 2003).

[422] WIPO Case No D2006-1039 (21 Oct 2006).

[423] See, eg, *Shell Co (Aust) Ltd v Esso Standard Oil (Aust) Ltd* (1961) 109 CLR 407, 414–5 per Windeyer J.

[424] eRes Case No AF-0104 (20 Mar 2000).

> Mere identicality of a domain name with a registered trademark is sufficient to meet the first element of the test, even if there is no likelihood of confusion whatsoever.

The distinction was significant in *Chernow Communications, Inc v Jonathan D Kimball*,[425] as the complainant, who owned the 'C-COM' mark, alleged that the <ccom.com> domain name was identical but, apparently because of an oversight, failed to allege that the domain name was confusingly similar. The majority in that decision held that a dehyphenated domain name was identical, rather than confusingly similar, to a hyphenated domain name, essentially on the basis that to hold otherwise would create the possibility of future potential cybersquatters escaping the policy by merely adding or removing punctuation.

As the comparison under paragraph 4(a)(i) is not primarily directed at the extent of source confusion, however, and as most complainants will allege both identicality and confusing similarity, the distinction between the two standards is not especially relevant in practice. Moreover, as pointed out at [5.31], most panels hold that 'essential or virtual identity' is sufficient. If, despite this, a distinction is considered desirable, the following statement from *Kerly's Law of Trade Marks and Trade Names* as to what amounts to identicality may be helpful:

> A sign will be identical with the registered mark where it reproduces, without any modification or addition, all the elements constituting the mark or where, viewed as a whole, it contains differences so insignificant they may go unnoticed by the average consumer.[426]

[5.31] 'Essential or Virtual Identity is Sufficient'

UDRP decisions commonly incorporate statements similar to the following:

> Essential or virtual identity is sufficient for the purposes of the Policy: see *The Stanley Works and Stanley Logistics, Inc v Camp Creek Co, Inc* (WIPO Case No. D2000-0113), *Toyota Jidosha Kabushiki Kaisha d/b/a Toyota Motor Corporation v S&S Enterprises Ltd* (WIPO Case No D2000-0802) *and Nokia Corporation v Nokiagirls.com aka IBCC* (WIPO Case No. D2000-0102).

This statement means that merely adding minor or inconsequential elements to the complainant's mark does not remove the similarity between the domain name and the mark such that there is no confusing similarity. Moreover, merely removing minor or inconsequential elements does not dispel identicality or confusing similarity.

To begin with, the mere addition of generic or descriptive terms to the complainant's mark will not distinguish the domain name from the mark sufficiently to remove confusing similarity. The addition of generic or descriptive terms is dealt with further at [5.32–[5.33] below. Secondly, the deliberate misspelling of the complainant's mark in a domain name, or 'typosquatting', will not dispel confusing similarity. The practice of 'typosquatting' is dealt with further at [5.34] below. This paragraph deals with marks that incorporate other relatively minor alterations.

One of the earliest and most commonly cited decisions to consider this issue was *The Stanley Works v Camp Creek Co, Inc*[427] In that dispute, the complainant had trade mark rights in the names 'STANLEY' and 'BOSTICH' and the respondent had registered 10 similar domain names, including <stanleybostich.com>, <stanley-bostich.com>,

[425] WIPO Case No D2000-0119 (18 May 2000).
[426] Kitchin *et al*, above n 20, para [14-051].
[427] WIPO Case No D2000-0113 (13 Apr 2000).

<bostichnails.com> and <bostichtools.com>. The panellist had no difficulty in concluding that the disputed domain names were either identical or confusingly similar to the complainant's marks, on the basis that the:

> Respondent has appropriated the trademarks Stanley and Bostitch, sometimes rearranging them, sometimes adding a generic word such as nail, and has registered the names in various top level domains such as org., com. and net.

The following are examples of categories of minor changes to marks that do not dispel confusing similarity:

- The addition or removal of single letters or numbers, and especially the addition of the plural 's', will usually not remove confusing similarity.

  —*Toyota Jidosha Kabushiki Kaisha v S&S Enterprises Ltd*[428]—the domain names <itoyota.com> and <itoyotas.com>, which added an 'i' and an 's' to the complainant's mark, were held to be confusingly similar.
  —*EthnicGrocer.com, Inc. v LatinGrocer.com*[429]—the domain names <ethnicgroceries.com> and <quericogrocers.com> were held to be confusingly similar to the complainant's marks, 'ETHNICGROCER.COM' and 'QUERICO.COM'.
  —*i2 Technologies Inc v Richard Alexander Smith*[430]—the domain name <i2technology. com> was held to be confusingly similar to the complainant's mark, 'i2 TECHNOLOGIES'.
  —*Verio Inc v Sunshinehh*[431]—the domain name <1verio.com> was held to be confusingly similar to the complainant's mark 'VERIO'.
  —*Bank for International Settlements v BIS*[432]—the domain name <bankforinternationalsettlement.com> was held to be confusingly similar to the complainant's mark 'BANK FOR INTERNATIONAL SETTLEMENTS'.
  —*Intuit Inc v NOLDC, Inc.*[433]—the domain name <innovativemerchantsolution.com> was held to be confusingly similar to the complainant's mark 'INNOVATIVE MERCHANT SOLUTIONS'.

- The addition or removal of a punctuation mark, such as the addition or removal of a hyphen, or the removal of an apostrophe, will not remove confusing similarity.

  —*Stella D'oro Biscuit Co, Inc v The Patron Group, Inc*[434]—the domain name <stelladoro.com> was held to be confusingly similar to the complainant's mark 'STELLA D'ORO'.
  —*TPI Holdings, Inc v Jason Guida*[435]—the domain names <boatandrvtrader.com> and <boatrvtrader.com> were held to be confusingly similar to the complainant's mark 'BOAT & RV TRADER'.
  —*Chi-Chi's, Inc v Restauran Commentary (Restaurant Commentary)*[436]—the domain name <chichis.com> was held to be confusingly similar to the complainant's mark 'CHI-CHI'S'.

[428] WIPO Case No D2000-0802 (9 Sept 2000).
[429] NAF Case No FA94384 (7 July 2000).
[430] WIPO Case No D2001-0164 (4 June 2001).
[431] WIPO Case No D2003-0255 (24 June 2003).
[432] WIPO Case No D2003-0986 (2 Mar 2004).
[433] WIPO Case No D2006-1217 (6 Dec 2006).
[434] WIPO Case No D2000-0012 (17 Feb 2000).
[435] WIPO Case No D2000-0269 (23 May 2000).
[436] WIPO Case No D2000-0321 (29 June 2000).

—*Nintendo of America, Inc v Gray West International*[437]—the domain name <pokemon-games.com> was held to be confusingly similar to the complainant's mark 'POKEMON'.
—*Formula One Licensing BV v HyeongJeon, Gim*[438]—the domain name <formula-1.net> was held to be confusingly similar to the complainant's mark 'FORMULA 1'.
—*Ruth's Chris Steak House, Inc v Texas Internet*[439]—the domain name <ruthchris.com> was held to be confusingly similar to the complainant's mark 'RUTH'S CHRIS'.
—*Edmunds.com, Inc v The Registrant of <edmounds.com>*[440]—the domain name <edmounds.com> was held to be confusingly similar to the complainant's mark 'EDMUND'S'.
—*Las Vegas Sands, LLC v Full Spiral—NSW*[441]—the domain name <sands-casino.com> was held to be confusingly similar to the complainant's mark 'SANDS'.

- The addition of the letters 'www' in the domain name will not remove confusing similarity. This will usually be regarded as a form of 'typosquatting'.

—*Bank of America Corp v InterMos*[442] —<wwwbankofamercia> was confusingly similar to 'BANK OF AMERICA'.
—*Cable News Network LP v Stonybrook Investments LTD*[443]—<wwwcnn.com> was confusingly similar to 'CNN'.
—*Viacom International Inc v Erik Peterson*[444]—<www-mtv.com> was confusingly similar to 'MTV'.
—*NIKE, Inc v Alex Nike*[445]—<wwwnike.com> was confusingly similar to 'NIKE'.
—*Deutsche Telekom AG v Luis Javier Collazos*[446] —<wwwtonline.com> was confusingly similar to 'T-ONLINE'.

[5.32] Addition of Terms does not Dispel Confusion

UDRP panels commonly refer to the following statement made by the panellist in *Britannia Building Society v Britannia Fraud Prevention*:[447]

> the incorporation of a trademark in its entirety is sufficient to establish that a domain name is identical or confusingly similar to the complainant's registered mark. *See, eg, Quixtar Investments, Inc v Dennis Hoffman*, Case No D2000-0253 (WIPO May 29 2000) (finding that QUIXTAR and <quixtarmortgage.com> are legally identical); *Nikon, Inc and Nikon Corporation v Technilab, Inc.*, Case No. D2000-1774 (WIPO Feb. 26, 2001) (incorporation of Nikon mark into domain name <nikondealers.com> satisfies first prong of Policy). The addition of other terms in the domain name does not affect a finding that the domain name is identical or confusingly similar to the Complainant's registered trademark *Wal-Mart Stores, Inc v Richard MacLeod d/b/a For Sale*, Case

[437] WIPO Case No D2000-1219 (10 Jan 2001).
[438] WIPO Case No D2004-0210 (14 June 2004).
[439] WIPO Case No D2005-0292 (17 May 2005).
[440] WIPO Case No D2006-0741 (25 Sept 2006).
[441] WIPO Case No D2006-0981 (19 Oct 2006).
[442] NAF Case No FA95092 (1 Aug 2000).
[443] NAF Case No FA96282 (22 Jan 2001).
[444] WIPO Case No D2000-0346 (3 June 2001).
[445] WIPO Case No D2001-1115 (7 Nov 2001).
[446] WIPO Case No D2003-0371 (9 July 2003).
[447] WIPO Case No D2001-0505 (6 July 2001).

No. D2000-0662 (WIPO Sept 19 2000) (for purposes of the first prong of the Policy, <walmartsucks.com> is confusingly similar to Wal-Mart trademark).

The statement really stands for two, more precise propositions. First, a domain name will be identical or confusingly similar to the complainant's mark if the mark is included in its entirety in the domain name without any changes that may serve to distinguish the domain name. Secondly, a domain name that merely adds terms to the complainant's mark that do not distinguish the domain name from the mark—namely, generic or descriptive terms—will be confusingly similar to the complainant's mark. As the panellist, citing a number of relevant decisions, put it in *Las Vegas Sands, LLC v Full Spiral–NSW*:[448]

> The combination of a trademark with a non-distinctive, descriptive word in a domain name does not necessarily change the overall impression of the trademark. In fact, there may even be a greater risk of confusion where the additional words are descriptive of the wares or services with which the trademark is ordinarily used, because there is an increased chance that Internet users will believe that the domain name resolves to a website that is owned or sponsored by, or affiliated with, the trademark owner. See *Bayerische Motoren Werke AG v(This Domain is For Sale) Joshuathan Investments, Inc*, WIPO Case No. D2002-0787; *Harrods Limited v Pierre*, WIPO Case No D2001-0456; *Dr Ing hc FPorsche AG v VasiliyTerkin*, WIPO Case No. D2003-0888; *Diageo North america, Inc and United Distillers Manufacturing Inc, Diageo North America Inc,Diageo Brands BV v iVodka.com aka Alec Bargman*, WIPO Case No. D2004-0627; *Pivotal Corporation v Discovery Street Trading Co Ltd*, WIPO Case No. D2000-0648; *Advanced Micro Devices Inc v Dmitry*, WIPO Case No D2000-0530; *Caterpillar Inc v Roam the Planet Ltd*, WIPO Case No D2000-0275; *Telstra Corporation Limited v Ozurls*, WIPO Case No. D2001-0046; and *Caixa D´Estalvis I Pensions de Barcelona ("La Caixa") v Awenebi David*, WIPO Case No. D2006-0632.

The general principles are illustrated by the following representative examples:

- *Nokia Corporation v Nokiagirls.com.*[449] The complainant, who had registered rights in the 'NOKIA' mark, brought a complaint in relation to the domain name, <nokiagirls.com>, which resolved to sites that were unrelated to the complainant, including sites with images of girls. The panellist held that the addition of the word 'GIRLS' to the mark did not remove confusing similarity as it was a 'neutral addition' that did not distinguish the domain name.
- *Britannia Building Society v Britannia Fraud Prevention.*[450] The complainant building society had registered rights in the mark 'BRITTANIA' and brought a complaint in relation to the <britanniabuildingsociety.org> domain name. The panellist essentially held that, as the words 'BUILDING SOCIETY' were generic, the domain name was confusingly similar to the mark.
- *Ansell Healthcare Products Inc v Australian Therapeutics Supplies Pty, Ltd.*[451] The complainant, who had registered rights in the mark 'ANSELL' in connection with the sale of condoms, brought a complaint against the registrant of the <ansellcondoms.com> domain name. The panellist held that the addition of the generic term, 'CONDOMS', did not avoid confusion. Moreover, as the term described the complainant's business, use of the term was likely to increase confusion.

[448] WIPO Case No D2006-0981 (19 Oct 2006).
[449] WIPO Case No D2000-0102 (18 Apr 2000).
[450] WIPO Case No D2001-0505 (6 July 2001).
[451] WIPO Case No D2001-0110 (2 Apr 2001).

- *Nokia Corporation v Phonestop.*[452] The complainant had registered rights in the mark 'NOKIA' and brought a complaint against the registrant of the domain name, <nokiaonthe.net>. The panellist held that the additional terms did not affect a finding of confusing similarity.
- *Revlon Consumer Products Corporation v Brandy Farris.*[453] The complainant, who had registered rights in the 'REVLON' mark, brought a complaint against the registrant of the <iloverevlon.com> and <loverevlon.com> domain names. The panellist held that the addition of the generic terms 'I' and 'love' did not distinguish the domain names so as to avoid confusion.
- *Spencer Douglass, MGA v Bail Yes Bonding.*[454] The complainant had registered rights in the mark 'ALADDIN' and brought a complaint against the registrant of the <aladdinbailbonding.com> domain name. The panellist held that the addition of the generic terms 'BAIL BONDING' did distinguish the domain name from the complainant's mark. As the panellist pointed out, given that both complainant and respondent were in the business of bail bonding, the addition of the terms may have increased the likelihood of confusion.
- *Gerber Products Company v LaPorte Holdings.*[455] The complainant, who had registered rights in the mark 'GERBER', brought a complaint in relation to the <gerberfood.com>, <gerberinsurance.com> and <gerberproducts.com> domain names. The panellist held that the addition of the generic terms 'FOOD', 'INSURANCE' and 'PRODUCTS' did not eliminate the confusing similarity between the domain names and the complainant's marks on the basis that 'the addition of generic or descriptive words is not sufficient to distinguish a mark from its subsequent adopter'.
- *David Z, Inc v Admin c/o LaPorte Holdings.*[456] The complainant, who had registered rights in the mark 'DAVID Z', brought a complaint in relation to the <davidzinc.com> domain name. The panellist held that the domain name was confusingly similar as it incorporating the complainant's mark in its entirety, adding only the generic term 'INC'.
- *Red Bull GmbH v PREGIO Co, Ltd.*[457] The complainant had registered rights in the mark 'RED BULL' and brought a complaint in relation to the <redbull-jp.net> domain name. The panellist held that the addition of the geographical term 'jp' was insufficient to distinguish the domain name from the complainant's mark. In this respect, the panellist added that the geographical term contributed to the likelihood of confusion, as it could falsely suggest that the respondent sold 'RED BULL' in Japan.
- *International Organization for Standardization ISO v Capaccio Environmental Engineering, Inc.*[458] The complainant, an international standards body, had registered rights in marks that included 'ISO'. The disputed domain name, <iso14000.com>, resolved to a Web site headed 'ISO 14000 Information Center'. One of the most widely known of the ISO standards, the 'ISO 14000' family of standards, is concerned with environmental management. The panellist held that 'ISO' was the distinctive component of the disputed domain name and that the addition of the number '14000' did not remove confusing similarity. In fact,

[452] WIPO Case No D2001-1237 (6 Dec 2001).
[453] WIPO Case No D2003-0291 (4 June 2003).
[454] WIPO Case No D2004-0261 (1 June 2004).
[455] WIPO Case No D2005-1277 (8 Feb 2006).
[456] NAF Case No FA653829 (18 Apr 2006).
[457] WIPO Case No D2006-0909 (19 Sept 2006).
[458] WIPO Case No D2006-1190 (28 Nov 2006).

by referring to one of the complainant's 'products', the numbers reinforced the confusing similarity between the domain name and the complainant's mark.

[5.33] Addition of Generic 'Internet' Terms

A number of panel decisions have held that the addition of Internet-related generic terms, such as 'software', 'online' and 'download', will not adequately distinguish the domain name from the complainant's mark so as to dispel confusing similarity. The following are examples of these kinds of decision.

- The addition of the word 'SOFTWARE' is not sufficient to remove confusing similarity: *Pivotal Corporation v Discovery Street Trading Co, Ltd*[459] (<pivotalsoftware.com>); *Altavista Corporation v Brunosousa*[460] (<altavistasoftware.com>); *Ahead Software AG v Leduc Jean*[461] (<nero-neu.com> and <nero-software.com>); *Ikarus Software GmbH v Longo*[462] (<ikarus-software.com>); *Sharman License Holdings Ltd v Dustin Dorrance*[463] (<kazaa-lite-downloads.com>, <kazaa-media.net>, <kazaa-music.net> and <kazaa-software-download.com>); *Sharman License Holdings Ltd v Domains at Fabulous*[464] (<download-kazaa.com>, <freekazaa.com>, <kazaaalite.com>, <azaadesktop.com>, <kazaa-download.com>, <kazaa-downloads.com>, <kazaafree.com>, <kazaalie.com>, <kazaa-media.com>, <kazaa-music.com>, <kazaarlite.com>, <kazaars.com>, <kazaaupdate.com>, <kazaa-virus.com>, <kazacom.com>, <kazadownload.com>, <kazammusic.com>, <kazars.com>, <kazasa.com>, <kazzadownload.com>, <kazzamusicdownloads.com>, <kazzasoftware.com>, etc.); *Symark International, Inc v Gary McCurty*[465] (<symarksoftware.com>).
- The addition of the word 'ONLINE' (or similar) is not sufficient to remove confusing similarity: *Dixons Group Plc v Mr Abu Abdullaah*[466](<dixons-online.net> and <dixons-online.org>); *America Online, Inc. v. Anthony Peppler*[467] (<aol7.com>, <aolnick.com>, <aolpogo.com>, <aolmediaplayer.com>, <aolprofile.com>, <mailaol.com>, <mapquesst.com>, <mapcrest.com>, <icqprograms.com>, <icqdownloads.com>, <americans-online.com> and <digitacity.com>); *Encyclopaedia Britannica, Inc v null John Zuccarini, Country Walk*[468] (<britannicaonline.com>); *Dr Grandel GmbH v Drg Randel Inc*[469] (<drgrandel-online.com>); *Confederation Nationale du Credit Mutuel v Josh Self*[470] (<creditmutuelonline.com>); *F Hoffmann-La Roche AG v George McKennitt*[471] (<buy-online-valium.com>, <buyonlinevalium.com> and <online-valium.com>); *F Hoffmann-La Roche AG v Pinetree Development, Ltd*[472] (<buy-xenical-cheap-

[459] WIPO Case No D2000-0648 (14 Aug 2000).
[460] WIPO Case No D2002-0109 (3 Apr 2003).
[461] WIPO Case No D2004-0323 (29 June 2004).
[462] WIPO Case No D2004-0646 (5 Oct 2004).
[463] WIPO Case No D2004-0659 (29 Oct 2004).
[464] WIPO Case No D2004-0915 (24 Dec 2004).
[465] WIPO Case No D2005-0235 (15 June 2005).
[466] WIPO Case No D2001-0843 (23 Aug 2001).
[467] NAF Case No FA103437 (22 Feb 2002).
[468] WIPO Case No D2002-0487 (12 Aug 2002).
[469] WIPO Case No D2002-0829 (30 Sept 2005).
[470] WIPO Case No D2005-1057 (25 Nov 2005).
[471] WIPO Case No D2005-1300 (6 Feb 2006).
[472] WIPO Case No D2006-0049 (3 Mar 2006).

online.com>); *Banca Intesa SpA v Roshan Wickramaratna*[473] (<bancaintesa-online.com>); *Hoffmann-La Roche Inc v Dmitri Zoubkov*[474] (<tamiflu-buy-online.com>).

- The addition of Internet-related generic terms, such as 'my' or 'e-', will not dispel confusing similarity: *Infospace.com, Inc v Delighters, Inc*[475] (<myinfospace.com>); *Busy Body, Inc v Fitness Outlet Inc*[476] (<efitnesswarehouse.com>); *Sony Kabushiki Kaisha v. Sin, Eonmok*[477] (<mysony.com>); *Hugo Boss AG v Dr Yang Consulting*[478] (<myhugoboss.com>); *Société des Bains de Mer et du Cercle des Etrangers à Monaco v. International Lotteries et al*[479] (<e-casinomontecarlo.com>); *Prada SA v Nevis Industries (HK) Limited*[480] (<myprada.com>); *Mattel, Inc v Equipment Outlet Express*[481] (<myhotwheels.com>); *American Express Company v Lee HyunGyu*[482] (<myamericanexpress.com>); *Microsoft Corporation v Woo Seungchul*[483] (<myxbox.com>).

Typosquatting

[5.34] 'Typosquatting'

UDRP panels have defined 'typosquatting' as:

> taking advantage of common misspellings made by Internet users who are looking for a particular site of a particular provider of goods or services, in order to obtain some benefit therefrom.[484]

In *Shields v Zuccarini*,[485] an action brought against the notorious 'typosquatter', John Zuccarini, under the US ACPA, the US court referred to 'typosquatting' as:

> the intentional registration of domain names that are misspellings of distinctive or famous names, causing an Internet user who makes a slight spelling or typing error to reach an unintended site . . .

In that case, the court held that a domain name that included an intentional misspelling of a mark would be 'confusingly similar' to the misspelled mark for the purpose of the ACPA. It is unsurprising that UDRP panels have also held that domain names that are the result of 'typosquatting' are confusingly similar to a complainant's misspelled mark. As the panellist pointed out in *Marriott International, Inc v Seocho:*[486]

> Respondent's efforts constitute 'typosquatting,' whereby a registrant deliberately introduces slight deviations into famous marks in order to commercially benefit. Respondent's typosquatting, by its definition, renders the domain name confusingly similar to Complainant's mark.

Similar comments were made by the panellist in *Dell Computer Corp v Clinical Evaluations:*[487]

473 WIPO Case No D2006-0215 (14 Apr 2006).
474 WIPO Case No D2006-0732 (13 Sept 2006).
475 WIPO Case No D2000-0068 (1 May 2000).
476 WIPO Case No D2000-0127 (22 Apr 2000).
477 WIPO Case No D2000-1007 (16 Nov 2000).
478 WIPO Case No D2000-1109 (18 Dec 2000).
479 WIPO Case No D2000-1327 (8 Jan 2000).
480 WIPO Case No D2002-0161 (9 May 2002).
481 NAF Case No FA114459 (12 July 2002).
482 NAF Case No FA126649 (25 Nov 2002).
483 NAF Case No FA601455 (20 Jan 2006).
484 *Expedia, Inc v Alvaro Collazo,* WIPO Case No D2003-0716 (30 Oct 2003).
485 254 F 3d 476, 483 (3d Cir, 2001).
486 NAF Case No FA149187 (28 Apr 2003).
487 WIPO Case No D2002-0423 (30 June 2002).

> Typo-squatting seeks to take advantage of errors by users in typing domain names into their Web browser's location bar. Domain names which constitute typo-squatting are confusingly similar by definition; it is this similarity which makes them attractive.

At the same time, however, there is no absolute rule that the addition or deletion of a single letter in a domain name necessarily means that there is confusing similarity, as there are many legitimate words that differ from other words by one letter more or less. As the panellist in *Dell Computer Corp v Clinical Evaluations* went on to state:

> It is not necessarily the case that alteration of a single letter will constitute typo-squatting; there may be legitimate trademark or domain name uses that happen coincidentally to differ from another's legitimate use by a single letter.

Moreover, 'typosquatting' may well involve the alteration of more than one letter in a complainant's mark. The extent to which the spelling of a domain name differs from the spelling of the complainant's mark must therefore be considered as part of the broader inquiry into whether the domain name and the complainant's mark are confusingly similar. As the panellist in *Sharman License Holdings, Limited v IcedIt.com*[488] cautioned:

> The Panel considers that it should approach each of the domain names and consider whether each is confusingly similar on their face. Whether or not, letters differ between the mark KAZAA and the domain name complained upon is merely part of the test which the Panel has to decide, ie is the domain name confusingly similar? The Panel does however agree with the Complainant that confusing similarity must turn 'on the similarity of sound, appearance and idea suggested by the mark and the domain name' among other factors.

In cases where there is room for doubt whether a domain name that incorporates minor changes to the complainant's mark may be confusingly similar, the degree of distinctiveness of the mark may be taken into account. As the panellist in *Reuters Limited v Global Net 2000, Inc*[489] maintained:

> A domain name which differs by only one letter from a trademark has a greater tendency to be confusingly similar to the trademark where the trademark is highly distinctive.

Special cases involving the addition of the single letter 's' as a plural, or the 'www' prefix, are dealt with at [5.31] and [5.28]. These practices will also usually fall within the definition of 'typosquatting'. The following is a selection of representative decisions that illustrate the principle that the deliberate misspelling of a complainant's mark in a domain name does not remove confusing similarity:

- *Neuberger Berman Inc v Alfred Jacobsen*[490]—<newbergerberman.com> was confusingly similar to 'NEUBERGER BERMAN'.
- *Reuters Limited v Global Net 2000, Inc*[491]—<wwwreuters.com>, <reters.com>, <ruters.com>, <reuers.com> and <reutersnews.com> were confusingly similar to 'REUTERS'.
- *Dow Jones & Company, Inc v John Zuccarini*[492]—domain names <wallstreetjounal.com> and <wallstreetjournel.com> were confusingly similar to 'THE WALL STREET JOURNAL'.

[488] WIPO Case No D2004-0713 (17 Dec 2004).
[489] WIPO Case No D2000-0441 (13 July 2000).
[490] WIPO Case No D2000-0323 (12 June 2000).
[491] WIPO Case No D2000-0441 (13 July 2000).
[492] WIPO Case No D2000-0578 (28 Aug 2000).

- *Telstra Corporation Ltd v Warren Bolton Consulting Pty Ltd*[493]—<bigpons.com> was confusingly similar to 'BIG POND'.
- *Cimcities, LLC v John Zuccarini*[494]—<accesatlanta.com> was confusingly similar to 'ACCESS ATLANTA'.
- *Google, Inc v Abercrombie 1*[495]—<googld.com> was confusingly similar to 'GOOGLE'.
- *Bank of America Corporation v BanofAmerica*[496]—<banofamerica.com> was confusingly similar to 'BANK OF AMERICA'.
- *Go Daddy Software, Inc v Daniel Hadani*[497]—<fodaddy.com>, <gidaddy.com>, and <gpdaddy.com> were confusingly similar to 'GO DADDY'.
- *Marriott International, Inc v Seocho*[498]—<marrriott.com> was confusingly similar to 'MARRIOTT'.
- *Yurtici Kargo Servisi AS v Yurticicargo Yurticikargo*[499]—<yurticicargo.com> was confusingly similar to 'YURTICI KARGO'.
- *Sharman License Holdings, Limited v IcedIt.com*[500]—<jazaa.com>, <kaazar.com>, <kadaa.com>, <karzaa.com>, <kavaa.com>, <kawaa.com>, <kaxaa.com>, <kazaal.com>, <kazuu.com>, <kezaa.com>, <kszaa.com>, <zakaa.com> and <zazaa.com> were confusingly similar to 'KAZAA'.
- *ESPN, Inc v XC2*[501]—<espnnews.com> was confusingly similar to 'ESPNEWS'.
- *Humana Inc v Cayman Trademark Trust*[502]—<humanna.com> was confusingly similar to 'HUMANA'.
- *Express Scripts, Inc v Kal*[503]—<epxress-scripts.com>, <espress-script.com>, <esxpress-scripts.com>, <expess-scripts.com>, <exprescriptions.com>, <express-cripts.com>, <express-scipts.com>, <express-scritps.com>, <express-scrpts.com>, <express-sript.com>, <express-sripts.com>, <expresssripts.com>, <exrpess-scripts.com> and <rxpress-scripts.com> were confusingly similar to 'EXPRESS SCRIPTS', with the exception of <exprescriptions.com>, on the basis of 'typosquatting'. <exprescriptions.com> was held to be phonetically and conceptually similar to the complainant's mark and, consequently, confusingly similar.
- *Amazon.com, Inc v Steven Newman a/k/a Jill Wasserstein a/k/a Pluto Newman*[504] — <amazoh.com>, <qmqzon.com> and <smszon.com> were confusingly similar to 'AMAZON' (the panel pointing out that 'all of the substituted letters are proximately located to the original letters on a standard "qwerty" keyboard, and the domain names are predictable mistypings of the Complainant's mark').
- *Edmunds.com Inc v Ed Munski*[505]—<edmuns.com> was confusingly similar to 'EDMUND'S'.

[493] WIPO Case No D2000-1293 (21 Nov 2000).
[494] WIPO Case No D2001-0491 (31 May 2001).
[495] NAF Case No FA101579 (10 Dec 2001).
[496] NAF Case No FA105885 (12 Apr 2002).
[497] WIPO Case No D2002-0568 (1 Aug 2002).
[498] NAF Case No FA149187 (28 Apr 2003).
[499] WIPO Case No D2003-0707 (24 Oct 2003).
[500] WIPO Case No D2004-0713 (17 Dec 2004).
[501] WIPO Case No D2005-0444 (28 June 2005).
[502] WIPO Case No D2006-0073 (7 Mar 2006).
[503] WIPO Case No D2006-0301 (12 May 2006).
[504] WIPO Case No D2006-0517 (28 June 2006).
[505] WIPO Case No D2006-1094 (3 Oct 2006).

Gripe Sites

[5.35] Gripe Sites and Confusion

Some domain names add a negative or derogatory term, such as 'sucks', to a trade mark. Other examples of 'sucks'-type domain names include those that take forms such as <ihate[trade mark].com>, <[trade mark]isbadforyou.com> and <fuck[trade mark].com>. A 'sucks'-type domain name may, or may not, resolve to a web site that includes material that is critical of the trade mark owner. Web sites that include critical material, especially material that is critical of a commercial enterprise, are known as 'gripe' sites. Apart from resolving to a 'gripe' site, a 'sucks'-type domain name may be registered with the intention of demanding payment from the trade mark owner, or of attracting Internet users for a commercial purpose, such as advertising or obtaining click-through revenue. 'Sucks'-type domain names that resolve to sites that are used for business purposes that are unrelated to any criticism of a trade mark owner are known as 'sham speech' domain names.

Whether a 'sucks'-style domain name that incorporates the complainant's trade mark is confusingly similar to the mark has been one of the most controversial issues in the application of the first element of the UDRP. At the time of writing, no consensus view has emerged, with UDRP panellists continuing to express divergent opinions on the status of 'sucks'-type domain names. In the absence of a consensus, the *WIPO Overview* sets out a majority view and a minority view. The majority view is as follows:

> A domain name consisting of a trademark and a negative term is confusingly similar to the complainant's mark. Confusing similarity has been found because the domain name contains a trademark and a dictionary word; or because the domain name may not be recognized as negative; or because the domain name may be viewed by non-fluent English language speakers, who may not recognize the negative connotations of the word that is attached to the trademark.[506]

The WIPO minority view, on the other hand, is that:

> A domain name consisting of a trademark and a negative term is not confusingly similar because Internet users are not likely to associate the trademark holder with a domain name consisting of the trademark and a negative term.[507]

[5.36] Treatment of Gripe Sites under US Law

The division between UDRP panellists can be best understood against the background of the treatment of gripe sites and 'sucks'-type domain names by US courts. The starting point for this analysis is the decision of District Judge Pregerson in *Bally Total Fitness Holding Corp v Faber.*[508] In that case, the owner of the 'BALLY' trade mark brought actions for trade mark infringement, unfair competition and trade mark dilution against the owner of the 'BALLY sucks' gripe site. While the term 'BALLY SUCKS' was used in a sub-domain, and was not used directly as the gripe site's domain name, the court made it clear that it would have applied the same analysis if the domain name took the form <ballysucks.com>.

[506] WIPO, above n 57, para 1.3.
[507] *Ibid.*
[508] 29 F Supp 2d 1161 (CD Cal, 1998).

The key legal issue in the trade mark infringement action was whether there was a 'likelihood of confusion' between the term 'BALLY SUCKS' and Bally's marks. Applying the '*Sleekcraft* factors', the court held that there was no likelihood of confusion between the name and the marks. In reaching this conclusion, the court stated in a footnote that 'no reasonably prudent Internet user would believe that "ballysucks.com" is the official Bally site or is sponsored by Bally'.[509] The court also dismissed the action for trade mark dilution on the basis that there was no commercial use of the mark and that, even if there was a commercial use, use of the mark for the purposes of legitimate criticism did not amount to tarnishment.

In a subsequent case, *Lucent Technologies, Inc v Lucentsucks.com*,[510] the owner of the 'LUCENT' mark brought an action against the registrant of the <lucentsucks.com> domain name under the ACPA. Although the action was dismissed for lack of jurisdiction, District Judge Brinkema addressed an argument that the action should have been dismissed on the ground that the defendant's free speech rights outweighed the claimant's trade mark rights. On this point, the court observed that, for the purposes of trade mark law, a parody or criticism site diminishes the risk of consumer confusion. In relation to the instant case, the court found the defendant's contention that the average consumer would not confuse the <lucentsucks.com> web site with a site sponsored by the claimant to be persuasive. The court concluded that, if it could be established that the <lucentsucks.com> site was for effective parody or critical commentary, this would 'seriously undermine' the claimant's action, as it would be difficult for the claimant to make out a likelihood of confusion and bad faith intent.

Reflecting the balance to be struck between freedom of expression and trade mark rights, other US decisions dealing with trade mark actions against gripe sites have, in general terms, drawn a distinction between non-commercial sites and sites with a commercial element. The use of a trade mark in the domain name for non-commercial gripe sites has been held not to amount to trade mark infringement on the ground that a 'sucks'-type domain name creates no likelihood of confusion, there is no commercial use of the mark or the mark is not used in connection with goods or services.[511] On the other hand, gripe sites with some commercial element, such as the attraction of users for the purpose of advertising or linking to the site of a competitor of the trade mark owner, have been held to infringe on the ground of 'initial interest confusion' and use of the mark in commerce.[512] As yet, however, the US courts have not had to consider the position where a 'sucks'-type domain name is used in connection with a commercial gripe site.

[5.37] Panel Views on 'Sucks'-type Domain Names

The source of the division between panellists concerning the status of 'sucks'-type domain names can be traced to fundamentally different views regarding the purposes of the comparison between the disputed domain name and the complainant's mark under paragraph

509 *Ibid*, n2.

510 95 F Supp 2d 528 (ED Va, 2000).

511 See, eg, *Taubman Co v Webfeats*, 319 F 3d 770 (6th Cir, 2003) (<taubmansucks.com>); *Bosley Medical Institute, Inc v Kremer*, 403 F 3d 672 (9th Cir, 2005) (<www.bosleymedical.com>).

512 See, eg, *Planned Parenthood Federation of America, Inc v Bucci*, 42 USPQ 2d 1430 (SDNY, 1997) (<plannedparenthood.com>); *OBH, Inc v Spotlight Magazine, Inc*, 86 F Supp 2d 176 (WD NY, 2000) (<thebuffalonews.com>).

4(a)(i). As explained at [5.25] above, the comparison required under the UDRP differs from the comparison between marks under national trade mark laws, as the UDRP is aimed at preventing abusive, bad faith registration, and does not directly address trade mark infringement. As explained at [5.26] above, while there is agreement among panellists that the comparison must focus on the disputed domain name and the complainant's mark, irrespective of use factors, there is disagreement about whether a strict literal comparison is required or whether the comparison is aimed at protecting against source confusion. In any case, as explained at [5.27], the content of a web site is irrelevant to the comparison between the domain name and the complainant's mark under the UDRP. These issues demand further analysis in the context of the treatment of 'sucks'-type domain names.

To begin with, if the comparison under paragraph 4(a)(i) is directed mainly at determining whether there is source confusion, then it is extremely unlikely that a 'sucks'-type domain name would be confusingly similar to a mark included in the domain name. The addition of a pejorative or derogatory term would usually be sufficient to dispel any confusion as to source. In this sense, the position would be similar to that reached by the court in the *Bally* case, in that no 'reasonably prudent' Internet user would believe that there is a source association. If, however, some lesser form of association between the domain name and the complainant's mark is required, then the addition of a pejorative or derogatory term may not be sufficient to remove confusion.

There appear to be two key issues that need to be resolved in order to determine the approach taken as to whether or not a 'sucks'-type domain name may be confusingly similar to a complainant's mark. First, given that some degree of association is necessary, then how much association is required? If the standard to be applied is whether a 'reasonably prudent' person would be confused as to source, then 'sucks'-type domain names are unlikely to be confusingly similar. Whether this is an appropriate standard to apply in a context in which Internet users make instantaneous decisions about whether or not to access a web site may, however, be questioned. A form of confusion analogous to the 'initial interest confusion' recognised by US law, whereby a similarity between marks attracts users to a web site even though the association is later dispelled, would seem to be more readily applicable. Moreover, there is a need to consider the standard to be applied to the hypothetical person who must form an association: should it be a 'reasonably prudent' person, or should some lesser standard be adopted? As explained at [5.38] below, some panels have considered the extent to which Internet users may be non-English speakers to be relevant to this question.

Secondly, there is an underlying need to take into account the conduct sought to be caught by the UDRP. In this respect, it is important to understand that a negative association between a domain name and a mark may facilitate certain conduct just as much as a positive association or confusion as to source. For example, a person who registers a 'sucks'-type domain name may have a similar degree of leverage over a trade mark owner, in terms of an ability to threaten harm to the mark unless payment is made to transfer the name, as does someone who registers a domain name that creates source confusion. Additionally, a person who registers a 'sucks'-type domain name may attract users by virtue of the negative connotations of the name, such as users who are critical of the trade mark owner. If the domain name registrant attracts such users for his or her own commercial purposes, such as to re-direct users to other sites or to obtain advertising revenue, the question arises whether this conduct should fall within the scope of the UDRP. Is it, for example, conduct that harms the trade mark owner in a way that is analogous to conventional cybersquatting? In each of these instances, while there is no source confusion in the traditional trade mark sense, it could be said that the 'sucks'-type domain name is 'confusingly similar' to the mark, in the sense that the use of the mark in the name creates some association with the complainant, even if it is a negative association, that results in confusion.

These issues are revisited at [5.41], following an analysis of the divergent views expressed in panel decisions concerning the status of 'sucks'-type domain names.

[5.38] Justifications for the Majority View

The appropriate standard to apply when comparing a disputed domain name with the complainant's mark was at the centre of the reasoning in the first major UDRP decision to consider the issue, *Wal-Mart Stores, Inc v Richard MacLeod d/b/a For Sale.*[513] In that dispute, which concerned the allegedly infringing domain name <wal-martsucks.com>, the complainant argued that 'sucks' was simply a generic term which, when added to a trade mark, would not remove confusion. The panellist dismissed this argument without difficulty, concluding that:

> No reasonable speaker of modern English would find it likely that Wal-Mart would identify itself using wal-martsucks.com . . . The Panel specifically rejects Complainant's argument that consumers are likely to be confused as to the sponsorship or association of a domain name that combines a famous mark with a term casting opprobrium on the mark.

In other words, the panellist concluded that, if the purpose of comparing the domain name and the mark is to avoid source confusion, a 'sucks' domain name cannot be confusingly similar. The problem in the dispute, however, was that, after registering the domain name, the respondent had offered to sell it to the complainant for a substantial sum, conduct that resembled cybersquatting. Quite clearly, an implied threat to use a domain name to damage a complainant's mark may be just as credible if made by the registrant of a 'sucks'-type domain name as a threat made by the registrant of other confusingly similar names. Taking into account the extent to which the UDRP is directed against extortionate registration and use of a domain name, the panellist concluded that the comparison was not concerned with source confusion, and held that the domain name was confusingly similar to the complainant's mark. In this respect, the panellist stated that:

> the Panel understands the phrase 'identical or confusingly similar' to be greater than the sum of its parts. The Policy was adopted to prevent the extortionate behavior commonly known as 'cybersquatting,' in which parties registered domain names in which major trademark owners had a particular interest in order to extort money from those trademark owners. . . . Thus, the Panel concludes that a domain name is 'identical or confusingly similar' to a trademark for purposes of the Policy when the domain name includes the trademark, or a confusingly similar approximation, regardless of the other terms in the domain name. In other words, the issue under the first factor is not whether the domain name causes confusion as to source (a factor more appropriately considered in connection with the legitimacy of interest and bad faith factors), but instead whether the mark and domain name, when directly compared, have confusing similarity.

If this reasoning is accepted, then domain names that incorporate a complainant's mark may be held to be confusingly similar, even if the domain name includes terms that potentially distinguish it from a complainant's mark. The emphasis in the policy therefore shifts from 'confusing similarity' to whether the respondent has rights or legitimate interests, and

[513] WIPO Case No D2000-0062 (19 Sept 2000).

whether there is bad faith registration and use. At the same time, it must be borne in mind that a domain name that includes a negative or derogatory term is more likely to be used to threaten a complainant than a name that includes other potentially distinguishing terms.

A narrow majority of panel decisions have concluded that 'sucks'-type domain names are confusingly similar to marks incorporated in the name, some extending the reasoning in *Wal-Mart Stores, Inc v Richard MacLeod,* while others have introduced new considerations. The following discussion identifies the main decisions to conclude that a 'sucks'-type domain name is confusingly similar, and outlines the reasoning in each case.

In *Wal-Mart Stores, Inc v Walsucks and Walmarket Puerto Rico,*[514] a decision reached before *Wal-Mart Stores, Inc v Richard MacLeod,* the panellist decided that the domain names <wal-martcanadasucks.com>, <wal-martcanadasucks.com>, <walmartuksucks.com>, <walmartpuertorico.com> and <walmartpuertoricosucks.com> were confusingly similar to the 'WAL-MART' mark. In reaching this conclusion, the panellist applied the '*Sleekcraft* factors, referred to at [5.25] above, emphasising the strength of the complainant's well-known mark and the intention of the respondent in registering the domain name, which was to seek payment from the complainant. The panellist also concluded that the domain name was sufficiently similar to the mark to attract Internet users, in this sense distinguishing the required confusion from source confusion:

> Internet users with search engine results listing Respondent's domains are likely to be puzzled or surprised by the coupling of Complainant's mark with the pejorative verb "sucks". Such users, including potential customers of Complainant, are not likely to conclude that Complainant is the sponsor of the identified websites. However, it is likely (given the relative ease by which websites can be entered) that such users will choose to visit the sites, if only to satisfy their curiosity. Respondent will have accomplished his objective of diverting potential customers of Complainant to his websites by the use of domain names that are similar to Complainant's trademark.

Given the questionable status of the multi-factor *Sleekcraft* test under the UDRP, as explained at [5.25] above, the <wal-martcanadasucks.com> decision may be of limited use in this respect. Nevertheless, the observations that Internet users might be confused by search engine results and attracted to a web site, despite the absence of source confusion, have been influential. In *Cabela's Incorporated v Cupcake Patrol,*[515] for example, the panellist referred to the <wal-martcanadasucks.com> decision to support a finding of confusing similarity, stating:

> By using Complainant's marks in its domain names, Respondent makes it likely that Internet users entering 'Cabela's' into a search engine will find 'CABELASSUCKS.COM' in addition to the Complainant's site <cabelas.com>. Respondent's domain name is sufficiently similar to Complainant's marks that the search engine results will confusingly list the Respondent's domain name when searching for Complainant's mark.

Similarly, the <wal-martcanadasucks.com> decision was expressly approved by the panellist in *Direct Line Group Ltd v Purge IT,*[516] who held that the domain name <directlinesucks.com> was confusingly similar to the 'DIRECT LINE' mark. In reaching this conclusion, the panellist stated:

[514] WIPO Case No D2000-0477 (20 July 2000).

[515] NAF Case No FA95080 (29 Aug 2000).

[516] WIPO Case No D2000-0583 (13 Aug 2000). See also *Dixons Group PLC v Purge IT and Purge IT Ltd,* WIPO Case No D2000-0584 (13 Aug 2000).

The first and immediately striking element in the Domain Name is the Complainants' name and adoption of it in the Domain Name is inherently likely to lead some people to believe that the Complainants are connected with it. Some will treat the additional 'sucks' as a pejorative exclamation and therefore dissociate it after all from the Complainants; but equally others may be unable to give it any very definite meaning and will be confused about the potential association with the Complainants.

In subsequent decisions, the potential for non-English speakers, who may not understand the negative connotations of terms such as 'sucks', to be confused into believing there is some association with the trade mark owner has been emphasised. In *Vivendi Universal v Mr Jay David Sallen*,[517] for example, the majority of the panel determined that the domain name <vivendiuniversalsucks.com> was confusingly similar to the 'VIVENDI UNIVERSAL' mark largely on this basis, justifying their conclusion in the following terms:

> being satisfied that certain members of the public in general and 'Internauts' in particular, not being English speakers and/or aware of the meaning of the word 'sucks' in the Internet world, would be likely to understand 'sucks' as a banal and obscure addition to the reasonably well-known mark VIVENDI UNIVERSAL and that, accordingly, <vivendiuniversalsucks.com> refers to goods or services provided by the Complainant, this Panel, by majority, finds the requirements of paragraph 4(a)(i) of the Policy to have been met.

This reasoning was taken up by other panels, in decisions such as *Koninklijke Philips Electronics NV v In Seo Kim*,[518] in which the panellist held that the <philipssucks.com> was confusingly similar to the 'PHILIPS' trade mark. In that decision, the panellist stated:

> Not all Internet users are English speaking or familiar with the use of 'sucks' to indicate a site used for denigration. Furthermore, it is not unknown for companies to establish complaint or comment sites or areas of sites to obtain feedback on their products; accordingly, some people might suppose that a website of this nature at the Domain Name was operated by the Complainant.

In *Bayer Aktiengesellschaft v Dangos & Partners*,[519] the panellist, who held that the domain names <bayersucks.org>, <bayersucks.biz> and <bayersucks.info> were confusingly similar to the 'BAYER' mark, referred to the minority of decisions that had held that 'sucks'-type domain names were not confusingly similar, before providing a helpful summary of the majority of decisions to that point:

> in the majority of cases, the opposite conclusion was reached mainly because not all the Internet users are English speakers or familiar with the use of 'sucks' to indicate a site used for denigration or criticisms. Furthermore, when the Internet users enter a word in a search engine, the engine identifies websites of potential relevance and includes the term that the users are searching. As such, the search engine also would certainly retrieve the domain names consisting of the search term and the term 'sucks'. The domain name holder will thus have reached its objective to divert potential consumers to its website through the adoption of a domain name that is similar to a third party's trademark (See *Koninklijke Philips Electronics NV v In Seo Kim*, Case No. D2001-1195; *Vivendi Universal v Mr Jay David Sallen and GO247.COM, INC*, Case No D2001-1121 (with dissenting opinion); *The Salvation Army v Info-Bahn, Inc*, Case No D2001-0463; *ADT Services AG v ADT Sucks.com*, Case No D2001-0213; *Société Accor contre M Philippe Hartmann*, Case No D2001-0007; *TPI Holdings, Inc v AFX Communications a/k/a AFX*, Case No D2000-1472; *Diageo plc v John Zuccarini, Individually and t/a Cupcake Patrol*; Case No D2000-0996; *Standard Chartered PLC v*

[517] WIPO Case No D2001-1121 (7 Nov 2001).
[518] WIPO Case No D2001-1195 (12 Nov 2001).
[519] WIPO Case No D2002-1115 (3 Feb 2003).

Purge IT, Case No. D2000-0681; *Wal-Mart Stores, Inc v Richard MacLeod d/b/a For Sale*, Case No D2000-0662; *National Westminster Bank PLC v Purge IT and Purge IT Ltd*, Case No D2000-0636; *Freeserve PLC v Purge I.T. and Purge IT Ltd*, Case No D2000-0585; *Dixons Group PLC v Purge IT and Purge IT Ltd*, Case No D2000-0584; *Direct Line Group Ltd, Direct Line Insurance plc, Direct Line Financial Services Ltd, Direct Line Life Insurance Company Ltd, Direct Line Unit Trusts Ltd, Direct Line Group Services Ltd vPurge IT, Purge IT Ltd*, Case No D2000-0583; *Wal-Mart Stores, Inc v Walsucks and Walmarket Puerto Rico*, Case No. D2000-0477).

Given the continued dominance of English on the Internet and elsewhere, the argument that non-English speakers may be confused does not seem to be especially persuasive. As the dissenting panellist pointed out in *Vivendi Universal v Mr Jay David Sallen:*

> the word 'sucks' is used quite frequently within domain names (and elsewhere) to criticize, disparage, and ridicule—probably more frequently than any other single word in any language. The mere fact that some people may not fully comprehend the colloquial meaning of 'sucks' is thus immaterial, at least absent evidence that many people understand the word to mean something that, when combined with the Complainant's trademark, is confusingly similar to the trademark.

Nevertheless, it has been suggested that this consideration may be more of a factor where a trade mark has a strong reputation, or customer base, among non-English speakers. In *Berlitz Investment Corp v Stefan Tinculescu*,[520] for example, the panellist took into account the fact that the Complainant's core business catered almost exclusively to individuals who did not speak English as a first language in determining that the <berlitzsucks.com> domain name was confusingly similar to the complainant's 'BERLITZ' mark. At the same time, the panellist noted that he was 'not entirely convinced' by the conclusion reached by the majority of panels that 'sucks'-type domain names are confusingly similar.

[5.39] Justifications for the Minority View

By the time of the *Berlitz* decision, a substantial minority of panels had concluded that 'sucks'-type domain names should not be considered confusingly similar to marks included in the name. The early decisions to reach this conclusion were strongly influenced by the US cases dealt with at [5.36] and tended towards adopting a per se rule, under which 'sucks'-type domain names could never be held to be confusingly similar. In this regard, this line of decisions equated confusing similarity under the UDRP with the source confusion applied in trade mark infringement actions.

The first decision to conclude that a 'sucks'-type domain name was not confusingly similar was *Wal-Mart Stores, Inc v wallmartcanadasucks.com*,[521] which concerned the domain name <wallmartcanadasucks.com> (which is not to be confused with the <walmartcanadasucks.com> decision dealt with at [5.37] above). The domain name resolved to a web site that included very little critical material, and the respondent had requested payment from the complainant. In this decision, the panellist engaged in an extensive discussion of US cases, including the *Bally* and *Lucentsucks.com* cases, simply assuming that principles derived from US trade mark law, including the multi-factor *Sleekcraft* test, could be directly applied to the UDRP. Essentially adopting the analysis of the court in *Bally*, the panellist decided that a 'reasonably prudent user' would not confuse the

[520] WIPO Case No D2003-0465 (22 Aug 2003).
[521] WIPO Case No D2000-1104 (23 Nov 2000).

<wallmartcanadasucks.com> domain name with the 'WALMART' mark. The panellist concluded his discussion with the following statement:

> I do not see how a domain name including 'sucks' ever can be confusingly similar to a trademark to which 'sucks' is appended. But whether or not a per se privilege for use of 'sucks' is appropriate, the record in this case does not support a finding that the ICANN Policy has been violated.

In the next decision to conclude that a 'sucks'-type domain name was confusingly similar, *Lockheed Martin Corporation v Dan Parisi*,[522] the issue divided the majority of the panel and the dissenting panellist. In that dispute, the respondent used the domain names <lockheedsucks.com> and <lockheedmartinsucks.com> for sites that provided a forum for criticism of the trade mark owner. The majority followed the <wallmartcanadasucks.com> decision in relying on the *Bally* and *Lucentsucks.com* cases, pointing out that the products associated with the complainant's marks were quite different from the respondent's service of providing a forum for criticism. The majority also observed that, while the disputed domain names might well be returned by a search engine, any confusion would be removed once the searcher saw the 'sucks' domain names in the search results.

The minority panellist in *Lockheed* dissented on the basis that concluding that a 'sucks'-type domain name was not confusingly similar would mean that the UDRP could not apply to a respondent who registered a 'sucks'-type domain name with the intention of extorting payment from a trade mark owner. To ensure that the policy applies to abusive, bad faith registration of 'sucks'-type domain names, the dissenting panellist concluded that:

> The outcome requires each case to be considered on its own facts and precludes a per se conclusion that there can be no 'confusing similarity' where 'sucks' is added to the trademark.

In a decision reached around the time of the *Lockhead Martin* decision, *McLane Company, Inc v Fred Craig*,[523] the panellist concluded that the disputed domain name <mclanenortheastsucks.com> was not confusingly similar to the 'MCLANE NORTHEAST' mark. As in other panel decisions to come to this conclusion, the panellist referred to the *Lucentsucks.com* case and expressed the view that the addition of a 'sucks'-type term would avoid source confusion. In this respect, the panellist stated:

> This domain name includes the term 'sucks.' While often the inclusion of a generic term will not serve to distinguish a domain name from a trademark, in the case of the term 'sucks,' the addition of the generic term does reduce the likelihood of confusion. It is unlikely that a viewer would confuse the domain name 'mclanenortheastsucks.com' with the trademark mclane northeast. Instead, it should be evident to a viewer that any site reached using 'mclanenortheastsucks.com' is not sponsored or endorsed by the trademark owner. In a recent decision regarding registration of the domain name 'lucentsucks.com' the judge accepted the argument that 'the average consumer would not confuse lucentsucks.com with a web site sponsored by [Lucent].' See *Lucent Technologies, Inc v Lucentsucks.com*, 95 F Supp 2d 528, 535 (ED Va, May 3, 2000).

The link between the view that the comparison required by paragraph 4(a)(i) is concerned with the same sort of source confusion as trade mark infringement actions, and the minority view that 'sucks'-type domain names are not confusingly similar, was clearly drawn by the panellist in *America Online, Inc v Johuathan Investments, Inc*.[524] As pointed out at [5.26] above, in that decision the panellist said that confusing similarity under the UDRP

522 WIPO Case No D2000-1015 (26 Jan 2001).
523 WIPO Case No D2000-1455 (11 Jan 2001).
524 WIPO Case No D2001-0918 (14 Sept 2001).

refers to 'confusion as to trade origin', in other words, to trade mark source confusion. In relation to the dispute itself, the panellist held that the <fucknetscape.com> domain name was not confusingly similar to the 'NETSCAPE' mark. In reaching this conclusion, the panellist stated:

> The Panel regards it as inconceivable that anyone looking at this Domain Name will believe that it has anything to do with a company of such high repute as the Complainant. It is manifestly, on its face, a name, which can have nothing whatever to do with the Complainant. It is a name, which, by its very nature, declares that it is hostile to Netscape.

[5.40] 'Sucks'-type Domain Names: no Universal Rule

The division between the majority and minority views on the status of 'sucks'-type domain names is illustrated by the division between the panellists in *Société Air France v Virtual Dates, Inc*,[525] in which the majority held that the <airfrancesucks.com> domain name was confusingly similar to the 'AIR FRANCE' mark. The majority began their analysis by stressing that 'the standard for finding confusing similarity under the first element of the policy, which is a rule specifically designed to address cybersquatting, is not the same standard as for finding trademark infringement'. Pointing out that each dispute is to be assessed on its own merits, the majority considered the fact that international customers are not necessarily familiar with the pejorative nature of the term 'sucks' to be decisive in this case. The dissenting panellist, on the other hand, referred to the *Bally* case in deciding that no 'reasonably prudent' Internet user would believe that there was an association between <airfrancesucks.com> and 'AIR FRANCE'. The dissenting panellist also rejected the argument that non-English speakers might be confused by the domain name, on the basis that a substantial percentage of Air France customers would be familiar with English.

Despite the abiding differences between panellists, there is an emerging consensus on at least one issue, namely that a per se approach, that would automatically find that a 'sucks'-type domain is not confusingly similar, has been rejected. The per se approach to finding a 'sucks'-type domain name not confusingly similar derives from the US court decisions in the *Bally* and *Lucentsucks.com* cases. In the early decision in *Cabela's Incorporated v Cupcake Patrol*,[526] the panellist referred to the *Bally* and *Lucentsucks.com* precedents, before concluding that:

> The Panel does not infer that '-sucks' domain names are immune from scrutiny as being confusingly similar to trademarks to which they are appended. Each case must be considered in light of the facts presented.

There are sound reasons for rejecting a per se approach. To begin with, as the majority pointed out in *Vivendi Universal v Mr Jay David Sallen*, much may depend on the context in which a term such as 'sucks' is used, as there are clearly examples where the use of such a term is not entirely negative or derogatory. In that decision, the majority gave the following two examples of uses of the term 'sucks' that would not dispel an association with the trade mark owner:

[525] WIPO Case No D2005-0168 (24 May 2005).
[526] NAF Case No FA95080 (29 Aug 2000).

(1) the use of the words 'sucks' purely descriptively, as in the advertising slogan 'Nothing sucks like Electrolux' (If there were a website at <electroluxsucks.com>, it would be unlikely to be taken as unaffiliated with the company Electrolux); and
(2) the website of the band Primus, < primussucks.com>, so named after the album Suck on This (1990). (The website of the band's lead singer, Les Claypool, at <lesclaypool.com>, has a link to the <primussucks.com> website).

In *Wachovia Corporation v Alton Flanders*,[527] the panellist also rejected the per se approach, on the basis that the *Bally* and *Lucentsucks.com* cases were concerned with trade mark infringement, which 'is not the same as proof of the first element under the UDRP Policy'. In what has been a consistent theme among those who have adopted the majority view, the panellist, in effect, considered that *sui generis* principles should be developed under the UDRP, which is specifically designed to address cybersquatting, rather than resorting to national trade mark laws. In the instant dispute, the panellist concluded that the domain names <wachovia-sucks.com>, <wachoviabanksucks.com> and <wachoviasucks.com> were confusingly similar to the 'WACHOVIA' family of marks. In coming to this conclusion, the panellist referred to the trade mark owner's international reputation, which made it relevant to consider the extent to which international customers might not recognise the meaning of the word 'sucks'.

In a comprehensive review of previous panel decisions, the panellist in *Medimmune, Inc v Jason Tate*[528] confirmed the now accepted view that there is no universal rule and each case must be decided on its merits. In that dispute, the complainant claimed that the domain names <synagisisbadforyou.com> and <synagisisnotsafe.com> were confusingly similar to the 'SYNAGIS' mark, arguing that the words 'isbadforyou' and 'isnotsafe' were descriptive, dictionary words that did not distinguish the domain names from the mark. Rejecting this argument, the panellist maintained that, where a domain name combines a mark with words that indicate the domain name is critical of the complainant's products, such as in this case, it would be unlikely for Internet users to be confused about an association between the domain name and the complainant. At the same time, the panellist pointed out that adopting a per se rule, that a 'sucks'-type domain name can never be confusingly similar, would prevent the policy applying to 'sham speech' domain names that resolve to sites that are used for purposes such as obtaining click-through revenue. As the panellist observed, it was really concerns such as this that influenced some panels to conclude that there is less need to show confusing similarity where there is overwhelming evidence of a lack of legitimate interests and of bad faith registration.

The panellist in *Medimmune* proceeded to address the main justifications presented by panels that have concluded that 'sucks'-type domain names are confusingly similar. To begin with, the panellist referred to the argument that non-English speakers would be confused about an association with the trade mark owner as they might not understand terms such as 'sucks'. In this dispute, the panellist simply held that this justification had no application as, first, 'isbadforyou' and 'isnotsafe' were common English words and, secondly, the complainant did not have a significant non-English-speaking customer base. The panellist then turned to a consideration of the required standard of confusion, pointing out that those supporting the majority view had applied a form of 'initial interest confusion', under which confusion could be found even where someone using a search engine understands a derogatory

[527] WIPO Case No D2003-0596 (19 Sept 2003).
[528] WIPO Case No D2006-0159 (14 Apr 2006).

meaning but, nevertheless, is curious enough to follow a link to the 'sucks'-type domain name. On the facts before him, the panellist concluded that there was insufficient evidence of 'initial interest confusion', pointing out that the disputed domain names did not appear in the first 20 pages of returns from an Internet search. While the panellist remained concerned about the problem of 'sham speech' sites, then, he felt compelled to conclude that there was no confusing similarity.

An interesting, and nuanced, approach to the issue of 'sucks'-type domain names has been adopted in two related decisions, *Wal-Mart Stores, Inc v xc2*[529] (<walmartblows.com>) and *Wal-Mart Stores, Inc v Traffic Yoon*[530] (<boycottwalmart.com>), in which the panellist applied consistent principles to reach opposing conclusions. The careful reasoning in the decisions confirms the trend of UDRP panellists rejecting a rigid approach to 'sucks'-type sites, in favour of a close attention to the facts of the particular dispute. As the panellist put it in <walmartblows.com>:

> blanket propositions in relation to whether a derogatory term appended to a Complainant's mark constitutes a phrase which is confusingly similar to the Complainant's mark should be avoided. Instead, careful consideration needs to be given to whether the particular disparaging term combined with the Complainant's mark may confuse members of the public into believing that there exists an affiliation between the Complainant or the Respondent, or that the Complainant operates or owns the domain name.

The decisions are notable, however, mainly because the panellist essentially proposed a new test for determining whether a domain name is confusingly similar in the context of 'sucks'-type domain names. As pointed out at [5.26] above, the panellist in the <boycottwalmart.com> decision generally approved the approach to confusing similarity adopted by the panellist in *America Online, Inc v Johuathan Investments, Inc., and AOLL-NEWS.COM*[531], maintaining that, in addition to similarity, a complainant must establish confusion as to the trade origin of products, services or proprietorship of the site to which the disputed domain name resolves. As also pointed out, the panellist rejected the wholesale adoption of the *Sleekcraft* factors, while claiming that some of the factors might be relevant to the comparison under the UDRP. Finally, as further pointed out, the panellist, concerned at the evidential burden imposed by a comprehensive analysis of source confusion, considered that the inquiry should be limited to a side by side comparison of the domain name and the mark 'through the eyes of the likely viewers of the two'.

These conclusions do not, however, dispose of the kind of confusion that needs to be established. On this issue, the panellist proposed that:

> confusing similarity will be established where those persons who are mostly likely to want to access a complainant's website will be confused as to whether the complainant is the owner and operator of the website to which the disputed domain name resolves. As previously mentioned, this is determined with no regard to the content of the website but merely comparing the domain name (without the generic top-level domain) with the trademark. This is quite important to appreciate, since the Policy is only intended to deal with cases where the domain name is identical or confusingly similar to a complainant's trademarks. In this Panel's view the Policy is not intended to, and cannot, be responsive to all variants of confusing conduct by a respondent, and that is why in this Panel's view resort to the body of a respondent's website is not helpful when considering the test on the first aspect of the Policy.

[529] WIPO Case No D2006-0811 (29 Aug 2006).
[530] WIPO Case No D2006-0812 (20 Sept 2006).
[531] WIPO Case No D2001-0918 (14 Sept 2001).

In other words, the panellist proposed a side-by-side comparison of the disputed domain name and the complainant's mark with a view to establishing whether or not there is confusion that the domain name is associated with a web site operated by the complainant. This approach clearly rules out other sorts of confusion regarding an association between the domain name and the complainant, including confusion potentially arising from a negative association such as might arise from 'sham speech' domain names.

Applying this approach to the <boycottwalmart.com> domain name, the panellist took into account three factors in deciding there was no confusing similarity. First, as most of the complainant's customers would be familiar with English, the panellist concluded there would be no confusion regarding the use of the term 'boycott'. Secondly, the panellist distinguished this case from the 'sucks' cases on the basis that, unlike 'boycott', the term 'sucks' relies on an understanding of American slang. Thirdly, the panellist applied a side by side comparison to point out that there is less likelihood of confusion where a derogatory term is used as a prefix, rather than as a suffix. Therefore, the panellist concluded that members of the public would not be confused about whether the complainant owned the web site to which the <boycottwalmart.com> domain name resolved.

The outcome in the <boycottwalmart.com> decision can be contrasted with that in the <walmartblows.com> dispute, where the panellist held that the disputed domain name was confusingly similar to the 'WAL-MART' mark. In <walmartblows.com>, the panellist held that the domain name could cause confusion as to whether or not the complainant owned or operated the Web site to which the domain name resolved. In reaching this conclusion, the panellist took into account the extent to which the term 'blows' is an idiomatic usage of American English, and that the appended derogatory term was a suffix and not a prefix.

[5.41] Conclusions on 'Sucks'-type Domain Names

The extent to which panellists remain divided on whether or not 'sucks'-type domain names are confusingly similar is indicated by the divergent outcomes and reasoning in recent UDRP decisions. The following is a representative sample of recent decisions that have considered the issue:

- *Medimmune, Inc v Jason Tate*[532]—<synagisisbadforyou.com> and <synagisisnotsafe.com> were not confusingly similar to the 'SYNAGIS' mark.
- *Ameridream, Incorporated v The Owners Alliance*[533]—the <fuckameridream.net>, <fuckameridream.org>, and <fuckameridream.com> domain names were confusingly similar to the 'AMERIDREAM' mark.
- *Buw Holdings GmbH, Jens Bormann, Karsten Wulf v Anon-Web.com*[534]—the <buwsuckz.com> domain name was confusingly similar to the 'BUW' mark.
- *Booz Allen Hamilton Inc v Marc Buchard c/o Domains Unlimited EU*[535]—the <boozallen-sucks.com> domain name was confusingly similar to the 'BOOZ ALLEN HAMILTON' mark.

532 WIPO Case No D2006-0159 (14 Apr 2006).
533 NAF Case No FA676671 (26 May 2006).
534 WIPO Case No D2006-0462 (7 June 2001).
535 NAF Case No FA706689 (30 June 2006).

- *The Works at Wyomissing, LLC v Youth Fellowship & Rescue c/o Guillermo Jalil*[536]—the <theworkssucks.com> and <gearedforfunsucks.com> domain names were not confusingly similar to the marks 'THE WORKS AT WYOMISSING' and 'GEARED FOR FUN!'.
- *Wal-Mart Stores, Inc v xc2*[537]—the <walmartblows.com> domain name was confusingly similar to the 'WAL-MART' mark.
- *Wal-Mart Stores, Inc v Traffic Yoon*[538]—<boycottwalmart.com> was not confusingly similar to the 'WAL-MART' mark.
- *Countrywide Financial Corporation v Bluehost.com*[539]—<countrywidehomeloanssuck.com> was confusingly similar to the 'COUNTRYWIDE HOME LOANS' mark.

Despite the differences in outcomes, the majority of panellists now seem to have accepted the view that the confusion between the disputed domain name and the complainant's mark required under the first element of the UDRP is not the source confusion required under trade mark law. This conclusion necessarily follows from an understanding that the UDRP is directed against extortionate and other opportunistic practices involving the registration and use of a mark as a domain name, and not at protecting against source confusion. Consequently, the confusion required under the first element of the UDRP is that which is necessary for a domain name registrant to engage in such extortionate and opportunistic practices. In other words, the degree of confusion must be such as to enable the domain name registrant to make a credible threat to harm the trade mark owner, or to engage in other opportunistic practices involving the use of the mark for commercial gain. Therefore, while the use of the domain name in a way that creates source confusion is certainly relevant to determining whether the domain name has been registered in bad faith, as is acknowledged by paragraph 4(b)(iv) of the UDRP, the confusion that is required under the first element is a lesser standard, requiring only that it is such as to create the potential to harm the trade mark owner, such as through source confusion.

While there is considerable agreement that the confusion referred to in paragraph 4(a)(i) is not source confusion, there is disagreement about the precise level of confusion that is required. This disagreement is related to the two different approaches to the interpretation of the act of registering a domain name identified at [3.15]: that registration of a domain name that includes someone else's mark is automatically suspect, or that domain name registrants should be given the benefit of the doubt whether the registration of a domain name that includes a mark is for a legitimate use. At the same time, it should be borne in mind that the role of the first element of the UDRP is primarily that of a gatekeeper, in that it sets a minimum threshold to be satisfied before turning to the issues of whether the respondent has rights or legitimate interests in the domain name, and of whether the complainant has sufficiently established the respondent's bad faith. Moreover, the extent to which Internet users make immediate, often intuitive decisions about accessing web sites, rather than reasoned decisions, suggests that a relatively undemanding standard for determining the degree of confusion should be applied.

Over all, however, the test for determining the required level of similarity between the disputed domain name and the complainant's mark should be determined by reference to the objective of preventing extortionate and opportunistic practices, which harm the trade

[536] NAF Case No FA720831 (7 July 2006).
[537] WIPO Case No D2006-0811 (29 Aug 2006).
[538] WIPO Case No D2006-0812 (20 Sept 2006).
[539] NAF Case No FA804739 (7 Nov 2006).

mark owner or enable the domain name registrant to free ride on the goodwill in the mark. In this sense, the question to be asked in disputes involving 'sucks'-type domain names is whether the degree of similarity between the domain name and the complainant's mark is such that the domain name registrant can use the domain name to make a credible threat to harm the trade mark owner, or otherwise to take advantage of the goodwill in the mark. As the panellist pointed out in the <boycottwalmart.com> decision, this will often involve an assessment of whether potential visitors to a web site operated by the complainant trade mark owner would be attracted to the web site to which the disputed domain name resolves in the belief that there is some association with the trade mark owner. It is, however, probably not strictly necessary for potential visitors to be confused as to whether the trade mark owner is the owner or operator of the web site to which the disputed domain name resolves, as some lesser form of association may be sufficient.

At the same time, the test proposed in the <boycottwalmart.com> decision only addresses the question of what the confusion should be about—which is self-evidently confusion between the disputed domain name and the complainant's mark—and not the degree of confusion required between the two. Given the sorts of spontaneous decisions made by Internet users, it is suggested that a liberal approach should generally be adopted in determining the degree of confusion required under the first element of the UDRP. This approach is reinforced by the consideration that it is perfectly feasible for domain names with negative connotations, including 'sucks'-type domain names, to be used either to make extortionate demands of a trade mark owner or opportunistically to exploit the goodwill in the complainant's mark. In this respect, it is also arguable that the use of a domain name that includes negative connotations—such as a 'sucks'-type domain name—intentionally to attract Internet users that are looking for a cybergripe site, for purely commercial purposes—such as redirecting Internet users to pornographic web sites—is the kind of opportunistic practice that should be prohibited by the policy. In other words, the use of the domain name for a 'sham speech' site should be prohibited, as this still involves the use of the complainant's mark, albeit in a negative sense, to attract Internet users for the domain name registrant's own commercial purposes. The contrary argument, presented by the panellist in the <boycottwalmart.com> decision, is that the UDRP is targeted at a narrower range of behaviour and does not extend to 'all variants of confusing conduct' by domain name registrants.

While acknowledging the accuracy of this assertion, however, it would seem reasonable for a domain name registrant who knowingly incorporates someone else's trade mark in the domain name in its entirety to bear the onus of establishing that the domain name has been registered for a legitimate purpose. That said, in the event of a negative domain name, such as a 'sucks'-type domain name that includes a mark in its entirety, being found to have breached the UDRP, it would seem that the appropriate remedy should be cancellation of the domain name, rather than transfer to the complainant trade mark owner.

6

Rights or Interests in the Disputed Domain Name

Overview and Policy

[6.1] Overview

The second element of abusive, bad faith registration, set out in paragraph 4(a)(ii) of the UDRP, requires the complainant to prove that:

> the respondent (domain name holder) has no rights or legitimate interests in respect of the domain name.

This element implements the recommendation of the WIPO Final Report that the scope of the procedure be confined to disputes between a complainant with trade mark rights and a respondent with no rights or interests in the domain name, and not include disputes on the merits between parties with competing rights in the domain name. In this respect, the report stated:

> the scope of the procedure is limited so that it is available only in respect of deliberate, bad faith, abusive, domain name registrations or "cybersquatting" and is not applicable to disputes between parties with competing rights acting in good faith.[1]

The report further stated that:

> good faith disputes between competing right holders or other competing legitimate interests over whether two names were misleadingly similar would not fall within the scope of the procedure.[2]

[6.2] Policy Considerations

The policy underlying the limitation established by the second element of abusive registration is that complex issues arising in disputes between parties with competing rights or interests in a domain name are properly dealt with by courts, and not by the expedited UDRP procedure. The ICANN staff report submitted to the board in October 1999 explained the limitation in the following terms:

[1] World Intellectual Property Organization (WIPO), *The Management of Internet Names and Addresses: Intellectual Property Issues,* Report of the WIPO Internet Domain Name Process (30 Apr 1999), available at http://www.icann.org/amc/en/processes/process1/report/, para [135].

[2] *Ibid,* para [172].

> the fact that the policy's administrative dispute-resolution procedure does not extend to cases where a registered domain name is subject to a legitimate dispute (and may ultimately be found to violate the challenger's trademark) is a feature of the policy, not a flaw. The policy relegates all 'legitimate' disputes—such as those where both disputants had longstanding trademark rights in the name when it was registered as a domain name—to the courts; only cases of abusive registrations are intended to be subject to the streamlined administrative dispute-resolution procedure.[3]

This policy has been echoed by panel decisions. In *Action Sports Videos v Jeff Reynolds*,[4] for example, the panellist stated:

> The Panel does not believe it is the function of the Policy to serve to disentangle the conflicting rights of parties to a mark once it is apparent both parties have colorable claims to a mark underlying a disputed domain name.

While the policy intent behind the introduction of the second element is clear, there are two major difficulties with the application of the second element:

- First, under paragraph 4(a) of the UDRP, the complainant bears the onus of proving each of the three elements of abusive, bad faith registration. The second element of the UDRP, however, requires that the complainant establish a negative. In recognition of the difficulties involved in proving a negative, UDRP panels have held that a complainant need only establish a prima facie case, whereupon the burden shifts to the respondent. This is dealt with in more detail at [6.4].
- Secondly, there is obviously considerable ambiguity concerning precisely what may qualify as 'rights or legitimate interests' in a domain name. Determining whether or not a respondent has 'rights or legitimate interests' may involve the application of national legal principles, but there are significant differences between national laws in relation to what might qualify as a 'right'. For example, freedom of expression is regarded as a strong 'right' in some national jurisdictions, but not treated as a legal 'right' at all in others. The differences between national laws, combined with uncertainty concerning what is meant by 'legitimate interests', creates the conditions for inconsistencies in the application of the second element of the UDRP. Most of the difficulties in applying the second element derive from these uncertainties.

A distinction may be drawn between what amounts to a 'right' and what amounts to a 'legitimate interest'. A 'right' has been interpreted to mean a legal right or entitlement, whereas a 'legitimate interest' is something less than a strict legal entitlement that essentially arises from bona fide use, or preparations to use, the disputed domain name. The rationale for distinguishing between a 'right' and a 'legitimate interest' was explained by the panellist in *International E-Z UP, Inc v PNH Enterprises, Inc*[5] in the following terms:

> The intention was that, first, if the registrant had an actual legal right to the domain name, that would defeat the trademark owner's claim by itself. But, secondly, the Policy went further and added another criterion, that of legitimate interest. Those words have to be given some work to do or they serve no purpose. The intention here was to cover cases where the registrant may not have a legal right, but where it nevertheless has a bona fide association or connection of some sort with the domain name, where for example '. . . the Respondent has not registered the domain name for merely speculative reasons . . .' (*MAHA Maschinenbau Haldenwang GmbH & Co KG v Deepak*

[3] ICANN, *Second Staff Report on Implementation Documents for the Uniform Dispute Resolution Policy* (24 October 1999), ('*Second Staff Report*'), para [4.1(c)].

[4] WIPO Case No D2001-1239 (13 December 2001).

[5] NAF Case No FA808341 (15 November 2006).

Rajani, D2000-1816 (WIPO March 2, 2001), but has a 'real interest in the domain name' and is not making 'an abusive domain name registration . . .' (*Madonna Ciccone, p/k/a Madonna v Dan Parisi and "Madonna.com"*, D2000-0847 (WIPO Oct. 12, 2000).

In determining what amounts to a 'legitimate interest', it is important to bear in mind that the policy refers to a legitimate interest 'in respect of' the domain name. It necessarily follows from this that a respondent does not necessarily have a 'legitimate interest' merely because it is engaged in a 'legitimate business' undertaking. As the panellist explained in *International Data Group, Inc v K McGovern:*[6]

> the proper enquiry on this issue is not whether a respondent is conducting a legitimate business, but whether he has a right or legitimate interest in the domain name itself; thus, a respondent might have a legitimate business and yet be obliged to conduct it under a different name from the complainant's trademark. For that reason, it has long been held that the mere fact that a respondent has used the domain name to set up a business is not sufficient to show that it has a right or legitimate interest in the domain name.

Adopting a purposive approach to interpreting the UDRP, the scope of the second element should be interpreted in light of the objectives of the policy. As the UDRP is aimed at preventing 'abusive, bad faith' registration and use of a domain name, then a respondent who engages in such practices in relation to a domain name will be unable to establish that it has 'legitimate interests' in the domain name. There is therefore a link between the second and third elements of the UDRP. Given the difficulties in determining what is meant by a 'legitimate interest', a useful approach is to consider whether the respondent's activities disentitle it from claiming that an 'interest' is 'legitimate'. For example, registration of a domain name with the intention of trading off the reputation of the complainant's mark by attracting users to the respondent's web site, or diverting users to the sites of the complainant's competitors, cannot amount to a legitimate interest in the domain name. In these circumstances, the respondent may clearly have a commercial 'interest' in the domain name, but that interest will not be 'legitimate'. Ultimately, however, the difficulties experienced in determining the scope of the second element of the UDRP hinge on the fundamental differences in approach to interpreting the UDRP dealt with at [3.15].

While the UDRP does not attempt the impossible task of exhaustively defining what is meant by the terms 'rights' or 'legitimate interests', paragraph 4(c) of the policy provides considerable guidance by listing three non-exclusive circumstances of what may amount to a 'right' or 'legitimate interest'. The first recourse of many decisions concerning 'rights or legitimate interests' is often to examine the applicability of the circumstances set out in paragraph 4(c). These three circumstances, which can be thought of as affirmative defences, are dealt with at [6.3] immediately below.

Affirmative Defences

[6.3] Affirmative Defences in Paragraph 4(c)

As explained at [3.5], the August 1999 ICANN board resolution required that a number of factors that had not been identified in the WIPO Final Report be considered in determining

[6] WIPO Case No D2005-0960 (31 Oct 2005).

whether there was abusive, bad faith registration. Two of the additional factors were: whether the domain name holder was making a legitimate non-commercial or fair use of the mark; and whether the domain name holder was commonly known by the domain name. These factors were incorporated into paragraph 4(c) of the UDRP, which identifies three circumstances in which a domain name holder has 'rights' or 'legitimate interests' in a domain name.

In general terms, paragraph 4(c) operates in the following manner. The second element of abusive, bad faith registration requires a complainant to establish that the domain name holder has no rights or legitimate interests in respect of the domain name, obviously begging the question of what amounts to a 'right or legitimate interest'. Paragraph 4(c) provides some guidance by setting out three circumstances in which a registrant will have 'rights' or 'legitimate interests' in the disputed domain name. At the same time, it should be borne in mind that, as paragraph 4(c) is expressed to be 'without limitation', the three circumstances are non-exhaustive. As the ICANN Second Staff Report of 24 October 1999 expressed it:[7]

> the three circumstances described in paragraph 4(c) are not intended to constitute an exhaustive list of legitimate registration and usage of domain names. Even if none of the three listed circumstances is present, the administrative procedure would still not apply to a dispute where the domain-name holder can show that its activities are otherwise legitimate.

As explained in more detail at [6.4], for all practical purposes, the three circumstances, which are sometimes referred to as 'safe harbours', effectively act as affirmative defences.

The policy background to the three circumstances in which a domain name registrant is taken to have rights or legitimate interests in the disputed domain name is dealt with at [3.8]. The three affirmative defences are:

> (i) before any notice to the domain name holder of the dispute, use of, or demonstrable preparations to use, the domain name or a name corresponding to the domain name in connection with a bona fide offering of goods or services.

As explained at [3.8], the first affirmative defence set out in paragraph 4(c) was not one of the additional factors identified by the ICANN board in August 1999. The circumstances that fall within this defence and the issues that have arisen in interpreting the defence are explained at [6.7]ff.

> (ii) the domain name holder (as an individual, business, or other organization) has been commonly known by the domain name, even if he or she has acquired no trademark or service mark rights.

The second affirmative defence closely reflects the wording of the second factor identified in the August 1999 ICANN board resolution. The policy background to this defence, sometimes known as the 'pokey' clause, is set out at [3.8]. The circumstances that fall within this defence are dealt with at [6.8]ff.

> (iii) the domain name holder is making a legitimate non-commercial or fair use of the domain name, without intent for commercial gain to misleadingly divert consumers or to tarnish the trademark or service mark at issue.

The third affirmative defence closely reflects the wording of the first factor identified in the August 1999 ICANN board resolution. The policy background to this defence, and the problems posed by the awkward wording of the defence, are dealt with at [3.8]. As

[7] See *Second Staff Report*, above n 3, para [4.6].

explained, concerns relating to the interpretation of the term 'tarnishment', and the potential for this limitation on the defence to deprive legitimate criticism sites of protection, were raised during the drafting of the provision. The application of the defence and the difficulties experienced by UDRP panels in interpreting and applying the defence are dealt with at [6.9]ff.

In general terms, the objectives of the three affirmative defences set out in paragraph 4(c) can be explained as follows. First, paragraph 4(c)(i) is aimed at protecting rights or legitimate interests of the respondent arising from commercial use of the disputed domain name that are analogous to trade mark rights, in that they are related to use of the name as a source identifier. Secondly, paragraph 4(c)(ii) is aimed at protecting other rights or legitimate interests of the respondent in the domain name that arise from the name being used to identify the respondent. Thirdly, paragraph 4(c)(iii) is aimed at other non-trade mark, non-commercial rights or interests of the respondent in the domain name, including free speech rights.

Burden of Proof

[6.4] Burden of Proof

Paragraph 4(a) of the UDRP expressly provides that the complainant bears the onus of establishing each of the three elements of the UDRP. The difficulties in proving a negative proposition—such as establishing that a respondent has no rights or legitimate interests—are, however, considerable. As the respondent necessarily has more knowledge of the facts relevant to determining whether there are 'rights or legitimate interests' than does the complainant, the respondent is clearly better placed to produce evidence relevant to this issue than is the respondent. Moreover, the wording of paragraph 4(c) of the UDRP, which sets out the three circumstances in which a respondent can demonstrate 'rights or legitimate interests' in a domain name, and which is headed 'How to Demonstrate Your Rights to and Legitimate Interests in the Domain Name in Responding to a Complaint', clearly places an evidentiary burden on the respondent.

These two factors have led UDRP panels to adopt the rule that a complainant needs only to make out a prima facie case that the respondent has no rights or legitimate interests, whereupon the burden shifts to the respondent, who must produce evidence establishing rights or legitimate interests in the domain name, such as evidence of one of the three circumstances set out in paragraph 4(c). In this sense, the three circumstances set out in paragraph 4(c) operate as affirmative defences.

The approach taken to the shift in the burden of proof and the reasons given for this approach were explained by the panelist in *Croatia Airlines d d v Modern Empire Internet Ltd*,[8] in the following terms:

> Since it is difficult to prove a negative (i.e. that Respondent lacks any rights or legitimate interests in the mark)—especially where the Respondent, rather than the Complainant, would be best placed to have specific knowledge of such rights or interests—and since Paragraph 4(c) describes how a Respondent can demonstrate rights and legitimate interests, a Complainant's burden of proof on

[8] WIPO Case No D2003-0455 (21 Aug 2003).

> this element is light (See e.g.: *De Agostini SpA v Marco Cialone*, WIPO Case No. DTV2002-0005 <deagostini.tv>). Hence, Complainant must make at least a *prima facie* showing that Respondent has no rights or legitimate interests in the mark. After Complainant has met its initial burden of proof, if Respondent fails to submit a response Complainant will be deemed to have satisfied Paragraph 4 (a) ii of the Policy.

The reasons for the approach were expanded upon by the panellist in an earlier decision, *Intocast AG v. Lee Daeyoon*,[9] who explained that:

> For methodical reasons it is very hard for the Complainant to actually prove that Respondent does *not* have rights or legitimate interests in respect of the domain name, since there is no strict logical means of verifying that a fact is not given. Falsification does not equal positive proof for the *lack* of a fact. Many legal systems therefore rely on the principle *negativa non sunt probanda*. If a rule contains a negative element it is generally understood to be sufficient that the complainant, by *asserting* that the negative element is not given, provides *prima facie* evidence for this negative fact. The burden of proof then shifts to the respondent to rebut the complainant's assertion.

UDRP panels have universally adopted the view that the burden of proving that the respondent has no rights or interests in the domain name shifts to the respondent once the complainant establishes a prima facie case. The WIPO Consensus View on this issue presents a good summary:

> While the overall burden of proof rests with the complainant, panels have recognized that this could result in the often impossible task of proving a negative, requiring information that is often primarily within the knowledge of the respondent. Therefore a complainant is required to make out an initial prima facie case that the respondent lacks rights or legitimate interests. Once such prima facie case is made, respondent carries the burden of demonstrating rights or legitimate interests in the domain name. If the respondent fails to do so, a complainant is deemed to have satisfied paragraph 4(a)(ii) of the UDRP.[10]

It is important to understand that the shift in the burden of proof in establishing the second element of the UDRP does not alter the evidentiary burden in relation to the other elements of the UDRP, which as explained at [4.14], remains with the complainant

[6.5] What Amounts to a Prima Facie Case?

Although the complainant need only make out a prima facie case that the respondent has no rights or legitimate interests in the disputed domain name, the complainant must still discharge this burden. The burden placed on the complainant is, however, far from onerous, with some panels stating that the burden is 'light'. The panellist in *Educational Testing Service v Netkorea Co*,[11] for example, expressed the initial burden in the following terms:

> the burden on a complainant in respect of this element must, by necessity, be relatively light. In any event, Paragraph 4(c) of the Policy gives respondents ample opportunity to rebut any evidence tendered by complainants.

[9] WIPO Case No D2000-1467 (17 Jan 2001) (emphasis in original).

[10] WIPO, *WIPO Overview of WIPO Panel Views on Selected UDRP Questions* (23 Mar 2005), available at http://www.wipo.int/amc/en/domains/search/overview/index.html, para 2.1.

[11] WIPO Case No D2000-0087 (4 Apr 2000).

The two different formulations of the complainant's initial burden—the need to establish a prima facie case or to discharge a 'light' burden— have no practical significance. Moreover, requiring a low initial burden is consistent with the UDRP Rules. As explained by the panellist in *Levi, Ray & Shoup, Inc v Capitol Systems Corporation*:[12]

> Establishing that Respondent has no rights or legitimate interest in a domain name—proving a negative—can be difficult. Some panels have held that the burden of establishing this element is light because the evidence lies within a respondent's knowledge. Others have held that after a complainant makes a *prima facie* showing, the burden shifts to the respondent to rebut the complainant's evidence. These approaches are consistent with the general structure of Rules. Rules, Paragraph 3(b)(ix)(2), requires a complainant to show why the respondent 'should be considered' as having no rights or interests in the domain name (not why the respondent *has* no rights or interests in the name), while Rules, Paragraph 5(b), requires the respondent to '[r]espond specifically to the statements and *allegation*' of the complaint.

It is possible to derive a number of principles from UDRP decisions about what will amount to a prima facie case, which are best illustrated by examples drawn from the decisions. To begin with, a mere allegation that a respondent has no rights or legitimate interests in the domain name is not sufficient to establish a prima facie case. As the panellist pointed out in *Pomellato SpA v Richard Tonetti*,[13] 'simply stating the reasons is not normally sufficient to meet the Complainant's burden of proof'. This view necessarily follows from the wording of paragraph 4(a), which provides that the complainant must *prove* each of the three elements of the UDRP.[14] This requires a complainant to support its allegations with some actual evidence. As the panellist put it in *Do The Hustle, LLC v Tropic Web*,[15] the complainant must produce 'concrete evidence', meaning more than bald assertions that the elements of abusive, bad faith evidence have been made out. Moreover, some complainants have limited themselves to mere allegations that none of the circumstances set out in paragraph 4(c) have been made out. Complainants would be well advised to produce more evidence than mere assertions drawn from the wording of the UDRP.[16]

In *Stream International Inc v dotPartners LLC*,[17] for example, the complainant claimed that the respondent had no legitimate interest in the domain name <stream.biz>, alleging that a search of databases and of the Internet indicated that the respondent was not commonly known by the name 'STREAM', and that a further search indicated that it had not applied to register 'STREAM' as a trade mark. The panellist held that, as the complainant had merely made allegations and produced no evidence to support the allegations, it had failed to satisfy its initial evidentiary burden. Moreover, as the respondent provided evidence of an application for the mark 'MAINSTREAM TV' in connection with an Internet broadcasting business, the panellist held that even if the burden had shifted, the respondent had established a prima facie case of legitimate interests in the domain name.

A similar outcome was reached in *GA Modefine S A v Sparco P/L*,[18] where the complainant simply made the following allegations:

[12] WIPO Case No D2005-0700 (31 Aug 2005) (emphasis in original).

[13] WIPO Case No D2000-0493 (7 July, 2000). See also *EFG Bank European Financial Group SA v Jacob Foundation*, WIPO Case No D2000-0036 (22 Mar, 2000).

[14] See, eg, *Janine Turner v Mercedita Kyamko*, WIPO Case No D2004-1036 (27 Jan 2005).

[15] WIPO Case No D2000-0624 (21 Aug 2000).

[16] See, eg, comments made in *Commerce LLC v Charles M Hatcher*, NAF Case No FA105749 (9 Apr 2002).

[17] NAF Case No FA112428 (12 July 2002).

[18] WIPO Case No D2000-0419 (13 Aug 2000).

> Sparco P/L certainly knew at the moment of the registration of the Domain Name in issue the importance and the notoriety of all the ARMANI trademarks, the company certainly registered the Domain Name in bad faith to the purpose of taking advantage from an unlawful use of the Modefine world famous trademarks.

In deciding that the complainant had failed to make out a prima facie case the panellist pointed out that it was not sufficient simply to allege that the respondent knew of the complainant's mark.

The principle that mere allegations are insufficient is subject to the qualification that, in certain circumstances, the amount of evidence required to establish a prima facie case is minimal. First, as explained further below, if the complainant alleges circumstances surrounding the registration and use of a disputed domain name that could not result in the respondent having rights or interests in the domain name, and the respondent fails to rebut the allegations, then panels will readily find that a prima facie case has been made out. In such circumstances, panellists may be able to verify the complainant's allegations by visiting the web site associated with the disputed domain name. At the same time, complainants should provide some evidence of the circumstances alleged, such as shots taken from a web site. Secondly, in practice, the failure of a respondent to respond to a complaint, while not removing the obligation on the respondent to produce evidence will, in certain circumstances, influence a panel to find that a prima facie case has been made out. While a complainant must support any assertions with evidence before the burden shifts to the respondent, as explained at [4.16], once some evidence has been adduced, pursuant to paragraph 14(b) of the UDRP Rules, failure of a respondent to respond enables a panel to draw negative inferences. The combination of evidence of bad faith registration or use of a domain name and the negative inferences that may be drawn from a failure of a respondent to rebut a complainant's allegations will often lead to the conclusion that the respondent has no rights or interests.

Certain allegations, even if accompanied by evidence, are not capable of establishing a prima facie case. First, merely alleging that a respondent has not been authorised or licensed to use the complainant's trade mark is not, without more, sufficient to show an absence of rights or interests. In *VeriSign Inc v VeneSign C A*,[19] the complainant, the owner of the 'VERISIGN' mark, alleged that the respondent had no rights or legitimate interests in the <venesign.com> domain name, mainly on the basis that the respondent was not a licensee of the complainant and was not otherwise authorised to use the mark. In undefended proceedings, the complainant simply stated that it had no information about whether the respondent's name, 'VeneSign C A', was a valid company name. From this, the panellist concluded that the complainant had failed to make out a prima facie case that the respondent's company name did not confer rights or interests in the corresponding domain name. Consequently, although the complainant was able to establish confusing similarity and bad faith, the complaint was dismissed.

Secondly, merely showing that the disputed domain name resolves to an inactive Web site is, without more, insufficient for the complainant to make out a prima facie case. Certainly this may go some way to establishing that there has been no use of the domain name under paragraph 4(c)(i) but this, in itself, is not enough to shift the burden of proof. For example, in *Love Your Neighbor Corp v EMC a/k/a Dennis Cail*[20] the complainant, who had rights in

[19] WIPO Case No D2000-0303 (5 July 2001).
[20] NAF Case No FA97345 (12 July 2002).

the 'LOVE THY NEIGHBOR' mark, claimed that the respondent had no rights or interests in the <loveyourneighbor.net> domain name, simply asserting that there was no 'website page up at this time'. The panellist held that this was not enough to establish a prima facie case that the respondent had no rights or interests.

The circumstances surrounding the registration and use of the disputed domain name are relevant to, and well may be determinative of, whether there is a prima facie case of no rights or interests. Where the circumstances indicate that it is unlikely that the respondent could have rights or interests in the domain name, the burden will shift to the respondent to establish that it has rights or interests. These circumstances can commonly be independently verified by the panellist visiting the web site concerned. The following circumstances will commonly lead panels to conclude that a prima facie case has been established.

- Evidence that the web site associated with the disputed domain name is being used to redirect Internet users to commercial web sites that are not related to the complainant is usually sufficient to establish a prima facie case. In such circumstances, it is difficult to envisage how a respondent could establish rights or interests in the domain name under paragraph 4(c), and seemingly impossible for the respondent to establish that it is making a legitimate non-commercial use under paragraph 4(c)(iii). Accordingly, if the web site is used for these purposes, the burden will be readily shifted to the respondent.

In *Janine Turner v Mercedita Kyamko*,[21] for example, the complainant, an American actress, claimed that the respondent had no interests in the <janineturner.com> domain name. The complainant further alleged that the web site associated with the disputed domain name directed Internet users to a pornographic site, and advertised the domain name as 'for sale'. After independently verifying these claims, and drawing negative inferences from the respondent's default, the panellist held that the complainant had discharged its evidentiary burden, pointing out that using a domain name merely to attract and redirect traffic is insufficient to amount to a legitimate interest.

Similarly, in *Sports Holdings, Inc v Whois ID Theft Protection*[22] the disputed domain name <hibbitts.com> was used for a site that included a commercial search engine and direct links to sites run by the complainant's competitors. In undefended proceedings, the panellist pointed out that using the domain for these purposes could not constitute a legitimate non-commercial or fair use under paragraph 4(c)(iii) of the UDRP. As the other two circumstances set out in paragraph 4(c) were not relevant, the panellist concluded that the complainant had established a prima facie case.

- *A fortiori*, if it is clear that the disputed domain name is an example of 'typosquatting', then evidence that the associated web site is being used to divert Internet users to web sites that are not related to the complainant is usually sufficient for a prima facie case to be made out.

In *Matrix Group Limited, Inc v LaPorte Holdings, Inc*,[23] for example, the complainant, who owned rights in the 'WRESTLING ONE' mark and sold wrestling equipment, asserted that the respondent had no legitimate interest in the <wrestlerone.com> domain name on the basis that the respondent had never been known by the name 'WRESTERONE' and that there was no evidence of legitimate use of the domain name. The evidence indicated that

[21] WIPO Case No D2004-1036 (27 Jan 2005).
[22] WIPO Case No D2006-1146 (14 Nov 2006).
[23] WIPO Case No D2005-0059 (7 Apr 2005).

the respondent's web site redirected Internet users to another site that contained links to web sites run by the complainant's competitors. Pointing to the respondent's default, the panellist inferred that the disputed domain name was being used to divert Internet users, and held that the complainant had established a prima facie case. In reaching this conclusion, the panellist stated:

> It would be a remarkable coincidence if the Respondent also happened to have a legitimate trade in wrestling equipment alongside its typosquatting activities; the more reasonable inference is that this is just one more instance of the Respondent's registering a domain name in which it lacks any rights or legitimate interests within the meaning of the Policy.

Similarly, in *Google Inc v Sergey Gridasov*,[24] the complainant alleged that the respondent had no interest in a number of misspelled domain names that were clear examples of 'typosquatting'. The web sites associated with the domain names contained links to web sites that offered goods and services that were not related to the complainant. Moreover, the complainant asserted that the respondent's web site attempted to install malicious software on the computers of Internet users. In undefended proceedings, the panellist held that the respondent had no rights or legitimate interests, pointing out that 'the use of domain names that are confusingly similar to Complainant's mark to attract Internet users to a website that may annoy or harm the users in some way, and that appears to serve no legitimate function, is harmful to Complainant as the users may assume that Complainant has some affiliation with the harmful content'.

- Other circumstances indicating that it is unlikely for a respondent to have a legitimate interest in the disputed domain name may be sufficient for a prima facie case to be made out. Such circumstances may include evidence of an unusually extended period of non-use of a domain name, evidence of improper commercial use of the domain name, or evidence that a confusingly similar domain name is being offered for sale. While, in isolation, these circumstances may not be sufficient, in the context of all of the surrounding circumstances, they may contribute to a panel finding that the burden has shifted to the respondent.

Although evidence of an inactive web site is not, in itself, sufficient for the burden to shift, in some circumstances evidence of extended inactivity may be. In *Clerical Medical Investment Group Limited v Clericalmedical.com (Clerical & Medical Services Agency)*,[25] for example, the complainant claimed that the respondent had no legitimate interest in the <clericalmedical.com> domain name mainly on the basis that there was no evidence of any use or preparation for use of the domain name in the two years since it had been registered. Moreover, the complainant asserted that a search of the respondent's name produced no evidence of a legitimate business being conducted under that name. The panel held that, in the circumstances where the respondent had no obvious connection with the disputed domain name and the domain name had been inactive for a significant period, the mere assertion that the respondent had no rights or legitimate interests was sufficient to establish a prima facie case.

Evidence that a respondent is making a commercial use of a web site associated with a disputed domain name is obviously not, in itself, sufficient for the burden to shift. At the same time, evidence of an improper commercial use of the web site, such as using it to pass off the

[24] NAF Case No FA474816 (5 July 2005).

[25] WIPO Case No D2000-1228 (28 Nov 2000).

respondent's goods as those of the complainant, or to offer 'counterfeit' goods, may be sufficient to establish a prima facie case. In *Wellquest International, Inc v Nicholas Clark*,[26] for example, the complainant alleged that the respondent's <bloussantadvanced.com> web site offered 'breast enhancement products' in competition with the complainant, and promoted 'counterfeit' herbal supplements. Although the panellist could not determine whether the respondent was selling counterfeit products from the disputed web site itself, the respondent's default allowed the panellist to infer that this was happening. As there can be no legitimate interest in the sale of 'counterfeit' products, the panellist held that the complainant had satisfied the burden of proof.

Similarly, in *F. Hoffmann-La Roche AG v Web Marketing Limited*,[27] the disputed domain name, <valium-net.com>, resolved to a web site that offered something described as 'generic valium' and contained a link to complainant's web site. The panellist held that, in the circumstances, it was impossible for the respondent's use of the domain name to be bona fide. Given that the respondent was offering a product using the complainant's 'VALIUM' trade mark, it was clear that the respondent was making an illegitimate commercial use of the domain name misleadingly to offer a product in competition with the complainant.

In all the surrounding circumstances, use of the web site to which the disputed domain name resolves to advertise that a non-generic domain name is 'for sale' may contribute to the establishment of a prima facie case that the respondent had no interest in the domain name. In *Compañía Anónima Cigarrera Bigott Sucesores v Wang Yi-Chi*,[28] for example, the disputed domain name, <bigott.com>, included a notice on the web site to which the domain name resolved advertising the domain name for sale. In undefended proceedings, the panellist held that there was no basis for believing that the respondent was known by the domain name, that the respondent was making a legitimate non-commercial use of the domain name, or that he was using it in good faith for the sale of goods or services. Similarly, in *Gorilla Trades, Inc v Clark Cochran*,[29] when the panellist reviewed the disputed domain names, four of the web sites advertised the domain names for sale. Given the respondent's default, the panellist held that the complainant had not authorised the respondent to use its mark, that there was nothing to indicate that the respondent made any use of the domain names other than attempting to sell them and there was no evidence of use of the domain names for criticism of the complainant. In the circumstances, the panellist was able to conclude that the respondent had no rights or interests.

If the circumstances indicate that the respondent is making a legitimate use of the disputed domain name, however, then even in undefended proceedings the complainant will not have made out a prima facie case. In *Leo Quinones, Jr v William Chambers*,[30] for example, the complainant alleged that the respondent had no rights or legitimate interests largely on the basis that the disputed domain name <filmfreakcentral.net> was registered one and a half years after the complainant commenced using the mark 'THE FILM FREAK' and after the complainant had registered the <filmfreak.com> domain name. The complainant presented copies of the respondent's web site, showing use of the 'FILM FREAK' mark, and the respondent defaulted. The panellist held that, taking into account the generic nature of the mark, the complainant had failed to make out a prima facie case that the respondent had no

[26] WIPO Case No D2005-0552 (19 July 2005).
[27] WIPO Case No D2006-0005 (14 Mar 2006).
[28] WIPO Case No D2003-0252 (28 May 2000).
[29] WIPO Case No D2006-0530 (12 June 2006).
[30] WIPO Case No D2004-0461 (17 Aug 2004).

rights or interests. On the contrary, the panellist considered that the material submitted by the complainant, which showed use of the respondent's site for film reviews, went some way to establishing a prima facie case that the respondent had legitimate interests in the disputed name. This decision also illustrates the principle that merely alleging that the complainant is aware of the complainant's mark is not sufficient to make out a prima facie case.

Although there is no obligation on a complainant to present evidence of a trade mark search, in some circumstances a trade mark search can assist in establishing a prima facie case. In *Educational Testing Service v. Netkorea Co*,[31] for example, the complainant trade mark owner, based in the United States, conducted a trade mark search of the trade mark office in the respondent's home jurisdiction, which indicated that the respondent had not registered or applied to register the 'TOEIC' trade mark. Together with evidence indicating that the respondent intended to use the domain name to mislead consumers for commercial gain, this led the panellist to conclude that the respondent had no rights or interests in the domain name.

Similarly, in *Klein-Becker IP Holdings, LLC v Symblic Systems Pty Ltd*,[32] the complainant asserted that, as the respondent was not an authorised dealer, the disputed domain name <strivectin-explained.com> was being used illegitimately to sell the complainant's 'STRIVECTIN' product and advertise the products of competitors. To support its case, the complainant presented evidence that a search of Australian trade marks indicated that the respondent had no registrations or applications to register marks that included the term 'STRIVECTIN'. The panellist took this into account in finding that the complainant was not making a legitimate use of the domain name and, therefore, had no interests in the name.

If the complainant is not able to establish a prima facie case there is a possibility that a finding of reverse domain name hijacking could be made out. Reverse domain name hijacking is dealt with at [4.17].

[6.6] The Respondent's Burden of Proof

After the complainant has established a prima facie case, the burden shifts to the respondent to rebut the case by producing evidence that it has rights or legitimate interests. If the respondent fails to respond, the panel may draw negative inferences from the respondent's default. The respondent may rebut the case made by the complainant by producing evidence that one of the circumstances set out in paragraph 4(c) of the UDRP applies. Once the burden has shifted, the respondent must establish that it has rights or interests in the domain name to the same standard as applies to all elements of the UDRP, that is, on the balance of probabilities.

In *Document Technologies, Inc v International Electronic Communications Inc*,[33] the panel stated that the respondent must rebut the initial case by providing 'concrete evidence' of rights or legitimate interests. In *Do The Hustle, LLC v Tropic Web*,[34] the panellist observed that 'concrete evidence' requires more than 'mere personal assertions'. A further gloss was provided on the kind of evidence required by the panellist in *ARAG Allgemeine Rechtsschutz-Versicherungs-AG v Seung Nam Kim*,[35] who observed that:

[31] WIPO Case No D2000-0087 (4 Apr 2000).
[32] WIPO Case No D2006-0498 (14 June 2006).
[33] WIPO Case No D2000-0270 (6 June 2000).
[34] WIPO Case No D2000-0624 (21 Aug 2000).
[35] WIPO Case No D2006-1001 (29 Sept 2006).

> assertions are clearly a form of evidence. It is obvious to observe that assertions, in the form of written or oral testimony, routinely constitute evidence in domestic legal proceedings. The difference under the Policy is that such evidence is not subject to cross examination.

These comments are useful in that they reflect the extent to which it is necessary for evidentiary requirements to be adapted to the expedited nature of UDRP proceedings.

The evidentiary burden faced by respondents is best illustrated by representative examples drawn from UDRP decisions. The following are examples of decisions in which the respondent failed to satisfy its evidentiary burden.

- *Pivotal Corporation v Discovery Street Trading Co Ltd.*[36] The complainant alleged that it had not authorised or licensed the respondent to use the 'PIVOTAL' mark, that the respondent could not have been unaware of the complainant's mark and that the respondent had registered the <pivotalsoftware.com> domain name for the purpose of selling it to the complainant. Taking into account evidence that the domain name was not being used and that the respondent had registered a number of other domain names, the panellist held that there was a prima facie case that the respondent had no interest in the domain name. In reply, the respondent argued that it had acquired the domain name in good faith through proper legal channels, the domain name was registered for an unspecified future use, the domain name had not been used largely because of the actions of the complainant and that the complainant had failed to make out a prima facie case. The panellist held that these assertions failed to rebut the complainant's case. Even if substantiated, the factors identified by the respondent could not establish an interest in the domain name.
- *Tribeca Film Center, Inc v Lorenzo Brusasco-Mackenzie.*[37] The complainant trade mark owner claimed that the respondent had no rights or interests in the <tribecafilmcenter.com> domain name on the following grounds: the respondent had not used, and had not made plans to use, the domain name in connection with the offering of goods or services; the respondent had never been known by the domain name; and the respondent's proposal to use the domain name in connection with a 'fan' web site was not a legitimate non-commercial use. The panellist evidently considered the allegations sufficient to constitute a prima facie case. In response, the respondent presented evidence of an unpublished home page to support the claim that he intended to make a legitimate non-commercial use of the domain name for a 'fan' site. The panellist concluded that the respondent had failed to satisfy its evidentiary burden, giving a number of reasons. First, the panellist pointed out that mere assertions of intent are not enough to establish a legitimate, non-commercial use is being made of the domain name. Secondly, as the home page had never been published, there had been no use of the domain name. Thirdly, as the home page looked confusingly like the trade mark owner's home page, rather than a criticism or 'fan' site, it would not be a legitimate use of the domain name even if it were to be published.
- *International Data Group, Inc v K McGovern.*[38] The complainant alleged that the respondent had no interests in the <pc-world-online.com> domain name on the basis that it used the domain name to allow consumers to purchase the complainant's 'PC WORLD' magazines without permission and to advertise the goods and services of complainant's competitors. The panellist held that the allegations were sufficient to establish a prima facie case. In reply, the respondent contended that the domain name and the associated

[36] WIPO Case No D2000-0648 (14 Aug 2000).
[37] WIPO Case No D2000-1772 (10 Apr 2001).
[38] WIPO Case No D2005-0960 (31 Oct 2005).

web site were part of a legitimate online publishing business. The panellist correctly pointed out that the issue was whether the respondent had a right or legitimate interest in the domain name, not whether he was using the disputed domain name to conduct a legitimate business. The substantive issues in the dispute concerned rival assertions made by the complainant and the respondent. On the one hand, the complainant argued that the respondent was using the confusingly similar domain name to mislead consumers to believe there was a connection with the respondent and thereby illegitimately attract consumers to its web site. On the other, the respondent claimed that the complainant's mark was not well known in the respondent's home jurisdiction, the Republic of Ireland, and that it had not been aware of the complainant's trade mark rights. The panellist concluded that there was no bona fide use of the domain name as, on the balance of probabilities, the respondent knew of the complainant's mark when the domain name was registered and also that there was a likelihood of consumers being confused. In this respect, the evidence indicated that the complainant's magazines were sold in the Republic of Ireland. Consequently, the respondent's assertions had failed to rebut the prima facie case made by the complainant.

The following are representative examples of decisions in which the respondent produced sufficient evidence to rebut a prima facie case.

- *Deutsche Welle v DiamondWare Limited.*[39] The complainant, a German broadcasting company, sent a letter to the American respondent demanding transfer of the <dw.com> domain name, to which the respondent replied with an offer to sell the domain name. The complainant contended that this reply was evidence that the respondent had no rights or legitimate interests in the domain name. In reply, the respondent pointed out that it had been using the terms 'DiamondWare' and 'DW' in connection with its software since early 1994 and that it had continuously used the domain name in connection with the bona fide offering of goods and services since its registration at that time. The panel pointed out that the domain name corresponded to the initials of the respondent's trade name and accepted that the domain name was registered prior to the respondent becoming aware of the complainant. As the evidence indicated that the domain name was being used in connection with a bona fide offering of goods or services, the respondent had established that it had rights or legitimate interests in the domain name, and the complainant had failed to satisfy its ultimate burden that the respondent had no rights or interests.
- *Commerce LLC v Charles M Hatcher.*[40] The complainant alleged that none of the circumstances set out in paragraph 4(c) had been established, including that the respondent had never been known by the domain name and had not used the domain name for any purpose. While pointing out that the unsupported allegations were 'sparse', the panellist held that they were sufficient to raise a prima facie case that there had been no legitimate use of the domain name. In response, the respondent claimed that it was using the domain name as the host name of its e-mail address, but provided no evidence that it was doing so. In the absence of any evidence, the panellist pointed out that it could not presume that the respondent had a legitimate interest in the domain name. Nevertheless, given the weakness of the complainant's case and the inconclusive battle between the 'conclusory allegations' of the parties, the panellist was inclined to find that the complainant had failed

[39] WIPO Case No D2000-1202 (2 Jan 2001).
[40] NAF Case No FA105749 (9 Apr 2002).

to satisfy its ultimate burden. The implication is that, where the complainant's case consists of mere assertions, assertions made by the respondent may be sufficient to rebut a prima facie case.

- *Windsor Fashions, Inc v Windsor Software Corporation.*[41] The complainant alleged that the respondent had no rights or interests in the disputed domain name <windsor.com> on the ground that it had no registered trade mark rights in the name, there had been no use of the domain name and it was not commonly known by the domain name. In response, the respondent produced evidence that it was formally incorporated under the name 'Windsor Software Corporation', entered into a contract with NASDAQ under that name, filed documents under that name with the Greater Washington Board of Trade and paid taxes to the IRS under that name. While conceding that it was possible for a registrant to concoct a false business name for the purpose of registering a domain name, in this case, given the preponderance of evidence, the panel observed that this was highly improbable. Accordingly, the panel held that the respondent had established rights and legitimate interests in the domain name, and that the complaint was an example of reverse domain name hijacking.
- *Deutsche Post AG v NJDomains.*[42] The complainant, the privatised German postal authority, alleged that the respondent had no rights or interests in the <post.com> domain name as it 'makes no sensible use of the domain name but uses it for committing internet fraud' and uses the domain name 'to harm other people and steal their money'. In reply, the respondent alleged that it had been making the domain name available to subscribers for personal e-mail addresses, and that this amounted to use of the domain name in connection with a bona fide offering of goods or services. The fraud referred to by the complainant consisted of use of the domain name to send fraudulent e-mails, which appeared to originate from a subsidiary of the complainant, designed to induce recipients to pay for non-existent goods. The respondent denied that it was involved with the fraud, and produced evidence that it had terminated the e-mail addresses once the fraud was brought to its notice. The panel concluded that being involved in providing the subscription e-mail service was sufficient to establish that the respondent had legitimate interests in the domain name. Moreover, while not making a finding of reverse domain name hijacking, the panel cautioned the complainant against making serious allegations, such as allegations of fraud, without sufficient supporting evidence.

Bona Fide Offering of Goods or Services

[6.7] Bona Fide Offering of Goods or Services

The first of the circumstances set out in paragraph 4(c), which, if established, means that the respondent has rights or legitimate interests in the disputed domain name, provides that such rights or interests arise where:

[41] WIPO Case No D2002-0839 (14 Nov 2002).
[42] WIPO Case No D2006-0001 (1 Mar 2006).

(i) before any notice to the domain name holder of the dispute, use of, or demonstrable preparations to use, the domain name or a name corresponding to the domain name in connection with a bona fide offering of goods or services.

By specifying that a respondent will have rights or legitimate interests if it is preparing to use, or using, the domain name in connection with the bona fide offering of goods or services, the policy is, in effect, protecting legitimate preparations to use, or use, of the domain name as a kind of source identifier for goods or services. In effect, this excludes from the UDRP not only disputes among competing trade mark rights in the domain name, but disputes involving potentially competing trade mark rights. The exclusion of genuine preparations to use the disputed domain name from the scope of the UDRP clearly reflects the extent to which the policy is directed against extortionate registration and use of a domain name, and not at plans to use a domain name in connection with a bona fide business. Complex issues relating to whether plans to use a domain name may infringe the rights of a trade mark owner are best left to the courts.

[6.7.1] What is Meant by 'Notice' of the Dispute?

Paragraph 4(c)(i) provides that use, or demonstrable preparations to use, the domain name must occur before the respondent has notice of the dispute. It is important to note that the use or preparations must occur before the respondent has notice of the dispute, and not before the complaint is initiated. The relevant point for determining whether there has been bona fide use or preparations to use the domain name will therefore usually be the time that the respondent receives a cease and desist letter. As the panellist expressed it in *D.C.C. Wine Exchange Inc v Giumarra Farms Inc:*[43]

> The issue of rights to and legitimate interests in the domain name must be determined, as it existed before any notice to the Respondent of the dispute. That time frame is the time from when the Respondent registered the domain name in May 1996 and when the Complainant notified Respondent of its position that use of the domain name 'winesdirect.com' infringed on its trademark.

In *Corinthians Licenciamentos LTDA v. David Sallen,*[44] the panellist considered what was meant by 'notice' and 'dispute'. In that dispute, the complainants, the exclusive licensee of a famous Brazilian soccer team, sent a cease and desist letter on 16 November 1999 in response to an e-mail from the respondent indicating that the <corinthians.com> domain name was for sale. At an indeterminate time after this, the respondent posted text from the biblical book of Corinthians to the Web site associated with the domain name. A notification of the UDRP complaint was sent to the respondent on 30 May 2000. The respondent asserted that he had been using the domain name to offer a bona fide service to the religious public prior to the UDRP complaint. Construing the text of the UDRP as a whole, the panellist held that the term 'dispute' is not the same as the term 'proceeding', and that a domain name dispute is necessarily previous to UDRP proceedings. As the biblical text was posted after the respondent received notice of the 'dispute', the panellist concluded that the respondent had not established a bona fide use of the domain name under paragraph 4(c)(i).

A similar conclusion was reached in *The Sun Chronicle v Web Solutions, Inc.*[45] In that dispute, the evidence indicated that, after receiving a cease and desist letter, the respondent

[43] NAF Case No FA95065 (28 Aug 2000).
[44] WIPO Case No D2000-0461 (17 July 2000).
[45] NAF Case No FA94895 (18 July 2000).

modified the web site associated with the <sunschronicle.com> domain name to suggest an association with 'Sun' computer workstations. As there was no evidence to suggest that any preparations had been made to use the domain name until after the cease and desist letter, the panellist held that the respondent had produced no evidence to rebut the complainant's case that the respondent had no recognised interest in the domain name.

In *Sanlam Limited v Selat Sunda Incorporated*[46] the complainant trade mark owner sent a letter to the respondent indicating that it would like to register the disputed domain name, <sanlam.com>, asking what rights the respondent had in the name and what the respondent intended to use the domain name for. The complainant lodged its complaint on 28 July 2000. In the interim, the complainant alleged that the respondent had constructed a makeshift web site solely to create the impression that it had rights in the domain name. In undefended proceedings, the panellist was required to determine the date on which the respondent received notice of the dispute, the date on which the respondent could reasonably be said to have received the complainant's letter, or the date that the UDRP complaint was notified to the respondent. The panellist held that the correct date had to be the date that the respondent received the complainant's letter, as the competing interpretation would allow the respondent to make preparations to use the domain name long after it had notice that there was a competing claim to the name. This would readily allow respondents to defeat the claims of trade mark owners. As there was no evidence of any use of the domain name before the respondent received the complainant's letter, the panellist concluded that the respondent had no rights or interests in the domain name.

It has been suggested that notice of the dispute may include constructive notice. In *Entre-Manure, LLC c/o Craig B Stern v Integriserv*,[47] the complainant applied to register the mark 'ENTRE-MANURE' on 17 August 2005, and sent a letter of demand to the respondent, who had registered the domain name <entremanure.com> on 23 February 2006. The panellist held that the respondent had made a bona fide, if limited, use of the name in connection with goods and services as early as June 2001. In the course of the decision, however, the panellist assumed that constructive notice of a dispute would be sufficient to put the respondent on notice for the purpose of paragraph 4(c)(i). Moreover, referring to federal US trade mark law, which provides that registration on the Principal Register is 'constructive notice of the registrant's claim to ownership',[48] the panellist claimed that registration of a trade mark could amount to constructive notice. As paragraph 4(c)(i) refers to 'notice of the dispute', and not notice of the mark, the reasoning of the decision on this point is clearly wrong: registration of a trade mark cannot be notice of a dispute. Despite the suggestions made in this decision, in most circumstances 'notice' under paragraph 4(c)(i) should be interpreted as actual notice of the dispute.

If, however, the respondent has constructive notice of the complainant's mark prior to receiving notice of the dispute, then the preparations to use the domain name or the actual use of the domain name may not be bona fide. This approach to the issue, which would seem to be preferable, was adopted by the panellist in *Cosmos European Travels AG v Eurotech Data Systems Hellos, Ltd.*[49] In that dispute, the respondent had registered the domain name <cosmos.com>, which it proposed to use as an e-commerce portal to offer travel services which would compete with the services offered by the complainant, which

[46] WIPO Case No D2000-0895 (10 Oct 2000).
[47] NAF Case No FA741534 (16 Aug 2006).
[48] Lanham Act § 22, 15 USC § 1072.
[49] WIPO Case No D2001-0941 (8 Oct 2001).

owned rights in the 'COSMOS' mark. The evidence indicated that, before receiving notice of the dispute, the respondent was aware that the web site associated with the domain name was attracting would-be customers of the complainant. The panellist held that this put the respondent on notice that continued use of the domain name would be contrary to the complainant's trade mark rights and, consequently, that the respondent's preparations to use the domain name were not bona fide.

[6.7.2] 'Use' of the Domain Name

Paragraph 4(c)(i) provides that 'use' of a domain name in connection with the bona fide offering of goods or services means that the respondent will have rights or legitimate interests in the domain name. This circumstance cannot be made out unless the domain name has been 'used' or there have been demonstrable preparations to 'use' the domain name prior to the respondent receiving notice of the dispute. In determining whether this circumstance has been made out it is therefore necessary to understand what is meant by 'use' of a domain name. As the 'use' required is in connection with the '*bona fide* offering of goods or services', however, UDRP panels do not usually separate out considerations relating to whether the domain name has been 'used' from considerations relating to whether there is a 'use' in connection with the '*bona fide* offering of goods or services'.

UDRP panels have held that there is no 'use', or no '*bona fide* use' of the disputed domain name in the following circumstances.

- Where the web site, to which the domain name resolves, has no more than a notice that the site is 'under construction' there is no 'use' of the domain name.

In *Sears, Roebuck and Co v Hanna Law Office*,[50] before the respondent received a cease and desist letter from the complainant, the only content on the <searsroebuck.com> web site was a message reading:

> The new website for 'www.searsroebuck.com' is currently under construction. Please check back soon!

It was only after receiving the letter that the respondent created a 'complaint' site at the domain name. The panellist held that the respondent had not used the domain name in connection with a bona fide offering of goods or services before receiving notice of the dispute and that, given the fame of the complainant's mark, could not have had a legitimate contemplated use for offering goods or services.

Similarly, in *The Hamlet Group, Inc v James Lansford*[51] the disputed domain name, <hamburgerhamlet.com>, resolved to a web site that contained only a standard form 'under construction' page. In deciding that the respondent had no rights or interests in the domain name, the panellist stated:

> The Panel does not regard this limited use as of legal significance for the purpose of deciding this proceeding. Using the domain name registration materials routinely provided by Network Solutions, the posting of an 'Under Construction' web page is effectively no more than a part of the registration process, and for the purpose of this proceeding the Panel treats it as such. Registration, standing alone, does not establish rights or legitimate interests in a domain name in the sense of paragraph 4(a)(ii) of the Policy.

[50] WIPO Case No D2000-0669 (8 Sept 2000).
[51] WIPO Case No D2000-0073 (31 Mar 2000).

In *AlohaCare v PPI*,[52] the respondent asserted that the disputed domain names, <alohacare.com>, <alohacare.net> and <alohacare.org>, were being used as 'a marketing tool to demonstrate our software prototype and business model'. The panellist held that this assertion was contradicted by the fact that the domain names were attached to 'under construction' web-pages at all relevant times. Consequently, the panellist concluded that the respondent had not produced concrete evidence sufficient to rebut the prima facie case that had been made out by the complainant.

- Where the web site, to which the domain name resolves, contains no more than links to other sites that are unrelated to a bona fide business of the respondent there is no 'use' of the domain name, no 'demonstrable preparations' to use the domain name or no use in connection with a bona fide offering of goods or services.

In *Easyjet Airline Company Ltd v Andrew Steggles*,[53] the disputed domain name, <easyjet.net>, resolved to a web site that contained links to the site of a Hudson, New York, vacuum cleaner business. The panellist held that the link to the web site did not demonstrate any serious preparations to use the domain name as the web site did not reveal any use of the domain name whatsoever. Moreover, there was not even evidence that the link was made with the permission of the third party business concerned.

In *SunFest of Palm Beach County, Inc v Electronic System Technologies, Inc*,[54] the disputed domain name <sunfest.com> was used for a web site that contained little more than links to sites associated with the respondents. The panellist held that this did not amount to serious preparations to use the domain name in connection with a bona fide offering of goods or services.

In *NetLearning, Inc v Dan Parisi*,[55] the sole use of the disputed domain name, <netlearning.com>, was to link to another web site operated by the respondent, <whitehouse.com>, which was a repository for adult entertainment links and pornographic images. The panel held that this was not a bona fide use of the domain name in connection with goods or services.

Similarly, in *Janine Turner v Mercedita Kyamko*,[56] the <janineturner.com> domain name resolved to a web site that redirected Internet users to a site containing pornographic material at <clubpink.com>. The panellist in this dispute stated that:

> in the current state of affairs, the Respondent cannot claim to 'use' the disputed Domain Name in connection with a bona fide offering of goods and services, as it is merely attempting to attract and divert online traffic to its other website and this cannot constitute a bona fide offering of goods and services.

In *Network Solutions, LLC v Click Search*,[57] the disputed domain name, <networksoluions.com>, resolved to a web site that appeared to have links to other commercial sites. The panellist held that merely having links to commercial sites cannot be viewed as a bona fide offering of goods or services.

The question of the extent to which a web site that links to other sites can amount to a bona fide offering of goods or services is dealt with further at [6.7.4] ff below.

[52] WIPO Case No D2002-1160 (12 Feb 2003).
[53] WIPO Case No D2000-0024 (17 Mar 2000).
[54] WIPO Case No D2000-0631 (3 Oct 2000).
[55] NAF Case No FA95471 (16 Oct 2000).
[56] WIPO Case No D2004-1036 (27 Jan 2005).
[57] WIPO Case No D2005-0453 (7 June 2005).

- Where the web site to which the domain name resolves contains no more than an offer to sell the domain name there is no 'use' of the domain name, no 'demonstrable preparations' to use the domain name or no use in connection with a bona fide offering of goods or services.

In *Caterpillar Inc v Roam the Planet, Ltd*,[58] the disputed domain name, <catmachines.com>, resolved to a Web site that announced that the domain name was for sale and that contained a link to the respondent's site that offered a range of domain names for sale. The respondent admitted that it acted as a 'domain name broker'. The panellist held that the complainant's 'activity is not and has not been directed to the use of the domain name for the offering of goods or services, other than the domain name itself'. Finding that this did not fall within paragraph 4(c)(i), the panellist concluded that the respondent had no rights or interests in the domain name.

In *No Slippy Hair Clippy v Damage Sports, Inc*,[59] the Web site at <noslippyhairclippy.com> resolved to a banner which said, 'interested in buying this domain? Inquire at: [email address]'. The respondent offered no goods or services from the disputed domain name, but sold sporting goods from another, unrelated domain name. The panellist held that the respondent had failed to rebut the prima facie case made out by the complainant.

In *Gorilla Trades, Inc v Clark Cochran*,[60] the respondent, an apparently disgruntled customer of the complainant, registered a number of domain names that were confusingly similar to the complainant's mark. Although the respondent suggested that the domain names might be used to criticise the complainant, the only content on each of the web sites was a notice that '[t]his domain name is for sale'. In concluding that the respondent had no rights or interests, the panellist observed that there was 'nothing in the record to indicate Respondent made any use at all of any of the disputed domain names other than attempting to sell all but one of them'.

The rule that offering a domain name for sale from a web site is not a 'use' of the domain name must be distinguished from the use of the domain name for redirecting users to the site of a generic domain name reseller. As explained at [6.11], the majority of panels have held that the registration of a generic domain name as part of the business of a generic domain name broker or reseller amounts to use of the domain name in connection with a bona fide offering of goods or services.

- Where the web site is 'warehoused', with only an intention to use the domain name, then there may be no 'use' of the domain name.

A completely inactive web site does not amount to a 'use' of the domain name. In *The J Paul Getty Trust v Domain 4 Sale & Company*,[61] for example, the disputed domain names <gettymuseum.com> and <gettysmuseum.com> were inactive. The panellist held that the respondent had made no use of the domain names and, in the absence of a response, concluded that the domain names had been registered with the intention of selling them for a profit.

Where there is no actual use of the disputed domain name, the question under paragraph 4(c)(i) will usually be whether the respondent can establish that it has made 'demonstrable

[58] WIPO Case No D2000-0631 (25 May 2000).
[59] WIPO Case No D2004-0757 (26 Oct 2004).
[60] WIPO Case No D2006-0530 (12 June 2006).
[61] NAF Case No FA95262 (7 Sept 2000).

preparations' to use the domain name. As explained below, mere assertion of an intention to use the domain name is not sufficient to constitute 'demonstrable preparations'.[62]

[6.7.3] 'Demonstrable Preparations' to Use a Domain Name and the Status of Business Plans

Paragraph 4(c)(i) of the UDRP provides that a legitimate interest in a domain name may be established if, before notice of the dispute, the respondent made 'demonstrable preparations' to use the domain name in connection with a bona fide offering of goods or services. The nature of this provision means that whether or not it is made out will often turn on the credibility of the evidence presented by the respondent. As the panellist pointed out in *Trygg-Hansa AB v Arild Rosenberg*,[63] 'it is sometimes easy to offer *post facto* [*sic*] explanations for certain actions or inactions in a given context'.

To begin with, once the complainant has established a prima facie case, the respondent must produce some credible evidence of actual preparations to use the domain. Mere uncorroborated assertions of an intention to use the domain name will not amount to demonstrable preparations to use the name. For example, in *World Wrestling Federation Entertainment, Inc v Ringside Collectibles*,[64] the respondent alleged that it intended to use the domain names <wwfauction.com> and <wwfauction.net> in connection with an auction site for World Wrestling Federation action figures. In finding that the respondent had failed to make out its case, the panel stated:

> Respondents seeking to show preparations to make a legitimate use must give Panels some evidence; acceptable evidence may include business plans or documented expenses, but will of course vary with the nature of the use and the particulars of the domain name. . . . the failure to present any credible evidence of demonstrable preparations to offer auction services is fatal to Respondent's claim of a legitimate interest in the domain names.

Similarly, in *Del Monte Corporation v. David Crumpacker*,[65] the respondent claimed that he had registered the domain name <orchardselect.com> in connection with a business plan to sell fresh fruit over the Internet. He further claimed to have consulted with experts in the area of Internet sales, and to have arranged with orchardists to sell fruit from the proposed web site. The respondent, however, failed to provide any documentary support for his allegations. In the absence of documentary proof, the panellist found that the claims made by the respondent were not credible and that, consequently, there were no demonstrable preparations to use the domain name in connection with a bona fide offering of goods or services.

In *LIBRO AG v NA Global Link Limited*,[66] the respondent argued that it intended to use the domain name <libro.com> in connection with a virtual book shop. It further argued that the word 'LIBRO', which means 'books' in Spanish and Italian, is a generic term. The panellist, however, pointed out that there was no evidence of any preparations to use the domain name, it being used solely to redirect users to a Web site at <restaurant.com>. Accordingly, the panellist concluded that the 'mere speculative idea for a bona fide business application of a generic domain name' will not confer rights or legitimate interests.

[62] See also *Document Technologies, Inc v International Electronic Communications Inc*, WIPO Case No D2000-0270 (6 June 2000).

[63] WIPO Case No D2003-0171 (11 Apr 2003).

[64] WIPO Case No D2000-1306 (24 Jan 2001).

[65] WIPO Case No D2000-0498 (21 Aug 2000).

[66] WIPO Case No D2000-0186 (16 May 2000).

In *Société Nouvelle du Journal L'Humanité v Cyril Havas,*[67] the respondent contended that, although a web site had not been established for the <pifgadget.net> domain name, he had undertaken 'a lot of preparation work', and had a business plan to advertise and sell gadgets online. As the respondent failed to submit any evidence of the alleged preparation work, however, the panellist concluded that he had no rights or interests in the domain name. Similarly, in *GoPets Ltd v Edward Hise,*[68] the respondent stated that it had 'big plans' for the <gopets.com> web site, and referred to a 'full report' allegedly written in 1999 in connection with the plans. The respondent did not, however, produce a copy of the report. The panellist was 'unconvinced' by the reasons given by the respondent for registering the domain name, commenting that the 'failure by the Respondent to develop a website using the Disputed Domain for over seven years and the fact that the Respondent does not provide copies of the supposed event in which the alleged plans were said to be recorded speaks volumes'. As the complainant failed to establish bad faith registration, however, the panellist did not need to come to a conclusion about whether the respondent had a legitimate interest in the domain name.

The 'demonstrable preparations' must be preparations to 'use' the disputed domain name, and not simply preparations to establish a business. In *Medisite SA RL v Intellisolve Limited,*[69] for example, the respondent asserted that it was in the early stages of planning a business, and that, after carrying out market research, business planning and financial statistics and contacting an accountant, it had decided to reserve the <medisite.com> domain name. Although the panellist was prepared to accept that the respondent had been considering an online business connected with medical or pharmaceutical services, there was no evidence that the particular domain name was to be used in connection with such a business. On balance, the panellist therefore held that there were no demonstrable preparations to use the particular domain name in connection with the bona fide offering of services.

At the same time, the 'demonstrable preparations' need not include actual 'use' of the domain name. In *IKB Deutsche Industriebank AG v Bob Larkin,*[70] for example, the respondent registered the disputed domain name <ikb.com> in 1996, but there was little evidence that the domain name had been used by the time the complaint was lodged in 2002. The respondent presented evidence of preparations to use the domain name, including evidence of the formation of a company, 'IKB Investments Limited' in 1997, evidence of the development of IKB ('information knowledge base') software from a third party investor and the respondent's diary entries, evidence of limited evaluation trials conducted on the <ikb.com> Web site and a design logo prepared for use with the software. The respondent also presented evidence explaining the delay in using the domain name, including difficulties in obtaining venture capital after the dot com bust and unexpected difficulties in using Java software. The panel took the investor's evidence and the respondent's diary entry into account in finding that, despite the long period of non-use, the respondent had made demonstrable preparations to use the domain name.

A genuine business plan that shows the respondent is preparing to use the domain name may be sufficient evidence of demonstrable preparations to use the domain name in connection with the bona fide offering of goods or services. In *Physik Instrumente GmbH & Co*

[67] WIPO Case No D2006-0018 (6 Mar 2006).
[68] WIPO Case No D2006-0636 (26 July 2006).
[69] WIPO Case No D2000-0179 (19 May 2000).
[70] WIPO Case No D2002-0420 (23 July 2002).

v Stefan Kerner,[71] for example, the respondent, a new media and Internet consultancy, developed a proposal for an 'office support' Internet portal using the domain name <pi.com>. The proposal was evidenced by a document stored on a computer. The complainant contended that the respondent might have fabricated the computer records, but submitted no evidence to support this claim. The panellist, however, concluded that the evidence of the business plan, together with evidence that the respondents had paid a substantial amount for the disputed domain name in the month following their commission of a business plan, was sufficient to establish that the respondents had made demonstrable preparations. The panellist therefore held that the complainant had failed to establish that the respondent had no legitimate interest in the domain name, even though the respondent subsequently decided not to pursue the business plan.

Sometimes relatively minimal evidence of a business plan is sufficient. In *Smart Design LLC v Carolyn Hughes*,[72] the respondent, who had registered the domain name <smartdesign.com>, did not produce a business plan, but claimed that the domain name was to be used as a 'thinkubator' to research and develop patentable inventions to be used in business. In support of this, the respondent produced a letter from a patent attorney indicating that the respondent, trading as 'Smart Design', had requested assistance with a patent application for an e-commerce business method. Although the respondent had not used the domain name at all, and there was little evidence of business activity, the panellist held that the respondent had rights or interests in the domain name.

The more detailed and extensive a business plan is, the more likely it is that it will be sufficient evidence of demonstrable preparations to use the domain name. In *The Jolt Company v Digital Milk, Inc*,[73] for example, the respondent, who was the registrant of the disputed domain names <jolt.com> and <jolt.net>, presented a range of evidence of preparations to use the domain names. The evidence indicated that the respondent: had hired consultants to conduct a search regarding use of the 'JOLT' name in connection with online entertainment; acquired the domain names following the search; prepared a detailed business plan, identifying a management team and including financial projections; and had changed the nature of the proposed business from online entertainment to online game shows. Rejecting the complainant's contention that the respondent had no rights because it had not used the domain name for approximately two years, the panel held that the cumulative weight of the evidence was sufficient to amount to demonstrable preparations to use the domain name in connection with goods or services. Similarly, in *Saltworks, Inc v Tom Graham*,[74] the respondent claimed that he had registered the domain name <saltworks.com> for use in connection with a business of producing salt and selling luxury bath salts. In support of this contention the respondent produced a business plan presented to the government of Anguilla and evidence of a feasibility study for a saltworks conducted by a university botanist. The panellist held that this established that the respondent had made demonstrable preparations to use the domain name.

The evidence of demonstrable preparations to use the domain name, such as a business plan, must be evidence of genuine preparations, and not merely fabrications. In particular, in assessing whether a 'business plan' is evidence of genuine preparations to use the domain name, UDRP panels will examine the plan to determine whether it is a genuine plan or a

[71] WIPO Case No D2000-1001 (3 Oct 2000).
[72] WIPO Case No D2000-0993 (18 Oct 2000).
[73] WIPO Case No D2001-0493 (1 Aug 2001).
[74] WIPO Case No D2005-0874 (5 Oct 2005).

mere 'sham'. In *Pharmacia & Upjohn Ab v Gabriel Sipa,*[75] for example, in response to a procedural order from the panellist, the respondent produced an expurgated version of a purported business plan for use of the <i-pharmacia.com> domain name in connection with an Internet pharmacy. After examining the plan, however, the panellist concluded that it was a 'sham', stating:

> The document appears to have been produced without any serious attention to detail and provides no useful information of any kind. It is undated and contains nothing to suggest that the Respondent had any serious intention of using the Domain Name for the purpose given.

These factors, taken together with the fact that it was clear that the domain name was for sale, led the panellist to conclude that the respondent had no rights or legitimate interests in the domain name.

In *Deutsche Telekom AG v Kerstin Ice,*[76] the respondent alleged that the domain name, <tele-kom.com>, was an abbreviation of 'telepathic komponents', and that it was intended to be used to market 'telepathic music'. Although the respondent produced a business plan, the panellist concluded that it had been fabricated for the purpose of the dispute. In reaching this conclusion, the panellist made the following comments:

> Although the so-called business plan is addressed to a banking institute, the Panel doubts that a bank in fact would consider such paper for the purpose of advancing money. The business plan does not describe real activities that are intended to be made over the disputed website or any other websites mentioned therein. The expected yearly turnovers which are listed for each website—including the disputed one—are without any commercial support. Furthermore, it is difficult to believe that the business concept, which allegedly is to sell music, can be achieved through websites designated as 'www.SexTo.Biz', 'www.CloneAid.de', 'www.Alt-Gold.com', and 'www.GenTechnics.com' that are all mentioned (among others) in the business plan. Therefore, a bona fide offering of goods and services, or demonstrable preparations thereto, have not been established by the Respondent.

A business plan that is prepared for the purpose of the dispute clearly cannot amount to evidence of demonstrable preparations to use a disputed domain name, as it will have been produced after the respondent has received notice of the dispute. For example, in *Australian Trade Commission v Matthew Reader,*[77] the respondent, who had registered the domain name <austrade.com>, claimed that he had been pursuing a business based on inter-related domain names, and presented evidence consisting of registration of the domain names, the existence of a business plan and employment of a consultant. The business plan, however, appeared to have been created for the purpose of defending the domain name, as it contained pages of argument directed at the instant dispute. As the plan was obviously created following notice of the dispute, the panel held that the plan could not be admitted as evidence of demonstrable preparations to use the domain name.

[6.7.4] Bona Fide Offering of Goods or Services: General Principles

Under paragraph 4(c)(i) of the UDRP, rights or interests may be established by demonstrable preparations to use the domain name only if the preparations to use the domain name are in connection with the bona fide offering of goods or services. Furthermore, actual use of a domain name can establish rights or interests in the domain name under paragraph

[75] WIPO Case No D2000-0309 (7 Aug 2000).
[76] WIPO Case No D2003-0688 (21 Oct 2003).
[77] WIPO Case No D2002-0786 (12 Nov 2002).

4(c)(i) only where the use is in connection with the bona fide offering of goods or services. In each case, not only must there be an offering of goods or services, but the offering must be bona fide. Consequently, the question of what amounts to a bona fide offering will often be the most important issue in determining whether or not the defence applies.

As explained at [6.7.2], UDRP panels usually do not separate out considerations relating to whether there has been a 'use' of the domain name from considerations relating to whether there has been a '*bona fide* use'. As further explained, where a web site to which the domain name resolves contains no more than links to sites that are unrelated to a bona fide business, the defence will not apply because there is no recognised 'use' of the domain name, there are no 'demonstrable preparations' to use the domain name, or there is no '*bona fide* use'. This paragraph explains the general principles that apply in determining whether there is a bona fide offering of goods or services, while the succeeding paragraphs give examples of the operation of the principles.

The following are the general principles that apply in determining whether there is a bona fide offering of goods or services.

- The domain name must not be used intentionally to attempt to attract, for commercial gain, Internet users to a web site to which the domain name resolves by creating a likelihood of confusion with the complainant's mark.
- In particular, the domain name must not be used to 'bait' Internet users through a suggested association with the trade mark owner, in an attempt to 'switch' the users to other goods or services.
- The domain name must not be used for commercial gain by intentionally attempting to divert Internet users to other sites that are not related to the complainant.
- In particular, the domain name must not be used to divert users to a pornographic web site that is not related to the complainant, regardless of whether goods or services are offered from that web site.
- The respondent must not try to corner the market in all domain names, thus depriving the trade mark owner of the opportunity to reflect its own mark in a domain name.

[6.7.5] Bona Fide Offering of Goods or Services: Trading Off the Reputation of the Trade Mark Owner to Attract or Divert Internet Users

Commercial use of the disputed domain name intentionally to trade off the reputation of the trade mark owner is not a bona fide offering of goods or services.

The panel in *Madonna Ciccone v Dan Parisi*[78] summed up the general approach to this issue, when it stated that the <madonna.com> domain name:

> was selected and used by Respondent with the intent to attract for commercial gain Internet users to Respondent's web site by trading on the fame of Complainant's mark. We see no other plausible explanation for Respondent's conduct and conclude that use which intentionally trades on the fame of another can not constitute a 'bona fide' offering of goods or services. To conclude otherwise would mean that a Respondent could rely on intentional infringement to demonstrate a legitimate interest, an interpretation that is obviously contrary to the intent of the Policy.

In *Advance Magazine Publishers Inc v Lisa Whaley*,[79] the panel pointed out that there will be no bona fide offering of goods or services where the domain name is not used in good faith, adding that 'intentionally infringing use should not be viewed as *bona fide* use'.

[78] WIPO Case No D2000-0847 (12 Oct 2000).
[79] WIPO Case No D2001-0248 (7 May 2001).

The use of the domain name intentionally to attempt to attract, for commercial gain, Internet users to the respondent's web site by creating a likelihood of confusion with the complainant's mark is not a bona fide offering of goods or services. In particular, the use of the domain name to 'bait' Internet users by suggesting an association with the trade mark, in an attempt to 'switch' consumers to other goods or services, is not a bona fide offering. The following decisions are representative of the many examples of decisions that illustrate these principles.

- *Advance Magazine Publishers Inc v Lisa Whaley.*[80] The complainant was the owner of the 'GQ' mark and the respondent had registered a number of domain names, including <gqkids.com>, which incorporated the 'GQ' mark. The respondent used the <gqkids.com> domain name to sell children's clothing from a web site that featured a logo that was similar to the complainant's 'GQ' logo. Taking into account the similarity in the design of the logo, the panel held that the domain names had been adopted with an intention to benefit from an association with the 'GQ' mark and, consequently, gave rise to no legitimate interests.
- *Amphenol Corporation v Applied Interconnect, Inc.*[81] The complainant was the owner of the 'AMPHENOL' mark, which it used in connection with the sale of electrical components, including cables, while the respondent was the registered holder of <amphenolcables.com>, which it used to sell hybrid cable assemblies. The panellist concluded that the respondent was using the complainant's mark as a 'bait' to attract Internet users, and that this use was not a bona fide offering of goods or services.
- *Chanel, Inc v Estco Technology Group.*[82] The respondent, who had registered the domain names <chanelstore.com> and <chanelfashion.com>, produced an elaborate business involving use of the domain names in connection with web sites providing fashion information. The panel held that the respondent could not use the complainant's famous mark to entice Internet users to its fashion database and that the respondent had no legitimate interests in the domain name.
- *America Online, Inc v AOL Logistics Co, Ltd.*[83] The respondent used the disputed domain name <aologis.com> on a commercial web site that displayed the 'AOL' mark with a globe and with a computer screen. The 'AOL' mark was used both in the visible text and the metatags of the web site. The panellist held that this was not a bona fide offering of goods or services.
- *Eldrick 'Tiger' Woods v Shay Kostiner.*[84] The respondent used the <tigerwoodsdesign.com> domain name to advertise its e-commerce business at the disputed domain name, which was not affiliated with the complainant in any way. The panellist held that this diversionary use of the disputed domain name for the respondent's commercial gain was not a bona fide offering of goods or services.
- *Edmunds.com, Inc v The Registrant of <edmounds.com>, Free Domains Parking, Andrey Vasiliev.*[85] The respondent used the <edmounds.com> domain name in connection with a web site that contained advertisements for car price quotations. The panellist held that

[80] WIPO Case No D2001-0248 (7 May 2001).
[81] WIPO Case No D2001-0296 (8 May 2001).
[82] WIPO Case No D2000-0413 (18 Sept 2000).
[83] NAF Case No FA166019 (20 Aug 2003).
[84] NAF Case No FA772899 (25 Sept 2006).
[85] WIPO Case No D2006-0741 (25 Sept 2006).

the domain name was being used improperly to capitalise on the reputation and goodwill of the complainant's mark and, as such, was not a bona fide offering of goods or services.

Similarly, the use of the domain name for commercial gain by intentionally attempting to attract and divert Internet users to other sites that are not related to the complainant is not a bona fide offering of goods or services. In particular, using the domain name to divert users to a pornographic web site is not a bona fide use, despite the fact that goods or services may be offered from that web site. The following is a sample of the many decisions to implement these principles.

- *Reuters Limited v Global Net 2000, Inc.*[86] The complainant alleged that the domain names <wwwreuters.com> and <reutersnews.com> linked to sites hosting pornographic material. As the complainant provided no evidence to support the assertions, however, the panellist held that the assertions could not be established as facts. Nevertheless, in the absence of evidence from the respondent, and given that the 'REUTERS' mark was 'not one that the Respondent would legitimately choose in the context of provision of goods or services via a web site unless seeking to create an impression of an association with the complainant', the panellist held that the respondent had no rights or interests in the domain names.
- *Women in Military Service for America Memorial Foundation, Inc v Russian Web Marketing.*[87] The disputed domain name, <wimsa.org>, redirected Internet users to a site requiring users to click on an 'Enter' icon. When the user clicked on 'Enter', the user was redirected to pornographic Web sites. If the user did not click on 'Enter', the user was taken to other pornographic sites. The panellist held that redirecting Internet users to sites that offer goods or services, and especially redirecting users to pornographic sites, was not a bona fide offering of goods or services.
- *Reed Elsevier Inc & Reed Elsevier Properties Inc v Fast Break Information.*[88] The respondent admitted that it intended to use the disputed domain name, <lexisnexis.org>, to operate an 'adult' site, claiming that it based the name 'Lexis' upon a female adult entertainer, and the name 'Nexis' on a feature of the Star Trek films. The panellist held that using the domain name to divert users to an 'adult' site does not constitute a legitimate interest.
- *DLJ Long Term Investment Corporation v BargainDomainNames.com.*[89] The disputed domain name <dlj.org> was used to divert users to the <visual.com> Web site that contained links to Web sites, including those offering services related to investing, stocks, personal finance and financial services. The panellist held that using the complainant's mark to divert Internet users to a web site with goods that compete with the complainant's is not a bona fide offering of goods or services.
- *US Franchise Systems, Inc v Thurston Howell III.*[90] The respondent used the disputed domain name, <microtels.com>, to redirect Internet users to the <travelnow.com> web site, which hosted a travel-related search engine. The domain name was later used to redirect users to a web site providing information about a computer services company. The panellist held that using the complainant's mark and goodwill to attract users to an unrelated business was not a bona fide offering of goods or services.

[86] WIPO Case No D2000-0441 (13 July 2000).
[87] WIPO Case No D2001-0610 (11 July 2001).
[88] NAF Case No FA97691 (25 July 2001).
[89] NAF Case No FA104580 (9 Apr 2002).
[90] NAF Case No FA152457 (6 May 2003).

- *American Civil Liberties Union of New Mexico v Vilma Morales.*[91] The disputed domain name, <aclunm.org>, was used to 'mousetrap' Internet users, by automatically diverting them to pornographic web sites, such as <pinkflicks.com> and <gothfuck.com>, and opening pornographic 'pop-up' advertisements. The panellist held that redirecting Internet users in this way was not a bona fide offering of goods or services, even though the web sites to which users were redirected clearly offered goods or services.
- *Expedia, Inc v Compaid.*[92] The respondent used the <expediate.com> domain name to redirect Internet users to a web site featuring links to travel services that directly competed with the complainant's business. The panellist held that the use of a domain name that is confusingly similar to the complainant's mark to redirect Internet users to web sites that offer products that compete with the complainant's is not a use in connection with a bona fide offering of goods or services.
- *Humana, Inc v Unasi Inc a/k/a Domaincar.*[93] The respondent used the disputed domain name <huumana.com> to redirect users to a portal site offering links to the sites of the respondent trade mark owner's competitors. The panellist held that using the complainant's mark to divert traffic in this way was not a bona fide offering of goods or services.
- *Württembergische Versicherung AG v Emir Ulu.*[94] The disputed domain name <xn-wrttembergische-versicherung-16c.com> redirected users to a web site that offered domain names for sale. The panellist held that, as the respondent was 'baiting' Internet users to its own web site by using a domain name that was confusingly similar to the complainant's mark, paragraph 4(c)(i) was not made out.
- *Electronic Arts Inc v Cayman Trademark Trust aka Generic Search Terms.*[95] The disputed domain name <eastore.com> resolved to a web site that consisted of a list of sponsored links relating to sites promoting products offered by competitors of the complainant, including computer and video games software. The panellist held that use of the domain name to promote competitors' products was not a legitimate use, and could confer no rights or interests in the domain name.
- *Victoria's Secret Stores Brand Management, Inc v Diana Abeyta.*[96] The respondent used the disputed domain name, <victoriaslilsecret.com>, to redirect Internet users to the respondent's web site, which featured an escort service, including a list of rates. The panellist held that this was not a bona fide offering of goods or services.
- *Goldman, Sachs & Co v Lis Wevers c/o Goldman Advertising Services BV.*[97] The disputed domain name, <goldmensex.com>, was used automatically to divert Internet users to the <cityoflove.com> site, which contained an adult entertainment directory and provided pornographic films for sale. The panellist held that use of a disputed domain name in connection with adult-oriented goods or services is not a bona fide offering of goods or services.
- *The Royal Bank of Scotland Group plc v Caribbean Online International Ltd.*[98] The disputed domain name, <rbsjobs.com>, featured links to web sites that offered financial and bank-

[91] WIPO Case No D2004-0473 (23 Aug 2004).
[92] NAF Case No FA520654 (30 Aug 2005).
[93] WIPO Case No D2006-0199 (6 Mar 2006).
[94] WIPO Case No D2006-0278 (4 May 2006).
[95] WIPO Case No D2006-0570 (12 July 2006).
[96] NAF Case No FA747975 (23 Aug 2006).
[97] NAF Case No FA812109 (22 Nov 2006).
[98] NAF Case No FA849147 (8 Jan 2007).

ing services in competition with the services offered by the complainant. The panellist inferred that the respondent was generating click-through fees from diverted consumers and concluded that the domain name was not being used in connection with a bona fide offering of goods or services.

A considerable number of panel decisions have held that use of a disputed domain name that is confusingly similar to a pharmaceutical manufacturer's mark to redirect Internet users to an online pharmacy is not a bona fide offering of goods or services. Decisions that have reached this conclusion include: *Pfizer, Inc v Seocho and Vladimir Snezko* (<wwwpfizer.com>);[99] *AVENTIS Pharma SA and Merrell Pharmaceuticals Inc v Rx USA* (<anzemet.net> and <taxotere.net>);[100] *G D Searle & Co v Fred Pelham* (<doctorslink-viagra-xenical-propecia-celebrex.com>);[101] *Pfizer Inc v lipidor.com DNS Services* (<lipidor.com>);[102] *Hoffman-La Roche Inc v #1 Viagra Propecia Xenical & More Pharmacy* (<viagra-propecia-xenical.com>);[103] *Aventis, Aventis Pharma SA v VASHA Dukes* (<buy-aventis.com>);[104] *Lilly ICOS LLC v Dan Eccles* (<cialis-drug-online-buying-guide.com>);[105] *Pfizer Inc v jg a/k/a Josh Green* (<pfizerforwoman.com>; <wwwpfizerforwomen.com>);[106] *Lilly ICOS LLC vTudor Burden d/b/a BM Marketing/Burden Marketing* (domain names including <buy-cialis-generic-viagra.com> and <cheap-viagra-cialis-cheap.com>);[107] *F Hoffmann-La Roche AG v Direct Response Marketing Ltd* (<xenical-sales.com>).[108]

[6.7.6] Bona Fide Offering of Goods or Services: Depriving the Complainant of the Opportunity to Reflect its Mark in a Domain Name

The disputed domain name will not be used to make a bona fide offering of goods or services if the respondent has attempted to 'corner the market' in domain names that incorporate, or are similar to, the complainant's mark, thereby depriving the complainant of the opportunity to reflect its mark in a domain name. Registration of a large number of domain names may be indicative of domain name speculation, which will not confer legitimate interests in the domain name.

In *Nabisco Brands Company v The Patron Group, Inc*,[109] for example, the respondent admitted to registering 40 domain names, including names that were identical, or confusingly similar, to the complainant's marks. In holding that the respondent had no legitimate interests in the disputed domain names, the panellist stated that 'it is highly unlikely that Patron could legitimately use each of the 40 domain names it assertedly owns in providing goods and services in commerce'.

In *Ermenegildo Zegna Corporation v Estco Enterprises Ltd*,[110] the respondent registered a range of fashion-related domain names, including <zegnafashion.com>, as part of a business plan of linking the domain names to a fashion information portal site at

[99] WIPO Case No D2001-1188 (28 Nov 2001).
[100] WIPO Case No D2002-0290 (23 May 2002).
[101] NAF Case No FA117911 (19 Sept 2002).
[102] WIPO Case No D2002-1099 (8 Apr 2003).
[103] WIPO Case No D2003-0793 (30 Nov 2003).
[104] WIPO Case No D2004-0276 (24 May 2004)
[105] WIPO Case No D2004-0750 (8 Nov 2004).
[106] WIPO Case No D2004-0784 (6 Dec 2004).
[107] WIPO Case No D2005-0313 (21 June 2005).
[108] WIPO Case No D2006-1370 (29 Dec 2006).
[109] WIPO Case No D2000-0032 (23 Feb 2000).
[110] WIPO Case No D2000-0395 (27 July 2000).

<designerglobe.com>. In rejecting the contention that the plan was sufficient for the respondent to have a legitimate interest in the domain name, the panel held that there was no need for the respondent to register all of the domain names to meet its business needs.

In *Magnum Piering, Inc v Mudjackers,*[111] the respondent registered 14 domain names that were identical, or confusingly similar, to the complainant's mark in the .com and .net gTLDs, arguing that it had registered the domain names to pre-empt the activities of third parties that might register the domain names to damage its business interests. The panellist rejected this contention, finding that the respondents were 'mere officious interlopers', who had no legitimate interests in the domain names. In reaching this conclusion, the panellist observed that 'a single distributor is extremely unlikely to have a legitimate interest in precluding others from using numerous variants on a mark'.

In *Ferrari SPA, Fila Sport SPA, v Classic Jack,*[112] the respondent registered 11 domain names that incorporated the 'FERRARI' mark, including <ferrari-fila.com> and <ferraristuff.com>. The respondent claimed that it had 'legitimate interests' as the domain names were used only to offer the complainant's products and the web sites included a disclaimer of affiliation with the complainant. The panellist, however, concluded that the respondent had registered the 11 domain names with an intention to corner and block the complainants from using the domain names. As with similar decisions, such as *Nabisco Brands,* the panellist's conclusion that the respondent had no legitimate interests was reinforced by the fact that the respondent had registered a range of other related domain names, indicating that it had engaged in 'multiple domain name grabbing'.

As explained at paragraph [6.7.7], the principle that a respondent who registers domain names in an attempt to 'corner the market' does not make a bona fide offering of goods or services has been incorporated into the test for determining whether a reseller is using a domain name to make a bona fide offering of goods or services.

[6.7.7] Bona Fide Offering of Goods or Services: Can a Respondent that Resells a Complainant's Goods or Services have a Right or Legitimate Interest in the Disputed Domain Name?

The most controversial issue in determining whether a respondent has used the disputed domain name to make a bona fide offering of goods or services has been whether a respondent who uses the domain name to resell the complainant's goods or services is making a bona fide offering. The essential question is whether the respondent's use of the disputed domain name in connection with the offering of the complainant's goods or services can be bona fide.

In recognition of a continuing division of views on this issue, the *WIPO Overview* sets out a majority view and a minority view. The WIPO majority view is that a reseller may make a bona fide offering, provided that it complies with a set of conditions aimed at minimising source confusion. The majority view is as follows:

> A reseller can be making a bona fide offering of goods and services and thus have a legitimate interest in the domain name if the use fits certain requirements. These requirements include the actual offering of goods and services at issue, the use of the site to sell only the trademarked goods and the site accurately disclosing the registrant's relationship with the trademarked owner. The respondent must also not try to corner the market in domain names that reflect the trademark.[113]

[111] WIPO Case No D2000-1525 (29 Jan 2001).
[112] WIPO Case No D2003-0085 (15 Apr 2003).
[113] WIPO, above n 10, para 2.3.

The minority view, on the other hand, is that use of the disputed domain name by a reseller is not bona fide unless the complainant has expressly authorised the use of its trade mark as a domain name. Accordingly, the WIPO minority view is that:

> Without the express permission of the trademark holder the right to resell the trademark holder's products does not create a right to use the trademark as the basis for a domain name.[114]

In applying these views, a particular issue that has arisen is whether the general position applies only to resellers, such as express licensees or exclusive dealers, that are in a contractual relationship with the complainant, or whether it extends to all resellers of the complainant's products, regardless of whether they are in a such a contractual relationship. This issue is dealt with at [6.7.8].

The multi-factor test set out in the majority view was first formulated by the panellist in *Oki Data Americas, Inc v ASD, Inc.*[115] The panellist in that dispute held that the following minimum requirements must be met for a reseller to be making a bona fide offering of goods or services:

1. The respondent must actually be offering the goods or services at issue.
2. The respondent must use the site to sell only the trademarked goods; otherwise, it could be using the trade mark to bait Internet users and then switch them to other goods.
3. The site must accurately disclose the registrant's relationship with the trade mark owner; it may not, for example, falsely suggest that it is the trade mark owner, or that the Web site is the official site, if, in fact, it is only one of many sales agents.
4. The respondent must not try to corner the market in all domain names, thus depriving the trade mark owner of reflecting its own mark in a domain name.

The respondent in the *Oki Data* dispute itself, who was the registrant for the <okidataparts.com> domain name, was listed on the complainant's web site as an authorised 'OKI DATA' dealer. Moreover, the respondent used the domain name only on 'OKI-DATA' goods and services, prominently disclosed that it was merely a repair centre and had not registered numerous confusingly similar domain names. As the respondent satisfied the test identified by the panellist, it had a legitimate interest in the domain name.

Prior to the formulation of the *Oki Data* test, the division between those panel decisions concluding that a reseller could have legitimate interests in the disputed domain name and those holding that a reseller could only have a legitimate interest if the domain name was registered with the express permission of the trade mark owner had already emerged. One line of decisions held that resellers could establish legitimate interests in certain circumstances. In *Weber-Stephen Products Co v Armitage Hardware,*[116] for example, the respondent, who was a distributor of the complainant's products, registered a number of domain names that incorporated the complainant's mark, including <webergrills.com>. The respondent used the web site to sell the complainant's goods under the complainant's trade mark, while expressly stating that it was the respondent's dealer. Although the respondent had used the web site to sell products other than those of the complainant, the panel accepted that those instances were minor and that the respondent had removed the products. In these circumstances, the panel held that the domain name was used in connection with the bona fide offering of the complainant's goods or services.

[114] *Ibid.*
[115] WIPO Case No D2001-0903 (6 Nov 2001).
[116] WIPO Case No D2000-0187 (11 May 2000).

Similarly, in *Columbia ParCar Corp v S Brustas GmbH,*[117] the respondent, who had registered the <columbiaparcar.com> domain name, had an exclusive distribution agreement with the complainant. The complainant, however, claimed that this did not extend to use of its mark as a domain name. The panellist held that, as the respondent was an exclusive dealer who had, with the knowledge and consent of the complainant, used the mark to promote the complainant's products, it had legitimate interests in the domain name.

The competing line of decisions held that the express permission of the complainant trade mark owner is required for a reseller to acquire a legitimate interest. In *Motorola, Inc v NewGate Internet, Inc,*[118] the respondent registered the domain name <walkabout.com>, and claimed that it was authorised by one of Motorola's distributors to sell 'TALKABOUT' products. The complainant did not dispute the claim, but the majority of the panel maintained that whether or not the respondent was a reseller is irrelevant, 'since the right to resell products does not create the right to use a mark more extensively than required to advertise and sell the product'. As the majority of the panel regarded use of the mark as a domain name as going further than the rights to advertise and sell the goods or services, it held that the respondent had no interests in the domain name. The dissenting panellist, by way of contrast, held that the respondent had acquired legitimate interests as its proposed use of the domain name as an 'adult sex site' did not compete with the complainant's goods or services.

Similarly, in *The Stanley Works and Stanley Logistics, Inc v Camp Creek Co, Inc,*[119] the respondent registered a number of web sites that included the complainant's mark, including <stanleybostitch.com> and <stanley-bostitch.com>, and claimed that it used the sites to sell the complainant's products. The panellist maintained that even if the respondent were an authorised retail seller of the complainant's products, this would not confer rights to use the complainant's mark as a domain name. Consequently, the panellist held that retail sellers of a complainant's products have no legitimate interests in a disputed domain name without something more, such as authorisation in a licensing agreement or other special circumstances.

Since the *Oki Data* test was first enunciated it has been followed in a majority of panel decisions. The following are examples of decisions that have applied the *Oki Data* decision and that serve to illustrate the operation of the multi-factor test.

- *Experian Information Solutions, Inc v Credit Research, Inc.*[120] The respondent, who had been an authorised reseller of the complainant's credit reports, used the domain names <experiancredit.com>, <experiancredit.net> and <experiancredit.org> to offer for sale 'triple merge' credit reports. The 'triple merge' reports combined information regarding specific consumers from three major credit reporting agencies, including the complainant. As the reports incorporated the 'products' of the complainant's competitors, the panel held that the respondent failed the second element of the *Oki Data* test. Moreover, as the respondent did not disclose its relationship with the complainant, the third element was not satisfied. The respondent therefore had no legitimate interests in the domain names.
- *Volvo Trademark Holding AB v Auto Shivuk.*[121] The respondent used the domain name <volvo-auto-body-parts-online.com> to promote 'VOLVO' auto parts. The panellist held

[117] WIPO Case No D2001-0779 (23 Aug 2001).
[118] WIPO Case No D2000-0079 (20 Apr 2000).
[119] WIPO Case No D2000-0113 (13 Apr 2000).
[120] WIPO Case No D2002-0095 (7 May 2002).
[121] WIPO Case No D2005-0447 (8 June 2005).

that, as each of the elements of the *Oki Data* test was satisfied, the respondent was making a bona fide offering of goods or services. First, the disputed domain name was used only in connection with 'VOLVO' parts. Secondly, the respondent's web site disclosed that it was merely a distributor of the complainant's goods, indicating that it was not intentionally using the domain name illegitimately to benefit from the goodwill in the complainant's mark. Thirdly, although the respondent had registered a number of domain names that included famous car brands, such as <audi-auto-body-parts-online.com> and <bmw-auto-body-parts-online.com>, there was no evidence that it had registered the domain names to corner the market or prevent the complainant from reflecting its mark in a domain name. Having passed the *Oki Data* test, the panellist held that the respondent reseller had legitimate interests in the domain name.

- *Toyota Motor Sales, USA, Inc v Pick Pro Parts Inc.*[122] The respondent used the domain names <lexus-parts.com> and <toyota-parts.com> to direct Internet users to a web site that advertised automotive parts for a variety of makes of vehicles, including those of the complainant's competitors. The web site included a notice disclaiming affiliation with the complainant. As the respondent's web site offered goods or services in competition with the complainant's, the panellist held that it failed the second element of the *Oki Data* test. While the panellist observed that the respondent could conceivably have satisfied the second element if it had produced evidence that the domain names were not being used to 'bait and switch' consumers, in the absence of a response, the respondent was not making a bona fide offering of the complainant's goods.
- *Lance Armstrong Foundation v Chris Angeles.*[123] The complainant sold yellow 'LIVESTRONG' bracelets to promote awareness of cancer survivorship. The respondent used the disputed domain names, <livestrongbracelets.net> and <talk-livestrong.com>, to advertise 'LIVESTRONG' bracelets, as well as other wrist bands and decorative bracelets. The 'LIVESTRONG' bracelets were purchased from the complainant and resold at a higher price. The panellist held that, as the domain names were used to sell a variety of other bracelets, the respondent failed to satisfy the second element of the *Oki Data* test. Concluding that the panellist had not made a bona fide offering, the panellist observed that the respondent's activities presented 'a distinct danger of "bait and switch" tactics, as the Respondent trades on the goodwill associated with the Complainant's "LIVESTRONG" mark to attract Internet users and then attempt to sell them other products and services'.
- *Toyota Motor Sales, USA, Inc v Fleetrates.com.*[124] The respondent used the disputed domain names, <lexusus.com> and <joelexus.com>, to redirect Internet users to sites offering the sale of new and used motor vehicles. The panellist held that the second element of the *Oki Data* test was not made out, as the domain names were used to sell the products of the complainant's competitors, as well as the complainant's products. Moreover, as there was no disclaimer of a relationship with the complainant, the respondent was clearly not making a bona fide offering of goods or services.
- *Grundfos A/S v King Pumps Inc.*[125] The respondent registered the domain name <grundfospumpsupply.com>, intending to use it to sell the complainant's products and to direct users to the respondent's site at <waterpumpsupply.com>. As the panellist held

[122] WIPO Case No D2005-0562 (20 July 2005).
[123] WIPO Case No D2005-0888 (13 Oct 2005).
[124] NAF Case No FA568488 (21 Nov 2005).
[125] WIPO Case No D2005-1073 (7 Jan 2006).

that the respondent intended to link the domain name to web sites that sold products of the complainant's competitors, he concluded that the second element of the *Oki Data* test had not been satisfied.

- *Allen-Edmonds Shoe Corporation v joseph c/o joseph scorsone.*[126] The respondent, a reseller of 'ALLEN EDMUNDS' shoes, used the <allenedmondshoes.com> domain name to direct Internet users to sites offering the products of the complainant's competitors. Applying the *Oki Data* test, the respondent clearly could not be making a bona fide offering. The respondent, however, claimed that the site was 'under construction', and that the links had been chosen by the domain name registrar. The panellist rejected this contention, holding that the respondent was responsible for the links and that, as they included links to web sites of the complainant's competitors, the second element of *Oki Data* was not made out.
- *Control Techniques Limited v Lektronix Ltd.*[127] The respondent used the domain names, <controltechniques-repairs.com> and <controltechniques-spares.com>, to promote the repair of and spare parts for 'CONTROL TECHNIQUES' products. As the domain names redirected Internet users to a web site that also provided spare parts for the products of the complainant's competitors, however, the respondent did not satisfy the second of the *Oki Data* principles. The panellist also held that the respondent failed to satisfy the third element of *Oki Data*, as the disclaimer on the web site was inadequate, in that it did not expressly disclaim a connection with 'CONTROL TECHNIQUES'. The respondent was therefore unable to show a bona fide offering of goods or services.
- *Fleming Sales Company, Inc v David Marketing Group.*[128] The respondent was the registrant of the <patiohearth.com> domain name and a reseller of the complainant's 'PATIO HEARTH' brand goods, including portable fireplaces. The respondent purchased the goods from a licensee and authorised distributor of the complainant's products. In addition to using the domain name to sell the complainant's goods, however, the respondent used it to sell goods sourced from the complainant's competitors. Moreover, the respondent's web site did not contain a disclaimer disclosing its relationship with the complainant. As the respondent had not complied with the second and third factors of the *Oki Data* test, the panellist held that it had not used the domain name to make a bona fide offering of goods or services.
- *Seiko Epson Corporation and Epson America, Inc v AOS Web Com, Inc.*[129] The respondent claimed that it planned to use the <epsonusa.com> domain name solely for the sale of 'EPSON' products, and placed a disclaimer on the web site. The plan was not, however, put into effect, and the domain name was used to redirect users to a web site that also sold the products of the complainant's competitors. Applying the *Oki Data* test, the panellist held that the respondent was not making a bona fide offering as it was using the complainant's mark to attract customers and sell competitive products.

Provided that the site associated with the domain name is used only to sell the complainant's goods or services, the second element of the *Oki Data* test will be satisfied, even if the site includes links to sites that sell products of the complainant's competitors. In *Dr Ing hc F Porsche AG v Del Fabbro Laurent,*[130] for example, the domain names

[126] NAF Case No FA624511 (28 Feb 2006).
[127] WIPO Case No D2006-1052 (11 Oct 2006).
[128] WIPO Case No D2006-1174 (8 Nov 2006).
[129] NAF Case No FA823033 (27 Nov 2006).
[130] WIPO Case No D2004-0481 (20 Aug 2004).

<porsche-buy.com> and <porschebuy.com> were used to offer for sale used Porsche cars. The web sites, however, contained links to sites that offered used cars of competitors of the complainant. In holding that the links did not jeopardise the respondent's rights or interests in the domain names, the panellist stated:

> Clearly designated links to websites, under which other brands or products are offered, cannot be considered as the offering of other brands or products under the Domain Names, provided they take up only a minor part of the site and the overall impression of that site remains that of a site offering corresponding goods.

Similarly, in *Inter-Tel, Inc v Michael Marcus,*[131] the respondent used the disputed domain name <intertelphonesystems.com> to sell the complainant's products. The web site contained statements distinguishing the respondent from the complainant trade mark owner. At the same time, the web site contained links to other sites that sold products manufactured by the complainant's competitors, and advertised two telephones not produced by the complainant. The panellist applied the reasoning in the *Porsche* decision to hold that the links to the other sites were irrelevant. Moreover, he held that the use of the site to advertise the competitor's products was *de minimus,* as the respondent's site was 'directed overall toward selling Complainant's products'. Consequently, the domain names were being used to make a bona fide offering of the complainant's products.

By way of contrast, in *PRL USA Holdings, Inc v Philip Chang,*[132] the respondent used the <polousa.com> domain name to offer the complainant's 'POLO' goods for sale, claiming that the fact that 10 to 15 per cent of the sales from the web site were the goods of competitors did not undermine its claim to a 'legitimate interest' in the domain name. In concluding that the second element of the *Oki Data* test had not been satisfied, however, the panel observed that the sales of the competitor's goods 'is not great but neither is it insignificant'.

As indicated by the *WIPO Overview,* a minority of panellists hold the view that a distributor or reseller cannot acquire rights or interests in a disputed domain name in the absence of express permission from the trade mark owner. Prior to the *Oki Data* decision, this appeared to be the view of the majority of panel decisions.[133] The minority view, however, is now becoming less prominent. Moreover, those panels who continue to apply the minority view now also commonly apply the *Oki Data* test in the alternative. The following are examples of decisions that apply the minority view or refer to it in positive terms.

- *Nokia Corporation v Nokia Ringtones & Logos Hotline.*[134] The respondent used the disputed domain name <worldnokia.com> to offer logos, ringtones, software and other products and services to owners of 'NOKIA' mobile phones. The panellist held that the 'mere fact that Respondent sells logos and ringtones compatible with NOKIA mobile phones is not sufficient for Respondent to claim a legitimate interest'. The decision was, however, also influenced by the fact that the respondent offered illegally copied 'NOKIA' logos and used the web site to sell products that were compatible with mobile phones other than 'NOKIA' phones.
- *Enzymatic Therapy, Inc v VNF Nutrition.*[135] The respondent was a reseller of the complainant's dietary supplements, and used the <enzymatictherapy.com> web site to

[131] NAF Case No FA727697 (27 July 2006).
[132] WIPO Case No D2006-0093 (20 Apr 2006).
[133] See, eg, the comments in *Easy Heat, Inc v Shelter Products,* WIPO Case No D2001-0344 (14 June 2001).
[134] WIPO Case No D2001-1101 (18 Oct 2001).
[135] WIPO Case No D2001-1310 (17 Dec 2001).

promote the sale of the complainant's products, as well as products of the complainant's competitors. The panellist held that, as the complainant had not licensed or authorised the respondent to use its mark as a domain name, the respondent could have no rights or legitimate interests in the domain name.

- *Pfizer Inc v Alex Schreiner/ Schreiner & Co.*[136] The respondent used the disputed domain name <blu-viagra.com> to link to an online pharmacy that offered 'genuine Viagra from Pfizer' for sale. The panellist applied both the minority view, that a distributor or reseller has no rights in the disputed domain name in the absence of the trade mark owner's consent, and the majority view, that distributors can have legitimate interests under the *Oki Data* conditions. In this dispute, regardless of the approach adopted, the respondent was not making a bona fide offering of goods or services. On the minority view, the respondent was not licensed or authorised to use the complainant's mark as a domain name. Moreover, as the respondent failed to disclaim an affiliation with the complainant, it did not satisfy the third element of the *Oki Data* test.
- *Merrell Pharmaceuticals Inc and Aventis Pharma SA v Lana Carter.*[137] The respondent used the disputed domain name, <allegra-top-pharmacy.net>, for a site that advertised a wide range of brand name prescription medications, including 'ALLEGRA', and that linked to sites that supplied pharmaceuticals. The use of the domain name in connection with the sale of a competitor's products clearly failed the second element of the *Oki Data* test. Nevertheless, the panellist expressly referred to the minority of panel decisions that have 'rejected out of hand any right on the part of a distributor or reseller to register a domain name consisting of or containing the manufacturer's trademark, at least in the absence of the trademark owner's consent'.

The difference between the majority and minority views is ultimately attributable to different views concerning the scope of the UDRP. The majority view is essentially that, unless a reseller is engaging in conduct that would disentitle it from having a legitimate interest in a domain name, such as trading off the complainant's reputation in a mark to sell competing products, then there should be no obstacle to it registering and using the complainant's mark in a domain name. After all, resellers are able legitimately to use trade marks in advertising and marketing the products of their suppliers. On this view, the UDRP is narrowly targeted at preventing 'abusive, bad faith' registration and use. The rationale for the majority view was expressed by the panellist in the *Oki Data* decision itself, where he said:

> It is important to keep in mind that the Policy was designed to prevent the extortionate behavior commonly known as cybersquatting. It cannot be used to litigate all disputes involving domain names. *The Thread.com, LLC v. Poploff*, Case No. D2000-1470 (WIPO Jan. 5, 2001). If trademark owners wish to prevent the use of their marks by authorized sales and repair agents in domain names, they should negotiate such protections through appropriate contractual language or, when permitted under the relevant law, seek recovery in classic trademark infringement or dilution litigations. In the absence, however, of some element of illegitimacy, they should not use the Policy to prevent uses that ICANN deemed to be legitimate, including the use of domain names in connection with the bona fide offering of goods and services.

The minority view, on the other hand, is based on the assumption that a trade mark owner has rights to reflect the mark in corresponding domain names that are good against all the world except for those who can establish competing rights or interests. On this view,

[136] WIPO Case No D2004-0731 (24 Oct 2004).
[137] WIPO Case No D2004-1041 (25 Jan 2005).

a reseller's rights in a domain name that reflects a complainant's mark are necessarily derivative of the complainant's rights, and therefore depend upon the consent or permission of the trade mark owner to registration of the domain name. As the panellist in the *Motorola* decision expressed this point:

> the right to resell products does not create the right to use a mark more extensively than required to advertise and sell the product.

Similarly, the panellist in *Volvo Trademark Holding AB v Peter Lambe*[138] maintained that:

> It is a well-known principle of trademark law that the manufacturer of 'unoriginal' spare parts is entitled to a certain limited use of the trademark of the manufacturer of the original products in connection with the *bona fide* offering of these goods, but this principle does not entitle the said manufacturer to incorporate the trademark in his business name or in any other type of business identifier such as a domain name.

As the UDRP is narrowly targeted at preventing 'abusive, bad faith' registration, and not at conferring broader rights on trade mark owners in relation to domain names, the majority view is to be preferred. This conclusion is reinforced by the decision to extend the protection conferred by the second element of the UDRP beyond legal 'rights' to include 'legitimate interests'. In other words, even if any legal rights of a reseller in relation to registration and use of a domain name must be acquired from the trade mark owner, it can obtain 'legitimate interests' in the domain name quite independently of the trade mark owner. In that event, as explained by the panellist on *Oki Data,* the onus is on the trade mark owner to protect its interests, either by means of contract or through the courts.

[6.7.8] Bona Fide Offering of Goods or Services: Can a Reseller that is Not in a Contractual Relationship with the Complainant have Rights or Interests in the Disputed Domain Name?

UDRP panels have differed on the question whether a reseller that is not in a contractual relationship with the complainant can have rights or interests in the disputed domain name as a result of merely offering the complainant's goods or services for sale. In general, the division of views on this issue tracks the differences between the majority and minority views on whether resellers, including those who have a contractual relationship with the complainant, can establish rights or interests in the domain name.

Adherents to the minority view must hold the view that a reseller that is not in a contractual relationship with the complainant and that otherwise has no express permission to register and use the domain name can never make a bona fide offering of goods or services. On this view, the question whether a respondent that is in a contractual relationship, such as a licensee or exclusive distributor, has express or implied permission to register and use the domain name depends upon the construction of the contract. Given that UDRP panels are not well-equipped to make complex determinations on the construction of contracts, there are clearly limitations with the minority view.

The majority of panellists, on the other hand, have held that resellers who do not have a contractual relationship with the complainant are to be treated in precisely the same way as those who have a contractual relationship. The majority of panels therefore apply the *Oki Data* factors to determine whether a reseller who does no more than sell the complainant's

[138] WIPO Case No D2001-1292 (20 Jan 2002).

products is using the domain name to make a bona fide offering of goods or services. The following panel decisions have considered the issue.

- *Ferrero SpA v Fistagi Srl.*[139] The respondent, who was a wholesaler of the complainant's products, registered the <kinderferrero.com> domain name. It argued that it had a legitimate right to use the domain name because it had been authorised to sell the complainant's products. In the absence of evidence of an agreement, the panel held that the respondent was a mere reseller of the complainant's products. Adopting the minority view, the panel further held that a mere right to resell a product does not create a right in a mark that is more extensive than a right to advertise and sell the product. On the basis that the registration of the disputed domain name suggested a closer relationship with the complainant than that of a mere reseller, the panel concluded that the respondent was not using the domain name to make a bona fide offering of goods or services. This conclusion was reinforced by the fact that the domain name merely linked to a test page and there was scant evidence of preparations to use the domain name.
- *DaimlerChrysler AG v Donald Drummonds.*[140] The respondent was the registrant of the disputed domain name <mercedesshop.com>, which it used to offer genuine Mercedes parts and accessories for sale. The web site included a clear disclaimer of affiliation with the complainant. The majority of the panel took this into account in finding that the respondent was making a bona fide offering of the complainant's products, despite the absence of a contractual relationship with the complainant. The dissenting panellist, however, held there was no bona fide offering, largely on the basis that the disclaimer was not sufficient to dispel the 'initial interest confusion' of Internet users.
- *Whirlpool Properties, Inc & Whirlpool Corporation v Ace Appliance Parts and Service.*[141] The respondent used the <whirlpoolparts.com> and <kitchenaidparts.com> domain names to sell and advertise the complainant's products and spare parts, but had no contractual relationship with the complainant. The panellist pointed out that this case differed from *Oki Data* in that the respondent in this dispute was a mere reseller, and not an authorised 'repair facility'. Despite this difference, the panellist applied the *Oki Data* test to conclude that the respondent had no legitimate interests in the domain names as the web sites were used to sell parts for competitors of the complainants.
- *Philip Morris Incorporated v Alex Tsypkin.*[142] The respondent, who was not related to the complainant, used the disputed domain name <discount-marlboro-cigarettes.com> to sell 'MARLBORO' cigarettes from a web site that featured the complainant's distinctive logo. Although there was no contractual relationship, the panel applied the *Oki Data* test. As the web site was used to sell discount cigarettes produced by the complainant's competitors, the panel held that the respondent was illegitimately trading off the complainant's mark. From the sale of the competitors' products, when combined with the use of the complainant's logo and the absence of a disclaimer, it was clear that the respondent had failed to satisfy the *Oki Data* test.
- *Dr Ing hc F Porsche AG v Del Fabbro Laurent.*[143] The respondent, who was not in a contractual relationship with the complainant, used the disputed domain names to sell used 'PORSCHE' cars. The panellist pointed out that the facts in this dispute differed from

[139] WIPO Case No D2001-0262 (23 May 2001).
[140] WIPO Case No D2001-0160 (18 June 2001).
[141] NAF Case No FA109386 (24 May 2002).
[142] WIPO Case No D2002-0946 (13 Feb 2003).
[143] WIPO Case No D2004-0481 (20 Aug 2004).

those in the *Oki Data* decision, where the parties were in an existing distributor relationship. Nevertheless, the panellist applied the *Oki Data* test on the basis that there were no facts in the dispute that would render the test inappropriate. Finding each of the *Oki Data* factors satisfied, the panellist held that the respondent was making a bona fide offering of the complainant's goods or services.

If the ability of a respondent to establish rights or legitimate interests in a domain name that uses the complainant's mark does not depend upon obtaining the complainant's permission, there seems no reason whatsoever to distinguish resellers who are in a contractual relationship with the complainant from mere resellers of the complainant's products. This means that the *Oki Data* test should be applied to mere resellers and, provided that the *Oki Data* factors are satisfied, there should be no difficulty in establishing rights or legitimate interests in a domain name that uses the complainant's mark. There is absolutely nothing to suggest that the UDRP should be used to confer rights on trade mark owners over and above what is needed to prevent 'abusive, bad faith' registration and use.

[6.7.9] Bona Fide Offering of Goods or Services: Can a Reseller have Rights or Interests in the Disputed Domain Name Where a Contractual Relationship with the Complainant has been Terminated?

An issue that has given rise to some difficulty has been whether a reseller that has been in a contractual relationship with the complainant trade mark owner retains 'rights or legitimate interests' in a domain name once the contract has been terminated. The approach taken to this issue necessarily depends upon the view taken of the basis of a reseller's rights or interests in the disputed domain name. The following decisions have dealt with the position of a reseller following termination of a contractual relationship with the complainant.

- *UVA Solar GmbH & Co KG v Mads Kragh.*[144] The respondent, who had registered the <sunmaxx.com> domain name, was the sole distributor of 'SUNMAXX' products in Denmark at the time the domain name was registered, but the distribution arrangement was subsequently discontinued. In the absence of evidence, the panellist pointed out that it was usual for distribution agreements to contain a clause limiting the use of the trade mark owner's mark to the duration of the contractual relationship. Although the respondent initially had rights or interests in the domain name, the panellist held that any rights or interests ceased once the distributorship came to an end.
- *Geckodrive Inc v Jon Hollcraft.*[145] The respondent, the registrant of the <geckodrives.com> domain name, was a reseller of the complainant's products until the complainant changed its marketing strategy, whereupon it ceased to be a reseller. The panellist held that, as the respondent's claims to the domain name arose only from the resale of the complainant's products, any rights or interests in the domain name were revoked with the termination of the resale arrangement.
- *Jerome Stevens Pharmaceuticals, Inc v Watson Pharmaceuticals.*[146] The respondent, who had registered <unithroid.com>, had a marketing and distribution agreement with the complainant which granted the respondent limited rights to use the complainant's mark during the term of the agreement. The panellist adopted the minority view that a reseller does not

[144] WIPO Case No D2001-0373 (7 May 2001).
[145] eRes Case No AF-1047 (20 Nov 2001).
[146] WIPO Case No D2004-1029 (26 Mar 2004).

have rights to a domain name that includes the trade mark owner's mark unless the trade mark owner has given permission to use the mark as a domain name. The panellist accepted that, as the UDRP complaint was not lodged until after the distribution agreement was terminated, the respondent had rights or interests in the domain name during the term of the agreement. As the distribution agreement had been discontinued, however, the panellist held that the respondent no longer had rights or interests in the domain name.

- *International E-Z UP, Inc v PNH Enterprises, Inc.*[147] The respondent, who was the complainant's Canadian distributor, registered the domain name <ezup-canada.com> and used it to offer the complainant's products for sale. The distribution agreement was, however, discontinued. The complainant argued that the termination of the distribution agreement meant that the respondent no longer had any rights or interests in the domain name. The respondent, however, claimed that it retained an interest in the domain name as it retained some of the complainant's stock, and was required to provide warranty and after sales services. The panellist first held that, in these circumstances, the *Oki Data* test applied to determine whether the respondent was using the domain name to make a bona fide offering. Applying this test, it was clear that the respondent had rights or interests in the domain name prior to the termination of the distribution agreement. In examining the position following termination of the agreement, the panellist distinguished between legal 'rights' and 'legitimate interests' in the disputed domain name. As a business has an 'interest' in servicing its customers, the panellist held that the respondent retained 'legitimate interests' in the disputed domain name even after the termination of the distribution agreement with the complainant. The panellist distinguished *Geckodrive* and *Jerome Stevens* on the facts.

Where a trade mark owner has a contractual relationship with a reseller, then obviously the contract can expressly deal with the reseller's use of the mark. If the contract clearly provides that use of the mark as a domain name must be discontinued upon termination of the contractual arrangement, then the reseller must lose all claims to the domain name upon the agreement coming to an end. The clearest case would be where the contract expressly provided that the use of the mark as a domain name had to be discontinued upon termination of the agreement. If the contract is ambiguous or unclear, then the panel must make a decision, taking all factors into account and bearing in mind that contractual disputes are best dealt with by the courts.

As the policy is not confined to protecting the legal 'rights' of respondents in a disputed domain name, but extends to protect 'legitimate interests', then where a contract with a trade mark owner does not clearly govern the use of the mark as a domain name, the analysis of the panellist in *International E-Z UP* should be applied. It seems that, just as a reseller can acquire interests in a domain name from its resale activities, provided that the *Oki Data* multi-factor test is satisfied, then those interests will not cease to exist merely because the contractual relationship has come to an end. This must be the position, as the respondent's 'legitimate interests' in the domain name are independent of the contract with the trade mark owner. Provided that the reseller continues to sell or service the complainant's products legitimately, and provided all elements of the *Oki Data* test continue to be complied with and that the contract does not clearly prohibit the use of the mark as a domain name following termination, then the respondent must therefore retain rights or interests in the domain name.

[147] NAF Case No FA808341 (15 Nov 2006).

'Commonly Known by' Domain Name

[6.8] Holder 'Commonly Known by' Domain Name

The second of the circumstances set out in paragraph 4(c) which, if established, means that the respondent has rights or legitimate interests in the disputed domain name, provides that such rights or interests arise where:

> (ii) the domain name holder (as an individual, business, or other organization) has been commonly known by the domain name, even if he or she has acquired no trademark or service mark rights.

This circumstance clearly excludes personal, trading and other names of organisations from the scope of the policy on the basis that individuals, corporations and other organisations should not be deprived of the ability to reflect a name by which they have become known as a domain name. Regardless of whether the respondent is able to establish legal 'rights' in the name, this circumstance effectively provides that a respondent has a 'legitimate interest' in a domain name if it has been 'commonly known' by the domain name.

The key issue in establishing the scope of the defence is clearly determining what amounts to being 'commonly known' by the name. The policy does not attempt to define what is meant by 'commonly known', leaving this to be determined by UDRP panellists. The two other important issues that have arisen in the application of paragraph 4(c)(ii) are the point in time at which the respondent must be 'commonly known' by the name and whether the defence applies to nicknames.

[6.8.1] Domain Holder has been Commonly Known by the Domain Name: the Time at which the Respondent must be 'Commonly Known'

Panel decisions have held that a respondent must have been 'commonly known' by the domain name before the domain name is registered. The reason for this rule is obviously that a respondent's rights or interests in the domain name must arise independently of registration of the domain name, and cannot be derived solely from registration. To hold otherwise would open the door to 'abusive' registrations that are subsequently ratified by use of an illegitimately registered domain name.

The following decisions illustrate the operation of the general rule.

- *RMO, Inc v Andy Burbidge.*[148] The respondent acquired the domain name <rmo.com> for use as a personal site for information about the respondent's family, and for establishing e-mail addresses for friends and family members. The panellist held that, as the respondent was not known by the name 'RMO' prior to registration of the domain name, paragraph 4(c)(ii) had not been satisfied.
- *Rothschild Bank AG v Rothchild Corporation.*[149] The respondent registered the domain name <rothchild.com> and subsequently changed his surname from 'LUHTALA' to 'ROTHCHILD'. The panel held that, as the respondent was not known by the 'ROTHCHILD' name at the time the domain name was registered, he had no rights or legitimate interests in the domain name.

[148] NAF Case No FA96949 (16 May 2001).
[149] WIPO Case No D2001-1112 (15 Jan 2002).

[6.8.2] Domain Holder has been Commonly Known by the Domain Name: When is the Respondent 'Commonly Known' by the Name?

UDRP panels have not formulated a general test for determining whether a respondent has been 'commonly known' by a name, but have generally looked to evidence of the extent to which the respondent has used the name in deciding whether the respondent has been sufficiently known by the name. Panels have been especially prepared to assume that a respondent is 'commonly known' by a name when the evidence indicates that the name has been used in a business context. Nevertheless, a respondent that presents sufficient evidence of being known by a genuine personal name may establish rights or legitimate interests in the domain name.

The following are examples of decisions that have dealt with the question of when a respondent has rights or legitimate interests in a domain name as a result of the respondent being 'commonly known' by the name which illustrate the practical issues that may arise.

- *Gordon Sumner, p/k/a Sting v Michael Urvan.*[150] The respondent registered the <sting.com> domain name and claimed that he was 'commonly known' by the name as he used the name 'STING' as his user-name for online gaming. The panelist held that this use of the name did not mean that the respondent was 'commonly known' by the name, pointing out that the evidence was weak and that use of the name in the context of an online user-name was more consistent with anonymity than with the respondent being 'known' by the name.
- *GA Modefine SA v AR Mani.*[151] The respondent registered the domain name <armani.com> on the basis of his personal name, 'A R MANI'. He presented evidence that his name was 'ANAND RAMNATH MANI', including a listing in a telephone directory and a copy of a baptismal certificate. The panellist concluded that, as the domain name reflected the respondent's real name, he had established a legitimate interest in the name.
- *AST Sportswear, Inc v Steven R Hyken.*[152] The respondent registered the domain name <johnnyblaze.com> and presented evidence that he had used the name 'JOHNNY BLAZE' as a stage name for 25 years. The evidence included evidence of a television appearance and use of the name on a t-shirt. The panelist held that, regardless of whether the respondent was using the name at the time of the dispute, the evidence established that he had been commonly known by the name, and this was sufficient to establish legitimate interests in the domain name.
- *IKB Deutsche Industriebank AG v Bob Larkin.*[153] The respondent registered the domain name <ikb.com> and had registered the company name 'IKB INVESTMENTS LIMITED', and there was evidence from an investor of use of the name in connection with a proposed business. Despite the relatively weak evidence, the panel held that it was sufficient to establish that the respondent was commonly known by the name.
- *Royal Bank of Canada v RBC Bank.*[154] The respondent registered the domain name <rbcalliance.com> and claimed rights or interests in the name on the basis of registration of the company name 'RBC ALLIANCE BANK' in the Republic of Montenegro. The respondent, however, presented no evidence of use of the name before the domain name

[150] WIPO Case No D2000-0596 (19 July 2000).
[151] WIPO Case No D2001-0537 (20 July 2001).
[152] WIPO Case No D2001-1324 (26 Mar 2002).
[153] WIPO Case No D2002-0420 (23 July 2002).
[154] WIPO Case No D2002-0672 (20 Nov 2002).

had been registered. The panelist pointed out that evidence of mere registration of a company name cannot establish that the respondent is 'commonly known' by the name, as this would open the door to the policy being circumvented. In this respect, the panellist observed:

> If the intention of the Policy were otherwise, every cybersquatter would be able to avoid the operation of the Policy by the simple expedient of: (i) quietly registering someone else's trademark as a corporation name (possibly in some jurisdiction having no connection with either the trademark owner or the cybersquatter); (ii) waiting some decent interval of time before registering the corporation name as a domain name; and (iii) resisting the trademark owner's challenge under the Policy by claiming that the fact of the registration of the corporation proves that the corporation has been 'commonly known by' the corporation name/trademark/domain name, and therefore has a legitimate interest in the domain name.

In the absence of evidence of actual use of the name, the panellist held that the respondent was not commonly known by the name.

- *CITGO Petroleum Corporation v Matthew S Tercsak.*[155] The respondent registered the domain name <mystic.com> and was the president of Mystik, Inc, which conducted business under the 'MYSTIK' name. The evidence indicated that the respondent had been doing business under the 'MYSTIK' name before the domain name was registered. As the respondent had used the name in the context of his business, he had been commonly known by that name by his business customers prior to registration of the domain name, and therefore had legitimate interests in the domain name.
- *Toyota Motor Sales USA Inc v J Alexis Productions.*[156] The respondent registered the domain name <lexusmichaels.com>, and presented evidence that he had used the name 'LEXUS MICHAELS' for some time as a stage name in the adult entertainment industry. In the absence of evidence to the contrary, the panellist held that the respondent had legitimate interests in the domain name.

[6.8.3] Domain Holder has been Commonly Known by the Domain Name: Can the Respondent be 'Commonly Known' by a Nickname?

There has been controversy concerning whether a respondent who is commonly known by a nickname can establish legitimate interests in a domain name that reflects the nickname. As the terms of paragraph 4(c)(ii) do not distinguish between personal names and other names by which the respondent is 'commonly known', there is no reason why the defence cannot apply to nicknames in the same way that it applies to other names, including personal names. The key issue in these disputes is whether the evidence is sufficient to establish that the respondent has been 'commonly known' by the nickname. Panels should review such evidence carefully, to guard against the opportunistic adoption of a name purely for the purpose of establishing rights or interests in a famous mark.

A number of decisions have held that a respondent that is 'commonly known' by a nickname has rights or legitimate interests in a domain name that reflects the nickname. In *Penguin Books Limited v The Katz Family,*[157] for example, the respondent registered <penguin.org> and presented evidence that he was known by the nickname 'PENGUIN' long before the domain name was registered. In the absence of evidence to the contrary, the panel

[155] WIPO Case No D2003-0003 (28 Feb 2003).
[156] WIPO Case No D2003-0624 (16 Oct 2003).
[157] WIPO Case No D2000-0204 (20 May 2000).

held that it was reasonable for someone to register a domain name based on a nickname and that the respondent had established legitimate interests in the disputed domain name.

Other decisions have held that the respondent has failed to present sufficient evidence that he or she has been commonly known by the nickname. For example, in *The Clorox Company v Marble Solutions,*[158] the respondent registered the domain name <clorox.org>, claiming that he used the name 'CLOROX' as an email address and was known by that name by friends. As the respondent did not provide evidence of widespread use of the alleged nickname, the panel concluded that the respondent was not commonly known by the name. Similarly, in *E Remy Martin v Ramy Fahel,*[159] the respondent registered the domain names <remyxo.com>, <remyxo.net> and <remyxo.org>, and claimed that he was known by the nickname 'REMYXO', asserting that 'remy' was the common pronunciation of his first name, and 'XO' stood for hugs and kisses. In support of these assertions, the respondent presented a list of the names and e-mail addresses of people who allegedly knew him by the nickname, and a picture of him in a hat bearing the word 'REMYXO'. The panellist, however, held that, in the absence of persuasive documentary evidence such as testimonies or correspondence, the respondent had not presented sufficient evidence to establish that he was commonly known by the name.

The way in which the domain name is being used may influence the panel to conclude that an alleged nickname is not genuine. In particular, if the domain name is being used to trade off the complainant's mark unfairly, the respondent will find it difficult to establish legitimate interests in the domain name on the basis of an alleged nickname. In *Kraft Foods North America Inc v The Pez Kiosk,*[160] for example, the respondent registered the domain name <sugus.com>, and claimed that close friends knew him by the nickname 'SUGUS'. The domain name was used to promote the products of the complainant trade mark owner's competitors. The panellist held that, even if the respondent was known by the nickname by some friends, given the way in which the name was used, this was insufficient to establish that the respondent had rights or interests in the domain name. Presumably, the panellist held that the evidence was insufficient to establish that the respondent was 'commonly known' by the name.

Given the wording of paragraph 4(c)(ii), which refers to the 'domain holder' being commonly known by the domain name, decisions that have held that rights or interests can be established on the basis of a nickname of a family member or the name of a family pet should be viewed with circumspection. The following are examples of decisions that have dealt with this issue.

- *Jollibee Foods Corporation v Graham Chrystman.*[161] The respondent registered the domain name, <jollibee.com>, and claimed legitimate interests in it on the basis that 'JOLLIBEE' was his wife's long-standing nickname. The panellist held that the respondent was making a legitimate non-commercial or fair use of the domain name under paragraph 4(c)(iii), as the name was a family nickname and the domain name had been used only for the purposes of the family.
- *Vanguard Medica Limited v Theo McCormick.*[162] The respondent registered the domain name <miguard.com>, claiming that 'MIGUARD' was the name of his guard dog. Despite

[158] WIPO Case No D2001-0923 (20 Nov 2001).
[159] WIPO Case No D2001-1026 (11 Oct 2001).
[160] WIPO Case No D2001-1191 (8 Nov 2001).
[161] NAF Case No FA95561 (31 Oct 2000).
[162] WIPO Case No D2000-0067 (3 Apr 2000).

the suspicion aroused by the timing of the registration of the domain name, which coincided with the announcement of the complainant's 'MIGUARD' product, the panellist held that, as there was no basis to discredit the respondent's contention, he had established legitimate interests in the domain name.

- *Finter Bank Zurich v Gianluca Olivieri.*[163] The respondent registered the domain names <finter.com> and <finter.org>, and claimed that the names were based on his dog's name, 'RUBY DER FINTER', and presented vaccination certificates in support of this assertion. Taking into account that the sole use of the domain names by the respondent was to offer them for sale to the complainant, the panellist held that the respondent had no legitimate interests in the domain name.
- *Tower Laboratories Ltd v Eric Seltzer.*[164] The respondent registered the domain name <bromoseltzer.com>, claiming that it was a combination of his surname, 'SELTZER', and the name of his father's deceased dog, 'BROMO'. Prior to notification of the dispute, the respondent's web site displayed an image of the complainant's product that, upon notification of the dispute, was replaced by an image of a child with the statement 'Upset Stomach'. Taking into account the use of the domain name, the panellist concluded that the respondent had registered the domain name because of its commercial notoriety, and not because it reflected a name by which the respondent was commonly known.

As can be seen from these examples, the decisions dealing with the status of nicknames by which the respondent is allegedly known are not entirely satisfactory. It is suggested, however, that greater consistency could be achieved by the application of two basic principles. First, UDRP panels must scrutinise with great care the evidence presented to support a contention that a respondent has been commonly known by a nickname in order to ensure that the nickname has not been fabricated for opportunistic reasons. Secondly, paragraph 4(c)(ii) is, by its terms, concerned with whether the respondent is commonly known by the name, and not with whether a family member or family pet is commonly known by the name. Consequently, if rights or legitimate interests are to be successfully asserted in a name other than one by which the respondent is commonly known, it must be on the basis that the respondent is making a legitimate non-commercial or fair use of the domain name under paragraph 4(c)(iii), and not on the basis of paragraph 4(c)(ii).

Non-commercial or Fair Use

[6.9] Legitimate Non-commercial or Fair Use

The third of the circumstances set out in paragraph 4(c) which, if established, mean that the respondent has rights or legitimate interests in the disputed domain name provides that such rights or interests arise where:

> (iii) the domain name holder is making a legitimate non-commercial or fair use of the domain name, without intent for commercial gain to misleadingly divert consumers or to tarnish the trademark or service mark at issue.

[163] WIPO Case No D2000-0091 (23 Mar 2000).
[164] NAF Case No FA791325 (16 Oct 2006).

This defence recognises the importance of non-commercial considerations, in that it is aimed at balancing trade mark rights against other non-commercial rights and interests, including 'fair use' rights and rights to (or interests in) freedom of expression. The recognition of non-commercial rights and interests was foreshadowed by the WIPO Final Report, which stated that:

> Domain name registrations that are justified by legitimate free speech rights or by legitimate non-commercial considerations would . . . not be considered to be abusive.[165]

The protection of non-commercial rights or interests, in part, does no more than recognise the necessarily limited scope of trade mark rights. Since trade mark law confers rights over words, a complete monopoly over use of a mark would deplete the language. As the US Court of Appeals for the Ninth Circuit put it in *New Kids on the Block v News America Publishing Inc:*[166]

> the primary cost of recognizing property rights in trademarks is the removal of words from (or perhaps non-entrance into) our language.

The balance between protecting rights in a trade mark, on the one hand, and non-commercial rights and interests in a domain name that reflects the mark, on the other, raises different issues from those raised by the use of marks more broadly. The issues, which arise most clearly in connection with domain names that take the form <trademark.TLD>, are different because the use of a trade mark as a domain name is different from other uses of a trade mark. On one view, preventing others from using a mark as a domain name does not unduly restrain freedom of expression, as it is perfectly possible for commentary on or criticism of a trade mark owner to be made on a site identified by a domain name that is not confusingly similar to the mark. The other view, however, is that freedom of expression necessarily includes the ability to reflect a mark as a domain name. The different approaches to freedom of expression on the Internet are illustrated by the different views adopted by UDRP panels in relation to criticism sites, which are dealt with at [6.9.7]–[6.9.11].

There are two main underlying difficulties that have arisen in the application of paragraph 4(c)(iii). First, associated with the different views of free speech rights on the Internet, there are considerable differences in the extent to which national laws recognise non-commercial legal 'rights', including rights to freedom of expression. This invariably results in inconsistent outcomes, depending upon the choice of law. In particular, UDRP panels that apply US law accord greater protection to a respondent's rights to freedom of expression than do panels applying other national laws. Secondly, as explained at [3.8], the wording of the defence is not especially helpful. In particular, the wording adopted leaves it open for respondents that operate criticism sites to be denied rights or interests in a domain name on the basis of an intention to use the domain name to 'tarnish' the trade mark. This is discussed further at [6.9.6].

A number of panels have stated that the following needs to be established for paragraph 4(c)(iii) to be made out:

1. the respondent's use is legitimate non-commercial or fair use of the domain name;
2. this use is without intent for commercial gain;

[165] WIPO, above n 1, para [172].
[166] 971 F 2d 302, 306 (9th Cir, 1992).

3. this use is without intent to misleadingly divert consumers; and
4. this use is without intent to tarnish the trademark of the complainant.[167]

This formulation does not, however, adequately explain the relationship between the separate elements of the defence. The preferred approach to establishing the elements of paragraph 4(c)(iii) was adopted by the panellist in *Howard Jarvis Taxpayers Association v Paul McCauley*.[168] On this approach, the panel first looks at whether the respondent is making a legitimate non-commercial or fair use of the domain name. In considering this question, the panel must assess the bona fides of the respondent, as if the domain name has been registered in bad faith, such as to trade off the goodwill of the complainant's mark, there can be no legitimate non-commercial or fair use. A finding that there is a legitimate non-commercial or fair use is not, however, sufficient for the defence to be established, as the panel must go on to consider whether there is an intent for commercial gain misleadingly to divert consumers or, in the alternative, an intent for commercial gain to tarnish the complainant's mark. For each of the elements, it is for the complainant to establish a prima facie case and, once that has been satisfied, for the respondent to establish that the defence has been made out on the balance of probabilities.

The scope of the defence set out in paragraph 4(c)(iii) should be understood against the background of the US legal doctrines of trade mark fair use, the treatment of 'parody' under US trade mark law, and the US doctrine of trade mark tarnishment. While the approach taken to these issues under US law is not controlling of the approach adopted under the UDRP, an explanation of the doctrines helps to elucidate the interpretation of paragraph 4(c)(iii). For example, an understanding of the US fair use doctrines indicates that a commercial use of a mark may still amount to a fair use, thereby assisting in distinguishing what is meant by a 'legitimate non-commercial use' from what is meant by a 'fair use'.

[6.9.1] US 'Classic Fair Use' Doctrine

The term 'fair use' is sometimes loosely used to refer to all non-infringing uses of a trade mark, but under US trade mark law it has a narrower, technical meaning. The technical meaning given to the term under US law is not determinative of the meaning of the term as used in paragraph 4(c)(iii) of the UDRP. Reference to US law, however, helps to give the term some meaning beyond merely all and every non-infringing use of a mark.

US trade mark law distinguishes between two, quite different, forms of fair use, known as 'classic fair use' and 'nominative fair use'. 'Classic fair use' provides a defence to trade mark infringement where a descriptive mark is used not in a trade mark sense, but in a descriptive sense, including as a description of the defendant's goods or services. As the *Restatement (Third) of Unfair Competition* puts it:

> Fair use is a reasonable and good faith use of a descriptive term that is another's trademark to describe rather than to identify the user's goods, services or business.[169]

'Nominative fair use', on the other hand, means the use of the claimant's mark to refer to the claimant's goods or services, which has been held to amount to a defence where there is no likelihood of confusion. 'Nominative fair use' is dealt with at [6.9.2] below.

[167] See *Mission KwaSizabantu v Benjamin Rost*, WIPO Case No D2000-0279 (7 June 2000); *Legal & General Group Plc v Image Plus*, WIPO Case No D2002-1019 (30 Dec 2002).
[168] WIPO Case No D2004-0014 (22 Apr 2004).
[169] *Restatement (Third) of Unfair Competition*, § 28, comment c (1995) .

The common law 'classic fair use' defence has been codified by § 33(b)(4) of the federal Lanham Act, which provides:

> use of the name, term, or device charged to be an infringement is a use, otherwise than as a mark, . . . of a term or device which is descriptive of and used fairly and in good faith only to describe the goods or services of such party, or their geographic origin . . .[170]

In *Seaboard Seed Co v Bemis Co*[171] a federal district court held that the statutory defence requires the defendant to establish the following three elements:

1. the term is not used as a trade mark or service mark;
2. the defendant has used the term 'fairly and in good faith'; and
3. the term has been used 'only to describe' the defendant's goods or services.

The need for the defence arises because a descriptive term that has acquired a secondary meaning has two uses: first, it retains its primary descriptive meaning and, secondly, it has an acquired meaning, as a mark that distinguishes the owner's goods or services'.[172] As trade mark law only protects against use of the term as a mark, anyone remains free to use the term in its descriptive sense.

The principal authority on the 'classic fair use' doctrine is the Supreme Court decision in *KP Permanent Make-Up, Inc v Lasting Impression I, Inc*,[173] which concerned whether a defendant relying upon the defence must establish that there is no likelihood of confusion in order for a descriptive term to be used 'fairly and in good faith'. In that case, Lasting Impression, which had registered the mark 'MICRO COLOR' in relation to cosmetic and medical make-up, brought an action against KP for use of the term on a colour chart. KP claimed that its use of the mark fell within the fair use defence, but the Ninth Circuit Court of Appeal held that there could be no fair use where there was a likelihood of confusion. On appeal, the Supreme Court unanimously held that a defendant seeking to rely on the fair use defence does not bear the burden of establishing that there is no likelihood of confusion. Moreover, the Supreme Court held that some degree of confusion was compatible with the fair use defence, explaining that:

> The common law's tolerance of a certain degree of confusion on the part of consumers followed from the very fact that in cases like this one an originally descriptive term was selected to be used as a mark, not to mention the undesirability of allowing anyone to obtain a complete monopoly on use of a descriptive term simply by grabbing it first.[174]

At the same time, the court held that the possibility of consumer confusion is a factor to be taken into account in determining whether a use is fair. Similarly, the extent to which an originally descriptive term has become so distinctive that a defendant's use cannot realistically be descriptive is relevant in determining whether there is a fair use.

From this, we can derive the principle that use of a descriptive mark as a domain name in a way that is descriptive of a respondent's goods or services may amount to a fair use of the mark. If we draw upon US case law, the extent to which the mark has been registered and used in good faith is a key factor in determining whether there is a fair use. For instance, use

[170] Lanham Act § 33(b)(4), 15 USC § 1115(b)(4).

[171] 632 F Supp 1133, 1138 (ND Ill, 1986).

[172] See JT McCarthy, *Trademarks and Unfair Competition*, 4th edn (Eagan, Minn, Thompson/West 2006) § 11:45.

[173] 543 US 111; 125 S Ct 542, (2004).

[174] 543 US 111, 122; 125 S Ct 542, 550 (2004).

of a mark with an intent to trade on the goodwill of the trade mark owner by creating source confusion will not amount to a fair use.[175] Thus, in *Washington Speakers Bureau, Inc v Leading Authorities, Inc,*[176] the court held that the use of the claimant's 'WASHINGTON SPEAKERS BUREAU' mark as the <washingtonspeakers.com> domain name was not a fair use as the defendant had chosen the domain name in bad faith to attract business from the claimant trade mark owner.

Some UDRP panel decisions have directly incorporated principles derived from the US fair use doctrine in considering whether a respondent has been making a fair use under paragraph 4(c)(iii). In *The Orange Bowl Committee, Inc v Front and Center Tickets, Inc,*[177] for example, the respondent registered the domain names <orangebowl.net> and <orangebowltickets.net>, which it used to redirect users to a site that resold event tickets. The panel held that this was not a bona fide offering of goods or services under paragraph 4(c)(i), as the mark was being used to promote the respondent's services. Moreover, as the use of the domain name was commercial, the availability of the paragraph 4(c)(iii) defence depended on whether or not the use was a fair use. The panel was split on this issue. In deciding that the use of the term 'ORANGE BOWL TICKETS' was not a fair use, the majority distinguished the *KP Permanent Make-Up* case on the ground that, in this dispute, the complainant's famous and distinctive mark was being used to attract Internet users to a web site that offered products in competition with those of the trade mark owner. The dissenting panellist, however, maintained that there was a fair use on the basis that the term 'ORANGE BOWL' was being used in a descriptive sense, as the name of an event, and not in a trade mark sense. Moreover, the dissenting panellist considered there was no diversion of users to competing products as the respondent was offering the trade mark owner's product, and not a competing product.

The outcome in the *Orange Bowl* dispute may be contrasted with the decision in *Dewey Ballantine LLP v Person Technology Group,*[178] which held that the use of a mark as a domain name in a descriptive sense was a fair use. In *Dewey Ballantine,* the complainant had registered rights in the 'DB' and 'DEWEY BALLANTINE LLP' marks and the respondent registered the <dbllplaw.com> domain name. The panellist held that, as the respondent was using the mark in a purely descriptive sense as the name for its law firm, the respondent was making a fair use of the domain name. This finding was reinforced by the absence of evidence that the respondent had registered the domain name with an intent misleadingly to divert consumers, and the fact that the complainant's mark had a relatively low level of distinctiveness.

[6.9.2] US 'Nominative Fair Use' Doctrine

'Nominative fair use' must be distinguished from 'classic fair use' in that it applies where the plaintiff's mark is used to identify the plaintiff's goods or services, such as in comparative advertising. Provided that the use does not create a likelihood of confusion, nominative fair use is a defence to an action for trade mark infringement.

[175] See McCarthy, above n 172, § 11:49, citing *EMI Catalogue Partnership v Hill, Holliday, Connors, Cosmopulos Inc*, 228 F 3d 56 (2d Cir, 2000).

[176] 33 F Supp 2d 488 (ED Va, 1999).

[177] WIPO Case No D2004-0947 (20 Jan 2005).

[178] WIPO Case No D2004-1032 (27 Jan 2005).

The nominative fair use defence was first recognised by the Ninth Circuit Court of Appeals in *New Kids on the Block v News America Publishing*.[179] In that case, *USA Today* conducted a telephone poll of the relative popularity of the five members of the 'New Kids' pop group, and published the results. In an action for trade mark infringement, the court held that this was a non-infringing nominative fair use of the mark, which did not imply an association with the trade mark owner.

Nominative fair use therefore involves a consideration of whether a use creates a likelihood of confusion, but the factors to be taken into account where there is a claim of nominative fair use are different from those that are usually taken into account in determining whether there is a likelihood of confusion. In the *New Kids* case, the court held that there would be a nominative fair use where the following three factors were satisfied:

1. The product or services in question cannot be readily identified without use of the trade mark;
2. The user only uses so much of the mark as is reasonably necessary to identify the product or service;
3. The user must do nothing that would suggest sponsorship or endorsement by the trade mark owner.[180]

Where there is a claim for nominative fair use, once the plaintiff has made out a prima facie case of a likelihood of confusion, the burden of proof shifts to the defendant to establish the above three factors. In the Internet context, the nominative fair use defence was held to apply where Terri Welles, a former Playmate of the Year, used the 'PLAYBOY' and 'PLAYMATE' marks to describe her title on her web site and in the web site metatags.[181]

The US doctrine of nominative fair use has been adopted and applied in a number of UDRP panel decisions. In *Pfizer Inc v Van Robichaux*,[182] for example, the respondent, an attorney specialising in medical injuries, registered the domain name <lipitorinfo.com>, claiming that he intended to use it to provide information about the complainant's 'LIPITOR' product. The complainant argued that there could be no fair use of the domain name, as the choice of the domain name created initial interest confusion concerning the sponsorship or endorsement of the respondent's web site. The respondent, however, argued that the use of the domain name was a nominative fair use of the complainant's mark, as the domain name was being used purely to identify the complainant's product. The panellist agreed that there might be some initial interest confusion arising from the use of the mark in the domain name, but observed that, if nominative fair use is to be permitted under paragraph 4(c)(iii), initial interest confusion cannot act as a per se preclusion. Although the respondent intended to make a commercial use of the domain name in conjunction with his practice as an attorney, the panellist pointed out that this did not prevent a finding of nominative fair use. The panellist therefore held that the respondent's proposed use of the domain name was sufficient in this instance to rebut the complainant's prima facie case that the respondent had no rights or legitimate interests in the domain name.

A number of decisions have held that the respondent is not making a fair use of the disputed domain name because one or more of the elements of nominative fair use established by the *New Kids* case have not been made out. In *Tacitica International, Inc v*

[179] 971 F 2d 302 (9th Cir, 1992).
[180] See 971 F 2d 302, 308 (9th Cir, 1992).
[181] *Playboy Enterprises, Inc v Welles*, 279 F 3d 796 (9th Cir, 2002).
[182] WIPO Case No D2003-0399 (16 July 2003).

YouCanSave.com, Inc,[183] for example, the respondent, who was not a licensee or authorised distributor of the complainant, registered the domain name <epilstopspray.com>, which was used on a web site that sold the complainant's products. The panellist held that there was no nominative fair use as the elements of the defence had not been satisfied. First, it was not necessary, in these circumstances, for the complainant's mark to be used in the domain name, the respondent being quite entitled to use its own business name. Secondly, the use of the domain name suggested a sponsorship or endorsement, which was not dispelled by an insufficient indication that the respondent was an 'independent' distributor contained deep within the respondent's web site. The panellist therefore held that the respondent had no rights or legitimate interests in the domain name.

A similar conclusion was reached in *Pepperdine University v BDC Partners, Inc*.[184] In that dispute, the respondent acquired the domain names <pepperdineuniversitywaves.com> and <pepperdineuniversitywaves.net>, together with over 23,000 other college, university and high school domain names. The respondent claimed that it was developing the domain names, which included the name of the university and the university's mascot name, 'WAVES', as part of a series of online social networking communities for students and alumni of universities and schools. The respondent argued that it was making a nominative fair use of the domain names as it could not identify the university without use of the complainant's marks. The panellist held that the use of the domain names was neither a classical fair use nor nominative fair use. In relation to nominative fair use, the panellist held that, as the web sites were inactive, there was no actual use of the domain names, and that, in any case, the names were not being used in a descriptive sense, but more as source indicators.

The panellist went on to find that none of the elements of nominative fair use had been made out. First, 'PEPPERDINE UNIVERSITY WAVES' was not the only way to describe the complainant's goods or services. Secondly, more of the mark had been used than was reasonably necessary, as 'WAVES' and possibly 'UNIVERSITY' were not necessary to describe the complainant's goods and services. Thirdly, although the respondent claimed that it would include a disclaimer on the web sites, the panellist held that the domain names themselves suggested sponsorship or endorsement by the university.

[6.9.3] Freedom of Expression and the Treatment of Parody under US Trade Mark Law

Under US law, the First Amendment may be a defence to an action for trade mark infringement where a defendant claims free speech rights in the communication of a message to the public. Where the mark is used in a commercial context, such as on merchandising, it is unlikely that the First Amendment defence will be made out. In such cases, the courts are likely to hold that there are other ways of communicating a message that do not involve use of the mark in association with goods or services.[185]

The non-commercial use of a mark to communicate a message, such as criticism of the trade mark owner, is quite a different matter. In such cases, the First Amendment ensures that trade mark law is not used to suppress criticism. As conventional trade mark infringement actions are available only in relation to use of a mark in a trade mark sense, as a source identifier, the First Amendment defence is much more relevant to actions for trade mark

[183] WIPO Case No D2002-0018 (12 Apr 2002).
[184] WIPO Case No D2006-1003 (25 Sept 2006).
[185] See, eg, *Mutual of Omaha Insurance Co v Novak*, 648 F Supp 905 (D Neb, 1986).

dilution, and especially to claims of tarnishment. In this respect, the *Restatement (Third) of Unfair Competition* states:

> Non-trademark uses, which do not involve a use to identify another's goods, service, or business . . ., are unlikely to have . . . [a] . . . diluting effect. In most instances such uses are intended to refer back to the original trademark owner and serve to confirm rather than undermine the associational significance of the mark.[186]

As explained further at [6.9.10] below, the Ninth Circuit in *Bally Total Fitness Holding Corp v Faber*[187] held that use of the plaintiff's 'BALLY' mark in the sub-domain 'BALLY SUCKS' did not amount to tarnishment of the mark, as consumer commentary is speech protected under the First Amendment.

Under US trade mark law, the treatment of parody is a particular application of the protection accorded to freedom of expression. In general, the use of a mark in a parody is a factor that is taken into account in determining whether there is a likelihood of confusion. As McCarthy points out:

> There are confusing parodies and non-confusing parodies. All they have in common is an attempt at humor through the use of someone else's trademark. A non-infringing parody is merely amusing, not infringing.[188]

In the analogous area of copyright protection, US law draws a distinction between 'parody', which makes fun of the original, and 'satire', which uses the original work in a humorous fashion to comment on something else. It is likely that the same distinction applies in the trade mark context. As a parody must bring to mind the object being parodied, the court in *Cliffs Notes, Inc v Bantam Doubleday Dell Publishing Group* held that a parody must communicate two contradictory messages:

> that it is the original, but also that it is *not* the original and is instead a parody.[189]

The basic principle in trade mark cases, however, remains that trade mark rights protect against confusion as to source, and not against criticism, including humorous criticism.

The treatment of a domain name that resolved to a web site that ostensibly parodied a trade mark owner was considered by the Fourth Circuit Court of Appeals in *People for Ethical Treatment of Animals (PETA) v Doughney.*[190] In that case, the defendant registered the domain name <peta.org>, and used it for a web site purporting to be for 'People Eating Tasty Animals', and containing links to web sites such as those offering meat, fur and leather products. The court held that the domain name itself was not a parody as it conveyed the message that it was related to the 'PETA' mark, but not the contradictory message required by *Cliffs Notes,* that the domain name was a parody. Rejecting the suggestion that the second meaning was conveyed by the contents of the web site, the court held that the two messages had to be communicated simultaneously and, as they were not in this case, there remained a likelihood of confusion.

Following the reasoning in the *PETA* case, some respondents in UDRP disputes have claimed rights or interests in a domain name on the basis that the domain name itself is a parody.

[186] *Restatement (Third) of Unfair Competition,* above n 169, § 25, comment i (1995).
[187] 29 F Supp 2d 1161 (CD Cal, 1998).
[188] McCarthy, above n 172, § 31:153.
[189] 886 F 2d 490, 494 (2d Cir, 1989) (emphasis in original).
[190] 263 F 3d 359 (4th Cir, 2001).

The most extensive consideration of the issue was in *A & F Trademark, Inc and Abercrombie & Fitch Stores, Inc v Justin Jorgensen*,[191] where the respondent registered the domain name <abercrombieandfilth.com>, claiming that it was a parody of the complainant's mark. The domain name was used for a gay-oriented pornographic web site. The panellist first accepted that, although not specifically mentioned in paragraph 4(c) of the UDRP, a parodic use of a mark as a domain name could establish rights in the domain name. The panellist then proposed the following two stage test for determining whether the respondent's mark amounts to a parody:

> In the first step, this Panel determines whether the domain name itself has the capacity to constitute parody. If the Panel can answer this question in the affirmative, it will then examine whether the Respondent's use of that domain name, on its web site or otherwise, is consistent with its claim of parody.

In the instant dispute, the panellist held that the domain name itself, <abercrombieandfilth.com>, satisfied the *Cliffs Notes* standard of communicating two contradictory messages. Turning to the content of the web site, however, the panellist found that there was no parodic content on the web site, and that the respondent's motive was purely commercial. Consequently, the respondent did not satisfy the second stage of the proposed test, and had no rights or interests in the domain name.

A more comprehensive discussion of the extent to which UDRP decisions have recognised rights or legitimate interests based on freedom of expression is found in the discussion of criticism sites at [6.9.7]–[6.9.11].

[6.9.4] Tarnishment

The term 'tarnishment' has a technical meaning under US anti-dilution law being, along with 'blurring', one of the two conventionally recognised forms of trade mark dilution. The US Lanham Act defines trade mark dilution to mean:

> the lessening of the capacity of a famous mark to identify and distinguish goods or services, regardless of the presence or absence of—
>
> (1) competition, between the owner of the famous mark and other parties, or
> (2) likelihood of confusion, mistake, or deception.[192]

Actions for trade mark dilution therefore protect against uses that erode the distinctiveness of a mark where there is no likelihood of confusion. Dilution by tarnishment refers to a lessening of distinctiveness of a famous mark through the use of the mark in a way that creates negative associations. In *LL Bean, Inc v Drake Publishers, Inc*, one of the most important US tarnishment cases, the First Circuit referred to tarnishment in the following terms:

> The threat of tarnishment arises when the goodwill and reputation of a plaintiff's trademark is linked to products which are of shoddy quality or which conjure associations that clash with the associations generated by the owner's lawful use of the mark.[193]

Tarnishment will often be found where a famous mark is used in connection with degrading or unwholesome material, such as illegal drugs or pornographic material. The following

[191] WIPO Case No D2001-0900 (19 Sept 2001).
[192] Lanham Act, § 45, 15 USC § 1127.
[193] 811 F 2d 26, 31 (1st Cir, 1987).

are examples of cases where the courts have held that particular uses of marks amount to tarnishment.

- *Coca-Cola Co v Gemini Rising, Inc*[194] ('ENJOY COCAINE' in script and colour identical to the 'COCA-COLA' mark).
- *Steinway & Sons v Demars & Friends*[195] ('STEINWAY' mark used on clip-on beer can handles).
- *Hasbro, Inc v Internet Entertainment Group*[196] ('CANDYLAND' mark used by pornographic Web site site <candyland.com>).
- *Toys 'R' Us, Inc v Akkaoui*[197] ('TOYS "R" US' mark used by <adultsrus.com> Web site in connection with the sale of adult products).
- *Polo Ralph Loren LP v Schuman*[198] ('POLO' mark used on 'POLO CLUB' adult entertainment business).
- *New York Stock Exchange, Inc v New York, New York Hotel LLC*[199] ('NEW YORK STOCK EXCHANGE' mark used by Las Vegas casino).

The US doctrine of trade mark tarnishment does not determine the meaning of the term in paragraph 4(c)(iii), but assists in the interpretation of the provision, especially given the potential for panellists without an understanding of the technical meaning of the term to regard any form of criticism as tarnishment. In this respect, it is significant that the free speech protection accorded by the First Amendment limits the extent to which actions for tarnishment can be used to prevent use of a mark for the purposes of genuine criticism or parody.

The balance struck under US law between protecting a mark against tarnishment and preserving freedom of expression is illustrated by the decision of the First Circuit Court of Appeals in *LL Bean, Inc v Drake Publishers, Inc.*[200] In that case, the defendant's adult magazine included an article that displayed the 'LL BEAN' mark with nude models in sexually explicit poses. As the defendant was not using the mark to promote its own products, but as part of a non-commercial parody, the Court of Appeal reversed the first instance finding of tarnishment. In reaching this conclusion, the court explained the relationship between the anti-dilution law and the protection of freedom of expression in the following terms:

> If the anti-dilution statute were construed as permitting a trademark owner to enjoin the use of his mark in a noncommercial context found to be negative or offensive, then a corporation could shield itself from criticism by forbidding the use of its name in commentaries critical of its conduct. The legitimate aim of the anti-dilution statute is to prohibit the unauthorized use of another's trademark in order to market incompatible products or services. The Constitution does not, however, permit the range of the anti-dilution statute to encompass the unauthorized use of a trademark in a noncommercial setting such as an editorial or artistic context.[200a]

In applying the balance between protection against tarnishment and freedom of expression in the context of the UDRP, it is important to bear in mind the need for any criticism or parody to be genuine. In *Microsoft Corporation v Paul Horner,*[201] for example, the respondent

[194] 346 F Supp 1183 (EDNY, 1972).
[195] 210 USPQ 954 (CD Cal, 1981).
[196] 40 USPQ 2d 1479 (WD Wash, 1996).
[197] 40 USPQ 2d 1836 (ND Cal, 1996).
[198] 46 USPQ 2d 1046 (SD Tex, 1998).
[199] 293 F 3d 550 (2d Cir, 2002).
[200] 811 F 2d 26 (1st Cir, 1987).
[200a] 811 F 2d 26, 33 (1st Cir, 1987)
[201] WIPO Case No D2002-0029 (27 Feb 2002).

used the domain name <microsoftsite.com>, on a website that contained pornographic material and links to pornographic sites. The respondent argued that the domain name was being used for the purpose of parody. In holding that the respondent was not making a legitimate non-commercial or fair use of the domain name, the panellist pointed out that the web site was not labelled as parody or humour. More importantly, the panellist distinguished *LL Bean* on the basis that, unlike the magazine in that case, the domain name in the instant dispute was used to attract users who would not expect to find pornographic material.

[6.9.5] Legitimate Non-commercial or Fair Use 'Without Intent for Commercial Gain': the Position of 'Commercial' Sites and 'Sham Speech' Domain Names

The affirmative defence set out in paragraph 4(c)(iii) applies only where a use is being made of the domain name without a commercial intent misleadingly to divert consumers or tarnish the complainant's mark.

In this respect, the *WIPO Overview* states:

> This section only concerns sites that practice genuine, non-commercial criticism. There are many UDRP decisions where the respondent argues that the domain name is being used for a free speech purpose, but the panel finds that, in fact, the domain name is being used for commercial gain.[201a]

To begin with, there must be a genuine intention to use the domain name for a legitimate non-commercial purpose, otherwise the door would be left open for opportunistic respondents to fabricate criticism sites in order to mask 'abusive, bad faith' registrations. The expedient of dressing up an abusive registration as an exercise of freedom of expression in order to take advantage of paragraph 4(c)(iii) is most clearly illustrated by the so-called 'sham speech' domain names. 'Sham speech' domain names are domain names selected not for the purpose of establishing a genuine criticism site, but to attract or divert Internet users for an ulterior commercial purpose. The following decisions illustrate the approach taken to domain names that purport to be for a 'legitimate' non-commercial use, but that have been registered for an ulterior motive.

- *Rolex Watch USA, Inc v Spider Webs, Ltd.*[202] The respondent registered the domain names <relojesrolex.com> and <erolexwatches.com>, which were used to redirect users to the respondent's <spintopic.com> web site. The <spintopic.com> site provided consumer information regarding a wide range of products and free forums for exchanging consumer information. The respondent had, in fact, registered a considerable number of domain names that directed users to the <spintopic.com> site, and obtained income from the number of 'hits' to that site. The complainant argued that, as there was no information whatsoever about 'ROLEX' watches, the site was no more than a Potemkin Village, that was designed solely to create the appearance of a 'legitimate interest' in the domain names. The panel held that, even if the use of the domain names was non-commercial, the diversion of users to a site that contained no information about 'ROLEX' products was clearly not a 'legitimate' or 'fair' use of the domain names. In this respect, the panel stated that:

> Accepting the Respondent's claim that its website is legitimate use would mean that cybersquatters could rightfully acquire domain names that incorporate famous trademarks by simply arranging for the domain names to resolve to generic and largely empty websites with no information or opportunities for commentary on the products known by the trademarks.

[201a] WIPO, above n 10, para 2.4.
[202] WIPO Case No D2001-0398 (2 July 2001).

- *National Collegiate Athletic Association v Dusty Brown.*[203] The respondent registered the domain names <ncaafootball2005.com> and <ncaafootball2006.com>. The first domain name initially resolved to a site that offered it for sale, but both sites later resolved to a site labelled 'NCAA Football 2005 Gripe Site'. The 'gripe site' included perfunctory criticism of the NCAA. The panellist found that the primary motive for registering the domain names was commercial, being to sell the names for more than the cost of registration, and that any intention to use the domain names as a source of criticism was subservient to the commercial purpose. The respondent therefore had no rights or legitimate interests in the domain names.
- *Medimmune, Inc v Jason Tate.*[204] The respondent registered a large number of domain names, including <synagisisbadforyou.com> and <synagisisnotsafe.com>. The domain names resolved to web sites that included disclaimers of an association with the complainant and statements of an intention to establish a criticism site for the 'SYNAGIS' product. The panellist held that the absence of any substantive comments on the sites indicated that the statements were no more than a pretext, and concluded that the respondent's likely motive was to use the domain names for 'pay-per-click' advertising revenue. This conclusion was reinforced by the fact that other domain names registered by the complainant resolved to sites that contained 'pay-per-click' advertising. The panellist concluded that this was not a 'legitimate' or 'fair' use of the domain names.

The requirement that the use be a 'legitimate non-commercial or fair use' means that the defence cannot apply where there is evidence of 'abusive, bad faith' registration. Where the evidence indicates that the domain name was registered in bad faith for a commercial purpose, a respondent's claims that there was a 'legitimate' purpose for the registration of the domain name, such as for use as a non-commercial criticism site, are clearly false and cannot be accepted. The bona fides of the respondent should be considered in determining whether there is a legitimate non-commercial or fair use, before the panel turns to consider the other elements of the defence.

For example, in *Wal-Mart Stores, Inc v Walsucks and Walmarket Puerto Rico,*[205] the respondent argued that the disputed domain names, including <walmartcanadasucks.com> and <walmartpuertoricosucks.com>, were being used as forums for criticism of the complainant. The respondent further claimed that transferring the domain names to the complainant would effectively silence criticism of the complainant, especially as the complainant had already acquired most other 'sucks' sites related to the trade mark. This view of the use of the domain names was, however, contradicted by other evidence. For instance, in communications with the complainant, the respondent called himself a 'domain name consultant' and offered to sell the domain names to the complainant. The panellist held that the intention of registering the multiple domain names was commercial, and therefore fell outside the scope of paragraph 4(c)(iii). In reaching this conclusion, the panellist stated:

> A demand for payment from the potential and actual subject of critical sites is fundamentally inconsistent with the right of free expression. It is as if a newspaper were to approach the potential subject of an adverse investigative report to propose that for an appropriate fee the report could be avoided. This would not be characterized as 'free speech' activity. It would rather be characterized as 'extortion'.

[203] WIPO Case No D2004-0491 (30 Aug 2004).
[204] WIPO Case No D2006-0159 (14 Apr 2006).
[205] WIPO Case No D2000-0477 (20 July 2000).

The defence set out in paragraph 4(c)(iii) will not apply where there is some intent to use the domain name for commercial gain. The respondent is therefore not entitled to rely on the defence where there is evidence of some commercial use of the domain name, even though the commercial use is not substantial. In *The New York Times Company v New York Internet Services,*[206] for example, the respondent registered the domain name <newyorktimes.com> and claimed that the site was to be used for commentary on the 'NEW YORK TIMES' newspaper. A commentary site was, however, only established after the complainant contacted the respondent. Moreover, the respondent had a web hosting agreement that provided free hosting in return for banner advertisements on the web site. The panellist held that, as the banner advertisements were part of a commercial arrangement, the respondent's use of the domain name was not a non-commercial or fair use.

Where the disputed domain name has been registered for a combination of commercial and non-commercial purposes, the respondent cannot rely on paragraph 4(c)(iii), even if the non-commercial purpose is the main reason for registering the domain name. In *Monty and Pat Roberts, Inc v J Bartell,*[207] for example, the respondent used the <montyroberts.org> domain name to divert users to a web site that contained information that was critical of the complainant, 'MONTY ROBERTS' (the famous 'horse-whisperer'). The site included links to other sites operated by the respondent that offered 'anti-Monty Roberts memorabilia' for sale, that advertised for sale a book that was critical of the complainant, and that advertised the services of competing horse trainers. The respondent claimed that he had legitimate interests in the domain name, as it was being used legitimately to criticise the complainant. The panellist, however, held that the respondent was using the domain name to divert Internet users to sites where merchandise was sold and where the services of the complainant's competitors were advertised. Consequently, even though the primary motive for establishing the site may have been to criticise the complainant, the fact that it was used for the commercial purpose of selling merchandise meant that paragraph 4(c)(iii) was not made out.

At the same time, where a genuine criticism site is used, in part, for the purpose of raising funds for a non-profit-making organisation, panels have held that this does not amount to an intent to use the domain name for commercial gain and that it does not prevent respondents from relying on paragraph 4(c)(iii). Decisions that have held that a non-profit-making respondent may use a domain name for fund raising and still make a legitimate non-commercial or fair use of the domain name include *Ahmanson Land Company v Save Open Space*[208] and *Covance, Inc v The Covance Campaign.*[209]

[6.9.6] Legitimate Non-commercial or Fair Use: Intent to 'Tarnish' the Complainant's Mark

Paragraph 4(c)(iii) provides that a legitimate non-commercial or fair use can only be made where there is no intent to 'tarnish' the complainant's mark. The wording adopted creates the potential for the defence to be emasculated by panels that interpret any criticism of a trade mark owner as a form of tarnishment.

[206] WIPO Case No D2000-0477 (5 Dec 2000).
[207] WIPO Case No D2000-0300 (13 June 2000).
[208] WIPO Case No D2000-0858 (4 Dec 2000).
[209] WIPO Case No D2004-0206 (30 Apr 2004).

The clearest example of a decision adopting a broad approach to what is meant by an intent to 'tarnish' the complainant's mark is *Mission KwaSizabantu v Benjamin Rost.*[210] In that dispute, the respondent registered the domain names <kwasizabantu.com>, <kwasizabantu.org> and <kwasizabantu.net>, and used them for criticism of the complainant, a Zulu religious organisation. The panellist held that the use of the sites to present critical views of the activities of the trade mark owner amounted to tarnishment and, on this basis, concluded that the respondent had no rights or legitimate interests in the domain names.

Some panel decisions have held that use of a domain name that is identical to the complainant's mark for a criticism site will invariably 'tarnish' the mark. In *The Royal Bank of Scotland Group and National Westminster Bank v Pedro Lopez,*[211] for example, the respondent registered domain names including <nationalwestminsterbank.com>, which resolved to a web site that provided a forum for criticism of the complainant bank. The panel held that by adopting domain names that were identical to the complainant's mark, and using them for uncontrolled criticism, the respondents were tarnishing the complainant's mark. In this respect, the panel stated:

> While one must concede that criticism per se is not a commercial use and that a criticism site may well be entirely legitimate, an argument can be made that a non controlled use can be anti-commercial when it has the effect of driving regular customers away from one's business. Having said that, the Panel also noted that the Respondents have used domain names that are identical and substantially similar to Complainants' trademarks to exercise their freedom of expression and this has the direct consequence of tarnishing Complainants' trademarks. Respondents' can very well achieve their objective of criticism by adopting a domain name that is not identical or substantially similar to Complainants' marks.

Decisions giving a broad scope to 'intent to tarnish' appear to be a misinterpretation of the terms of paragraph 4(c)(iii). If each of the components of the paragraph is to be given some meaning, then a separate role must be played by the requirement that the use be a legitimate non-commercial or fair use, and by the requirement that there be no intent to tarnish the complainant's mark. It is generally acknowledged that one of the main forms of legitimate non-commercial or fair use is the use of a mark to criticise the trade mark owner. To conclude that all criticism sites necessarily exhibit an intent to tarnish a complainant's mark would seem to render the inquiry into whether there is a legitimate non-commercial or fair use inoperative. Consequently, it is preferable for the meaning given to 'tarnishment' to be narrowly confined, and for the question whether a respondent who has established a criticism site has rights or legitimate interests in the domain name to be determined according to whether there is a legitimate non-commercial or fair use.

As explained at [6.9.4] above, the US doctrine of tarnishment applies to negative uses of a mark that create inferior or degrading associations with the mark, but not to uses of the mark to criticise the trade mark owner. This is the preferred interpretation to be applied to 'intent to tarnish' under paragraph 4(c)(iii). As the panellist put it in *Britannia Building Society v Britannia Fraud Prevention:*[212]

> Tarnishment in this context refers to such unseemly conduct as linking unrelated pornographic, violent or drug-related images or information to an otherwise wholesome mark.

[210] WIPO Case No D2000-0279 (7 June 2000).
[211] WIPO Case No D2002-0823 (3 Dec 2002).
[212] WIPO Case No D2001-0505 (6 July 2001).

On this approach, the question whether a respondent can have rights or interests in a domain name that is identical to a complainant's mark is best dealt with in determining what amounts to a legitimate non-commercial or fair use, and not in determining whether there is an intent for commercial gain to tarnish the mark.

The use of a confusingly similar domain name for the purpose of attracting or diverting users to pornographic material will clearly amount to an intent to tarnish the complainant's mark by establishing an association between the mark and the pornographic material. In *Nicole Kidman v John Zuccarini*,[213] for example, the respondent used the disputed domain name <nicolekidmannude.com> to redirect users to sexually explicit web sites that 'mouse-trapped' the use. The panel held that the sexual implications of the domain name and the use of the domain name for pornographic advertisements indicated an intent to tarnish the complainant's mark.

[6.9.7] Legitimate Non-commercial or Fair Use: Overview of the Treatment of Criticism Sites

There are persistent and continuing divisions among UDRP panellists concerning whether a respondent that uses a domain name that is confusingly similar to the complainant's mark for a genuine non-commercial criticism site has rights or legitimate interests in the domain name. The differences are deep-seated and reflect fundamentally different approaches to the balance between trade mark rights and freedom of expression in the context of registration and use of a confusingly similar domain name.

In the absence of the emergence of a consensus view, or even of clear majority and minority views, the *WIPO Overview* sets out two competing views on this issue. These are as follows.

- View 1: The right to criticize does not extend to registering a domain name that is identical or confusingly similar to the owner's registered trademark or conveys an association with the mark.
- View 2: Irrespective of whether the domain name as such connotes criticism, the respondent has a legitimate interest in using the trademark as part of the domain name of a criticism site if the use is fair and non-commercial.[214]

Since the publication of the *WIPO Overview*, some panellists have adopted a third view, which is a hybrid of the two other views and is known as the 'totality of circumstances' approach. The third view is as follows:

- View 3: Whether or not a criticism site is a legitimate non-commercial or fair use depends upon a careful examination of the facts and circumstances of the particular case, taking into account the nature of the criticism, the relationship of the parties, the existence and prominence of disclaimers and proof of actual confusion.

In a comprehensive review of the decisions dealing with criticism sites to that point, the majority of the panel in *Legal & General Group Plc v Image Plus*[215] analysed the distinction between the two views that were later identified by the *WIPO Overview*. As the majority explained, the first view, which it labelled the 'Domain Name itself is misleading' approach, focuses on the comparison between the complainant's mark and the disputed domain name. On this view, if the domain name appropriates the complainant's mark and it is not

[213] WIPO Case No D2000-1415 (23 Jan 2001).
[214] WIPO, above n 10, para 2.4.
[215] WIPO Case No D2002-1019 (30 Dec 2002).

apparent from the domain name itself that it is used for the purpose of criticism, then there can be no legitimate or fair use of the domain name. According to this view, the content of the web site, including whether or not it includes a disclaimer of association with the complainant, is irrelevant to whether or not there is a legitimate or fair use. The implication of this approach is that paragraph 4(c)(iii) can never be satisfied by a domain name that is identical to the complainant's mark (with the possible exception of a name that is a generic word), but that the use of domain names such as <trademarksucks.TLD> may amount to a legitimate non-commercial or fair use.

The second view, which the majority labelled the 'Complaints Site' approach, differs from the first view in that it focuses on the content of the respondent's web site, and not on the domain name. On this view, paragraph 4(c)(iii) will be satisfied if the web site is a genuine criticism site, regardless of whether it is clear from the domain name itself that it has been registered for the purpose of criticism. As the majority pointed out, many of the decisions that have adopted the second view have been influenced by the US approach to First Amendment protection of free speech.

The first view is explained in more detail at [6.9.8], the second view at [6.9.9] and the third view at [6.9.10]. The differences are ultimately referable to differences regarding the scope of the UDRP, which are dealt with at [6.9.11].

[6.9.8] First View: 'Domain Name Itself is Misleading'

The first view of the treatment of criticism sites under paragraph 4(c)(iii) is that there can be no legitimate non-commercial or fair use of a domain name that is identical or confusingly similar to the complainant's mark, as use of the mark will inevitably harm the trade mark owner by attracting users looking for the complainant's site. On this view, the content of the respondent's site is irrelevant, as it is not sufficient for the initial confusion to be later dispelled by the material on the site itself because the harm to the trade mark owner has already been incurred. Panellists who adopt this view consider that it does not abridge the respondent's freedom of expression, as the respondent is free to register a domain name that clearly indicates that it is not associated with the complainant's mark, but that it has been registered for the purpose of a criticism site.

The extent to which panellists who support the first view are concerned with the harm caused by users being attracted to the respondent's web site, despite the confusion later being dispelled by the content of the site, is analogous to the US doctrine of 'initial interest confusion', as applied to domain names by the US Court of Appeal for the Ninth Circuit in *Brookfield Communications, Inc v West Coast Entertainment Corp.*[216] In that case, the defendant registered the domain name <moviebuff.com>, which it later used to provide film reviews and commentary. The appeals court held that use of the plaintiff's mark in metatags resulted in initial interest confusion, as Internet users searching for the plaintiff's mark would be taken to the defendant's site, despite the fact that any actual confusion would later be dispelled. In reaching this conclusion, the court stated that:

> Although there is no source confusion in the sense that consumers know they are patronizing West Coast rather than Brookfield, there is nevertheless initial interest confusion in the sense that, by using 'moviebuff.com' or 'MovieBuff' to divert people looking for 'MovieBuff' to its web site, West Coast improperly benefits from the goodwill that Brookfield developed in its mark.[216a]

[216] 174 F 3d 1036 (9th Cir, 1999).
[216a] 174 F 3d 1036, 1062 (9th Cir, 1999).

The first view—that registration of a misleading domain name for a criticism site cannot be a legitimate non-commercial or fair use—has been adopted by a majority of non-US panellists. The following are examples of decisions that support the 'Domain name itself is misleading approach'.

- *Monty and Pat Roberts, Inc v J Bartell.*[217] The panellist held that the respondent was not making a legitimate non-commercial or fair use of the <montyroberts.com> domain name, adding that 'the right to express one's views is not the same as the right to use another's name to identify one's self as the source of those views'. In deciding against the respondent, the panellist explained that, by adopting a domain name that was identical to the complainant's mark, the respondent was illegitimately diverting Internet traffic to its own site.
- *Estée Lauder Inc v estelauder.com.*[218] The respondent registered the domain names <estelauder.com> and <estelauder.net>, and used them for criticism sites that included a consumer complaints forum. The sites contained clear disclaimers. In rejecting the respondent's fair use claim, the panellist stated that:

  > a distinction must be drawn between the domain name itself and the contents of the site which is reached through the domain name. Respondent has not denied, and cannot credibly deny, that its use of the domain names at issue was to attract internet users desirous of criticizing the Complainant or its products, and therefore likely to divert them from legitimate sites authorized by Complainant. The fact that the users, once so diverted or attracted, are confronted with numerous disclaimers does not cure the initial and illegitimate diversion.

- *Skattedirektoratet v Eivind Nag.*[219] The respondent's <skatteetaten.com> web site contained text in Norwegian that the respondent claimed to be satirical. The panel held that, although it is legitimate to operate a criticism site, it is not legitimate to use a domain name that is identical to the complainant's mark for this purpose. The panel added that 'anyone wishing to contact a trademark owner, has the right to contact the owner by addressing himself to the owner's exact identifier, followed by the top level suffix . . ., and to thereby reach the trademark owner, and not a third party, which itself does not have rights in that mark'.
- *Myer Stores Limited v Mr David John Singh.*[220] The respondent registered the domain name <myeronline.com> and a range of other domain names that included well-known Australian marks. The domain name resolved to a web site that included commentary on the financial performance of the trade mark owner and disclaimers of association with the complainant. The panellist held that the choice of the domain name falsely conveyed an association with the trade mark owner, which allowed the respondent to attract Internet users to his web site by trading off the complainant's goodwill in the mark. Holding that the respondent was not making a fair use of the domain name, the panellist added that:

  > By the time users reach the disclaimers on the website they have already been exposed to the misrepresentation inherent in the disputed domain name.

- *Gardere Wynne Sewell LLP v DefaultData.com.*[221] The respondent registered the domain names <garderewynne.com> and <gardereandwynne.com>, which resolved to a web site

[217] WIPO Case No D2000-0300 (13 June 2000).
[218] WIPO Case No D2000-0869 (25 Sept 2000).
[219] WIPO Case No D2000-1314 (18 Dec 2000).
[220] WIPO Case No D2001-0763 (10 July 2001).
[221] WIPO Case No D2001-1093 (16 Nov 2001).

that included material critical of the legal systems. The panellist held that it was not a fair use deliberately to register the name of a law firm in order to criticise the legal profession as a whole. In reaching this conclusion, the panellist stated that he:

> does not see why the internet public, seeking the mark holder and its products and services, should be surprised and detained by websites that criticize and protest but give no hint at what they are about until they are opened. The Panel believes the current way the internet operates requires that fair use protest sites under the Policy identify themselves with something in addition to the Complainant's trademark so the public can distinguish whether they want to visit the protest site or the Complainant's products and services site.

- *Triodos Bank NV v Ashley Dobbs.*[222] The respondent, who had been involved in a commercial dispute with the complainant, registered the <triodos-bank.com> domain name, which was used to air grievances against the complainant. The panellist held that the respondent's choice of domain name amounted to a misrepresentation of an association with the complainant, and that the nature of the respondent's site was irrelevant. In reaching this conclusion, the panellist stated:

> The fact that the nature of the site is apparent as soon as one arrives at it is too late. By then, the visitor has been deceived, the Respondent has achieved his objective and the damage is done.

- *The Royal Bank of Scotland Group plc v Pedro Lopez.*[223] The respondent registered the domain name <natwestbanksucks.com>, as well as a number of other domain names that were confusingly similar to the complainant's marks. The domain names resolved to a site that contained material that was highly critical of the complainant. Despite the fact that the disputed domain name included the term 'sucks', which could distinguish the domain name from the complainant's mark, the respondent followed the decision in the related dispute between the parties in *Royal Bank of Scotland Group and National Westminster Bank v Pedro Lopez,*[224] to hold that the use of a confusingly similar mark did not amount to a legitimate non-commercial or fair use.
- *Kirkland & Ellis LLP v DefaultData.com.*[225] The respondent registered the domain name <kirklandandellis.com> and a number of other domain names incorporating the names of law firms. The domain name resolved to a web site that included material critical of the legal profession. The panellist held that the respondent's purpose was to tarnish the legal profession as a whole, and the complainant in particular. In deciding that the respondent had no rights or legitimate interests in the domain name, the panellist stated that:

> the right to express one's views is not the same as the right to identify itself by another's name when expressing those views. Thus, while Respondent may express its views about the quality, or lack thereof, of the U.S. legal profession, in general, or any firm offering legal services, in particular, Respondent does not have the right to identify itself as that particular firm. And, there is nothing in the domain name to indicate that the site is devoted to criticism, even though that fact is apparent when visiting the site.

- *Justice for Children v R neetso.*[226] The respondent registered the domain names <justiceforchildren.net> and <justiceforchildren.org>, which resolved to a site that

[222] WIPO Case No D2002-0776 (3 Oct 2002).
[223] WIPO Case No D2003-0166 (9 May 2003).
[224] WIPO Case No D2002-0823 (3 Dec 2002).
[225] WIPO Case No D2004-0136 (2 Apr 2004).
[226] WIPO Case No D2004-0175 (4 June 2004).

included material that was critical of the complainant and of the operation of the judicial system in child custody matters. The panellist held that the use of the mark in identical domain names created initial interest confusion, and was therefore not a legitimate non-commercial or fair use. In reaching this conclusion, the panellist stated that:

> Decisions under the Policy focus upon a respondent's use of another's mark in a domain name to attract Internet users to respondent's site. This is true in typosquatting cases and in cases where a respondent selected his domain name in anticipation of subsequent sale to the mark owner. The content of Respondent's sites in these two categories of cases—cases in which respondents almost uniformly lose - is irrelevant to the harm to the mark owner and to the unwary consumer. That harm results from the confusion caused by the initial attraction to the site by means of borrowing the complainant's mark. And that is exactly the harm the Policy was adopted to address.

- *Texans For Lawsuit Reform, Inc v Kelly Fero.*[227] The complainant was formed to campaign for reform aimed at eliminating 'excessive' litigation in Texas. The respondent registered the domain name <texansforlawsuitreform.com>, which was used to criticise the complainant during a referendum campaign. The panellist held that the respondent was not making a fair use of the domain name as the choice of a name identical to the complainant's mark created initial interest confusion.
- *Gilead Sciences, Inc v Kumar Patel.*[228] The respondent registered the domain name <gilead-sciences.com> and numerous other domain names that incorporated pharmaceutical marks. The domain name resolved to a site that provided a forum for criticism of the complainant's 'GILEAD SCIENCES' product, and contained logos and scrolling images of other pharmaceutical products of the complainant. The panellist held that this was not a fair use as the domain name falsely conveyed an association with the complainant. The panellist further observed that it 'is not fair use of a criticism page to scroll [*sic*] multiple brand marks of the Complainant, together with their getup, while only criticising one brand name product'.
- *Moss and Coleman Solicitors v Rick Kordowski.*[229] The respondent registered the domain names <mossandcoleman.com> and <mossandcoleman.net>, which were used in relation to a customer 'grievance' site. The panellist referred to the two views of the treatment of criticism sites identified by the *WIPO Overview*, pointing out that few non-US panellists had adopted the second view. As the majority of panels had applied the first view to non-US parties, the panellist held that the choice of domain names that were identical to the complainant's mark was not a fair use, on the basis that it would create an association with the complainant and misleadingly divert Internet users to the respondent's web site.
- *InMed Diagnostic Services, LLC v James Harrison.*[230] The respondent registered the domain name <inmeddiagnostics.net> and, following a dispute with the complainant, published material critical of the complainant on the web site. After referring to the three views of the treatment of criticism sites, the panellist preferred to apply the first view. Applying this view, the panellist held that the respondent was not making a fair use of the domain name because choosing a name that was identical to the complainant's mark created initial interest confusion. The panellist maintained that the respondent could not increase the audience for his critical views by taking advantage of the goodwill in the complainant's mark to attract users to his web site.

227 WIPO Case No D2004-0778 (31 Oct 2004).
228 WIPO Case No D2005-0831 (17 Nov 2005).
229 WIPO Case No D2006-1066 (23 Oct 2006).
230 WIPO Case No D2006-1230 (11 Dec 2006).

[6.9.9] Second View: 'Complaints Site' Approach

The second view of the treatment of criticism sites under paragraph 4(c)(iii) is that the use of a mark as a domain name for a criticism site is a legitimate non-commercial or fair use as it is an exercise of the respondent's right of freedom of expression. The approach is similar to that adopted by the US Ninth Circuit Court of Appeals in *Bally Total Fitness Holding Corp v Faber*[231] to the claim that use of the term 'BALLY SUCKS' as a sub-domain to criticise the claimant's business amounted to tarnishment of the 'BALLY' mark. In rejecting the claim, District Judge Pregerson stated:

> Here . . . Faber is using Bally's mark in the context of a consumer commentary to say that Bally engages in business practices which Faber finds distasteful or unsatisfactory. This is speech protected by the First Amendment . . . As such, Faber can use Bally's mark to identify the source of the goods or services of which he is complaining. This use is necessary to maintain broad opportunities for expression.[231a]

On this view, provided that there are no indicia of bad faith indicating that the use is not 'legitimate', use of the domain name for a criticism site is sufficient to make out the defence. Underlying this approach is the view that the ability to register a domain name that is identical or confusingly similar to a trade mark for the purpose of identifying the trade mark owner as the object of a legitimate criticism site is a necessary incident of freedom of expression on the Internet. As the panellist in *Bridgestone Firestone, Inc v Jack Myers*[232] expressed the philosophical underpinnings of this view:

> The Internet is above all a framework for global communication, and the right to free speech should be one of the foundations of Internet law.

Once it is accepted that use of an identical or confusingly similar domain name for a criticism site is a legitimate non-commercial or fair use, then panels must turn to whether the domain name is being used, with intent for commercial gain, misleadingly to divert consumers or to tarnish the complainant's mark. As most clearly explained by the panellist in *Howard Jarvis Taxpayers Association v Paul McCauley*,[233] on this approach, genuine criticism sites will not be denied protection on these grounds. First, non-commercial criticism sites will not be used with intent for commercial gain. Secondly, such sites are not diverting consumers to products of the complainant's business competitors and, once the content of a criticism site is taken into account, any diversion is not 'misleading', as there is no source confusion with the trade mark owner. Thirdly, if tarnishment is confined, as it is in US law, to negative uses of a mark that create inferior or degrading associations then, as in the *Bally sucks* case, a genuine criticism site does not tarnish the complainant's mark.

The overwhelming majority of decisions dealing with US-based criticism sites have held that use of a domain name for a genuine criticism site is a legitimate non-commercial or fair use. The following are examples of the most important decisions to have adopted the second, 'Complaints site', view.

- *Bridgestone Firestone, Inc v Jack Myers*.[234] The respondent, a former employee of the complainant, registered the <bridgetown-firestone.net> domain name and used it for a complaint site. The panellist held that a 'reasonably prudent' Internet user would not be

[231] 29 F Supp 2d 1161 (CD Cal, 1998).
[231a] 29 F Supp 2d 1161, 1167 (CD Cal, 1998).
[232] WIPO Case No D2000-0190 (6 July 2000).
[233] WIPO Case No D2004-0014 (22 Apr 2004).
[234] WIPO Case No D2000-0190 (6 July 2000).

confused by the domain name as the respondent used only the .net gTLD and posted disclaimers to the web site. In deciding that the use of the domain name for a criticism site fell within paragraph 4(c)(iii), the panellist stated that:

> the Respondent's principal purpose in using the domain name appears not to be for commercial gain, but rather to exercise his First Amendment right to criticize the Complainants. The use of the <trademark.net> domain name appears to be for the communicative purpose of identifying the companies, which are the subject of his complaints. He is not misleadingly diverting users to his website, as he has not utilized the <.com> domain and has posted adequate disclaimers as to the source of the website. It does not appear that his actions are intended to tarnish, or have tarnished, the Complainants' marks.

The panellist further maintained that it was not necessary for the domain name to indicate clearly that it was not associated with the complainant's mark by, for example, adding a terms such as 'sucks' for the use of the domain name to be a legitimate non-commercial or fair use.

- *TMP Worldwide Inc v Jennifer L Potter.*[235] The respondent, who had been involved in a commercial dispute with the complainant, registered the domain names <tmpworldwide.net> and <tmpworldwide.org>, which she used as part of a campaign against the complainant. The panellist, referring to the *Bridgestone Firestone* decision, held that the use of the domain name by the respondent was an exercise of her First Amendment right to criticise the complainant and, consequently, was a legitimate non-commercial or fair use under paragraph 4(c)(iii).
- *Ahmanson Land Company v Save Open Space.*[236] The respondent registered the <ahmanson.org> domain name, which was used for criticism and commentary regarding environmental issues concerning the 'AHMANSON' ranch. The panel held that, as the web site was being used as a non-commercial platform to criticise the complainant's development activities, the domain name was being used for a legitimate non-commercial or fair use. Even though the respondent used the web site to solicit financial contributions, the panel held that the domain name was not being used for commercial gain, as the respondent was a non-profit-making organisation.
- *Britannia Building Society v Britannia Fraud Prevention.*[237] The respondent, consisting of a group of disgruntled members of the complainant building society, registered the <britanniabuildingsociety.org> domain name, which was used to criticise the management of the complainant. The panellist found that the use of the domain name fell within paragraph 4(c)(iii). In reaching this conclusion, the panellist held that there was no intent misleadingly to divert consumers to the web site as users were not diverted to a site promoting the products of a competitor of the complainant, and any initial confusion would be dispelled by the contents of the respondent's site. On this point, the panellist stated that a low level of initial confusion is 'a price worth paying to preserve the free exchange of ideas via the internet'. The panellist also held that 'fair use' criticism does not amount to tarnishment of the complainant's mark, even if the criticism includes defamatory material. On this point, the panellist pointed out that allegations of defamation must be pursued through the courts.

[235] WIPO Case No D2000-0536 (5 Aug 2000).
[236] WIPO Case No D2000-0858 (4 Dec 2000).
[237] WIPO Case No D2001-0505 (6 July 2001).

- *Legal & General Group Plc v Image Plus.*[238] The respondent registered the domain name <legal-and-general.com>, which it used for a criticism site. Following a review of the decisions dealing with criticism sites, the majority of the panel indicated that it preferred the second view. As the complainant was based in the United Kingdom, the majority adopted the second view on the basis of an analysis of the wording of paragraph 4(c)(iii), and not the application of US First Amendment law. In particular, the majority held that to regard the use of a domain name that is identical to a mark as per se impermissible would effectively nullify the defence set out in paragraph 4(c)(iii). Furthermore, the majority held that the use of the domain name was not 'misleading', as any initial confusion would be dispelled by the content of the respondent's site, the minimal initial confusion being the price to pay for freedom of expression on the Internet. The dissenting panellist, however, applied the first view of the treatment of criticism sites to conclude that the respondent did not have a legitimate interest in the domain name.
- *Howard Jarvis Taxpayers Association v Paul McCauley.*[239] In this decision, which is generally regarded as the best exposition of the second view, the respondent had been involved in a longstanding legal dispute with the complainant. The respondent registered the <hjta.com> domain name for use as a site to criticise the practices and management of the complainant. The panellist pointed out that, although panels had decided that use of a mark as a domain name on some US-based criticism sites was not a legitimate non-commercial or fair use under paragraph 4(c)(iii), in most of those decisions there was some evidence of bad faith. Applying the second view, which has been adopted by the majority of decisions relating to US-based sites, the panellist held that, as there was no evidence to suggest that the respondent's criticism site was not 'legitimate', the use of the domain name was a legitimate non-commercial or fair use. The panellist proceeded to hold that there was no intent misleadingly to divert consumers, as there was no diversion to the complainant's competitors and no confusion as the source or sponsorship of the respondent's web site. Finally, the panellist held that a genuine criticism site, even if it includes untrue or defamatory material, does not amount to tarnishment of the complainant's mark. The panellist concluded his analysis with the following observations on the scope of the UDRP:

> The Policy is designed to prevent abusive cybersquatting, but under United States law, it cannot extend to insulating trademark holders from contrary and critical views when such views are legitimately expressed without an intention for commercial gain. That is true even if the critical views are unfair, overstated, or flat-out lies, and even if they are posted at trademark.TLD websites. Use of the Policy to provide such insulation may undermine freedom of discourse on the Internet and undercut the free and orderly exchange of ideas that the First Amendment seeks to promote.

[6.9.10] Third View: the 'Totality of Circumstances' Approach

Since the publication of the *WIPO Overview*, some panels have adopted a third view of the treatment of criticism sites under paragraph 4(c)(iii). On this view, the 'totality of the circumstances' in each case must be examined in order to determine whether a criticism site is a legitimate non-commercial or fair use, with a particular emphasis being placed on the

[238] WIPO Case No D2002-1019 (30 Dec 2002).
[239] WIPO Case No D2004-0014 (22 Apr 2004).

nature of the criticism, the relationship of the parties, the presence of disclaimers and evidence of actual confusion.[240]

The roots of the third view may be traced to the approach adopted by the panellist in *Covance, Inc v The Covance Campaign*.[241] In that dispute, the respondent, an animal rights activist group, registered the domain name <covancecampaign.com>, and used it as part of a campaign against animal experiments conducted by the complainant. The panellist accepted that the domain name was being used for the non-commercial purpose of commenting on and criticising the complainant's activities. Emphasising the potential for an initial confusion of an association with the complainant, the panellist adopted the first view in relation to a domain name that is identical to the complainant's mark, namely, that such domain names do not qualify as legitimate non-commercial or fair uses. In this case, however, the domain name included the term 'campaign'. Where the domain name is not identical to the complainant's mark, but is 'confusingly similar', the panellist held that the 'degree of initial confusion needs to be considered in the particular factual circumstances of the case while keeping in mind the Policy's aims'.

As the panellist accepted the view that the policy is aimed at avoiding initial confusion, the analysis focused mainly on a comparison between the complainant's mark and the domain name. On this basis, the panellist held that, where a domain name includes a negative term such as 'sucks', there will usually be a legitimate non-commercial or fair use, but where the domain name includes a neutral term, the potential for initial confusion means that it is unlikely that the respondent will have legitimate interests in the domain name. In the instant dispute, the panellist held that, as the term 'campaign' was neither negative nor neutral, all of the circumstances of the case needed to be taken into account. Focussing on the potential for initial confusion, the panellist held that the respondent was making a legitimate non-commercial use of the domain name as, in this instance, it was clear from the results returned from a search engine that the respondent's site was not associated with the complainant, and any minimal confusion would be dispelled by the disclaimer on the respondent's web site. This clearly represented a departure from those decisions that adopted the first view, as it permitted the content of the respondent's web site to be taken into account in certain limited circumstances.

The following decisions have adopted and further developed the 'totality of the circumstances' approach that was first signalled in the *Covance* decision:

- *CBS Broadcasting Inc v Nabil Z aghloul*.[242] The respondent registered the <cbscares.com> domain name, which was inactive until the respondent was notified of the dispute, and thereafter included a disclaimer with links to sites that were critical of the complainant. The panellist adopted the approach first enunciated in *Covance*, of balancing the interests of trade mark owners against the interests of the public in freedom of expression, taking into account the 'totality of circumstances' of the case. The panellist first rejected the view that a domain name that is identical to a complainant's mark can never be a legitimate non-commercial or fair use, claiming that this would only be the case where there were some indicia of bad faith, such as where the respondent's site reproduced the 'look and feel' of the complainant's site. In the instant dispute, the panellist held that the

[240] The 'totality of circumstances' approach to criticism sites should be distinguished from the 'totality of circumstances' approach to determining bad faith, which is dealt with at [7.4].

[241] WIPO Case No D2004-0206 (30 Apr 2004).

[242] WIPO Case No D2004-0988 (5 Feb 2005).

respondent's passive holding of the domain name prior to receiving notice of the dispute was indicative of bad faith. In the totality of circumstances, therefore, the respondent was not making a legitimate non-commercial or fair use of the domain name.

- *Asset Loan Co Pty Ltd v Gregory Rogers.*[243] The respondent, who was involved in a legal dispute with the complainant, registered the domain name <assetloanco.net> and used it for a criticism site. Subsequently, the respondent offered to shut down the web site if outstanding disputes with the complainant were settled. The panel held that using the domain name as a form of leverage to bring litigation with the complainant to an end indicated that the domain name was being used for a commercial purpose, and that therefore there was no legitimate non-commercial or fair use. In reaching this conclusion, the panel took into account the absence of a disclaimer on the respondent's site which, together with the choice of a name that was identical to the complainant's mark, suggested an intention to create confusion with the complainant. From the decision, however, it appeared that at least one of the panellists believed that, in certain circumstances, use of an identical domain name could amount to a legitimate non-commercial or fair use.
- *The Highland Street Connection v Chris McGrath.*[244] The respondent, a member of the complainant's board of trustees, registered the domain names <highlandstreetfoundation.com> and <highlandstreetfoundation.org>, apparently as a protest against the complainant's change of name from 'HIGHLAND STREET CONNECTION' to 'HIGHLAND STREET FOUNDATION', and to block registration of the domain names, which remained inactive. In the totality of circumstances, the panel held that registration of the domain names was a mere pretext for blocking the complainant from registering the domain names, and therefore not a 'legitimate' use.
- *Ryanair Limited v Michael Coulston.*[245] The respondent registered the domain name <ryanaircampaign.org> and used it for a non-commercial criticism site. As the dispute involved European parties, the panellist, who had been responsible for the *Howard Jarvis* decision, applied the approach adopted in *Covance.* On this approach, the panellist held that any initial confusion caused by the similarity of the domain name to the complainant's mark would be dispelled by the contents of the respondent's web site, which was obviously a criticism site. The panellist therefore found that the respondent had a legitimate interest in the domain name as, in the totality of circumstances, the use to which the domain name was put counteracted any initial confusion.

The 'totality of circumstances' approach has been criticised by the panellist in *InMed Diagnostic Services, LLC v James Harrison.*[246] In that decision, the panellist, who adopted the first view, maintained that an approach based on an examination of all the circumstances of the dispute, including the content of the respondent's web site, necessarily involves an assessment of complex facts which UDRP panels are ill-equipped to analyse without the opportunity for live testimony and cross-examination. The panellist also considered that, given the nature of the UDRP process, it was inappropriate for panels to make determinations regarding the balance between trade mark rights and constitutional rights to freedom of expression. Consequently, the panellist preferred the 'bright line' rule that the use of a domain name that is identical to a complainant's mark can never amount to a legitimate non-commercial or fair use.

[243] WIPO Case No D2006-0300 (2 May 2006).
[244] WIPO Case No D2006-0516 (14 Aug 2006).
[245] WIPO Case No D2006-1194 (12 Dec 2006).
[246] WIPO Case No D2006-1230 (11 Dec 2006).

[6.9.11] Criticism Sites: Discussion

The different views about the treatment of criticism sites under paragraph 4(c)(iii) of the UDRP are ultimately referable to the fundamental tensions in the interpretation and application of the UDRP, dealt with at [3.15]. Additional difficulties have arisen because the objectives of paragraph 4(c)(iii) in particular were not specified with sufficient precision. These difficulties are not altogether surprising, as the balance between trade mark rights and freedom of expression in the context of domain name dispute resolution involves different considerations from the balance between trade mark rights and freedom of expression in the context of trade mark law.

The balance between trade mark rights and freedom of expression under trade mark law is conventionally struck by limiting a trade mark owner's rights to preventing infringing uses of a mark as a source identifier, while permitting use of the mark in a descriptive sense, including for criticising the trade mark owner. Nevertheless, as registration and use of a domain name do not necessarily involve use of a mark as a source identifier, but are more than a merely descriptive use, domain names require that a new balance be established between trade mark rights and freedom of expression. The different views on the treatment of criticism sites reflect differences in where this balance should be struck.

Proponents of the 'domain name itself is misleading' approach consider that a trade mark owner should be able to prevent the registration of a domain name that is identical, or 'misleadingly' similar, to the mark because of the harm incurred by an initial confusion with the complainant's mark. The harm is that users looking for the trade mark owner's site will be attracted to the site of someone who is not the trade mark owner, including the trade mark owner's critics. Proponents of the 'complaints site' view, on the other hand, acknowledge that there is some harm incurred by the adoption of a domain name that is identical, or 'misleadingly' similar, to the complainant's mark, but consider that the benefits in terms of freedom of expression outweigh the relatively minimal harm. In other words, supporters of the second view consider that freedom of expression on the Internet necessarily entails the freedom to use someone else's mark as a domain name for the purpose of criticising the mark owner.

While there are merits in both approaches, the starting point for analysis should be the limited scope of the UDRP, which is aimed only at preventing extortionate and opportunistic practices relating to the registration and use of marks as domain names. In particular, the policy does not confer on a trade mark holder pre-emptive rights over the registration of all domain names that reflect the owner's mark in its entirety. Furthermore, the stated objective of the second element of the UDRP is to exclude disputes where a respondent is able to establish rights or interests in a domain name from the scope of the policy. Thus, paragraph 4(c)(i) excludes disputes where a respondent has commercial trade mark-like rights in a domain name and paragraph 4(c)(ii) excludes disputes where a respondent has rights or interests that result from being commonly known by the domain name. By protecting rights arising from a 'legitimate non-commercial or fair use' of a domain name, paragraph 4(c)(iii) clearly envisages use of the domain name for the purpose of criticising a trade mark owner, which is reinforced by the statement from the WIPO Final Report, referred to at [6.9] above. Consequently, to conclude, as proponents of the 'domain name itself is misleading' approach have, that a respondent who uses an identical or 'misleadingly' similar domain name for a criticism site can never have rights or interests in the domain name is to treat rights or interests in freedom of expression as less than commercial or personal rights or interests in a domain name. In short, then, as the UDRP was designed to prevent only 'abusive, bad faith' registration and use, and not directed at suppressing criticism sites, the 'complaints site' view should therefore be preferred, as a matter of policy, over the 'domain name itself is misleading' view.

This does not mean that a trade mark owner is deprived of protection; merely that the policy is narrowly targeted, and that it does not give trade mark owners pre-emptive rights over domain names. In other words, the policy, on its terms, is not directed at preventing all forms of initial confusion between a mark and a domain name that results in users being attracted to the site of the domain name registrant as a result of the similarity between the mark and the domain name. Just as a degree of initial confusion is tolerated where the registrant has a legitimate commercial interest in the domain name, so a degree of initial confusion should be tolerated where the registrant has legitimate non-commercial rights or interests, including the right to exercise freedom of expression. This approach is consistent with the wording of the UDRP, which clearly does not confer pre-emptive rights on trade mark owners, and gives an appropriate role to paragraph 4(c)(iii) which, in terms, does not distinguish between domain names that are 'identical' to a complainant's mark and those that are merely 'confusingly similar'. The balance between protecting trade mark rights and freedom of expression is therefore established by the limitations inherent in paragraph 4(c)(iii): that the use to which the domain name is put is a 'legitimate' or fair use; that it is non-commercial; that there is no intent for commercial gain misleadingly to divert consumers; and no intent for commercial gain to tarnish the mark.

This approach necessarily involves an examination of whether or not a site to which the domain name resolves is a 'genuine' criticism site. By providing that a domain name holder may acquire rights or interests by making a 'legitimate non-commercial' use of a domain name, however, paragraph 4(c)(iii) expressly requires that account be taken of the use to which a domain name is put. Moreover, it is absolutely necessary for the bona fides of a respondent to be taken into account in determining whether a non-commercial use is 'legitimate'. How else, other than by visiting a site, is a distinction to be drawn between a 'sham speech' site or Potemkin Village, on the one hand, and a genuine criticism site, on the other?

As an analysis of the decisions that have adopted the 'totality of circumstances' approach reveals, most have concerned circumstances in which a respondent cannot rely on paragraph 4(c)(iii) because the use of the domain name was somehow not 'legitimate'. In practice, then, there is little to separate the 'complaints site' view from the 'totality of circumstances' view. The advantage of the 'complaints site' view, however, is that it adopts a clear rule as to where the balance between trade mark rights and freedom of expression is to be struck in the context of domain name dispute resolution.

As explained at [5.25], [5.26] and [5.41] a proper understanding of the objective of the UDRP requires the adoption of a new approach to comparing a complainant's mark with a domain name that is aimed at determining whether the two are sufficiently similar for a registrant to achieve the objectives of 'abusive, bad faith' registration and use of a domain name. Given the nature of a domain name, the objectives of 'abusive, bad faith' registration and use can be achieved by a relatively low level of similarity, which is analogous to initial interest confusion. Moreover, in comparing the mark and the domain name, the content of a site to which the domain name resolves is irrelevant, as the point in time at which the confusion is relevant is prior to the user accessing the respondent's site.

In assessing whether or not a respondent has rights or legitimate interests in a domain name, however, the content of a web site may well be determinative. For example, paragraph

4(c)(i) expressly refers to 'use' of a domain name in connection with an offering of goods or services. Similarly, paragraph 4(c)(iii) refers to a 'use' that is a 'legitimate non-commercial or fair use'. Provided that the activities of a respondent fit within these parameters, and the respondent is able to rebut any evidence of the disentitling factors set out in paragraph 4(c)(iii), then there can be no 'abusive, bad faith' registration and use. In short, absent bad faith, the UDRP is not aimed at preventing the use of a mark as the domain name for a genuine criticism site, nor should it be.

[6.9.12] Legitimate Non-commercial or Fair Use: 'Fan' Sites

The division of views between panellists concerning the treatment of criticism sites is mirrored by a division of views regarding the treatment of 'fan' sites under paragraph 4(c)(iii).

To begin with, for a respondent to be able to establish that a domain name for a 'fan' site is protected the domain name must actually be 'used' for a fan site and the use must be 'legitimate'. In relation to the treatment of fan sites under paragraph 4(c)(iii), the *WIPO Overview* states that:

> This section only deals with fan sites that are clearly active and non-commercial. There are many UDRP cases in which the respondent claims to have an active non-commercial fan site but the panel decides otherwise.[247]

In determining whether use of a mark as a domain name for a fan site is a legitimate non-commercial or fair use, it is important to bear in mind the differences in the wording between paragraphs 4(c)(i) and 4(c)(iii): while paragraph 4(c)(i) refers to 'demonstrable preparations to use' the domain name, paragraph 4(c)(iii) refers to 'making' a legitimate non-commercial or fair use. This has been interpreted to mean that, unless the respondent has not had an opportunity to establish a site, the domain name must resolve to an active fan site. In *Helen Fielding v Anthony Corbet*,[248] for example, the respondent registered the domain name <bridgetjones.com>, claiming that he intended to use it as a 'BRIDGET JONES' fan site, but later offered to transfer the domain name to the complainant for $15,000. At the time of the complaint, there had been no active use of the domain name in the two years since it was registered. Moreover, the respondent provided no evidence of any preparations to use the domain name as a fan site. Although the panellist acknowledged that there could be limited circumstances where a respondent might not be expected to make an immediate use of a domain name, yet still be considered to have registered the domain name for a legitimate non-commercial purpose, in the circumstances of this case, where the only use of the domain name was to offer it for sale, the panellist held that the respondent was clearly not 'making' any use at all of the domain name, let alone a legitimate non-commercial use.

Similarly, in *Seagate Technology LLC v Wang Zhanfeng*,[249] the respondent registered the domain name <seagate-maxtor.com> and claimed that he intended to use it as a 'fan' site for the complainant's products, but did not do so. As the respondent's site advertised the domain name for sale, the panellist held that the respondent was not making a legitimate non-commercial use of the domain name. Referring to *Helen Fielding*, the panellist held that, under paragraph 4(c)(iii), a respondent 'must show an actual and current use of the site for a fan club'. As the best that the respondent could establish was an intention to set up a fan site, paragraph 4(c)(iii) was clearly not made out.

[247] WIPO, above n 10, para 2.5.

[248] WIPO Case No D2000-1000 (25 Sept 2005).

[249] NAF Case No FA635276 (10 Mar 2006).

The rule that a mere assertion of an intention to establish a fan site is not sufficient for paragraph 4(c)(iii) to be satisfied is no more than an application of the general principle that once a complainant has established a prima facie case under paragraph 4(a)(ii), the respondent must produce 'concrete evidence' to rebut the allegations. Thus, in *Freddy Adu v Frank Fushille,*[250] the respondent registered the domain name <freddyadu.com> and claimed that he intended to create a non-commercial site for fans of the well-known soccer player, 'FREDDY ADU'. As the respondent presented no 'concrete evidence' of preparations to use the domain name for any purpose, let alone as a fan site, the panellist held that paragraph 4(c)(iii) was not made out and the respondent was not making a legitimate use of the disputed domain name.

A number of decisions have included definitive statements that the operation of a fan site either is or is not a legitimate non-commercial use. In *RAI Radiotelevisione Italiana SpA v Alessandro Pescetelli,*[251] for example, the panellist stated that:

> Operating a fan site has not generally being [*sic*] regarded as a 'legitimate interest' under the Policy, although it may be relevant for the discussion of whether there is a use in bad faith.

Against this, the panellist in *Freddy Adu v Frank Fushille*[252] stated that:

> Panelists in disputes under the Policy have consistently found that noncommercial fan sites qualified as legitimate use.

There is, in fact, a continuing division of views among panellists regarding the treatment of fan sites under paragraph 4(c)(iii). In recognition of this, the *WIPO Overview* sets out two views of UDRP panellists, neither of which has a majority of supporters. The two views are as follows.

- View 1: An active and clearly non-commercial fan site may have rights and legitimate interests in the domain name that includes the complainant's trademark. The site should be non-commercial and clearly distinctive from any official site.[253]
- View 2: Respondent does not have rights to express its view, even if positive, on an individual or entity by using a confusingly similar domain name, as the respondent is misrepresenting itself as being that individual or entity. In particular, where the domain name is identical to the trademark, the respondent, in its actions, prevents the trademark holder from exercising the rights to its mark and managing its presence on the Internet.[254]

The first view parallels the 'complaints site' approach to criticism sites in that an initial confusion created by a domain name that is confusingly similar to a complainant's mark is considered acceptable provided the confusion is dispelled by the contents of the respondent's site, including whether or not there is an adequate disclaimer. The second view, on the other hand, parallels the 'domain name itself is misleading' approach to criticism sites, in that it assumes that choosing a confusingly similar domain name for a fan site is impermissible on account of the initial confusion that the domain name will create among Internet users.

Proponents of the first view consider that a respondent can establish rights or legitimate interests in a genuine fan site, but only if two important conditions are satisfied. First, to fall within paragraph 4(c)(iii), the fan site must be 'non-commercial'. The clearest examples of

[250] WIPO Case No D2004-0682 (17 Oct 2004).
[251] WIPO Case No D2002-0716 (30 Aug 2002).
[252] WIPO Case No D2004-0682 (17 Oct 2004).
[253] WIPO, above n 10, para 2.5.
[254] *Ibid*, para 2.5.

decisions that have held that there can be no rights or interests in fan sites that are used for commercial purposes is a series of decisions involving the serial cybersquatter, Jeff Burgar, who registered a substantial number of domain names (between 75 and 200) based on the names of celebrities and redirected them to the <celebrity1000.com> web site. The web site provided information about celebrities and included advertising for services and products that were not related to the celebrities. In some cases, the domain names were later linked to sites dealing only with the celebrity concerned. In the vast majority of these decisions panels held that the respondent's activities amounted to a 'commercial' use of the domain name, as they were directed mainly at generating advertising revenues for the respondent, the inclusion of celebrity information being little more than a transparent attempt to bring the behaviour within the policy. In *Kevin Spacey v Alberta Hot Rods*,[255] for example, the panel stated that:

> Respondent's use of the <kevinspacey.com> domain name for an unauthorized fan site, as an apparent stratagem after previously using the site to direct Internet traffic to its own commercial site, demonstrates that Respondent is not making fair use of the domain name within the meaning of Policy ¶ 4(c)(iii).

Decisions rejecting claims to a legitimate interest in relation to alleged 'fan' sites operated by this particular respondent included: *Julie Brown v Julie Brown Club*,[256] *Celine Dion v Jeff Burgar*,[257] *Stephanie Seymour v Jeff Burgar*,[258] *Michael Andretti v Alberta Hot Rods*,[259] *Dr Michael Crichton v Alberta Hot Rods*,[260] *Bogart, Inc v Humphrey Bogart Club*,[261] *Jeffrey Archer v Alberta Hotrods*,[262] *Tom Cruise v Network Operations Center*.[263] In one decision, *Bruce Springsteen v Jeff Burgar*,[264] the majority held that the respondent's use of the domain name to divert users to the <celebrity1000.com> site did not indicate an intent for commercial gain misleadingly to divert users. On the facts, however, the *Bruce Springsteen* decision was clearly a strained interpretation of paragraph 4(c)(iii), with one of the majority later forming part of a unanimous panel in *Kevin Spacey v Alberta Hot Rods*[265] that concluded, on essentially indistinguishable facts, that there was no legitimate non-commercial or fair use of the disputed domain name. In that decision, in light of the pattern of activities of the respondent, the panel stated that 'two years after the *Springsteen* case, the persistent behavior of this Respondent in the registration of celebrity names militates against the "benefit of the doubt" which the Panelist afforded the Respondent in the previous case'.

Especially where there is evidence of bad faith, relatively minor commercial uses of a fan site may mean that there is no legitimate non-commercial or fair use of the disputed domain name. In *Universal City Studios v David Burns*,[266] for example, the respondents registered the domain name <adam-12.com>, and used it as a fan site for the 'ADAM-12' television series. The respondents posted solicitations for donations to cover potential legal costs to the web site, and offered an unauthorised 'ADAM-12' book for sale through the web site on

[255] NAF Case No FA114437 (1 Aug 2002).
[256] WIPO Case No D2000-1628 (13 Feb 2001).
[257] WIPO Case No D2000-1838 (13 Feb 2001).
[258] NAF Case No FA97112 (29 May 2001).
[259] NAF Case No FA99084 (4 Oct 2001).
[260] WIPO Case No D2002-0872 (25 Nov 2002).
[261] NAF Case No FA162770 (4 Aug 2003).
[262] WIPO Case No D2006-0431 (1 June 2006).
[263] WIPO Case No D2006-0560 (5 July 2006).
[264] WIPO Case No D2000-1532 (25 Jan 2001).
[265] NAF Case No FA114437 (1 Aug 2002).
[266] WIPO Case No D2001-0784 (1 Oct 2001).

a cost recovery basis. The respondents also posted offers to sell the domain name on the web site. The panel held that, although the name was being used for a fan site, the commercial use was sufficient for the respondents' paragraph 4(c)(iii) defence to fail.

The second condition that must be satisfied according to proponents of the first view is that, even where a fan site is non-commercial, the site must include an adequate disclaimer of association with the trade mark owner. On this approach, a disclaimer is considered necessary to ensure that the domain name does not misleadingly divert visitors to the respondent's fan site. The decision most commonly referred to in support of this principle is *Estate of Gary Jennings v Submachine.*[267] In that dispute, the respondent registered the domain name <garyjennings.com> and used it for a fan site that included reviews of books written by the author, 'GARY JENNINGS'. Although the site included links to sites that offered the author's books for sale, the panellist held that, as the respondent was not obtaining commercial gain, there was no 'commercial' use. The panellist also considered it significant that the site included a prominent disclaimer, as this indicated that it was not misleadingly diverting consumers.

In *Xuxa Promoções Produções Ltda v Ivo Conestabile Junior,*[268] the respondent registered the domain name <xuxaonline.com> and claimed it was used for a non-commercial fan site. The panellist observed that a fan site could be considered a legitimate interest in exceptional cases, but that in this dispute, as the respondent's site did not include an express disclaimer, there was no legitimate non-commercial use of the domain name.

The following is a representative sample of decisions that have adopted the first view.

- *Edward Van Halen v Deborah Morgan.*[269] The respondent registered the domain name <edwardvanhalen.com> and claimed that she intended to use it for a fan site. As the domain name was inactive, the panellist held that the respondent had no rights or interests in the domain name. In reaching this conclusion, the panellist stated that:

  > Her status as a fan might perhaps justify a finding of legitimate interests if she provided evidence of offline fan activity of which the website was an extension or if the website had been operational for a sufficient period.

- *Garth Brooks v Commbine.com, LLC.*[270] The respondent, who registered the domain name <garthbrooks.com> after the complainant inadvertently allowed registration to lapse, claimed that it was operating a non-commercial fan site for the singer. The respondent initially used the site for banner advertising, but the advertisements were later withdrawn. The panellist held that the respondent had been making a commercial use of the domain name and that, given the respondent's history of domain name brokerage, the explanation for registering the domain name was not plausible. At the same time, the panellist recognised that operating a non-commercial fan site may be a legitimate non-commercial use.
- *2001 White Castle Way, Inc v Glyn O Jacobs.*[271] The respondent registered the domain name <patbenatar.com> and used it for a fan site for the recording artist, 'PAT BENATAR', including detailed information about the artist and her albums. The web site included a recommendation for an Internet retailer that stocked the artist's work, there was no evidence that the respondent was obtaining a commercial benefit, and the site included a

[267] WIPO Case No D2001-1042 (25 Oct 2001).
[268] WIPO Case No D2005-0873 (1 Nov 2005).
[269] WIPO Case No D2000-1313 (20 Dec 2000).
[270] NAF Case No FA96097 (3 Jan 2001).
[271] WIPO Case No D2004-0001 (26 Mar 2004).

clear disclaimer. In these circumstances, the panellist held that the respondent was making a legitimate non-commercial use of the domain name.

- *Pearl Jam v Streaming Digital Media Dot Com.*[272] The respondent registered the domain name <pearljamlive.com>, which it used for a site dedicated to the band's music and fans. The panellist held that the respondent was operating a non-commercial fan site, and therefore making a legitimate non-commercial use of the domain name. In reaching this conclusion, the panellist held that the inclusion of a prominent link to the complainant's site sufficiently distinguished the respondent's site so that there was no intent misleadingly to divert consumers.
- *Blind Melon v KWI.*[273] The respondent registered the domain name <blindmelon.com> and used it as a fan site for the band, 'BLIND MELON'. The evidence indicated that fans had posted over 20,000 messages to the site's message board. The panellist held that the operation of the fan web site was a legitimate non-commercial or fair use.

The second view is that a respondent cannot, under any circumstances, establish rights or interests in a confusingly similar domain name that resolves to a fan site as, merely by choosing that domain name, the respondent has created a misleading association with the trade mark owner. This view was explained in the following terms by the panellist in *Daniel C Marino, Jr v Video Images Productions*:[274]

> in light of the uniqueness of the name DanMarino.com, which is virtually identical to the Complainant's personal name and common law trademark, it would be extremely difficult to foresee any justifiable use that the Respondent could claim. On the contrary, selecting this name gives rise to the impression of an association with Complainant, which is not based in fact.

The following are examples of decisions that support the second view.

- *David Gilmour v Ermanno Cenicolla.*[275] The respondent registered the domain name <davidgilmour.com>, together with a number of other celebrity domain names, and claimed that he intended to use it as a fan site for 'DAVID GILMOUR' and the band, Pink Floyd. The domain name was inactive and was initially offered for sale, along with the other celebrity domain names. The panellist held that the respondent had no rights or interests in the domain name as it had been selected with an intent to attract Internet users by trading on the goodwill of the complainant's mark. In reaching this conclusion, the panellist observed that, even if the domain name was used solely for favourable comments, the respondent could have no legitimate interests in the domain name as this would result in the respondent identifying himself with the trade mark owner.
- *Galatasaray Spor Kulubu Dernegi v Maksimum Iletisim AS.*[276] The respondent registered the domain name <galatasaray.com> with the intention of establishing a fan site for the complainant, a well-known Turkish football club. The respondent had also registered the domain names of other Turkish football clubs. Although the respondent was not making a commercial use of the domain name and the site included a prominent disclaimer, the panel held that the respondent was not making a legitimate non-commercial or fair use of the domain name. In reaching this conclusion, the panel stated:

[272] NAF Case No FA235831 (29 Mar 2004).
[273] NAF Case No FA741833 (21 Aug 2006).
[274] WIPO Case No D2000-0598 (2 Aug 2000).
[275] WIPO Case No D2000-1459 (15 Dec 2000).
[276] WIPO Case No D2002-0726 (15 Oct 2002).

the Respondent was well aware when it created the website and connected it to the Domain Name that a substantial proportion of visitors would assume it to be the Complainant's official site in the absence of the disclaimer. By then of course the Respondent has benefited from that initial interest confusion, whether the benefit be commercial or non-commercial.

- *The Estate of Marlon Brando v thewordbank twb.*[277] The respondent registered the domain name <marlonbrando.net> and offered books, films and other 'MARLON BRANDO' memorabilia from the web site. The respondent had also registered the names of other famous artists and celebrities as domain names. As the respondent was using the domain name to operate a commercial web site, the panellist held that it was not making a legitimate non-commercial use of the name. In any event, the panellist held that even if the respondent was operating a non-commercial fan site it could not have legitimate interests in the domain name, as this would open the door to other cybersquatters taking advantage of the exception.
- *Tia Carrere v Steven Baxt.*[278] The respondent registered the domain name <tiacarrere.com> and used it for a fan site that included advertisements and links to commercial sites, including sites for Hawaiian resorts and holidays. The panellist found that, as the commercial links on the web site were as prominent as the non-commercial material, the respondent was not making a fair use of the domain name. At the same time, the panellist held that using the complainant's name as a domain name could never be a legitimate or fair use of the domain name as 'the right to express views about a subject does not translate into a right to identify one's self as that subject'.

Generic Terms

[6.10] Rights or Interests in Generic Terms

A generic term is a name of a product or service itself. Given that, by definition, a generic term cannot operate to distinguish the source of goods or services, it might be thought that anyone should be free to register a generic term as a domain name. What is generic in one part of the world may, however, be distinctive in another part of the world. Moreover, a term that is descriptive rather than generic may become distinctive and acquire trade mark rights. Consequently, it is unfair to expect domain name registrants to be aware of trade mark rights that may exist in every generic or descriptive term. If a complainant can establish rights in a mark that includes a generic term, through establishing either registered or unregistered rights in the mark, the question arises whether the respondent has rights or legitimate interests under paragraph 4(a)(ii).

The general principle is that, where a complainant establishes a prima facie case that the respondent has no rights or legitimate interests in the domain name, the onus is on the respondent to show that one of the affirmative defences set out in paragraph 4(c) applies. In this sense, claims to rights or legitimate interests in a generic term are treated no differently from claims in respect of other domain names. The WIPO *Consensus View* reflects this approach and is as follows:

[277] NAF Case No FA505502 (16 Aug 2005).
[278] WIPO Case No D2005-1072 (28 Nov 2005).

> If the complainant makes a prima facie case that the respondent has no rights or legitimate interests, and the respondent fails to show one of the three circumstances under Paragraph 4(c) of the Policy, then the respondent may lack a legitimate interest in the domain name, even if it is a domain name comprised of a generic word(s). Factors a panel should look for when determining legitimate use would include the status and fame of the mark, whether the respondent has registered other generic names, and what the domain name is used for (a respondent is likely to have a right if it uses (*sic*) for a site for apples but not if the site is aimed at selling computers or pornography).[279]

As explained at [6.11], the consensus view is subject to an important qualification where there is no evidence of bad faith on the part of the respondent.

In disputes concerning generic domain names, panels often find that a prima facie case has been made out where the complainant has not authorised or licensed the use of its mark as a domain name. Where this is established, panels dealing with claims to rights or interests in a generic term, in essence, examine whether the respondent is able to make out any of the circumstances set out in paragraph 4(c). In most cases, the issue will be whether the respondent has used, or made demonstrable preparations to use, the domain name in connection with a bona fide offering of goods or services under paragraph 4(c)(i), or whether the respondent is making a legitimate non-commercial or fair use of the domain name under paragraph 4(c)(iii). Given that where there is evidence of bad faith registration or use of the domain name there cannot be a bona fide offering of goods or services, or a 'legitimate' use of a domain name, the question whether or not there are rights or interests in a generic domain name often centres on the bona fides of the respondent. For example, if the respondent has registered a generic domain name for the purpose of diverting users to a commercial site operated by the respondent, or to a competitor's site, then the respondent can have no rights or interests in the domain name.

The following are examples of decisions in which a respondent has claimed rights or interests in a domain name that consists of a generic term, but panels have found that the respondent has not made out any of the three circumstances set out in paragraph 4(c).

- *402 Shoes, Inc dba Trashy Lingerie v Jack Weinstock.*[280] The respondent registered the domain name <trashylingerie.com> and claimed rights or legitimate interests in the term 'TRASHY LINGERIE', arguing that it had become descriptive of provocative lingerie. The majority of the panel refused to entertain the defence that the complainant's registered mark had become descriptive or generic on the basis that this was properly a question to be determined by courts, as UDRP panels are ill-equipped to decide such complex questions. As the respondent had not presented evidence of any of the three circumstances set out in paragraph 4(c), the panel held it had no rights or interests in the domain name. In a concurring opinion, one member of the panel maintained that refusing to consider the defence that a mark had become descriptive or generic would discriminate against the respondent. On the facts, however, the concurring panellist held that the respondent had failed to establish that the mark was descriptive.
- *Classmates Online, Inc v John Zuccarini.*[281] The respondent, who had registered the domain names <classmat.com> and <classmats.com>, which were diverted to pornographic web sites that mousetrapped users, claimed that a respondent can have legitimate interests in a misspelt generic word. As the respondent had established rights in the 'CLASSMATES'

[279] WIPO, above n 10, para 2.2.
[280] WIPO Case No D2000-1223 (2 Jan 2001).
[281] WIPO Case No D2002-0635 (24 Sept 2002).

mark, the panel held that the respondent could not establish legitimate interests in the term, in the absence of evidence of any of the safe harbours set out in paragraph 4(c). Moreover, the commercial use of the domain name and the evidence of typosquatting and mousetrapping meant that use of the domain name was not 'legitimate'.

- *Emmanuel Vincent Seal v Ron Basset.*[282] The respondent registered the domain name <completesportsbetting.com> and argued that, as the words 'COMPLETE SPORTS BETTING' are descriptive, he was entitled to register the domain name in good faith. The panellist found that the respondent had registered the domain name knowing of the complainant's goodwill and reputation, and intending to use the domain name to compete with the complainant. In these circumstances, the panellist held that any preparations to use the domain name were not in connection with a bona fide offering of goods or services or making a fair use of the disputed domain name. Consequently, as the domain name had not been registered and was not being used in good faith, the respondent had no rights or interests in the domain name.
- *Owens Corning Fiberglass Technology, Inc v Hammerstone.*[283] The respondent registered the domain name <cultured-stone.com>, which it used to redirect Internet users to the <hammerstone.net> web site, which promoted the respondent's masonry products. Although the respondent claimed that the term 'CULTURED STONE' was generic, the panellist held that this use of the domain name was not in connection with a bona fide offering of goods or services and was not a fair use.
- *Société Générale v Kellerbauer.*[284] The respondent registered the domain name <societegeneral.com> and argued that the words 'SOCIETE GENERALE' are generic terms frequently used in many company names in French-speaking countries. The disputed domain name resolved to a web page containing links to commercial pages, including those of some of the complainant's competitors. In support of a claim that the respondent had made 'demonstrable preparations' to use the domain name under paragraph 4(c)(i), the respondent presented a printout of a web page for another of the respondent's sites. The panellist, however, rejected this evidence, as well as the respondent's explanation for the long delay in establishing an active web site. Consequently, the panellist held that the respondent had no rights or interests in the domain name.

[6.11] Registering a Generic Domain Name in Good Faith

Where a respondent registers a generic term as a domain name and there are no indications of bad faith, then the respondent will have rights or legitimate interests in the domain name. This is a qualification to the general position that a respondent who registers a generic term as a domain name must produce evidence that one of the circumstances set out in paragraph 4(c) has been satisfied. The *WIPO Overview* expresses this qualification in the following terms:

> If a respondent is using a generic word to describe his product/business or to profit from the generic value of the word without intending to take advantage of complainant's rights in that word, then it has a legitimate interest.[284a]

[282] WIPO Case No D2002-1058 (26 Jan 2003).
[283] WIPO Case No D2002-1058 (29 Dec 2003).
[284] WIPO Case No D2005-0816 (13 Oct 2005).
[284a] WIPO, above n 10, para 2.2.

Another way to state this principle is that the first person to register a generic term as a domain name in good faith will have legitimate interests in that domain name. This effectively focuses attention on the intention of the respondent in registering and using the disputed domain name. As the panellist expressed this point in *Jet Marques v Vertical Axis, Inc:*[285]

> mere registration of a generic word as a domain name will not be sufficient in and of itself to create a 'legitimate interest', if the respondent's substantial intention was to capitalize on the complainant's trademark interest.

A decision that is commonly cited in support of the principle that registration of a generic domain name may, absent bad faith, confer rights or interests in the domain name is *Target Brands, Inc v Eastwind Group.*[286] In that dispute, the respondent registered the domain name <target.org>, and used it as part of an affiliate advertising programme in which a third party advertising broker posted links to the respondent's web site. The links included links to web sites related to the dictionary meaning of the term 'TARGET', such as archery and goal-setting sites, and at one time had included a link to an adult web site. The panellist held that, as the term was generic, the first person to register the domain name in good faith had a legitimate interest in the domain name. Moreover, the panellist held that, in the absence of bad faith, using the domain name for an affiliate advertising programme was a bona fide offering of goods or services under paragraph 4(c)(i) of the UDRP.

A respondent who registers a generic domain name in good faith will have legitimate interests in the domain name even if it is offered for sale. In *Ultrafem Inc v Warren Royal,*[287] for example, the respondent registered the domain name <instead.com> and, although it had remained inactive for nearly three years, claimed that it planned to use it in connection with ecologically friendly products. The domain name was listed for sale on a domain name auction site. The panel held that, as the domain name was generic and there was no evidence of bad faith, the respondent had legitimate interests in the domain name. The panel also held that, in the absence of bad faith, an intent to sell a generic domain name can, in itself, amount to a legitimate interest in the domain name, provided there is no evidence that the domain name was registered to sell to a trade mark owner.

Quite a number of decisions dealing with rights or interests in generic domain names have raised the question whether a generic domain name broker or reseller, meaning a respondent who registers generic domain names with the intention of selling them, is using the domain names in connection with a bona fide offering of goods or services. The great majority of panels have held that generic domain name reselling is permissible, and amounts to a bona fide offering of goods or services under paragraph 4(c)(i). As the panellist put it in *The Landmark Group v Digimedia LP:*[288]

> as long as the domain names have been registered because of their attraction as dictionary words, and not because of their value as trademarks, this business model is permitted under the Policy.

There is, however, a continuing division of views as to whether or not the business of reselling domain names is a bona fide offering of goods or services, with a minority continuing to hold that it is not. This division of views is illustrated by the decision in *HP Hood*

[285] WIPO Case No D2006-0250 (26 May 2006).
[286] NAF Case No FA267475 (9 July 2004).
[287] NAF Case No FA97682 (2 Aug 2001).
[288] NAF Case No FA285459 (6 Aug 2004).

LLC v hood.com,[289] where the question of the proper treatment of a generic domain name broker in the circumstances of that case split the panel. In that dispute, the respondent, a bulk domain name reseller, used the domain name <hood.com> to obtain advertising revenue. The majority of the panel made a number of important points in the course of deciding that the respondent had rights or legitimate interests in the domain name.

First, the majority approved the following statement made by the panelist in *Shirmax Retail Ltd v CES Marketing, Inc*:[290]

> Where the domain name and trademark in question are generic—and in particular where they comprise no more than a single, short, common word—the rights/interests inquiry is more likely to favor the domain name owner. The ICANN Policy is very narrow in scope; it covers only clear cases of 'cybersquatting' and 'cyberpiracy', not every dispute that might arise over a domain name. See, eg, Second Staff Report on Implementation Documents for the Uniform Dispute Resolution Policy (Oct. 24, 1999).

Secondly, following the *Target Brands* decision, the majority held that using a common or generic word in connection with third party advertising services related to the descriptive meaning of the domain name will constitute use in connection with a bona fide offering of goods or services. Thirdly, referring to the evidence that the respondent had registered numerous generic and common words to generate advertising revenue, the majority held that this was sufficient to confer a legitimate interest. Fourthly, the majority held that generic domain name speculation could, in itself, amount to a bona fide offering of goods or services, concluding that:

> an established domain name resale enterprise that restricts its portfolio in a good faith effort to avoid misleading the public qualifies as a legitimate interest and a defense under Section 4(c)(i).

The dissenting panellist in that dispute took a quite different view. First, the dissenting panellist was concerned at an apparent attempt of the respondent to obscure its identity as a domain name reseller in the registration of the domain name. Secondly, the dissenting panellist held that simply registering a domain name, then offering to sell it, was insufficient to constitute a bona fide offering of goods or services.

A similar difference of views divided the panel in *Shoe Mart Factory Outlet, Inc v DomainHouse.com, Inc*,[291] where the majority related the differences to a different understanding of the purpose of the UDRP. In that dispute, the respondent registered the domain name <shoemart.com>, which it used solely to promote its domain registration services. The majority of the panel held that the complainant had failed to establish rights in the 'SHOEMART' mark, as it held no registered rights and was unable to establish unregistered rights in the generic or descriptive term. In this respect, the majority observed:

> a mark which is either generic or descriptive cannot be registered and cannot become a protected common law trademark unless it is conclusively shown to have become distinctive in the sense that it has achieved a secondary meaning such that consumers identify those common terms exclusively with the goods or services of the owner of the mark.

The majority went on to conclude that, as the UDRP was limited to providing summary proceedings for recovery of domain names that had been registered in bad faith, the

[289] NAF Case No FA313566 (20 Oct 2004).
[290] eRes Case No AF-0104 (20 Mar 2000).
[291] NAF Case No FA462916 (10 June 2005).

registration of a generic domain name for the purpose of selling it was not prohibited. The majority therefore concluded that:

> it is well established by UDRP decisions that the registration of common descriptive or generic terms and holding them for the purpose of sale is perfectly legal. What is prohibited is only the registration of protectible common law marks or registered marks for sale to the owners.

The dissenting panellist, however, held both that the complainant had established common law rights in the mark and that the respondent had no rights or interests in the domain name. In holding that the respondent was not using the domain name in connection with a bona fide offering, the dissenting panellist simply stated that:

> Reselling domain names does not constitute a *bona fide* offering of goods or services. To allow such an absurd construction would eviscerate the UDRP because *every* respondent could demonstrate rights.

The following decisions illustrate the principle that a respondent will generally have legitimate interests in a generic domain name in the absence of bad faith. The decisions also illustrate the general position that domain name resellers will be able to establish rights or interests in a generic domain name.

- *Shirmax Retail Ltd v CES Marketing, Inc.*[292] The respondent, a generic domain name reseller, registered the domain name <thyme.com> that redirected users to the <cesmarketing.com> site, from which it offered a considerable number of generic domain names for sale. The panellist held that the respondent had a tenable claim that 'its use on the web merely for the purpose of redirecting visitors to a different site constitutes a legitimate fair use, as long as this use is not misleading to consumers and does not tarnish a trademark'.
- *Allocation Network GmbH v Steve Gregory.*[293] The respondent, a domain name broker, registered the domain name <allocation.com>, but only commenced using it as a portal site after the UDRP complaint was initiated. The disputed domain name was one of approximately 400 descriptive or generic domain names that were offered for sale at the respondent's <rockbottomdomains.com> web site. The panellist held that, in principle, the practice of generic domain name brokerage could amount to a bona fide offering of goods or services under paragraph 4(c)(i), but added that 'this is different if it were shown that allocation.com has been chosen with the intent to profit from or otherwise abuse Complainant's trademark rights'. As the panellist concluded that the respondent was not aware of the complainant's German trade mark but had chosen domain names for registration at random from a dictionary, he held that the respondent had a legitimate interest in the domain name.
- *Lumena s-ka zo. o. v Express Ventures Ltd.*[294] The respondent registered the domain name <lumina.com> and claimed that it was in the business of registering generic and descriptive domain names for the purpose of developing them as 'productive' web sites. The panellist held that the admittedly perfunctory efforts to use the domain name were sufficient to amount to 'demonstrable preparations' to use the domain name under paragraph 4(c)(i). In deciding that the respondent had legitimate interests in the domain name, the panellist held that there was an arguable case that paragraph 4(c)(iii) was made out, observing that:

[292] eRes Case No AF-0104 (20 Mar 2000).
[293] WIPO Case No D2000-0016 (24 Mar 2000).
[294] NAF Case No FA94375 (11 May 2000).

given the generic nature of the domain name, Respondent has at least the arguable position that were it to use lumena.com merely for the purpose of redirecting visitors to a different site constitutes a legitimate fair use, as long as this use is not misleading to customers and does not tarnish a trademark.

- *Asphalt Research Technology, Inc v National Press & Publishing, Inc.*[295] The respondent registered the domain name <ezstreet.net> that was linked to a web site that advertised generic domain names, including the disputed domain name, for sale. The panellist held that, as generic terms may generally be registered on a 'first come, first served' basis, and as offers to sell generic, common or descriptive domain names are not acts of bad faith, the respondent had legitimate interests in the domain name.
- *Porto Chico Stores, Inc v Otavio Zambon.*[296] The disputed domain name <lovelygirls.com> was used in connection with a pornographic web site. As the domain name was purely descriptive, that there was no evidence that the respondent knew of the complainant's 'LOVELY GIRL' mark and that there was nothing 'illegitimate' in operating a pornographic web site, the panellist held that the complainant had not established that the respondent had no rights or interests in the domain name.
- *Micron Technology, Inc v Null International Research Center.*[297] The respondent, who was in the business of registering and reselling domain names, registered the domain name <crucialtechnology.com>, which remained inactive at the time of the dispute. The panellist held that 'domain name speculation can be a legitimate interest if the Respondent was not aware of Complainant's trademark rights'. As the complainant's mark was not well-known, the panellist concluded that the complainant had not established that the respondent had no rights or interests in the domain name.
- *Gorstew Limited v Worldwidewebsites.com.*[298] The nine disputed domain names, including <anguillabeaches.com>, <antillesbeaches.com> and <barbadosbeaches.com>, consisted of a geographic term relating to the Caribbean and the generic term 'BEACHES'. The domain names were diverted to a web site at <worldwidewebsales.com> that advertised a range of services and offered a large number of generic domain names for sale. The respondent indicated that it was willing to sell the disputed domain names to the complainant for a substantial amount. The panellist, however, was not satisfied that the domain names were registered primarily for the purpose of transfer to the complainant or to a competitor of the complainant. Finding that the policy does not prevent the registration of generic domain names with an intention to sell the domain names to third parties, the panellist concluded that the respondent had not registered the domain names in bad faith. In reaching this conclusion, the panellist approved the respondent's submission that while 'cybersquatting is precluded behavior, cyber speculating clearly is not'.
- *Jet Marques v Vertical Axis, Inc.*[299] The respondent acquired the domain name <jettour.com> after its previous registration had lapsed and used it to link to airline sites and other businesses offering airline travel. In its submission, the respondent claimed that as 'JET TOUR' was a generic term, the first person to register it in good faith was entitled to a legitimate interest in the domain name. The panel held that the respondent could establish legitimate interests in the domain name if the domain name was a generic

[295] WIPO Case No D2000-1005 (13 Nov 2000).
[296] WIPO Case No D2000-1270 (15 Nov 2000).
[297] WIPO Case No D2001-0608 (20 June 2001).
[298] WIPO Case No D2002-0744 (23 Oct 2002).
[299] WIPO Case No D2006-0250 (26 May 2006).

expression and, if so, the respondent had registered the domain name without intending to take advantage of the complainant's trade mark rights. On the first question, the panel held that there were contexts in which the term 'JET TOUR' was used in English in a purely descriptive sense, even if the term was not generic in other languages, including French. In relation to the second question, the panel considered that the key issue was whether the respondent was aware of the complainant's 'JET TOUR' mark at the time the domain name was acquired as, if the respondent was aware, it must have contemplated a likelihood of confusion. The panel did not, however, come to a conclusion on this point as the evidence was equivocal and, in any case, the panel held that the complainant had not established bad faith registration and use under paragraph 4(a)(iii). The considerations leading to this conclusion included the fact that the complainant's mark was registered only in Europe, while the respondent reseller was based in the Republic of Korea, and the extent to which the complainant's mark did not appear to be well-known outside France.

- *Shoulderdoc Limited v Vertical Axis, Inc.*[300] The respondent registered the domain name <shoulderdoc.com>, which originally provided sponsored links to a variety of sites offering medical services. The panel held that, as the term was generic or descriptive, the key issue was whether it was registered in bad faith. As in *Jet Marques,* this resolved to the question whether the domain name was registered with knowledge of the complainant's rights. Given the low level of distinctiveness of the 'SHOULDERDOC' mark, the panel held that the complainant had not established that the domain name had been registered in bad faith.

[300] WIPO Case No D2006-0625 (8 Sept 2006).

7

Bad Faith Registration and Use

Overview and Policy

[7.1] Overview

The third element of abusive, bad faith registration, set out in paragraph 4(a)(iii) of the UDRP, requires the complainant to prove that:

> the domain name has been registered and is being used in bad faith.

This element represents the gist of a complaint under the UDRP, as it effectively defines the behaviour that is proscribed by the policy. As the panellist put it in *Martha Stewart Living Omnimedia, Inc v Josh Gorton:*[1]

> The overriding objective of the Policy is to prevent abusive domain name registrations and use for the benefit of legitimate trademark owners, in circumstances where the registrant is seeking to profit from and exploit the trademark of another.

To the same effect, the panellist in *Match.com, LP v Bill Zag*[2] observed that:

> The Policy is designed to curb the abusive registration of domain names in the circumstances where the registrant is seeking to profit from and exploit the trademark of another ... The Policy was not intended to permit a party who elects to register as a trademark or service mark a common word to bar all others from using the common word in combination with other common words, unless it is clear that the use involved is seeking to capitalize on the goodwill created by the mark holder.

Given that the complainant bears the burden of making out the third element, registration and use of a domain name will be in good faith unless the complainant can establish that the domain name has been registered and is being used in bad faith. The WIPO Final Report explained this in the following terms:

> The cumulative conditions of the first paragraph of the definition make it clear that the behavior of innocent or good faith domain name registrants is not to be considered abusive.[3]

A wide variety of circumstances may give rise to a finding of bad faith registration and use. The *WIPO Overview,* in setting out the Consensus View as to what may constitute bad faith where the respondent makes no active use of a disputed domain name, provides that:

[1] WIPO Case No D2005-1109 (13 Dec 2005).
[2] WIPO Case No D2004-0230 (2 June 2004).
[3] World Intellectual Property Organization (WIPO), *The Management of Internet Names and Addresses: Intellectual Property Issues,* Report of the WIPO Internet Domain Name Process (30 Apr 1999) para [172].

> Examples of circumstances that can indicate bad faith include complainant having a well-known trademark, no response to the complaint, concealment of identity and the impossibility of conceiving a good faith use of the domain name.[4]

Other common circumstances that may result in a finding of bad faith include evidence that the respondent had notice of the complainant's mark at the time the domain name was registered, evidence of 'typosquatting', the use of the disputed domain name to divert Internet users to other web sites, the use of the disputed domain name to offer goods or services in competition with the complainant and the use of the disputed domain name to attract 'click-through' advertising revenue. As explained at [7.4], in determining whether a domain name has been registered and is being used in bad faith contrary to paragraph 4(a)(iii), UDRP panels will commonly look at the 'totality of circumstances'. This means that it is impossible to produce a comprehensive list of circumstances that will amount to bad faith for the purpose of the policy.

The relationship between the third element of the UDRP and the other two elements may be understood as follows: the first two elements effectively act as filters, excluding disputes that fall outside the scope of the policy, while the third element identifies the conduct prohibited by the policy. It is important to appreciate that, while it is more likely for there to be a finding of bad faith registration and use where the respondent has no rights or legitimate interests in the disputed domain name, as the second and third elements of the UDRP are separate and distinct, establishing that the respondent has no rights or legitimate interests is not sufficient to establish bad faith. As the panel explained in *Document Technologies, Inc v International Electronic Communications Inc.*[5]

> Bad faith is *not* proven by showing that Respondent lacks any rights to or legitimate interests in the Domain Name. That is a separate and distinct factor, with its own proof requirements. If the Panel adopted Complainant's approach, it would have the practical effect of conflating the second and third requirements. Such an approach is not supported by the clear wording of the Policy, which separates the requirements in Paragraph 4(a) into three distinct elements, and provides separate examples (Paragraphs 4(b) and 4(c)) of how to satisfy the second and third factors.

[7.2] Policy Considerations

Although the wording of paragraph 4(a)(iii) directly implements the wording proposed in the WIPO Final Report, it has been controversial. As explained at [3.5], as part of ICANN's process of reviewing WIPO's recommendations, trade mark interests proposed altering the wording of the third element of the UDRP to encompass bad faith registration *or* use of a domain name, arguing that this extension of the policy was needed to prevent bad faith 'passive warehousing'. As further explained, other submissions opposed this extension on the basis that this would effectively extend conventional principles of traditional trade mark law, according to which there is no infringement without commercial use of a mark.

This difference of views was never satisfactorily resolved, as ICANN decided against changing the policy, adopting the view that a change of this nature would be contrary to the August 1999 ICANN board policy resolutions. In this respect, the second ICANN staff

[4] WIPO, *WIPO Overview of WIPO Panel Views on Selected UDRP Questions* (23 Mar 2005), available at http://www.wipo.int/amc/en/domains/search/overview/index.html, para 3.2.

[5] WIPO Case No D2000-0270 (6 June 2000).

report on implementation of the UDRP, referring to the argument for extending the policy to cover 'passive warehousing', stated:

> While that argument appears to have merit on initial impression, it would involve a change in the policy adopted by the Board. The WIPO report, the DNSO recommendation, and the registrars-group recommendation all required both registration and use in bad faith before the streamlined procedure would be invoked. Staff recommends that this requirement not be changed without study and recommendation by the DNSO.[6]

In retrospect, it can be seen that the wording recommended by the WIPO Final Report and ICANN's decision to retain the wording were based on an inadequate understanding of the differences between traditional trade mark infringement actions and actions aimed at preventing cybersquatting. In short, it is readily apparent that there is absolutely no need for a domain name registrant actually to use a domain name before making an extortionate demand on a trade mark owner, seeking payment for the domain name. In other words, in order for a threat to be credible, there is no need for the threat to be acted upon. The lack of precision in satisfactorily identifying the objectives of the UDRP has created difficulties for panellists required to apply the policy to practices that do not clearly fit within the third element of the policy, especially those involving 'passive warehousing'.

Registration and Use in Bad Faith

[7.3] The Four Non-exclusive Circumstances

Paragraph 4(b) of the UDRP sets out four non-exclusive circumstances that amount to evidence of the registration and use of a domain name in bad faith. The *travaux préparatoires* for the four circumstances are explained in detail at [3.7]. This paragraph therefore simply outlines the circumstances set out in paragraph 4(b) and summarises the background to each of the circumstances. The four circumstances are as follows:

> (i) circumstances indicating that the domain name holder has registered or has acquired the domain name primarily for the purpose of selling, renting, or otherwise transferring the domain name registration to the complainant who is the owner of the trademark or service mark or to a competitor of that complainant, for valuable consideration in excess of the domain name holder's documented out-of-pocket costs directly related to the domain name.

The first circumstance is based on circumstance (a) proposed in the WIPO Final Report, but with two modifications. First, the wording of the paragraph was amended to apply to all circumstances indicating that the domain name was registered or acquired primarily for the purpose of selling, renting or transferring the domain name to the complainant, and not confined to actual offers to sell or transfer the domain name to the complainant. Secondly, following the August 1999 ICANN board resolution, the circumstance was confined to circumstances indicating an intention to seek payment from the complainant in excess of the registrant's 'out-of-pocket' costs, reflecting the pragmatic consideration that cybersquatters would not confine their claims to 'out-of-pocket' expenses.

[6] ICANN, *Second Staff Report on Implementation Documents for the Uniform Dispute Resolution Policy* (24 Oct 1999), available at http://www.icann.org/udrp/udrp-second-staff-report-24oct99.htm, para [4.5(a)].

(ii) the domain name holder has registered the domain name in order to prevent the owner of the trademark or service mark from reflecting the mark in a corresponding domain name, provided that the domain name holder has engaged in a pattern of such conduct.

The second circumstance accurately reflects circumstance (c) proposed in the WIPO Final Report. The wording adopted means that the second circumstance is not satisfied by the mere registration of a single domain name, but that there must be a 'pattern of conduct' indicating that the domain name has been registered to prevent the trade mark owner from reflecting its mark as a domain name. As explained at [3.7], proposals for expanding on what amounts to a 'pattern of conduct' were rejected by ICANN on the basis that each case must be assessed in accordance with its own particular facts.

(iii) the domain name holder has registered the domain name primarily for the purpose of disrupting the business of a competitor.

The third circumstance, which accurately reflects circumstance (d) proposed in the WIPO Final Report, was introduced in response to concerns that a competitor of a trade mark owner might register a trade mark for an anti-competitive purpose. Unlike the second circumstance, the third circumstance does not require a 'pattern of conduct', but is confined to registration of a domain name with the purpose of 'disrupting' the business of the trade mark owner.

(iv) by using the domain name, the domain name holder has intentionally attempted to attract, for commercial gain, Internet users to its web site or other on-line location, by creating a likelihood of confusion with the complainant's mark as to the source, sponsorship, affiliation, or endorsement of its web site or location or of a product or service on its web site or location.

The fourth circumstance reflects, with some changes in wording, circumstance (b) proposed in the WIPO Final Report. This circumstance differs from the other circumstances in that it applies to the 'use' of the domain name, whereas the other three circumstances refer to 'registration' of the domain name. In addition, this circumstance indicates that use of a domain name will be evidence of bad faith where it is such as to create a 'likelihood of confusion' as to the source, sponsorship, affiliation or endorsement of the respondent's web site or of a product or service on its web site. This circumstance is therefore aimed at behaviour that unfairly trades off the goodwill of a trade mark owner's mark by, for example, diverting users from the trade mark owner's site to the registrant's site, for the purpose of commercial gain.

It must always be borne in mind that the circumstances itemised in paragraph 4(b) are indicative, but not exhaustive. The ICANN staff report of October 1999 made this point in response to submissions that the illustrative circumstances set out in the UDRP were too restrictive:

> the delineation of evidence of bad-faith is not intended to be exhaustive, and other types of evidence may be presented to the panel in the administrative dispute-resolution procedure.[7]

The non-exhaustive nature of the circumstances set out in paragraph 4(b) is embodied in the *chapeau* to the paragraph, which refers to the four identified circumstances 'in particular but without limitation'. The non-exhaustive nature of the paragraph 4(b) circumstances has been commented on by many UDRP panels, such as the panellist in *Sony Kabushiki Kaisha v Richard Mandanice*,[8] who observed that the paragraph 4(b) circumstances:

[7] ICANN, *Second Staff Report on Implementation Documents for the Uniform Dispute Resolution Policy* (24 Oct 1999), available at http://www.icann.org/udrp/udrp-second-staff-report-24oct99.htm, para [4.5(b)].

[8] WIPO Case No D2004-1046 (20 Jan 2005).

are not exclusive and it is also open to a Complainant to rely on other matters to show what might be called general bad faith on the part of a Respondent domain name owner.

[7.4] 'Totality of Circumstances' Approach

As the list of circumstances that amount to evidence of bad faith registration and use under paragraph 4(b) is not comprehensive, other circumstances may be taken into account even where the specific circumstances set out in paragraph 4(b) are not made out. Where none of the paragraph 4(b) circumstances are established, panels adopt what is known as the 'totality of circumstances' approach. This simply means that bad faith registration and use can be inferred from the 'totality of circumstances', even where the circumstances set out in paragraph 4(b) are established. As the panel in *XM Satellite Radio Inc v Michael Bakker*[9] put it:

> While the Policy lists four circumstances demonstrating registration and use of a domain name in bad faith, this list merely illustrates possible situations that demonstrate bad faith. The Panel may look to the totality of the circumstances surrounding the registration and use of a domain name in determining whether bad faith is present.

While the term 'totality of circumstances' appears to have been initially confined to disputes involving 'passive holding' of a disputed domain name, which is dealt with at [7.10], it has acquired a wider use, with panels often simply stating that, in determining whether paragraph 4(a)(iii) is established, panels are required to look at the 'totality of circumstances'.

A good example of the application of the 'totality of circumstances' approach is the decision in *Twentieth Century Fox Film Corporation v David A Risser*,[10] which is commonly cited as authority for this approach. In that dispute, the respondent registered 21 domain names containing the complainant's 'FOX' mark, including <foxnetworknews.com>, <foxnewsnetwork.com> and <foxhomevideo.com>. The respondent claimed that he had registered the domain names as part of a new business of developing subscription web page link directories, but made no active use of the domain names. All but three of the disputed domain names were registered after the complainant contacted the respondent to request transfer of the domain names. In deciding that the domain names had been registered and were being used in bad faith, the panel stated that:

> The requirement in the ICANN Policy that a complainant prove that domain names are being used in bad faith does not require that it prove in every instance that a respondent is taking positive action. Use in bad faith can be inferred from the totality of the circumstances even when the registrant has done nothing more than register the names.

The circumstances that the panel took into account in determining that there was bad faith registration contrary to paragraph 4(a)(iii) in this particular dispute included an acknowledgement that one of the disputed domain names had been registered with the intent of selling it, and the following additional factors:

> his use of words that describe the Complainant's business activities, his lack of any legitimate interest in the name Fox, his employment with a Complainant subsidiary to which he owed a duty of loyalty, his registration of the names after notice of a dispute, his conflicting statements over when

[9] NAF Case No FA861120 (27 Feb 2007).
[10] NAF Case No FA93761 (18 May 2000).

> he supposedly decided on a use for one of the names, his contrived response to the complaint, and the world-wide knowledge that the Complainant does business under the name Fox . . .

[7.5] Reconciling Paragraphs 4(a)(iii) and 4(b)

There is clearly a tension between the wording of the third element of the UDRP, which refers to both registration and use of the disputed domain name, and the wording of the four circumstances set out in paragraph 4(b), the first three of which refer to registration but not use of the domain name, while the fourth circumstance refers to use only, but not to registration. In construing the relationship between paragraph 4(a)(iii) and the circumstances specified in paragraph 4(b), it is important to take into account the wording of the introductory words to paragraph 4(b), which provides that the four circumstances 'shall be evidence of the registration and use of a domain name in bad faith'. The panellist in *Shirmax Retail Ltd v CES Marketing, Inc*[11] reconciled the apparent inconsistency in the wording of paragraph 4(a)(iii) and the first three circumstances set out in paragraph 4(b) in the following terms:

> The requirement of bad faith registration and use in paragraph 4(a)(iii) is stated in the conjunctive. Registration in bad faith is insufficient if the respondent does not use the domain name in bad faith, and conversely, use in bad faith is insufficient if the respondent originally registered the domain name for a permissible purpose. The first three examples in paragraph 4(b) all refer to registration for various illegitimate purposes as evidence of registration and use in bad faith; but in each instance bad faith use may well be implicit in the act of registering a domain name, since all of the improper purposes mentioned can be accomplished merely by passively holding a domain name.

Applying this reasoning, sub-paragraphs (i) to (iii) of paragraph 4(b) can be seen as consistent with paragraph 4(a)(iii) if a broad interpretation is given to what amounts to 'use' of the disputed domain name. This issue is dealt with further at [7.6] below. It is, on the other hand, much more difficult to reconcile the wording of the fourth circumstance set out in paragraph 4(b), which refers only to bad faith 'use' of the disputed domain name and does not mention registration, with the wording of the third element of the UDRP. As the panellist in *Shirmax* explained:

> The fourth example (paragraph 4(b)(iv)), however, refers only to improper use, and does not appear to require that the domain name also have been registered in bad faith. This example thus appears to conflict with the rule set forth in paragraph 4(a)(iii). The language of paragraph 4(a)(iii) is clear, and the only reasonable interpretations are to regard the fourth example as a narrow exception to the preceding subparagraph's conjunctive rule, or to apply the conjunctive rule as it is written and disregard the example entirely.

An alternative suggestion for reconciling the wording in paragraph 4(a)(iii) with the fourth circumstance set out in paragraph 4(b) was made, in the following terms, by the panel in *Passion Group Inc v Usearch Inc:*[12]

> The panel is assisted in forming a view as to how to interpret paragraphs 4(a)(iii) and 4(b)(iv) by the contrast between the introductory words of paragraph 4(c): '*any of the following circumstances . . . shall demonstrate your rights or legitimate interests*' and the introductory words of paragraph

[11] eRes Case No AF-0104 (20 Mar 2000).

[12] eRes Case No AF-0250 (10 Aug 2000).

> 4(b): '*the following circumstances . . . shall be evidence of the registration and use of a domain name in bad faith*'.
>
> This contrasting language indicates that use of the kind described in 4(b)(iv) is to be taken as evidence of bad faith registration as well as evidence of bad faith use. But this evidence is not necessarily conclusive. Furthermore, the panel is not required to assign substantial weight to evidence of constructive bad faith registration furnished by paragraph 4(b)(iv), and the panel may have regard to other evidence in determining whether the requirements of 4(a)(iii) have been proved.
>
> This approach accords with the Policy by enabling a finding of bad faith registration to be made where bad faith use within 4(b)(iv) is the only evidence tending to show the purpose for which the domain name was registered. Where, however, there is other relevant evidence, such as evidence that the domain name was registered for a permissible purpose, it must be weighed against any evidence of bad faith registration constituted by evidence of bad faith use within 4(b)(iv). It is difficult to imagine circumstances in which, under this approach, subsequent bad faith use within 4(b)(iv) would suffice to prove that a domain name was originally registered in bad faith.

In other words, once the complainant satisfactorily establishes the circumstance set out in paragraph 4(b)(iv), then an inference can be drawn that the domain name was registered in bad faith, leaving it to the respondent to produce evidence that the domain name has been registered in good faith. This understanding of the operation of paragraph 4(b)(iv) gives an effective meaning to the terms used in paragraphs 4(a)(iii) and 4(b)(iv), and is to be preferred to the approach proposed by the panellist in *Shirmax*.

Bad Faith Use

[7.6] Registered *and* Used in Bad Faith

The precise wording of the third element of the UDRP is significant. It requires the complainant to establish that the domain name has been registered, and is being used, in bad faith.

As the panellist pointed out in *Shirmax*, the use of the conjunctive imposes the cumulative requirement that the disputed domain name be both registered *and* used in bad faith. Given the deliberate policy decision to retain the cumulative requirement, and not to amend the policy to allow the third element to be satisfied by either bad faith registration or bad faith use, then it is clear that a complainant must establish both that the domain name has been registered in bad faith and that it is being used in bad faith. From the very earliest decisions, UDRP panels have universally adopted this position. Once this is accepted, determining what amounts to a 'use' of the disputed domain name becomes a key issue.

The domain name does not need to resolve to an active web site in order for it to be 'used'. As explained at [7.8], it is necessarily the case that, in order to prevent bad faith registration, in certain circumstances passive warehousing (or passive holding) of a domain name must amount to 'using' the domain name in bad faith. Furthermore, an offer to sell the domain name will amount to a use of the domain name. This has been established since the first WIPO decision, *World Wrestling Federation Entertainment, Inc v Michael Bosman*.[13]

[13] WIPO Case No D1999-0001 (14 Jan 2000).

In that dispute, the domain name <worldwrestlingfederation.com> did not resolve to a web site but, soon after the registration of the domain name, the respondent offered to sell it to the owner of the 'WORLD WRESTLING FEDERATION' mark for US$1,000. After acknowledging that ICANN had decided to require both bad faith registration and use, the panellist referred to paragraph 4(b)(i) as support for the proposition that offering the domain name for sale for an amount in excess of out-of-pocket expenses constitutes use for the purpose of paragraph 4(a)(iii). In reaching this conclusion, the panellist referred to the decision of the US court in *Intermatic, Inc v Toeppen,*[14] which had held that registration of a domain name that incorporated a trade mark with an intention to sell it to the trade mark owner was a 'commercial use' of the mark for the purpose of the Lanham Act.[15]

At the same time, merely contemplating selling the disputed domain name is not sufficient evidence of bad faith registration and use. In this respect, it is important to note that paragraph 4(b)(i) requires evidence that the domain name has been registered or acquired *primarily* for the purpose of selling, renting or transferring the domain name. In *ZZounds Music, LLC v Zounds,*[16] for example, the domain name <zounds.com> did not resolve to an active web site, but there was evidence that the respondent had contemplated selling the domain to the complainant and to the predecessor of the complainant in response to offers made to purchase the domain name. The panel held that 'merely considering selling the domain name is not sufficient evidence of bad faith registration and use of the domain name'.

[7.7] Point in Time of Use in Bad Faith

Paragraph 4(a)(iii) refers to the use of a disputed domain name in the present tense, requiring that the domain name 'is being used' in bad faith. Despite the use of the present tense, the paragraph has been interpreted to mean that it is sufficient for the domain name to be used in bad faith at any time after registration.

In *Ingersoll-Rand Co v Frank Gully,*[17] for example, the respondent's only use of the domain names <ingersoll-rand.net>, <ingersoll-rand.org> and <ingersollrand.org> was as paid links to pornographic web sites. As a result of the dispute, however, the respondent had ceased using the domain names for this purpose by the time of the complaint. In concluding that the domain name was being used in bad faith, the panellist observed that:

> the term 'is being used' does not refer to a particular point in time (such as when the Complaint is filed or when the Panel begins deliberations), but refers to the period of time following registration of the domain name at issue . . . If at any time following the registration the name is used in bad faith, the fact of bad faith use is established. Thus, the fact that upon remonstrance by Complainant's counsel, Respondent ceased using the domain names at issue as links to pornographic sites, cannot alter the fact that the bad faith act had occurred during the period following registration.

Similarly, in *Valor Econômico SA v Daniel Allende,*[18] the panellist stated:

> The Policy should not be construed in such a manner that bad faith use should continue after the dispute has begun or that these proceedings have been commenced. To do that would amount to

[14] 947 F Supp 1227 (ND Ill, 1996).
[15] For a discussion of 'commercial use' under the Lanham Act see [5.24].
[16] NAF Case No FA817093 (24 Nov 2006).
[17] WIPO Case No D2000-0021 (9 Mar 2000).
[18] WIPO Case No D2001-0523 (28 June 2001).

> grant to any Respondent a protection against the Policy's consequences by simply allowing him to cease or cancel any posting of contents on the web site, and leaving the site without any use for all practical effects. Such would be an absurd interpretation of the Policy.

In *Caixa Economica Federal v Menterprises—Web Development,*[19] the respondent, a domain name reseller, offered the domain name <caixaeconomica.com> for sale, but removed the 'for sale' notice from the web site by the time of the complaint. The panellist held that he was 'entitled to consider acts of bad faith that occurred between the date of registration and the receipt of a notification from the Complainant, independently of whether they were "corrected"'.

[7.8] Good Faith Registration and Bad Faith Use

As paragraph 4(a)(iii) requires the complainant to establish both bad faith registration and bad faith use, it will not be satisfied if the domain name has been registered in good faith, even if it can be established that the domain name is being used in bad faith.

Statements in the WIPO Final Report might be interpreted to suggest that a finding of abusive, bad faith registration may be made where the domain name has been registered in good faith, but is subsequently being used in bad faith. In particular, paragraph 197 of the Final Report, which dealt with the possibility of introducing a limitation period, stated:

> The WIPO Interim Report recommended that a time bar to the bringing of claims in respect of domain names (for example, a bar on claims where the domain name registration has been unchallenged for a designated period of years) should not be introduced. It was considered that such a measure would not take into account that the underlying use of a domain name may evolve over time (with the consequence that the use of a domain name may become infringing through, for example, the offering for sale of goods of a different sort to those previously offered on the website); that any related intellectual property rights held by the domain name holder may lapse; and that a time bar would in any event be undesirable in cases of bad faith.[20]

As indicated by paragraph 198 of the Final Report, it is clear that the comments in the Interim Report were directed to proposals for a procedure that would have applied to all intellectual property disputes relating to domain name registrations, and not confined to abusive, bad faith registration and use. This context suggests that the reference to the use of the domain name becoming infringing over time refers to trade mark infringement, and not to a breach of the UDRP. Consequently, the comments in paragraph 197 should not be taken to mean that, where there is bad faith use, there is no need also to establish bad faith registration. This interpretation was confirmed by the following comments of the panellist in *Teradyne, Inc v 4Tel Technology:*[21]

> Read in conjunction with ¶198, it becomes clear that ¶197's reference to 'infringing', was intended to apply to the wider category of trademark infringement disputes, not to the narrow category of abusive registration. A conscious decision was made that UDRP proceedings be limited to abusive registrations. The question whether domain names registered in good faith become infringing is outside the scope of this inquiry. Further support for this interpretation is contained in the example of 'bad faith' contained in § 4(b)(i) of the Policy, which contemplates that the domain name

[19] WIPO Case No D2003-0254 (20 June 2003).
[20] WIPO, above n 3, para [197].
[21] WIPO Case No D2000-0026 (9 May 2000).

must have been registered or acquired 'primarily for the purpose of selling, renting or otherwise transferring' the domain name.

The need for a complainant to establish both bad faith registration and use is illustrated by one of the earliest UDRP decisions, *Telaxis Communications Corp v William E Minkle.*[22] In that dispute, the respondent registered the domain names<telaxis.com>, which he initially used to announce the impending launch of an estate agency business and applied for a business licence using the name 'TELAXIS'. The complainant subsequently registered the 'TELAXIS' mark in Morocco. Following this, the respondent redirected the domain name to the web sites of the complainant's competitors and to pornographic sites. The panellist found that the domain name had been registered in good faith and therefore that, even though the subsequent use of the domain name had been in bad faith, paragraph 4(a)(iii) was not satisfied.

Similarly, in *Miele, Inc v Absolute Air Cleaners and Purifiers,*[23] the respondent, an authorised dealer of 'MIELE' products, registered the domain name <miele.net> and used it to promote the dealership and sell 'MIELE' products. Subsequently, the dealership was discontinued and the complainant requested transfer of the domain name. The respondent apparently refused to transfer the domain name unless the dealership was re-established. The panellist held that, even though the domain name was being used in bad faith there was no breach of the policy, because it had not been registered in bad faith. In this respect, the panellist stated that:

> even though the Panel finds that the Respondent's continued maintenance of the domain name constitutes use in bad faith, the Panel can not find that the Respondent registered the domain name in bad faith within the narrow meaning and confines of the Policy as adopted. The Panel finds that it would be improper for it to impose equitable justice in this case in this forum, in light of the legislative history of the Policy. Rather, the Panel must conclude that there is a failure by the Complainant to meet this requirement of the Policy, because the Respondent's conduct at the time of registration was not bad faith conduct.

In *Leipa Pack Srl v Ulrich Brunhuber,*[24] the domain name <leipa.net> was registered by the respondent on behalf of the complainant. After the respondent left the complainant company, the domain name was used to provide information about the respondent's company. The panellist held that, as the domain name had not been registered in bad faith, the complainant had not satisfied the third element of the UDRP, stating:

> the Panel finds that Respondent did not have the requisite bad faith intentions, as meant by the Policy, when he registered the domain name at issue. In addition, the Panel finds that the circumstances mentioned by the Complainant, regarding Respondent's alleged use of the domain name in bad faith, being of a later date than the registration of the domain name, cannot, at least not in this case, lead to the conclusion that the original registration in good faith in retrospective has become a registration in bad faith.

As explained at [7.10], evidence that a disputed domain name is being used in bad faith may create an inference that the domain name has been registered in bad faith. Adopting the approach of the panel in *Passion Group Inc v Usearch Inc,*[25] however, the inference may be rebutted by evidence that the domain name has been registered in good faith. The applica-

[22] WIPO Case No D2000-0005 (5 Mar 2000).
[23] WIPO Case No D2000-0756 (11 Sept 2000).
[24] WIPO Case No D2005-0920 (8 Nov 2005).
[25] eRes Case No AF-0250 (10 Aug 2000).

tion of this approach may be illustrated by the decision in *Global Media Resources SA v Sexplanets*.[26] In that dispute, the respondent used the disputed domain name <sexplanets.com> to host 'adult' content, whereas the complainant's established web site at <sexplanet.com> was being used for the commercial sale of a wide range of 'adult entertainment' material. In these circumstances, the evidence may have indicated that the domain name was being used to divert Internet users from the respondent's web site contrary to paragraph 4(b)(iv), thereby giving rise to an inference of bad faith registration. The panellist, however, held that, as the respondent was not aware of the complainant's mark at the time the domain name was registered, the domain name had been registered in good faith, so that it was unnecessary to determine whether there was bad faith use. Accordingly, the complainant had failed to establish bad faith registration and use under paragraph 4(a)(iii).

Bad Faith Registration

[7.9] Is Renewal Registration?

In requiring bad faith registration, paragraph 4(a)(iii) refers to the initial registration of the domain name, and not to the renewal of the domain name. This means that provided the domain name has been registered in good faith, the third element will not be made out, even if the registration has been renewed in bad faith. At the same time, the transfer of a domain name to a third party is regarded as a registration of the domain name for the purpose of paragraph 4(a)(iii).

The WIPO Consensus View on this issue is as follows:

> While the transfer of a domain name to a third party does amount to a new registration, a mere renewal of a domain name does not amount to registration for the purposes of determining bad faith. Registration in bad faith must occur at the time the current registrant took possession of the domain name.[27]

A number of WIPO panel decisions illustrate the operation of the Consensus View.

In *Weatherall Green & Smith v Everymedia.com*,[28] the respondent registered the domain name <weatheralls.com> and offered it as a gift to his father, who was a partner in the complainant trade mark owner. The panellist held that, as the domain name was registered and immediately offered to the complainant, it was not registered in bad faith. After the domain name was renewed, however, it was used to redirect users to a web site that advertised 'WEATHERALLS' oilskins. On this basis, the panellist found that the domain name had been renewed, and was being used, in bad faith. As paragraph 4(a)(iii) refers to registration, and not to renewal, however, the panellist concluded that the third element of the UDRP had not been made out. In reaching this conclusion, the panellist referred to the following comments made in *Teradyne, Inc v 4Tel Technology*:[29]

[26] WIPO Case No D2001-1391 (24 Jan 2002).
[27] WIPO, above n 4, para 3.7.
[28] WIPO Case No D2000-1528 (19 Feb 2001).
[29] WIPO Case No D2000-0026 (9 May 2000).

A conscious decision was made that UDRP proceedings be limited to abusive registrations. The question whether domain names registered in good faith become infringing is outside the scope of this inquiry.

Similarly, in *Substance Abuse Management, Inc v SAMI,*[30] the complainant presented no evidence that the domain name <sami.com> had been registered in bad faith, but claimed that it had been renewed in bad faith, and that renewal should be regarded as equivalent to registration. In rejecting this contention, the panellist referred to the WIPO Final Report, stating that:

the authors of the Report could have extended its analysis of the UDRP to cover registrations obtained in bad faith either at the time of the original registration or at the time of renewal. It chose not to do so. Similarly, the Policy could have been drafted to address both bad faith registration and bad faith renewal. It does not, referring only to a registration having been obtained in bad faith. Thus, there is no support for Complainant's position.

The strict operation of the principle that the policy does not apply to bad faith renewals of registration, but is confined to bad faith registrations, is illustrated by the decision in *PAA Laboratories GmbH v Printing Arts America.*[31] In that dispute, the respondent, 'Printing Arts America', registered the domain name <paa.com>, but subsequently went into liquidation and offered the domain name for sale. While in liquidation, and after becoming aware of the complainant's interest in the domain name, the respondent renewed the registration of the domain name, thereby clearly blocking others from registering the domain name. Given that the domain name was initially registered in good faith, the panellist was reluctantly forced to conclude that paragraph 4(a)(iii) had not been satisfied. In reaching this conclusion, the panellist referred to the previous decisions to consider this issue before making the following observations:

the Panel wishes to clarify that its decision under this element is based on the need for consistency and comity in domain name dispute 'jurisprudence'. Were it not for the persuasive force of the cited decisions, this Panel would have expressed the view that paragraph 2 of the Policy demonstrates that references to 'registration' in the Policy were probably intended to be references to 'registration or renewal of registration.'. Absent the consistency of approach which has found favour with numerous earlier panels, this Panel would have seen no good reason for a renewal not to be considered as equivalent to 'registration' in the context of the objectives of the Policy. If the renewal had not been effected the disputed domain name would have lapsed and been available to others. The abusive refreshing of the original registration is an act which this Panel considers should be an act of a kind encompassed by paragraph 4(a)(iii) of the Policy. The benefit of an original good faith registration should not be perpetual to the point where it can cloak successors in title and successors in 'possession' long after the original registration would have expired.

[7.10] The *Nuclear Marshmallows* 'Inaction Doctrine'

'Passive holding' or 'passive warehousing' refers to the position where a domain name may have been registered in bad faith, but does not resolve to an active web site. The decision to require both bad faith registration *and* bad faith use of the disputed domain name clearly raises questions about the application of the policy where the domain name has been regis-

[30] WIPO Case No D2001-0782 (14 Aug 2001).
[31] WIPO Case No D2004-0338 (13 July 2004).

tered in bad faith but has not been actively used. Given that a domain name registrant does not need to make an active use of a domain name that is confusingly similar to a trade mark in order to make an extortionate demand on the trade mark owner, it is unsurprising that the issue of how to treat passive holding or passive warehousing arose in some of the earliest disputes to come before UDRP panels. From early on, it has been established that, in certain circumstances, passive holding or passive warehousing can amount to bad faith use of the disputed domain name.

The Consensus View set out in the *WIPO Overview* is as follows:

> The lack of active use of the domain name does not as such prevent a finding of bad faith. The panel must examine all the circumstances of the case to determine whether respondent is acting in bad faith. Examples of circumstances that can indicate bad faith include complainant having a well-known trademark, no response to the complaint, concealment of identity and the impossibility of conceiving a good faith use of the domain name. Panels may draw inferences about whether the domain name was used in bad faith given the circumstances surrounding registration, and vice versa.[31a]

The consensus view was first clearly established by the decision in *Telstra Corporation Limited v Nuclear Marshmallows*,[32] which remains the single most cited, and arguably most influential, UDRP panel decision. It is important to understand the reasoning adopted by the panellist in the *Nuclear Marshmallows* decision in order to appreciate correctly the treatment of passive holding under the UDRP. In that dispute, the respondent registered the domain name <telstra.org>, which was identical to the complainant's mark, but made no active use of the domain name and failed to file a response to the complaint. Furthermore, it appeared that the respondent had not provided accurate details of its identity to the domain name registrar.

The panellist, referring to the *travaux préparatoires* for the UDRP, first acknowledged that the use of the conjunction 'and' in paragraph 4(a)(iii) means that the policy necessarily requires both bad faith registration and bad faith use, such that bad faith registration alone is insufficient. The panellist went on to point out that this interpretation is reinforced by the use of the two tenses in paragraph 4(a)(iii): the past tense for registration and present tense for use. In relation to bad faith registration, the panellist considered that the steps taken by the respondent to obscure its identity were relevant. More importantly, the panellist considered that, given the strong reputation of the complainant's mark, 'it is not possible to conceive of a plausible circumstance in which the Respondent could legitimately use the domain name <telstra.org>'.

Nevertheless, the key question in the dispute was whether the domain name was being used in bad faith, there being no evidence of active use of the domain name whatsoever. The question was starkly raised in this instance, as the domain name did not resolve to an active web site, there was no evidence of preparations to use the domain name, there had been no advertising or promotion of the domain name, and there had been no offer to sell or transfer the domain name. In response to these circumstances, the panellist developed the 'inaction doctrine',[33] which effectively provides that all of the respondent's conduct, including inaction, may be taken into account in determining whether the respondent is acting in bad faith, which is equated with bad faith use. As the panellist explained this reasoning:

[31a] WIPO, above n 4, para 3.2.

[32] WIPO Case No D2000-0003 (18 Feb 2000).

[33] See G White, 'ICANN's Uniform Domain Name Dispute Resolution Policy in Action' (2001) 16(1) *Berkeley Technology Law Journal* 229.

> the relevant issue is not whether the Respondent is undertaking a positive action in bad faith in relation to the domain name, but instead whether, in all the circumstances of the case, it can be said that the Respondent is acting in bad faith. The distinction between undertaking a positive action in bad faith and acting in bad faith may seem a rather fine distinction, but it is an important one. The significance of the distinction is that the concept of a domain name 'being used in bad faith' is not limited to positive action; inaction is within the concept. That is to say, it is possible, in certain circumstances, for inactivity by the Respondent to amount to the domain name being used in bad faith.

In support of this conclusion, the panellist pointed out that the fact that the first three circumstances set out in paragraph 4(b) identify actions relating to the registration of the disputed domain name, but do not refer to positive actions relating to use of the domain name, suggests that there are circumstances in which passive holding may amount to bad faith use. Finally, the panellist made the important point that passive holding will not always amount to bad faith use, but whether or not it does so depends upon the circumstances of each case. In this dispute, the following circumstances led the panellist to conclude that the domain name was being used in bad faith:

- the strong reputation of the complainant's mark;
- the extent to which the respondent took active steps to conceal its identity, including acting under an unregistered business name and providing false contact details to the domain name registrar; and
- given these circumstances, it was 'not possible to conceive of any plausible actual or contemplated active use of the domain name by the Respondent that would not be illegitimate'.

As can be seen, the circumstances identified by the panellist in the *Nuclear Marshmallows* case form the basis for the circumstances that the WIPO Overview considers relevant in determining whether passive holding amounts to bad faith use. It is extremely important to appreciate that, while *Nuclear Marshmallows* decided that inaction may be bad faith use, UDRP panels cannot simply assume that passive holding invariably amounts to bad faith use, but should go through the process of identifying the circumstances in which passive holding amounts to bad faith. One obvious difficulty with the decision, however, is the assumption made by the panellist that it was impossible to conceive of any active use of the domain name that would not be illegitimate. Assuming that the use of a domain name that is identical to a complainant's mark as a criticism site may be legitimate in certain circumstances, the statement cannot be accepted in absolute terms. A better interpretation of the decision would be to interpret it as meaning that, in the particular circumstances of this case, including the absence of any indication that the domain name would be used for a criticism site, it would be impossible to conceive of any active use that would be legitimate.

UDRP panels do not always precisely identify the circumstances in which passive holding constitutes bad faith use. Nevertheless, identifying the particular circumstances is the preferred approach. The decisions in *DCI SA v Link Commercial Corporation*[34] and *Ladbroke Group Plc v Sonoma International LDC*[35] are good examples of how panels should approach this issue. In the *DCI* dispute, the respondent registered the domain name <5asec.com>, but made no active use of the domain name. Following the *Nuclear Marshmallows* doctrinc, the panellist was prepared to infer bad faith use on the basis of the following circumstances:

[34] WIPO Case No D2000-1232 (7 Dec 2000).
[35] WIPO Case No D2002-0131 (10 Apr 2002).

- the complainant's '5 A SEC' mark was used in the area of dry cleaning in France and other countries;
- the respondent had provided no evidence of any actual or contemplated good faith use;
- the domain name was not a domain name that any user would choose unless he or she knew of the complainant's marks, as the mark was not a generic or suggestive term;
- the respondent provided, and failed to correct, false contact details in breach of its registration agreement; and
- taking all of the above into account, it was not possible to conceive of any plausible actual or contemplated active use of the domain name by the respondent that would not be illegitimate.

Similarly, in the *Ladbroke Group* decision, the respondent registered a series of domain names that were confusingly similar to the complainant's well-known mark, but made no active use of them. The panellist adopted the *Nuclear Marshmallows* reasoning in full. In concluding that paragraph 4(a)(iii) had been satisfied, the panellist inferred that the domain names were being used in bad faith on the basis of the following circumstances:

- the domain names had been registered in bad faith, as the respondent must have been aware of the strong reputation of the complainant's mark and had registered a series of domain names in the .com, .net and .org gTLDs;
- any attempt actively to use any of the domain names would inevitably lead to a likelihood of confusion as to source, sponsorship, affiliation or endorsement of the registrant's web site;
- the respondent provided, and failed to correct, false contact details in breach of its registration agreement;
- there was no information about the respondent's business activities, and the complainant had made out a prima facie case that the respondent had no rights or legitimate interests in the domain names;
- the respondent failed to file a response; and
- taking all of the above into account, it was not possible to conceive of any plausible actual or contemplated active use of the domain name by the respondent that would not be illegitimate.

The list of relevant circumstances identified in the *WIPO Overview* is by no means exhaustive. In *Hotwire, Inc v Paris Willis*,[36] for example, the respondent registered the domain name <hotwi.com>, which was confusingly similar to the complainant's mark, but failed actively to use the disputed domain name. The panellist was prepared to infer bad faith use on the basis that the respondent knew of the complainant's mark as it became an affiliate of the complainant on the same day that it registered the domain name, and that it intended to use the domain name to profit from redirecting Internet traffic through typosquatting.

As illustrated by the *Ladbroke Group* decision, bad faith use may, in certain cases, be inferred from the circumstances surrounding bad faith registration. As explained at [7.4], this necessarily follows from the wording of the circumstances set out in paragraph 4(b), the first three of which refer to circumstances associated with registration as evidence of 'bad faith use and registration'. In *Jupiters Limited v Aaron Hall*,[37] for example, the respondent registered the domain names <jupiterscasino.com> and <jupiters-casino.com>, which

[36] NAF Case No FA741912 (11 Aug 2006).
[37] WIPO Case No D2000-0574 (3 Aug 2000).

were similar to the complainant's well-known registered Australian marks, but the domain names did not resolve to an active web site and the respondent did not file a response to the complaint. The respondent subsequently applied to register a mark identical to the complainant's mark in the United States in connection with services of the same description as those in relation to which the complainant's mark was registered.

The panellist first held that the circumstances set out in paragraph 4(b) had not been satisfied, pointing out that paragraph 4(b)(iv) differed from the first three circumstances in that it requires active use of the domain name. Nevertheless, given that the identified circumstances are not exhaustive, the panellist concluded that the domain name was registered in bad faith, taking into account the strong reputation of the complainant's mark, indicating that the respondent must have known of the complainant's rights and interests at the time the domain name was registered. From this, and absent an explanation of why the disputed domain names were chosen, the panellist was able to infer that the domain names were registered in bad faith in the hope of capitalising on the reputation of the complainant. This inference was reinforced by the respondent's actions in applying to register the complainant's mark. Affirming the *Nuclear Marshmallow* 'inaction doctrine', the panellist inferred that the circumstances surrounding registration, together with the US trade mark application, indicated that the respondent was holding the domain name for a future active use that would compete with, or otherwise be detrimental to, the complainant.

[7.11] Inferring Bad Faith Registration from Use

Just as an inference of bad faith use of a disputed domain name may be drawn from circumstances surrounding bad faith registration, so an inference that the domain name has been registered in bad faith can be drawn from circumstances indicating that the domain name is being used in bad faith. As explained at [7.5], this necessarily follows from the wording of paragraph 4(b)(iv), which identifies circumstances indicating bad faith use that will amount to evidence of both bad faith registration and use. As explained further at [7.5], the preferred approach to the interpretation of paragraph 4(b)(iv), which was adopted by the panel in *Passion Group Inc v Usearch Inc*[38] is that, while evidence of bad faith use may give rise to an inference of bad faith registration, the inference may be rebutted by evidence that the domain name has been registered in good faith.

Notice of Complainant's Mark

[7.12] Actual and Imputed Notice

A key issue in determining whether a domain name has been registered, or is being used, in bad faith is the extent to which the respondent knew of the complainant's mark at the time at which the disputed domain name was registered. Actual or imputed knowledge of the complainant's mark is compelling evidence of the respondent's bad faith. In a commonly cited decision, *Digi International Inc v DDI Systems*,[39] the panellist stated that:

[38] eRes Case No AF-0250 (10 Aug 2000).
[39] NAF Case No FA124506 (24 Oct 2002).

> there is a legal presumption of bad faith, when Respondent reasonably should have been aware of Complainant's trademarks, actually or constructively.

This statement should, however, be treated with some caution. First, it should be kept in mind that the complainant bears the onus of establishing bad faith. Secondly, as explained at [7.13], the doctrine of constructive notice is usually only applied where both parties are located in the United States. Nevertheless, knowledge of the complainant's mark at the time the domain name is registered may be decisive in finding that one of the circumstances set out in paragraph 4(b) has been made out, or that the respondent has otherwise acted in bad faith.

The fundamental principle in this area is that bad faith cannot be established where, at the time the domain name was registered, the respondent had no notice, either actual or imputed, of the complainant's mark. In *New Jersey Divorce Center, Inc v iGenesis Limited,*[40] for example, the panellist stated:

> If at the time of registration Respondent lacked actual knowledge or belief of Complainant's trademark rights . . . then a finding of bad faith registration is precluded . . . [A] finding of bad faith by its nature entails that there be some intentional act on the part of the party purported to be acting in bad faith such that that party reasonably believes or should believe it is offending the rights of another.

Lack of actual notice of a complainant's mark is not, however, a complete defence to an allegation of abusive, bad faith registration. As a respondent will customarily deny knowledge of the complainant's mark, the question of notice often depends upon the inferences that can be drawn from all of the circumstances of registration and use of the disputed domain name and, as explained below, in some circumstances knowledge of the complainant's mark will be imputed. On the other hand, even if the complainant is able to establish that the respondent had actual knowledge of the complainant's mark, this does not necessarily mean that abusive, bad faith registration and use will be established. The respondent may, for example, be able to establish that it has rights or legitimate interests in the disputed domain name.

To begin with, there is no positive obligation on a respondent to conduct a trade mark search prior to registering a domain name. It was never envisaged that this would be required under the policy. The WIPO Final Report, for example, included the following recommendation:

> It is not recommended that domain name registrations be made conditional upon a prior search of potentially conflicting trademarks, but it is recommended that the domain name application contain appropriate language encouraging the applicant to undertake voluntarily such a search.[41]

While there can be no bad faith in the absence of notice of the complainant's mark, wilful blindness as to the complainant's trade mark rights will not protect a respondent from a finding of bad faith. In *Mobile Communication Service Inc v WebReg, RN,*[42] for example, it appeared that the respondent engaged in the business of registering large numbers of domain names, the registration of which had expired, for the purpose of resale. Although the panellist pointed out that there is ordinarily no bad faith where a respondent registers

[40] NAF Case No FA584783 (14 Dec 2005).
[41] WIPO, above n 3, para [105].
[42] WIPO Case No D2005-1304 (24 Feb 2006).

generic terms as domain names without actual notice of a complainant's mark, this does not apply where a respondent deliberately closes its eyes to whether or not the domain names incorporate trade marks. In deciding that the respondent had engaged in bad faith registration and use, the panellist drew the following distinction:

> where a respondent has registered a domain name consisting of a dictionary term because the respondent has a good faith belief that the domain name's value derives from its generic qualities, that may constitute a legitimate interest and the offer to sell such a domain name is not necessarily a sign of bad faith. Where, in contrast, a respondent registers large swaths of domain names for resale, often through automated programs that snap up domain names as they become available, with no attention whatsoever to whether they may be identical to trademarks, such practices may well support a finding that respondent is engaged in a pattern of conduct that deprives trademark owners of the ability to register domain names reflecting their marks.

Similarly, in *Sprunk-Jansen A/S v Chesterton Holdings*,[43] the respondent used an 'automated registration program' to choose domain names for registration, including the disputed domain name, which was confusingly similar to the complainant's mark. Although the respondent could accurately claim that it did not have actual knowledge of the complainant's mark at the time the domain name was registered, the panellist held that wilful blindness, through the use of an automated program that does not include a mechanism to avoid trade marks, may be an indication of bad faith. Given that the disputed domain name was used to provide links to a web site that offered products that competed with those of the complainant, the panellist concluded that there was bad faith registration and use.

A pattern of registering domain names that correspond with marks and denying knowledge of the trade mark rights may lead to an inference of wilful blindness. In *Eurial Poitouraine v Compana LLC*,[44] for example, the respondent registered the domain name <soignon.com> as part of a pattern of registering expired domain names allegedly to provide e-mail services to customers with the corresponding surnames. Given that the respondent had been involved in a number of previous UDRP disputes and, in each case, had denied knowledge of the relevant marks, the panellist held that the respondent's assertion, in this instance, amounted to wilful blindness.

In the absence of evidence that a respondent had actual knowledge of the complainant's mark at the time the domain name was registered, in certain circumstances knowledge of the complainant's rights may be imputed. In particular, where the complainant's mark is famous or well-known, UDRP panels commonly hold that the respondent 'knew or should have known' of the complainant's mark. As the panel pointed out in *Banca Sella spa v Mr Paolo Parente*,[45] registration of a famous or well-known mark as a domain name suggests 'opportunistic bad faith'. 'Opportunistic bad faith' is dealt with further at [7.16].

Where a trade mark does not have a global reputation, the extent to which knowledge of the domain name may be imputed depends upon the location of the respondent. As the panellist pointed out in *Jupiters Limited v Aaron Hall*:[46]

> the fact that Complainant's name and trade mark is so well and widely known, coupled with the fact that Respondent's mailing address is in the same district as Complainant's business, makes it inevitable that Respondent registered the domain names in full knowledge of Complainant's rights and interests.

[43] WIPO Case No D2006-1080 (17 Nov 2006).
[44] WIPO Case No D2004-0270 (5 June 2004).
[45] WIPO Case No D2000-1157 (27 Nov 2000).
[46] WIPO Case No D2000-0574 (3 Aug 2000).

If the mark is well-known in the location of the respondent, it will be difficult for a respondent to establish that the domain name has not been registered for an illegitimate purpose.

The following are examples of decisions in which UDRP panels have imputed knowledge where a complainant has a famous or well-known mark.

- *America Online, Inc v Avrasya Yayincilik Danismanlik Ltd.*[47] The panellist held that the respondent, which had registered the disputed domain names <americaonline.net> and <americanonline.com>, 'had knowledge or reasonably should have had knowledge' of the complainant's famous mark. This was evidence of the respondent's bad faith.
- *Expedia, Inc v European Travel Network.*[48] The panellist held that the respondent, which had registered the <xpediatravel.com> domain name, 'knew of or should have known of' the complainant's 'EXPEDIA' mark, which had become famous in connection with e-commerce. This led to the inference that the domain name had been registered primarily for the purpose of diverting business from the complainant's web site.
- *SportSoft Golf, Inc v Hale Irwin's Golfers' Passport.*[49] The panellist held that the respondent, which had registered the <golferspassport.com> domain name, 'knew or should have known' of the registration and use of the complainant's 'HALE IRWIN GOLFERS' PASSPORT' and 'GOLFERS' PASSPORT' marks. This led to the conclusion that the domain name was being used to create a likelihood of confusion as to affiliation of the respondent's web site contrary to paragraph 4(b)(iv).
- *Sony Kabushiki Kaisha v Inja, Kil.*[50] The panellist held that the respondent, which had registered 19 domain names that included the term 'SONY', 'knew or should have known' of the registration and use of the complainant's famous mark prior to registering the domain name. On this basis, the panellist held that it was inconceivable that the respondent could make any active use of the domain names without creating a false impression of an association with the complainant.
- *Kevin Garnett v Trap Block Technologies.*[51] The panellist held that the respondent had engaged in bad faith when it registered the domain name <kevingarnett.com>, with full knowledge of the complainant's rights to 'KEVIN GARNETT, the personal name of a famous basketball player.
- *The Nasdaq Stock Market, Inc v Act One Internet Solutions.*[52] The panellist held that the respondent, which had registered the domain name <nasdasq.com>, was presumed to have known the complainant's famous 'NASDAQ' mark, and had engaged in 'opportunistic bad faith'.
- *Algemeen Nederlands Persbureau ANP BV v European Travel Network.*[53] The panellist held that it was highly likely that the respondent, which had registered the domain name <anpnieuws.com>, was aware of the complainant, a well-known Dutch news service, at the time the domain name was registered. Given the well-known mark, it was reasonable to expect that any use of the domain name would be associated with the complainant.
- *Sony Kabushiki Kaisha v Richard Mandanice.*[54] The panellist held that the respondent, who had registered the domain name <sony-z5.com>, must have known of the complainant's

[47] NAF Case No FA93679 (16 Mar 2000).
[48] WIPO Case No D2000-0137 (18 Apr 2000).
[49] NAF Case No FA94956 (11 July 2000).
[50] WIPO Case No D2000-1409 (9 Dec 2000).
[51] NAF Case No FA128073 (21 Nov 2002).
[52] WIPO Case No D2003-0103 (1 Apr 2003).
[53] WIPO Case No D2004-0520 (8 Sept 2004).
[54] WIPO Case No D2004-1046 (20 Jan 2005).

famous mark and the 'Z5' product at the time the domain name was registered. Accordingly, the panellist found that the respondent had acted in bad faith.
- *Eldrick 'Tiger' Woods v Whitsan Bay Golf Shop.*[55] The panellist held that the respondent had acted in bad faith by registered the domain name <tigerwoodscoursedesign.com> with full knowledge of the complainant's famous mark.

While a respondent may have imputed knowledge of a complainant's famous mark, this does not invariably mean that the respondent has acted in bad faith, as other evidence may be indicative of good faith registration or use. In *America Online, Inc v Anytime Online Traffic School,*[56] for example, the respondent registered the domain names <aoltrafficschool.com>, <aoldefensivedriving.com> and <aoldriverimprovement.com>, which were confusingly similar to the complainant's famous 'AOL' mark. The panellist, however, accepted that, as the respondent had legitimately used the domain names in connection with services offered by its 'ANYTIME ONLINE TRAFFIC SCHOOL' business, the complainant had failed to establish bad faith registration and use. Similarly, in *America Online, Inc v GO Technology Corporation,*[57] the panellist held that, although the respondent must have known of the complainant's famous 'AOL' mark when it registered the <laol.net> domain name, it had legitimately registered and used the domain name in connection with its 'Lake Area Online, Inc' business.

Apart from imputed knowledge of a complainant's famous or well-known mark, UDRP panels may infer knowledge of the complainant's mark in the following circumstances:

- *Typosquatting.* Where a respondent has obviously engaged in typosquatting, then knowledge of the complainant's mark will be readily inferred. Decisions in which typosquatting led to an inference of knowledge of the complainant's mark include: *Pfizer Inc v Re This Domain For Sale*[58] (registrant of <phiser.com> knew of complainant's 'PFIZER' mark); *The Nasdaq Stock Market, Inc v Act One Internet Solutions*[59] (registrant of <nasdasq.com> knew of complainant's 'NASDAQ' mark); *Paragon Gifts, Inc v Domain.Contact*[60] (registrant of <paragon-gifts.com> knew of complainant's PARAGON' mark); *Napster, Inc v John Gary*[61] (registrant of <napster4porn.info> and <sexnapster.info> knew of complainant's 'NAPSTER' mark).
- *Offering the domain name for sale.* An offer to sell the domain name to the complainant trade mark owner, especially an unsolicited offer, is clearly evidence that the respondent knew of the complainant's mark. In *Wal-Mart Stores, Inc v Longo,*[62] for example, the combination of typosquatting, the fame of the complainant's mark and an offer of sale indicated that the <www-wal-mart.com> domain was registered and was being used in bad faith. Similarly, in *Martha Stewart Living Omnimedia Inc v Josh Gorton,*[63] the respondent made an unsolicited offer to sell the domain names <marthastewartfoundation.com> and <marthastewartfoundation.org> on the day the domain name was registered. This was clearly evidence of opportunistic bad faith.

[55] NAF Case No FA772886 (26 Sept 2006).
[56] NAF Case No FA146930 (11 Apr 2003).
[57] NAF Case No FA403101 (9 Mar 2005).
[58] WIPO Case No D2002-0409 (3 July 2002).
[59] WIPO Case No D2003-0103 (1 Apr 2003).
[60] WIPO Case No D2004-0107 (20 Apr 2004).
[61] WIPO Case No D2005-0913 (11 Oct 2005).
[62] WIPO Case No D2004-0816 (29 Nov 2004).
[63] WIPO Case No D2005-1109 (13 Dec 2005).

- *Using the domain name to refer to the complainant's mark, or the complainant's products.* Where the web site to which the disputed domain name resolves refers to the complainant's marks or the complainant's products, it will be inferred that the respondent knew of the complainant's mark. In *Canon Kabushiki Kaisha v Price-Less Inkjet Cartridge Company,*[64] for example, the respondent used the domain names <canoninkjets.com>, <canoninkjet.com> and <canonink.net> to promote the sale of replacement ink cartridges for ink jet printers, including 'CANON' printers. The panellist held that this indicated that the respondent had chosen the domain names with knowledge of the complainant's mark. Unlike many other similar decisions, in *Kabushiki Kaisha Square Enix v NOLDC, Inc,*[65] the complainant was not able to establish that its 'VALKYRIE PROFILE' mark, which was used in connection with a computer game, had a strong reputation. Nevertheless, the panellist concluded that the respondent had knowledge of the mark on the basis that the web site to which the <valkyrieprofile.com> domain name resolved included a tab or 'button' with the name of another of the complainant's products.
- *Using the domain name to link to competitors' web sites.* Links from the web site to which the domain name resolves to the web sites operated by competitors of the complainant is compelling evidence that the respondent knew of the complainant's mark. In *PartyGaming Plc v WHOis Privacy Protection Service, Inc,*[66] for example, the disputed domain name, <patrypoker.com>, resolved to a web site that included a link to the complainant's web site, as well as the web sites of competitors of the complainant. This led the panellist to conclude that the respondent knew of the complainant's 'PARTYPOKER' mark at the time the domain name was acquired.
- *Being a competitor of the complainant in the same market.* Where the respondent is a direct competitor of the complainant trade mark owner, it will obviously be difficult to deny knowledge of the complainant's mark. In *Lubbock Radio Paging v Venture Tele-Messaging,*[67] for example, the complainant and the respondent were in competing lines of business in the same market area, and the respondent used the disputed domain names to advertise its services. The panellist held this to be evidence of bad faith. Similarly, in *Herbalife International of America, Inc v Ron Mauldin,*[68] the respondent used the <herbalifesurvivors.com> domain name to operate a web site that included a link to a site that promoted a multi-level marketing business model that competed with the complainant's business. This led to the conclusion that the domain name was registered in bad faith.
- *Adopting a design or layout for a web site that is similar to that of the complainant's web site, or referring to the complainant's mark on the web site.* Where the respondent adopts a design or layout for the web site to which the domain name resolves that is similar to the design or layout of a web site of the complainant, or actually refers to the complainant's mark, then knowledge of the complainant's mark may be inferred. In *Six Continents Hotels, Inc v Hotel Partners of Richmond,*[69] for example, the respondent reproduced the complainant's 'HOLIDAY INN' mark, including the trade mark registration symbol, on the web site to which the <holidayinnhotelreservations.com> domain name resolved. This was clearly evidence that the respondent knew of the complainant's mark.

[64] WIPO Case No D2000-0878 (21 Sept 2000).
[65] WIPO Case No D2006-0630 (14 July 2006).
[66] WIPO Case No D2006-0508 (26 June 2006).
[67] NAF Case No FA96102 (23 Dec 2000).
[68] NAF Case No FA862849 (23 Jan 2007).
[69] WIPO Case No D2003-0222 (14 May 2003).

Where a trade mark has a local reputation only, and the respondent is from a different part of the world, then knowledge of the complainant's mark will not be imputed. In *Rusconi Editore SPA v Bestinfo*,[70] for example, the complainant had registrations for the 'VITALITY' mark, which was used in connection with an Italian magazine, in non-English-speaking European countries. The panellist accepted that the respondent, which was based in Texas, was not aware of the complainant or its magazine prior to being notified of the complaint. Similarly, in *The Hebrew University of Jerusalem v Alberta Hot Rods*,[71] the complainant, a university located in Jerusalem, had registered rights in marks that included the term 'EINSTEIN' in France, Germany and Australia. The panel held that the Canadian respondent, which had registered the <alberteinstein.com> domain name, had no reason to believe that there were trade mark rights in the name 'ALBERT EINSTEIN'.

The general principle is therefore that where a mark is not well-known, or where it is unknown in the jurisdiction in which the respondent is located, knowledge of the mark will not be imputed. The following decisions illustrate the operation of this principle.

- *VZ VermögensZentrum AG v Anything.com*.[72] The complainant had registered rights in marks that included 'VZ' in Switzerland. The respondent, which was based in the Cayman Islands, registered a series of domain names that included combinations of two or three letter names, including <vz.com>, as part of a business of reselling generic domain names. The panel held that there was no reason to infer that the respondent knew, or should have known, of the complainant's Swiss mark.
- *KCTS Television Inc v Get-on-the-Web Limited*.[73] The complainant, a US public television broadcasting company, had registered rights in the 'KCTS' mark in the United States. The panel held that it could not infer that the respondent, which was based in the United Kingdom, knew of the complainant's mark when it registered the <kcts.com> domain name.
- *Seaway Bolt & Specials Corp v Digital Income Inc*.[74] The complainant had longstanding goodwill in the common law mark 'SEAWAY' in connection with the manufacture of pipes and drain plugs in the United States, but had no registered trade mark rights. The respondent, based in Canada, registered the domain name <seaway.com> as part of a business of registering generic domain names. As the panellist found that the complainant's mark was not well-known outside its field, she held that the complainant had not established that the respondent knew of the complainant at the time the domain name was registered.
- *Dollar Financial Group, Inc v Domainhiway.com*.[75] The complainant, one of the largest US originators of small consumer loans, had registered rights in the 'LOAN MART' mark in the United States. The panellist held that, as the respondent, which had registered the domain name <ezloanmart.com> as part of a business of registering domain names, was in a different line of business, it was not aware of the complainant at the time it registered the domain name.
- *Roco Modellspielwaren GmbH and Dipl-Wirt-Ing (FH) Peter Maedgdefrau v Plantraco Ltd*.[76] The complainants had registered rights in the 'MINITANK' mark in Austria and the

[70] WIPO Case No D2001-0656 (5 July 2001).
[71] WIPO Case No D2002-0616 (7 Oct 2002).
[72] WIPO Case No D2000-0527 (22 Aug 2000).
[73] WIPO Case No D2001-0154 (20 Apr 2001).
[74] NAF Case No FA114672 (5 Aug 2002).
[75] NAF Case No FA285123 (30 July 2004).
[76] WIPO Case No D2005-0112 (8 Apr 2005).

United States. As the respondent, which had registered the <minitank.com> mark, was based in Canada, and the Canadian mark was not registered until after the domain name had been registered, the panellist held that there was no evidence that the respondent knew, or should have known, of the complainant's mark.

- *Sasol Limited v Raymond Wong.*[77] The complainant, a South African petro-chemical company, had registered rights in the 'SASOL' mark in a number of countries, but had not applied to register the mark in the United States at the time the respondent registered the <esasol.com> domain name. The panellist found that, while the 'SASOL' mark was well-known in South Africa, and amongst oil and financial interests in the United States, it was not generally well-known in the United States. Accordingly, the panellist declined to infer bad faith registration of the domain name.

[7.13] Constructive Notice

Constructive notice of the complainant's mark, as opposed to actual or imputed notice, is generally not taken into account by UDRP panels in determining whether a respondent has acted in bad faith. The treatment of constructive notice under paragraph 4(a)(iii) has been complicated, however, by the US trade mark doctrine of constructive notice. The doctrine of constructive notice is established by § 22 of the Lanham Act, which provides that registration on the Principal Register is 'constructive notice of the registrant's claim of ownership'.[78] The doctrine means that the defendant in a trade mark infringement action cannot rely on the US common law defence of 'good faith' use of a mark without notice. The introduction of the doctrine, and the elimination of the common law defence, had the effect of conferring nationwide protection on marks registered in the Principal Register.[79] Some panellists have adopted the US doctrine of constructive notice to hold that, where a mark is registered in the United States and the respondent is located in the United States, the mere fact of registration means that the respondent has notice of the complainant's mark.

The *WIPO Overview* explains the approach that WIPO panels have taken to constructive notice of a complainant's mark in the following terms:

> Most panels have declined to introduce the concept of constructive notice into the UDRP. However, where a complainant had a United States registered trademark and respondent was located in the United States, this concept has been used in a few cases to support a finding of registration and/or use in bad faith. In those cases, where the complainant's trademark registration preceded the respondent's domain name registration, respondent was presumed to have notice of the trademark.[79a]

The introduction of a universal rule recognising constructive notice on the basis of registration of a trade mark would effectively undermine the policy, as it would mean that establishing registered rights in a mark would be sufficient to infect a respondent with bad faith. In this respect, the UDRP differs from trade mark infringement under US law, insofar as it permits good faith registration of a domain name that is confusingly similar to a complainant's mark provided that the respondent has no actual or imputed notice of the mark. To hold otherwise would, in effect, confer a veto on trade mark owners over the registration of confusingly similar domain names, unless the registrant could establish that it

[77] WIPO Case No D2005-0122 (10 May 2005).
[78] Lanham Act § 22, 15 USC § 1072.
[79] See, eg, *Dawn Donut Co v Hart's Food Stores, Inc*, 267 F 2d 358, 121 USPQ 430 (2d Cir, 1959).
[79a] WIPO, above n 4, para 3.4.

had rights or legitimate interests in a disputed domain name. The panel in *Advanced Drivers Education Products and Training, Inc v MDNH, Inc (Marchex)*[80] made this point when it observed that accepting a claim of constructive notice based on trade mark registration:

> would essentially establish a per se rule of bad faith any time a domain name is identical or similar to a previously-registered trademark, since constructive notice could be found in every such case. Such a result would be inconsistent with both the letter and the spirit of the policy, which requires actual bad faith.

The panellist in *The Way International, Inc v Diamond Peters*[81] elaborated on the objections to recognising the US doctrine of constructive notice in the following terms:

> As to constructive knowledge, the Panel takes the view that there is no place for such a concept under the Policy. The essence of the complaint is an allegation of bad faith, bad faith targeted at the complainant. For that bad faith to be present, the cybersquatter must have actual knowledge of the existence of the complainant, the trade mark owner. If the registrant is unaware of the existence of the trade mark owner, how can he sensibly be regarded as having any bad faith intentions directed at the complainant? If the existence of a trade mark registration was sufficient to give the Respondent knowledge, thousands of innocent domain name registrants would, in the view of the Panel, be brought into the frame quite wrongly.

Furthermore, as the panel pointed out in *AIB-Vincotte Belgium ASBL v Guillermo Lozada, Jr*,[82] purely from a practical point of view, domain name registrants do not have the resources to conduct trade mark searches in every jurisdiction in the world.

The following are examples of decisions that have refused to recognise constructive notice for the purpose of paragraph 4(a)(iii) of the UDRP.

- *Toronto Star Newspaper Ltd v Elad Cohen.*[83] The complainant, a Canadian cable broadcasting company, had registered the 'TSTV' mark in Canada. The panellist refused to recognise that the respondent, a resident of Israel who had registered the <tstv.tv> domain name, had constructive notice based on the Canadian registration.
- *Alberto-Culver Company v Pritpal Singh Channa.*[84] The complainant had registered the 'STATIC GUARD' mark in multiple jurisdictions, including the European Union and the United Kingdom. The panellist rejected the complainant's claim of constructive notice based on the trade mark registrations, pointing out that recognising constructive notice would be contrary to the 'clear intention' of paragraph 4(a)(iii).
- *Sterling Inc v Sterling Jewelers, Inc.*[85] The complainant had registered the mark 'JARED THE GALLERIA OF JEWELRY' in the United States and the respondent, which was based in Canada, registered the domain name <jaredjewelry.com>. Pointing out that the US doctrine of constructive notice was not designed to apply to domain name disputes, the panellist rejected the complainant's argument that the doctrine should be applied to a Canadian respondent that had purposely registered the domain name with a US registrar.
- *Olymp Bezner GmbH & Co KG v Olympus Access Service.*[86] The complainant had registrations for the 'OLYMP' mark in a number of jurisdictions, but not in the United States.

[80] NAF Case No FA567039 (10 Nov 2005).
[81] WIPO Case No D2003-0264 (29 May 2003).
[82] WIPO Case No D2005-0485 (29 Aug 2005).
[83] WIPO Case No DTV2003-0006 (22 Jan 2001).
[84] WIPO Case No D2002-0757 (7 Oct 2002).
[85] WIPO Case No D2002-0772 (13 Nov 2002).
[86] WIPO Case No D2003-0958 (17 Feb 2003).

While acknowledging that some UDRP panels had recognised constructive notice in limited circumstances, the panel refused to apply it in the circumstances of this dispute, where the respondent was based in the United States, concluding that a respondent was under no obligation to conduct a search prior to registering a domain name.

- *Sallie Mae, Inc v Chen Huang.*[87] The complainant had registered the 'SALLIE MAE' mark in the United States and the respondent, who was located in China, registered the <salliemaestudentloans.com> domain name. The panellist refused to apply the doctrine of constructive notice, holding that it was not applicable where the parties were not both from the United States.
- *Asset Marketing Systems, LLC v Silver Lining.*[88] The complainant had filed applications for registration of the common law mark 'GUEST SPEAKER' in the United States and Canada. Although the doctrine of constructive notice would not have applied to this dispute, the panellist indicated that there was no place for the concept under the policy.
- *Williams Electronics Games, Inc v Ventura Domains.*[89] The complainant owned the US registered mark 'WILLIAMS' for use in connection with pinball machines and the respondent, who was located in Sweden, registered the <williamspinball.com> domain name. Although the panellist held that the respondent had acted in bad faith, he expressly stated that the doctrine of constructive notice did not apply where the complainant was in the United States and the respondent in Sweden.
- *INVESTools Inc v KingWeb Inc.*[90] The complainant owned trade mark registrations in the United States for the 'INVESTOOLS' mark. The panel refused to accept that the Canadian respondent, who had registered the <investtools.com> domain name, had constructive notice of the complainant's mark on the basis of the US trade mark registrations.
- *Dr Mitchell Mandel and Ms Jill Slater v CSC Laboratories, Inc.*[91] The complainants owned the 'SKIN DOCTOR' mark, which had been registered in the United States and argued that the respondent, which was also based in the United States, had constructive notice of the registered mark. The panel held that the complainants could not rely on constructive notice to establish the necessary intent for bad faith registration and use.

In decisions involving trade marks registered in the United States where both parties are located in the United States, however, panellists have commonly applied the US doctrine of constructive notice. Despite the apparent inconsistency between the US doctrine of constructive notice and the balance between trade mark owners and domain name registrants sought to be struck by the UDRP, it now seems that constructive notice will generally be applied where both parties are located in the United States. Moreover, in some very limited circumstances the doctrine has been applied to a respondent that is not located in the United States. The following are examples of decisions that have held that the doctrine of constructive notice applies where the complainant's mark is registered in the United States.

- *J Crew International, Inc v crew.com.*[92] The complainant had registrations for the 'J CREW' mark in the United States and the respondent, who was also located in the United States, registered the domain name <crew.com> as part of a business of reselling generic domain

[87] WIPO Case No D2004-0880 (24 Dec 2004).
[88] WIPO Case No D2005-0560 (22 July 2005).
[89] WIPO Case No D2005-0822 (30 Sept 2005).
[90] NAF Case No FA598845 (9 Jan 2006).
[91] WIPO Case No D2006-0719 (6 Oct 2006).
[92] WIPO Case No D2000-0054 (20 Apr 2000).

names. The majority of the panel applied the doctrine of constructive notice to conclude that the domain name had been registered in bad faith. The dissenting panellist, however, rejected the application of the doctrine out of hand, claiming that it involved 'a gross over-simplification of the issues involved in preventing abusive domain name registrations'.

- *Kate Spade, LLC v Darmstadter Designs.*[93] The complainant had registered the 'JACK SPADE' mark with the USPTO and the respondent, which was based in Texas, registered the domain name <jackspade.com>. As the parties were both from the United States and the complainant's mark was registered with the USPTO, the panellist considered that it was 'fair to apply United States principles of constructive notice' in finding bad faith registration and use.
- *The Planetary Society v Salvador Rosillo Domainsforlife.com.*[94] The complainant had registered the mark 'THE PLANETARY SOCIETY' on the USPTO and the respondent, which was located in the United States, registered the domains names <planetarysociety.com> and <theplanetarysociety.com>. The panellist held that the respondent had constructive notice of the complainant's mark 'as a matter of law'.
- *The Sportsman's Guide, Inc v Modern Limited, Cayman Islands.*[95] The complainant owned US federal registration for 'THE SPORTSMAN'S GUIDE' mark and the respondent, which was based in the Cayman Islands, registered the <sportsmenguide.com> domain name. The panellist accepted that the principle of constructive notice was generally applicable to UDRP disputes, especially where the parties were both located in the United States. Even though the respondent was not based in the United States, the panellist held that it had constructive notice of the complainant's mark, taking into account the fact that the previous domain name registrant had been located in the United States, and that the domain name had been transferred to the respondent the day after the complainant requested transfer of the disputed domain name.
- *Staples, Inc v John Morgan.*[96] The complainant had registrations for marks containing the term 'STAPLES' in the United States, and the respondent, who was located in the United States, registered the <stpales.com> domain name. The panellist held that, as both parties resided in the United States, it was 'appropriate to apply the principle of constructive notice of Complainants' trademark registrations established by 17 USC § 1072'.
- *Sierra Suites Franchise, LP v International Marketing.*[97] The complainant had registered the 'SIERRA SUITES' mark in the United States, and the respondent, who was located in the United States, registered the <sierrasuites.net> and <sierrasuitesusa.com> domain names. The panellist held that the respondent had either actual or constructive knowledge of the complainant's mark by virtue of the complainant's prior registration of the mark.
- *Dermalogica, Inc v Paradigm Group.*[98] The complainant registered the 'DERMALOGICA' mark in the United States and the respondent, which was based in the United States, registered the <dermalogicadirectonline.com> domain name with either actual or constructive knowledge of the complainant's rights in the mark 'by virtue of Complainant's prior registration of that mark with the USPTO'.

[93] WIPO Case No D2001-1384 (3 Jan 2002).
[94] WIPO Case No D2001-1228 (12 Feb 2002).
[95] WIPO Case No D2003-0305 (18 June 2003).
[96] WIPO Case No D2004-0537 (20 Sept 2004).
[97] NAF Case No FA783218 (3 Oct 2006).
[98] NAF Case No FA874392 (30 Jan 2007).

It has been suggested that, in certain limited circumstances, the doctrine of constructive notice may be invoked where the complainant's trade mark is registered in a jurisdiction other than the United States. The principal decision to adopt this approach to constructive notice was *Red Nacional De Los Ferrocarriles Espanoles v Ox90*,[99] where the issue split the panel. In that dispute, the complainant, which was responsible for the Spanish railway system, had registered rights in the 'RENFE' mark in Spain. The complainant had been successful in a previous domain name dispute but, as a result of an 'administrative error', the registration of the domain name lapsed. The respondent, which engaged in a business that included registering lapsed or expired domain names, registered the <renfe.com> domain name, and claimed that it was unaware of the Spanish mark at that time. The respondent used software to identify the domain name for registration on the grounds that the term was a popular Internet search term and that registration of the domain name had lapsed. The majority of the panel held that, in these circumstances, the respondent was under an obligation to conduct a search, and had constructive notice of the complainant's mark. In reaching this conclusion, however, the majority were careful to point out that any obligation imposed on domain name registrants should be narrowly confined. In this respect, the majority stated:

> All members of this Panel are concerned that this decision not be misconstrued by subsequent Panels as holding that in every case a registrant is charged with conducting a detailed investigation into the reasons a domain name is available for registration. We do not believe that there is any such obligation on the part of a registrant, except in the unique circumstances and facts of this or other similar cases. Perhaps such an obligation might arise only where, as here, the name is available only because of its sudden expiration.

The dissenting panellist, on the other hand, refused to apply constructive notice on the basis that respondents have no obligation to conduct searches before registering a domain name. As, in this case, there was no evidence that the respondent had actual knowledge of the complainant's mark at the time the domain name was registered, the dissenting panellist concluded that it had not acted in bad faith. In reaching this conclusion, the dissenting panellist stated that:

> If a concept of constructive knowledge (i.e. 'he must have guessed and had he made the simplest of searches he would have known; he ought to have made the search, therefore he is to be taken as having known') is introduced into the UDRP, there is no knowing where this will all lead.

Even if the approach of the majority in this decision is adopted, however, it is clear that constructive notice can be applied to disputes where both parties are not located in the United States only in extremely limited circumstances, such as where the respondent is engaged in the business of registering lapsed domain names. In such circumstances, it may well be preferable to impute notice to the respondent on the basis that he or she has been wilfully blind to the complainant's mark, rather than resorting to constructive notice.

[7.14] Registration of Domain Name before Trade Mark

As explained at [5.15], the consensus view is that the time at which a complainant must establish rights in the mark is the time the complaint is made, and not the time at which the

[99] WIPO Case No D2001-0981 (21 Nov 2001).

domain name was registered. As further explained, the emergence of a consensus view means that the fact that a complainant acquired rights in the mark after the domain name was registered is irrelevant to the first element of the UDRP, but is to be dealt with in determining whether the respondent has registered the domain name in bad faith.

As a domain name registrant, in the usual course of events, will have no notice of a complainant's mark if the complainant has not acquired rights in the mark at the time the domain name is registered, in such circumstances the registration of the domain name cannot be in bad faith. The WIPO Consensus View on this issue is as follows:

> Normally speaking, when a domain name is registered before a trademark right is established, the registration of the domain name was not in bad faith because the registrant could not have contemplated the complainant's non-existent right.[100]

The following decisions are examples of the principle that registration of an identical or confusingly similar domain name prior to the complainant acquiring rights in the corresponding mark will usually not amount to bad faith.

- *John Ode v Intership Limited.*[101] The respondent registered the <ode.com> domain name on 18 May 1997, while the complainant did not use the service mark, 'ODE - OPTIMUM DIGITAL ENTERPRISES', until 28 May 2000, and did not file an application for trade mark registration until 28 September 2000. The panel adopted the view, since rejected by most UDRP panels, that the complainant must have rights in the mark that antedate the registration of the domain name in order to bring a complaint. Referring to a considerable number of decisions holding that there can be no bad faith where there were no rights in the mark at the time the domain name was registered, the panel held that, in this dispute, the domain name was not registered in bad faith.
- *Digital Vision, Ltd v Advanced Chemill Systems.*[102] The respondent registered the <digitalvision.com> domain name on 2 September 1997, while the complainant's US and European registrations for the 'DIGITAL VISION' mark post-dated this. The panellist held that the first element of the UDRP does not require rights in a trade mark to arise before the domain name is registered, but that this is relevant to the assessment of bad faith. In this dispute, as the complainant did not register its marks until four months after the domain name was registered, the panellist held that the domain name was not registered in bad faith.
- *PrintForBusiness B V v LBS Horticulture.*[103] The respondent registered the <print4-business.com> domain name on 4 March 2000, whereas the complainant had only applied for registration of a Community Mark for 'PRINT FOR BUSINESS'. As the domain name was registered before the complainant applied to register the trade mark, and as there was no evidence that the respondent should have known of the complainant's mark prior to registering the domain name, the panellist held that the domain name was not registered in bad faith.
- *Razorbox, Inc v Torben Skjodt.*[104] The respondent registered the <razorbox.com> domain name on 10 December 1999, while the complainant first used the 'RAZORBOX' mark in commerce on 15 January 2002. In these circumstances, the panellist held that it would be

[100] WIPO, above n 4, para 3.1.
[101] WIPO Case No D2001-0074 (1 May 2001).
[102] WIPO Case No D2001-0827 (23 Sept 2001).
[103] WIPO Case No D2001-1182 (21 Dec 2001).
[104] NAF Case No FA150795 (9 May 2003).

impossible for the complainant to establish that the domain name was registered in bad faith.

- *eeParts, Inc v E E All Parts Corp.*[105] The respondent registered the <eeallparts.com> and <eeallparts.net> domain names in July 2000, while the complainant obtained registration for the 'EEPARTS.COM' mark with the USPTO on 26 August 2003. As the domain names were registered before the complainant had any rights in the mark, and the respondent was unaware of the complainant's mark prior to receiving a 'cease and desist' letter, the panel held that the domain name was registered in bad faith.
- *Jazeera Space Channel TV Station v AJ Publishing.*[106] The respondent registered the <aljazeera.com> domain name on 12 April 1996, while the complainant was founded at some time in 1996 and first registered the <aljazeera.net> domain name on 30 August 1996. On the basis of the evidence submitted, the panel concluded that the domain name was registered before the complainant acquired rights in the 'AL JAZEERA' mark. Accordingly, the panel accepted that the domain name was not registered in bad faith.

In certain circumstances, however, a respondent may register a domain name in the expectation that the complainant is planning to acquire rights in a mark. Where a respondent registers a domain name with notice that a complainant is likely to acquire rights in a mark, such as where there has been a public announcement of a merger between two companies, then the disputed domain name may have been registered in bad faith, even though the registration of the domain name predates rights in the mark. The *WIPO Overview* expresses this qualification to the general principle that there can be no bad faith where the disputed domain name is registered before the complainant has acquired rights in the mark in the following terms:

> In certain situations, when the respondent is clearly aware of the complainant, and it is clear that the aim of the registration was to take advantage of the confusion between the domain name and any potential complainant rights, bad faith can be found. This often occurs after a merger between two companies, before the new trademark rights can arise, or when the respondent is aware of the complainant's potential rights, and registers the domain name to take advantage of any rights that may arise from the complainant's enterprises.[107]

The circumstances in which there may be bad faith registration, despite the complainant acquiring trade mark rights only after the domain name was registered, were set out by the panellist in *ExecuJet Holdings Ltd v Air Alpha America, Inc*[108] in the following terms:

> For the purposes of the Policy, a domain name registration can have been undertaken in bad faith even if the trade marks [*sic*] rights at which such bad faith was directed arose after the registration of the domain name. This might be the case where a registrant speculated on an impending merger between companies that would create a new name combining in whole or in part the names of the merger partners. Another example concerns the situation where a registrant based its registration of the domain name on its insider knowledge of the intentions of a (former) business partner or employer in relation to the latter's development of a new trademark. From the intent of the Policy, it is clearly irrelevant whether a registrant intended to abuse an existing trademark right or one which that registrant specifically knew would arise.

[105] NAF Case No FA481753 (14 July 2005).
[106] WIPO Case No D2005-0309 (19 July 2005).
[107] WIPO, above n 4, para 3.1.
[108] WIPO Case No D2002-0669 (7 Oct 2002).

In that particular dispute, the complainant was a Swiss company and the respondent was based in the United States. The complainant claimed that its Scandinavian operator had established a contractual relationship with the respondent's Danish subsidiary prior to the domain name being registered, even though the complainant acquired trade mark rights only after the domain name was registered. The panellist, however, held that, in this instance, there was insufficient evidence of the parties' relationship to establish bad faith registration.

The rationale for exceptions to the general principle that a domain name is not registered in bad faith when it is registered before the complainant acquires rights in a confusingly similar mark was clearly explained by the panellist in *Skype Limited v Benjamin Decraene*[109] as follows:

> There are . . . certain situations when a respondent is clearly aware of a complainant, and it is clear that the aim of the registration was to take advantage of the confusion between the domain name and any of the complainant's potential rights. In these cases, bad faith can be found. This could occur when a respondent is aware of a complainant's potential rights and registers the domain name to take advantage of any rights that may arise from a complainant's enterprises.

The following are representative panel decisions that have found that a domain name had been registered in bad faith where the complainant acquired rights in the mark after the domain name was registered.

- *Einstein Medical, Inc v Dennis Condon.*[110] The respondent registered the disputed domain names <plasticdocshop.com> and <cosmeticdocshop.com> on 1 April 1999, but the complainant's 'DOCSHOP' service mark was not registered until 9 December 1999. The complainant had, however, registered the domain names <plasticsdocshop.com> and <cosmeticsdocshop.com> in June 1998, and had informed the respondent of its confidential business strategies immediately prior to the registration of the disputed domain names. The panellist held that the respondents had used the domain names intentionally to attempt to attract, for commercial gain, Internet users to their web site by creating a likelihood of confusion with the complainant's mark and that, consequently, there was bad faith registration and use.
- *AP Møller v Web Society.*[111] The respondent registered the domain names <maersksealand.com> and <maersk-sealand.com> on 14 December 1998, while merger negotiations between AP Møller, the owner of the 'MAERSK' mark, and Sea-Land Service, Inc, the owner of the 'SEA-LAND' mark, had commenced earlier in 1998. The panellist held that the timing of the registration indicated a desire by the respondent to pre-empt registration of the disputed domain names by the merged entity and that, therefore, the registration was in bad faith.
- *eGalaxy Multimedia Inc v T1.*[112] The respondent registered the domain name <nakednew.com> on 6 October 2000, while the panellist held that the complainant acquired common law rights in the 'NAKED NEWS' mark only in April 2001. The evidence, however, established that the complainant had commenced its 'naked news' Internet programme in December 1999, and that the respondent had registered the domain name <victoriasinclair.com> at the same time as it registered the disputed

[109] WIPO Case No D2005-1112 (3 Jan 2006).
[110] NAF Case No FA92528 (2 Mar 2000).
[111] WIPO Case No D2000-0135 (15 Apr 2000).
[112] NAF Case No FA101823 (23 Dec 2001).

domain. As 'Victoria Sinclair' was the name of the lead anchor of the complainant's programme, the panellist held that the respondent had chosen the disputed domain name because of its similarity to the complainant's mark and that, consequently, the domain name had been registered in bad faith.

- *Kangwon Land, Inc v Bong Woo Chun.*[113] The respondent registered the disputed domain name <kangwonland.com> on 13 September 1998, while the complainant filed applications for registration of the <KANGWONLAND> mark only in 1999. As the complainant was established by a special law in June 1998 and had a monopoly over casino operations in the Republic of Korea, and as the respondent appeared to be involved in the online casino business, the panellist held that the respondent must have known of the reputation of the complainant as a state-run casino at the time the domain name was registered. In these circumstances, the panellist concluded that the respondent had registered the domain name in bad faith to trade off a mistaken affiliation with the complainant.
- *MADRID 2012, SA v Scott Martin-MadridMan Websites.*[114] The respondent registered the <madrid2012.com> domain name two days before the complainant filed its first application to register the 'MADRID2012' mark and before the complainant, which was established to promote Madrid's bid for the 2012 Olympic Games, was re-named 'MADRID 2012, SA'. The panellist pointed out that, although the respondent probably did not know of the imminent application for trade mark registration, the respondent was aware of Madrid's bid for the 2012 Games, which had been publicised months before the domain name was registered. On this basis, the panellist inferred that the domain name had been registered to hinder the complainant from registering the domain name. The panellist was also prepared to infer that the respondent knew of the complainant's applications to register the '2M12' mark at the time the respondent registered the <2m12.com> and <2m12.org> domain names, meaning that these domain names had also been registered in bad faith.
- *General Growth Properties v. Steven Rasmussen.*[115] The complainant began construction of 'PROVO TOWNE CENTRE', the largest shopping mall in Utah, on 5 June 1997 and the respondent registered <provotownecentre.com> domain name on the same day. The construction of the mall had been widely publicised. Referring to the decisions in *Kangwon Land* and *MADRID 2012*, the panellist concluded that the respondent knew of the 'pending widespread announcement' of the construction of the mall at the time he registered the disputed domain name. Concluding that the respondent had registered the <provotownecentre.com>, and subsequently registered and used the <provotownecentre.net> domain name, either to benefit from confusion with the complainant's mark or to sell the domain names to the complainant, the panellist held that the domain names had been registered in bad faith.
- *Aditya Birla Nuvo Limited v HJ Shin.*[116] The respondent registered the <adityabirlanuvo.com> domain name on 11 September 2005, while a restructuring and merger of existing Indian companies to form a new company, 'ADITYA BIRLA NUVO LIMITED', was publicly announced on or about the same date. The panellist held that the registration of the domain name at the time of the announcement of the merger was too much of a coincidence to be other than bad faith registration.

[113] WIPO Case No D2003-0320 (4 July 2003).
[114] WIPO Case No D2003-0598 (8 Oct 2003).
[115] WIPO Case No D2003-0845 (15 Jan 2004).
[116] WIPO Case No D2006-0224 (8 May 2006).

Disclaimer

[7.15] Relevance of Disclaimer on Respondent's Web Site

The inclusion of a disclaimer on a respondent's Web site will, in general, not reverse a finding of bad faith registration and use, as it will not be sufficient to dispel initial interest confusion. As the panel in *Société pour l'œuvre et la mémoire d'Antoine de Saint Exupéry—Succession Saint Exupéry—D'Agay v The Holding Company*[117] put it:

> A disclaimer may help to dispel confusion once the Internet user has reached the website, but it is very difficult to cure the problem created as to Complainant. Complainant may not recover the interest of the consumer. For this reason, the Panel is not inclined to place great weight on the introduction of the disclaimer.

Similarly, the panel in *Madonna Ciccone p/k/a Madonna v Dan Parisi and 'Madonna.com'*[118] stated that:

> Respondent's use of a disclaimer on its website is insufficient to avoid a finding of bad faith. First, the disclaimer may be ignored or misunderstood by Internet Users. Second, a disclaimer does nothing to dispel initial interest confusion that is inevitable from Respondent's actions. Such confusion is a basis for finding a violation of Complainant's rights.

Moreover, in combination with other factors, a disclaimer may well reinforce a finding of bad faith registration and use, as it clearly indicates that the respondent was aware of the complainant's mark. On the other hand, a disclaimer that is used on a criticism site may support a finding that the respondent has a legitimate interest in the domain name, as it makes it apparent to users that there is no association with the complainant.

As there is a consensus among UDRP panellists that the content of a web site is not relevant in determining whether the disputed domain name is confusingly similar to the complainant's mark, the presence of a disclaimer is not to be taken into account in considering the first element of the UDRP. As the panellist in *Myer Stores Limited v Mr David John Singh*[119] explained in relation to the first element:

> The test of confusing similarity under the Policy, unlike trademark infringement or unfair competition cases, is confined to a consideration of the disputed domain name and the trademark. Accordingly, disclaimers placed on the website are irrelevant.

In some decisions, however, a disclaimer of association on a web site has been accepted as evidence that the domain name is confusingly similar to the complainant's mark. In any case, the presence of a disclaimer can be taken into account in relation to either the second or third element of the UDRP.

The WIPO Consensus View on the role of a disclaimer on the web page of the disputed domain name is as follows:

> The existence of a disclaimer cannot cure bad faith, when bad faith has been established by other factors. A disclaimer can also show that the respondent had prior knowledge of the complainant's

[117] WIPO Case No D2005-0165 (9 June 2005).
[118] WIPO Case No D2000-0847 (12 Oct 2000).
[119] WIPO Case No D2001-0763 (10 July 2001).

trademark. However a disclaimer is sometimes found to support other factors indicating good faith or legitimate interest.[120]

The UDRP decisions that have considered the effect of a disclaimer fall into three broad categories. The first category adopts the position that a disclaimer can never dispel initial interest confusion. In this respect, some decisions have approved the following statement from a US decision:

> A disclaimer that purports to disavow association with the trademark owner after the consumer has reached the site comes too late; the consumer has already been misdirected.[121]

The second category of decisions have held that the particular disclaimer, in the circumstances of the case, does not effectively dispel confusion, and therefore does not cure a finding of bad faith. Finally, the third category of decisions have held that the use of a disclaimer on either a genuine criticism site or 'fan' site supports a finding that the respondent has legitimate interests in the disputed domain name.

The following decisions fall within the first category, in concluding that the disclaimer cannot reverse initial interest confusion.

- *Madonna Ciccone p/k/a Madonna v Dan Parisi and 'Madonna.com'.*[122] The disputed domain name <madonna.com> resolved to a pornographic web site and included a disclaimer of association with the complainant, 'MADONNA'. The panel held that the disclaimer did not avoid a finding of bad faith.
- *Estée Lauder Inc v estelauder.com, estelauder.net and Jeff Hanna.*[123] The disputed domain names, <estelauder.com> and <estelauder.net>, resolved to consumer criticism sites that contained several disclaimers of association with the complainant. The respondents argued that they had rights or interests in the domain names arising from use of the names for legitimate criticism sites. The panellist, however, found that the respondents had registered the domain names primarily for the purpose of disrupting the complainant's business. In deciding that the respondents had no rights or interests in the domain names, the panellist held that the initial illegitimate diversion of users, resulting from the confusingly similar domain names, could not be rectified by the disclaimers on the web sites. In this respect, the panellist stated that:

  > The fact that the users, once so diverted or attracted, are confronted with numerous disclaimers does not cure the initial and illegitimate diversion.

- *Arthur Guinness Son & Co (Dublin) Limited v Dejan Macesic.*[124] The disputed domain name <guiness.com> resolved to a web site that included a prominent disclaimer at the top of every page of the site. The panellist pointed to the deliberate misspelling of the complainant's mark in the domain name as evidence that the domain name was registered deliberately to divert Internet users. In finding that the respondent had no rights or interests in the domain name, the panellist followed the *Estée Lauder* decision to conclude that

[120] WIPO, above n 4, para 3.5.

[121] *Paccar, Inc v Telescan Technologies LLC,* 115 F Supp 2d 772, 778 (E D Michigan 2000), cited in *Alpha Xi Delta, Inc v Befriend Internet Services,* NAF Case No FA100681 (28 November 2001); *Seiko Epson Corporation and Epson America, Inc v Domain Administrator,* NAF Case No FA544995 (11 October 2005).

[122] WIPO Case No D2000-0847 (12 Oct 2000).

[123] WIPO Case No D2000-0869 (25 Sept 2000).

[124] WIPO Case No D2000-1698 (25 Jan 2001).

the content of a web site, including the use of disclaimers, could not cure the initial illegitimate diversion.

- *Alpha Xi Delta, Inc v Befriend Internet Services.*[125] The disputed domain name <alphaxidelta.com> resolved to a web site that included links to commercial sites, but included a disclaimer at the bottom of the homepage. The panellist held that the disclaimer did not dispel the initial interest confusion, which 'occurs when a domain name has the potential to misdirect consumers as they look for websites associated with the owner of a trademark'.
- *Bonneterie Cevenole SARL v Sanyouhuagong.*[126] The disputed domain name <montagut.com> resolved to a web site that was critical of the complainant, which had rights in the 'MONTAGUT' mark, and included disclaimers. The panellist held that the disclaimers did not dispel bad faith, stating that:

> The fact that when . . . [users] . . . reach the site there is a disclaimer is not enough. It is too late. By then, the damage has been done and/or the benefit to the Respondent has been achieved.

- *Seiko Epson Corporation and Epson America, Inc v Domain Administrator.*[127] The disputed domain names <epsoncart.com>, <picepson.com> and <qualityepson.com> were used to divert consumers for 'EPSON' products to sites that promoted the sale of non-'EPSON' replacement ink cartridges for ink jet printers, but included disclaimers. Referring to the *Alpha Xi Delta* decision, the panellist held that the disclaimers did not dispel initial interest confusion.

The following decisions, which have held that the particular disclaimer used on the respondent's web site does not dispel consumer confusion, fall within the second category of decisions to have considered the role of a disclaimer.

- *Pliva, Inc v Eric Kaiser.*[128] The disputed domain name <antabuse.net> resolved to a web site that marketed pharmaceutical products that competed with the complainant's product, but included a disclaimer at the bottom of the homepage. The panellist concluded that the domain name was registered and used in bad faith, and that the disclaimer could not cure the bad faith. In this instance, the panellist held that the disclaimer was ineffective for two reasons: first, it came after a full page of marketing where the complainant's mark appeared many times and, secondly, it appeared on the homepage after the point at which users were given the option of purchasing the competing product.
- *International Organization for Standardization ISO v ISO Easy.*[129] The disputed domain names, <isoeasy.com> and <isoeasy.org>, resolved to web sites that provided consultancy services and links to commercial sites related to standards, and included disclaimers of association with the 'ISO' trade mark. First, referring to the WIPO Consensus View, the panellist held that the disclaimers actually confirmed that the respondent knew of the complainant's mark. Secondly, in respect of a disclaimer that was located at the bottom of the homepage of the relevant sites, the panellist held that the placement gave the disclaimer a low profile. Thirdly, the panellist considered that the wording of another purported disclaimer did not unequivocally deny an association with the complainant. The disclaimers therefore did not dispel a finding that the domain names had been registered to divert Internet users.

[125] NAF Case No FA100681 (28 Nov 2001).
[126] WIPO Case No D2001-1309 (21 Jan 2002).
[127] NAF Case No FA544995 (11 Oct 2005).
[128] WIPO Case No D2003-0316 (9 June 2003).
[129] WIPO Case No D2005-0984 (8 Nov 2005).

- *Asset Loan Co Pty Ltd v Gregory Rogers.*[130] The disputed domain name <assetloanco.net> was used for a criticism site that included a disclaimer of connection with the complainant. The panel held that the presence of the disclaimer was negated because the first words confronting the user on reaching the site were not the disclaimer, but the title 'AssetLoanCo.net'. Consequently, instead of dispelling any initial confusion, the web site reinforced it.
- *SENSIS Pty Ltd v Kevin Goodall.*[131] The disputed domain name <tradingpost11.com> resolved to a web site, which was a portal for sites that competed with the complainant and included a disclaimer of association with the complainant's 'TRADING POST' mark at the bottom of the homepage. Adopting the reasoning in the *Pliva* decision, the panellist held that the disclaimer in this case did not have the effect of negating or diminishing the misleading effect of using the complainant's prominent mark. Accordingly, the panellist concluded that there was bad faith registration and use, as the domain names had been registered to disrupt the business of the complainant and to divert Internet users to the respondent's site.
- *Seiko Epson Corporation and Epson America, Inc v AOS Web Com, Inc.*[132] The disputed domain name <epsonusa.com> was used to sell the products of the complainant's competitors, as well as to resell the complainant's products, but included a fine print disclaimer. Finding that the disclaimer was both easy to overlook and equivocal, the panellist found that it could actually reinforce the view that the respondent's site was endorsed or approved by the complainant.

The following decisions fall within the third category of decisions, where panellists have held that the presence of a disclaimer on a criticism site or fan site may support a conclusion that the respondent has rights or legitimate interests in the disputed domain name. In a number of decisions in this category, panellists have held that the adoption of the particular domain name in question will result in a low level of initial interest confusion and that, consequently, the use of a disclaimer is more effective in dispelling confusion than in disputes involving a domain name that is identical to the complainant's mark.

- *Caterpillar Inc v Off Road Equipment Parts.*[133] The disputed domain name <catparts.com> was used by the respondent to resell the complainant's products, and included a disclaimer of association with the complainant. The panellist held that, by virtue of the disclaimer, there was no possibility that a visitor to the respondent's web site would be confused about a likelihood of affiliation.
- *Satchidananada Ashram v Domain Administrator.*[134] The disputed domain names, including <yogaville.com>, <yogaville.net> and <integralyoga.com>, directed users to a site that was critical of the complainant and included a disclaimer on every page of the web site. The panel held that this was a legitimate, non-commercial use of the domain names, and that the use of an explicit disclaimer, together with the content of the web site, meant that there was no likelihood of confusion with the complainant's mark.
- *Covance, Inc v The Covance Campaign.*[135] The panellist held that a prominent disclaimer on a criticism site supported the conclusion that the respondent had rights or legitimate

130 WIPO Case No D2006-0300 (2 May 2006).
131 WIPO Case No D2006-0793 (22 Aug 2006).
132 NAF Case No FA823033 (27 Nov 2006).
133 NAF Case No FA95497 (10 Oct 2000).
134 NAF Case No FA125228 (13 Dec 2002).
135 WIPO Case No D2004-0206 (30 Apr 2004).

interests in the disputed domain name, <covancecampaign.com>. In finding that the respondent was making a legitimate non-commercial or fair use of the domain name, the panellist stated that:

> In this case the Panel is satisfied that even where the reference to the Respondent's site comes up underneath the entry for the Complainant's site it is clear on the face of the search engine entries that the Respondent's site at 'www.covancecampaign.com' is not in any way run by, associated with or endorsed by the Complainant and is in fact concerned with criticism of the Complainant's activities. Further on reaching the Respondent's website a bold disclaimer makes it clear from the very outset that this site has no connection with the Complainant and is against the use of animal testing by the Complainant. In these circumstances it is extremely difficult to see how members of the public could be misled into thinking that the site is associated with, or has any connection with the Complainant.

- *Besiktas Jimnastik Kulubu Dernegi v Mehmet Tolga Avcioglu.*[136] The disputed domain name <besiktas.com> was used as a fan site for the 'BASIKTAS' football team, a well-known Turkish football club. The site included a number of disclaimers indicating that the site was not the official site of the 'BESIKTAS' football club. The panellist concluded that, as the site was a genuine 'fan site', the complainant had not established that the respondent had no legitimate interests in the domain name, nor that the domain name was registered in bad faith. In reaching this conclusion, the panellist held that the inclusion of appropriate disclaimers indicated that, even if the site was used for commercial gain, there was no likelihood of confusion with the complainant's mark.
- *America Online, Inc v Anytime Online Traffic School.*[137] The disputed domain names, <aoltrafficschool.com>, <aoldefensivedriving.com> and <aoldriverimprovement.com>, were used for services that did not compete with the complainant's services, and included disclaimers of association with the complainant. The panellist held that the use of a disclaimer supported the conclusion that the respondent was making a bona fide offering of services and was evidence of good faith.
- *CBS Broadcasting Inc, f/k/a CBS Inc v Nabil Z aghloul.*[138] The disputed domain name <cbscares.com> resolved to a site that was critical of 'CBS' that included a disclaimer of affiliation with the complainant. Adopting the approach of the panellist in the *Covance* decision, the panellist held that, given that the domain name was not identical to the complainant's mark, meaning that it would not automatically appear in response to an Internet search for the complainant's mark, the presence of the disclaimer could disabuse visitors to the site of an association with the complainant

'Opportunistic' Bad Faith

[7.16] 'Opportunistic' Bad Faith

The principle of 'opportunistic bad faith' is not an explicit part of the UDRP, but has been developed by UDRP panellists to deal with circumstances that strongly suggest that the

[136] WIPO Case No D2003-0035 (10 Mar 2003).
[137] NAF Case No FA146930 (11 Apr 2003).
[138] WIPO Case No D2004-0988 (5 Feb 2005).

disputed domain name has been registered and is being used in bad faith. As first formulated by the panellist in *Veuve Clicquot Ponsardin v The Polygenix Group Co*,[139] the principle applies where the disputed domain name:

> is so obviously connected with such a well-known product that its very use by someone with no connection with the product suggests opportunistic bad faith.

The principle essentially recognises that where a domain name is identical or confusingly similar to a well-known mark, it is extremely difficult for a respondent who has no connection with the mark credibly to deny bad faith registration or use. The application of the principle clearly depends upon the strength of the complainant's mark; the more well-known the mark is, the more difficult it is for the respondent to claim that the domain name has not been registered and used 'opportunistically'.

Where a panel finds that the principle of 'opportunistic bad faith' applies as a result of the strength of the complainant's mark, bad faith registration or use will generally be inferred unless the respondent can establish rights or legitimate interests in the disputed domain name. Moreover, given the strength of the complainant's mark, where the doctrine of opportunistic bad faith is applied, panellists will infer notice of the complainant's well-known marks.

The following decisions illustrate the application of the principle of 'opportunistic bad faith'.

- *Veuve Clicquot Ponsardin v The Polygenix Group Co.*[140] The disputed domain name <veuveclicquot.org> was identical to the complainant's famous 'VEUVE CLICQUOT' mark. Pointing out that the circumstances of bad faith are not limited to those identified in paragraph 4(b), and in the absence of a response, the panellist held that the domain name was so obviously connected with the complainant's well-known mark that its very use suggested opportunistic bad faith.
- *V&S Vin & Sprit Aktiebolag v Tyler Kownacki.*[141] The disputed domain name <absolutmandrin.com> was identical to the complainant's 'ABSOLUT MANDRIN' mark, which was used in connection with flavoured vodka. The panellist held that, given the reputation of the complainant's 'ABSOLUT' mark, the respondent knew or should have known of the mark, and that the domain name was so obviously connected with the complainant that there was opportunistic bad faith.
- *Banca Sella spa v Mr Paolo Parente.*[142] The disputed domain name <bancasella.com> was identical to the complainant's 'BANCA SELLA' mark. The complainant is one of the leading banks in Italy, and had operated under the 'BANCA SELLA' name since the nineteenth century. The panellist inferred that the respondent knew of the complainant's mark, and held that there was opportunistic bad faith, as the respondent could not possibly make any plausible legitimate actual or contemplated active use of the domain name.
- *Pavillion Agency, Inc v Greenhouse Agency Ltd.*[143] The disputed domain names, including <pavillionagency.com> and <pavillionagency.org> were confusingly similar to the complainant's 'PAVILLION' marks, which were well known for employment agency services. Given the distinctive nature of the complainant's marks, the panellist inferred that the

[139] WIPO Case No D2000-0163 (1 May 2000).
[140] WIPO Case No D2000-0163 (1 May 2000).
[141] NAF Case No FA95079 (8 Aug 2000).
[142] WIPO Case No D2000-1157 (27 Nov 2000).
[143] WIPO Case No D2000-1221 (4 Dec 2000).

respondent had notice of the marks, and that the domain names were so obviously connected with the complainant's marks that use by anyone other than the complainant was opportunistic bad faith.

- *Victoria's Secret et al v Model aka Lori Learmont.*[144] The disputed domain names <victoriassecretlingeriemodels.com> and <victoria-secret-models.com> were identical or confusingly similar to the complainant's famous 'VICTORIA's SECRET' mark. The panellist inferred that the respondent had notice of the complainant's mark, and held that the domain names were so obviously connected with the complainant that use by someone other than the complainant suggested opportunistic bad faith.
- *Boiron SA v José Antonio Paya Serer.*[145] The disputed domain name <boiron.net> was identical to the complainant's 'BOIRON' mark, which is famous in connection with homeopathics. Inferring that the respondent had notice of the complainant's mark, the panellist held that this was an example of opportunistic bad faith.
- *The Nasdaq Stock Market, Inc v Act One Internet Solutions.*[146] The disputed domain name <nasdasq.com> was identical to the complainant's famous 'NASDAQ' mark. The panellist inferred that the respondent was aware of the complainant's famous mark and that the domain name was so obviously connected to the complainant's mark that use by anyone other than the complainant was opportunistic bad faith.
- *America Online, Inc v D Brown.*[147] The disputed domain name <aolchat.com> was confusingly similar to the complainant's well-known 'AOL' mark. The panellist held that, given the well-known nature of the complainant's mark, the domain names were so obviously connected with the complainant that the use or registration of the mark as a domain name by anyone other than the complainant suggested opportunistic bad faith.

Registration for Purpose of Sale

[7.17] Registration for Purpose of Sale

The first non-exclusive circumstance that is evidence of bad faith registration and use, as set out in paragraph 4(b)(i), refers to:

> circumstances indicating that the domain name holder has registered or has acquired the domain name primarily for the purpose of selling, renting, or otherwise transferring the domain name registration to the complainant who is the owner of the trademark or service mark or to a competitor of that complainant, for valuable consideration in excess of the domain name holder's documented out-of-pocket costs directly related to the domain name.

Each of the parts of this circumstance has been subject to interpretation by UDRP panellists. The following sections explain the way in which this circumstance has been interpreted.

[144] NAF Case No FA96553 (2 Mar 2001).
[145] WIPO Case No D2001-0118 (2 May 2001).
[146] WIPO Case No D2003-0103 (1 Apr 2003).
[147] NAF Case No FA520497 (6 Sept 2005).

[7.17.1] Paragraph 4(b)(i): What Amounts to 'Circumstances Indicating' that the Domain Name has been Registered for the Purpose of Selling It to the Complainant or a Competitor?

To make out the first non-exclusive circumstance evidencing bad faith, the complainant need only establish 'circumstances indicating' that the domain name has been registered for the purpose of selling it to the complainant or a competitor. The 'circumstances' indicating that the domain name has been registered for this illegitimate purpose are not confined to formal, legal offers to sell the disputed domain name, but encompass the broad range of considerations that may lead to the conclusion that the respondent had the illegitimate purpose.

What may amount to 'circumstances' indicating that the domain name has been registered for the purpose of selling it to the complainant or a competitor was considered by the panel in *Channel 5 Broadcasting Limited v PT Pancawana Indonesia.*[148] In that dispute, the respondent claimed that the disputed domain name had not been registered in bad faith, as there had been no legally binding offer to sell the domain name to the complainant. In e-mails to the complainant, a representative of the respondent had insisted that the domain name was not for sale, but had invited the complainant to make an offer for the domain name. In rejecting the respondent's argument, the panel observed, in relation to paragraph 4(b)(i), that:

> Nowhere does this paragraph limit a panel's scrutiny to just those acts that constitute a legalistic "offer" and nothing else. 'Circumstances', as contemplated by this paragraph broadly encompasses those commercial negotiations, when viewed in their entirety and regardless of the specific negotiating nuance of who actually proposed what contractual term, through which a respondent's intention, as manifested by its actions, was to transfer a domain name to a complainant for consideration in excess of its costs of registration to its present holder.

[7.17.2] Paragraph 4(b)(i): Can an Offer to Sell a Domain Name in Settlement Negotiations Amount to Bad Faith?

A respondent may make an offer to sell the disputed domain name in the course of negotiations to settle the domain name dispute with the complainant. In general, offers made in these circumstances can be taken into account in determining whether paragraph 4(b)(i) has been satisfied. In common law jurisdictions, offers made as part of settlement negotiations in communications marked 'without prejudice' cannot usually be relied upon in court proceedings. The general view of panellists is that the rule precluding reliance on offers made 'without prejudice' does not apply to UDRP disputes. Some decisions have, however, applied the rule that offers that are made as part of a genuine effort to settle the dispute will not be taken into account, unless there is evidence of unambiguous impropriety. The position appears to be that if both parties agree that offers made in good faith negotiations to settle the dispute should not be taken into account, then they should be excluded from consideration by panellists, but that otherwise offers made in settlement negotiations can be evidence of registration for an illegitimate purpose.

The WIPO Consensus View on this issue is as follows:

> Evidence of offers to sell the domain name in settlement discussions is admissible under the UDRP, and is often used to show bad faith. This is because many cybersquatters often wait until a

[148] NAF Case No FA98415 (3 Oct 2001).

trademark owner launches a complaint before asking for payment and because panels are competent to decide whether settlement discussions represent a good faith effort to compromise or a bad faith effort to extort. Also the legal criteria for showing bad faith directly specifies that an offer for sale can be evidence of bad faith.[149]

The following decisions have held that offers to sell in settlement negotiations are evidence of bad faith under paragraph 4(b)(i):

- *Motorola, Inc v NewGate Internet, Inc.*[150] The respondent acquired the <talkabout.com> domain name for use as an adult sex site. In response to an offer by the complainant to acquire the domain name, the respondent indicated that it did not intend to relinquish the domain name in exchange for reimbursement of costs. The majority of the panel held that this was evidence that the domain name had been acquired primarily to sell the domain name to the complainant. The dissenting panellist, however, referring to US law, maintained that evidence of offers made in negotiations between the parties should be excluded from the consideration of UDRP panels.
- *CBS Broadcasting, Inc v Gaddoor Saidi.*[151] The complainant sent the respondent a letter requesting transfer of the disputed domain name and, in reply, the respondent offered to sell the domain name for a minimum of US$14,000. The panellist rejected the respondent's claim that, as the offer had been made in settlement negotiations, the panel was barred from taking it into account. In rejecting the respondent's contention, the panellist pointed out that paragraph 4(b)(i) expressly provides that an offer to sell the disputed domain name to the complainant is conclusive evidence of bad faith.
- *Magnum Piering, Inc v The Mudjackers.*[152] In negotiations over multiple disputed domain names, the respondents demanded US$2,500 per year from the complainant to lease the domain names. The panellist rejected the respondent's contention that the offer was not admissible on the basis that this would effectively insulate respondents who registered disputed domain names in the expectation that a complainant would make an offer to purchase the domain name from the UDRP. In this respect, the panellist stated:

  > The Panel is of the opinion that the Policy's goal of preventing cybersquatting would not be furthered by excluding evidence of a registrant's offer to sell or otherwise transfer the domain name for consideration in excess of out-of-pocket costs, even if the offer is made after the registrant is on notice of the dispute. Cybersquatters often wait until a trademark owner comes calling; they should not be able to avoid the Policy by being the second to speak.

  Furthermore, the panellist added that UDRP panels 'are fully capable of assessing whether an offer of sale reflects a good faith effort to compromise or part of a bad faith effort to extort'.

In the following decisions, panellists have held that there are circumstances in which good faith offers by a respondent to settle the dispute are not necessarily evidence of registration for an illegitimate purpose, and may be excluded from consideration by UDRP panels where the parties so agree.

- *The Vanguard Group, Inc v Emilio Sa.*[153] The complainant offered to purchase the disputed domain names for the respondent's out-of-pocket expenses, but in a series of negotiations

[149] WIPO, above n 4, para 3.6.
[150] WIPO Case No D2000-0079 (20 Apr 2000).
[151] WIPO Case No D2000-0243 (2 June 2000).
[152] WIPO Case No D2000-1525 (29 Jan 2001).
[153] WIPO Case No D2001-1453 (7 Apr 2002).

one of the disputed domain names was eventually sold for US$1,500. After a comprehensive review of UDRP decisions considering communications made in settlement negotiations, the panellist pointed out that the majority of decisions opposed the application of privilege to offers made 'without prejudice'. The panellist, nevertheless, considered that the desirability of promoting speedy resolution of disputes meant that, where the negotiations were a good faith attempt at settlement, the communications might be excluded. In this respect, the panellist approved the following comments made under the Nominet UK Dispute Resolution Service (DRS):

> It is desirable that any Dispute Resolution Service should be operated in a manner, which encourages the parties to settle between themselves if possible. In particular, in a case like this one, where the Complainant had the option of pursuing legal proceedings as an alternative to using the Nominet DRS, it cannot be equitable for a Respondent to be potentially subject to two distinct privilege regimes - on the one hand, if the Complainant opts for Court action, a Respondent can expect to have a settlement offer considered '*without prejudice*', whereas on the other, if the Complainant opts to use the DRS, a Respondent can only be confident that a settlement offer will not be used against him if he waits for proceedings to reach the 'Informal Mediation' stage. Such inconsistency cannot assist in promoting early resolution of a dispute.
>
> It would be incorrect to lay down a rule, which is absolute in its effect. The qualifying language in both . . . [the US and UK rules] . . . make it clear that under the common law, the principle applies only where the intent of the communication is to make a bona fide attempt to compromise a disputed claim and not where made to advance nefarious schemes.

Applying these principles to the instant dispute, the panellist found that the respondent had entered negotiations with the bad faith intent of inducing the complainant to purchase the domain names. Consequently, the panellist held that the respondent's communications were evidence of bad faith registration and use.

- *Pearl Jam v Streaming Digital Media Dot Com.*[154] The parties entered into settlement negotiations regarding the disputed domain name, and the respondent sought to introduce evidence from these communications. The panellist noted that there is a 'strong policy reason' for not considering such communications in UDRP proceedings, namely that parties should be encouraged to settle disputes amicably without fear that the negotiations will be used against them. At the same time, the panellist noted that UDRP decisions had taken 'without prejudice' communications into account in finding bad faith. In this dispute, the panellist held that it was not necessary to decide whether the complainant's communications were admissible, as there was no evidence of bad faith registration or use.
- *McMullan Bros, Limited and others v Web Names Ltd.*[155] Advisors for the complainant wrote to the respondent requesting transfer of the disputed domain name and solicitors for the respondent replied with a letter marked 'without prejudice' requesting that the complainant make an offer for the domain name. Despite subsequent negotiations, the parties failed to reach an agreement. After a comprehensive review of previous decisions, the panellist accepted the general principle that panellists should not be excluded from considering offers made in 'without prejudice' negotiations. At the same time, the panellist held that if both parties expressly or implicitly agreed that discussions with a view to settlement should not be brought before a panel, the negotiations should be excluded

[154] NAF Case No FA235831 (29 Mar 2004).
[155] WIPO Case No D2004-0078 (16 Apr 2004).

from consideration by the panel. In reaching this conclusion, the panellist stated that, if such material were not excluded from consideration:

> then a party might deliberately lure the other party into discussions, supposedly on a without prejudice basis, but in the hope that the other side might make a statement to its detriment during those discussions which could then be brought before a panel. This, the Panel believes, would be grossly unconscionable.

In the circumstances of the particular dispute, although the respondent marked its correspondence 'without prejudice', there was no evidence that the complainant agreed that the negotiations should be excluded, so the panellist concluded that the negotiations could be taken into account. In view of the fact that the respondent was willing to transfer the disputed domain name for IR£7,000, the panellist held that the respondent had registered the domain name primarily to sell it to the complainant.

- *HSBC Holding plc v Chaudry Cheema.*[156] The complainant wrote to the respondent and a series of negotiations ensued, in which the respondent sent an e-mail marked 'without prejudice' asking how much the complainant was prepared to pay for the disputed domain name. Adopting the reasoning in the *McMullan Bros* decision, the panellist held that the respondent's e-mail was not a bona fide attempt to settle the dispute and that there was no indication that the complainant had agreed that the correspondence should be treated as 'without prejudice'. The panellist therefore relied on the respondent's offers to sell the domain name to the complainant for £1 million, or £30,000, as evidence that the domain name had been registered for the purpose of selling it to the complainant.

[7.17.3] Paragraph 4(b)(i): 'Primarily' for the Purpose of Selling the Domain Name

Paragraph 4(b)(i) specifies that the disputed domain name must be registered or acquired 'primarily' for the purpose of selling it to the complainant or a competitor of the complainant. This means that the circumstance is not necessarily satisfied by merely offering the domain name for sale, without more, but that the primary purpose of registering the domain name must be to sell it to the complainant or a competitor.

To begin with, evidence that the respondent has merely considered offering the domain name for sale does not necessarily mean that paragraph 4(b)(i) has been satisfied. In *Mark Warner 2001 v Mike Larson,*[157] for example, the panellist concluded that the respondent only considered selling the disputed domain names after being contacted by the complainant. In finding that the domain names had not been registered 'primarily' for the purpose of selling them to the complainant or a competitor, the panellist stated:

> simply considering to sell or even offering to sell a domain name is insufficient to amount to bad faith under the ICANN Policy, because the domain name must be registered *primarily* for the purpose [*sic*] selling it to the owner of trademark for an amount in excess of out-of-pocket expenses.

Depending upon the circumstances of the particular dispute, doing no more than offering the domain name for sale does not mean that the domain name has been registered 'primarily' for the purpose of sale. In *LifePlan v Life Plan, C/O Relational Dynamics, Inc,*[158] the respondent offered to sell the disputed domain name in the course of settlement negotiations. In concluding that this, in itself, did not establish bad faith, the panellist stated that:

[156] WIPO Case No D2006-0295 (9 May 2006).
[157] NAF Case No FA95746 (15 Nov 2000).
[158] NAF Case No FA94826 (13 July 2000).

> the mere offering, without more, does not indicate circumstances suggesting that Respondent "registered the domain name primarily for the purpose of" selling, renting, or transferring the domain name to the Complainant as required under 4(b)(1).

Furthermore, given that the offer was made in settlement negotiations, the panellist concluded that evidence of the offer was inadmissible.

At the same time, however, it is clear that the fact that the complainant initiated negotiations to purchase the disputed domain name will not prevent a finding that the respondent has registered the domain name primarily for the purpose of selling it. As the panellist put it in *Magnum Piering, Inc v The Mudjackers*:[159]

> Cybersquatters often wait until a trademark owner comes calling; they should not be able to avoid the Policy by being the second to speak.

In a similar vein, the panellist in *Tracy Marrow p/k/a 'ICE-T' v iceT.com*,[160] observed that:

> Since it has become commonly known that an approach to a trade mark owner to sell a domain name containing its trade mark will be taken as evidence of bad faith, it is becoming increasingly common for registrants of domain names containing trade marks to sit and wait for an approach from the trade mark owner. I, therefore do not put much weight on the fact that it was the Complainant who contacted the Respondent to see if it was interested in selling the name.

In *SembCorp Industries Limited v Hu Huan Xin*,[161] for example, the complainant approached the respondent with an offer to purchase the disputed domain name, <sembcorp.com>, for the respondent's out-of-pocket expenses, but the respondent rejected the offer, stating that he would not accept less than the 'market value of the website'. In finding that the domain name had been registered in bad faith, the panellist held that it was immaterial that the complainant had initiated negotiations.

[7.17.4] Paragraph 4(b)(i): For the Purpose of 'Selling, Renting or Otherwise Transferring' for 'Valuable Consideration in Excess of Documented Out-of-pocket Costs'

Paragraph 4(b)(i) provides that the disputed domain name must have been registered for the purpose of 'selling, renting or otherwise transferring' the domain name for 'valuable consideration in excess of documented out-of-pocket costs'. Evidence that the disputed domain name has been registered for the illegitimate purpose of selling it to the complainant, or a competitor of the complainant, may be derived from a much wider range of circumstances than an explicit offer to sell the domain name for a set price. It may, for example, be clear from all of the surrounding circumstances that the domain name has been registered for the primary purpose of selling it to the complainant, even though the respondent has not directly communicated with the complainant. Moreover, what may amount to valuable consideration encompasses more than a fixed monetary amount, and will include other arrangements for payment in excess of out-of-pocket costs.

The following decisions illustrate the variety of circumstances in which paragraph 4(b)(i) has been made out.

- *World Wrestling Federation Entertainment, Inc v Michael Bosman*.[162] Three days after registering the disputed domain name, the respondent contacted the complainant by e-mail

[159] WIPO Case No D2000-1525 (29 Jan 2001).
[160] WIPO Case No D2000-1234 (22 Nov 2000).
[161] WIPO Case No D2001-1092 (26 Nov 2001).
[162] WIPO Case No D1999-0001 (14 Jan 2000).

and indicated an intention to sell the domain name, later making an express offer to sell the domain name. The panellist held that, as the circumstances clearly fell within paragraph 4(b)(i), this amounted to bad faith use, as well as bad faith registration, of the disputed domain name.

- *Metallica v Josh Schneider.*[163] In response to a 'cease and desist' letter from the complainant, the respondent, who had registered the <metallica.org> domain name, replied with a set of demands, including having dinner with the complainants (the musical group 'METALLICA'), conducting an interview and online chat session with the complainants, and requiring the complainants to telephone two of the respondent's friends and leave messages on their voicemails. The panellist held that this indicated that the respondent had registered the domain name to extract valuable concessions in contravention of paragraph 4(b)(i). In reaching this conclusion the panellist observed that:

  > The Respondent offered to transfer the domain name in exchange for a meeting with the Complainant, phone calls from the Complainant, and an interview with the Complainant. Given the Complainant's fame, these encounters with the Complainant would be of great value - certainly a value in excess of the minimal costs associated with registering and maintaining the domain name 'metallica.org'.

- *American Anti-Vivisection Society v 'Infa dot Ne' Web Services.*[164] The respondent registered the disputed domain name after the complainant's registration of the domain name was not renewed due to administrative error. The web site to which the domain name resolved was used to solicit bids to purchase the domain name. The panellist concluded that, although the respondent had not made an offer to sell the domain name directly to the complainant, paragraph 4(b)(i) was satisfied on the basis of the general offer of sale on the web site, which indicated that the domain name had been registered with the intent to make a profit.
- *Mr Severiano Ballesteros Sota v Patrick Waldron.*[165] Following the institution of UDRP proceedings, the respondent offered to transfer the <seveballesterostrophy.com> domain name to the complainant in return for four complimentary tickets for the 'SEVE BALLESTEROS TROPHY' each year or 'some such other gesture'. The panellist held that the offer showed an attempt to transfer the domain name for more than the expenses incurred by the respondent in registering the domain name.
- *WFBQ-FM/WRZX-FM, WNDE-AM v ecorp.com a/k/a ecorp.*[166] The respondent offered to sell the disputed domain names to the complainant in exchange for two years of discounted advertising of the respondent's business ventures on the complainant's radio stations. The panellist held that this was evidence that the respondent registered the disputed domain names in bad faith pursuant to paragraph 4(b)(i).
- *John David Lyon v James Casserly.*[167] The respondent offered to transfer the <southsidejohnny.com> domain name for compensation including concert tickets, backstage passes, press passes and a web site development contract. The panellist held that the respondent's requests suggested that the disputed domain name was acquired primarily for the purpose of transferring it to the complainant for valuable consideration in excess of the respondent's out-of-pocket expenses.

[163] NAF Case No FA95636 (18 Oct 2000).
[164] NAF Case No FA95685 (6 Nov 2000).
[165] WIPO Case No D2001-0351 (18 June 2001).
[166] NAF Case No FA105852 (23 Apr 2002).
[167] NAF Case No FA193891 (29 Oct 2003).

- *Compelling Content Creators, Inc v Pavel Ushakov.*[168] After acquiring the disputed domain name, which had previously been registered by the complainant, the respondent contacted the complainant offering to return the domain name in exchange for either US$5,000 or the right to publish the complainant's 'adult' videos on its own competing 'adult' web site. The respondent also threatened to sell the domain name to a third party. The panellist held that the offer was evidence that the respondent had registered the domain name primarily for the purpose of selling it back to the complainant.
- *Martha Stewart Living Omnimedia Inc v Josh Gorton.*[169] On the same day that the respondent registered the disputed domain names he offered to sell them to the complainant, saying that otherwise they would be put up for public auction. In subsequent negotiations, the respondent disclaimed any intent to profit from the domain names, and indicated that his intention was to establish a fan site. The panellist held that the respondent's unsolicited offer to sell the domain names to the complainant, with a threat to sell to the highest bidders, was compelling evidence that paragraph 4(b)(i) had been made out.
- *Anheuser-Busch, Incorporated v Wesley Niegro.*[170] After registering the disputed domain names the respondent contacted the complainant to 'pitch' an unsolicited business idea connected with the disputed domain names. The respondent refused to transfer the domain names unless the complainant agreed to meet to discuss the business idea. The panellist held that the respondent's attempt to extract either a business meeting or financial compensation in return for the domain names was evidence that paragraph 4(b)(i) had been satisfied.
- *Takaso Rubber Products Sdn Bhd v Selim Tasci and Tasci Dis Tic Ltd STI.*[171] In negotiations with the complainant, the respondent, who was the sole distributor of the complainant's condom products in Turkey, offered to transfer the disputed domain names in return for a 10-year exclusive distribution agreement. The complainant refused to entertain such an agreement, as it was inconsistent with its usual business practices. The panellist held that the demand for a business concession amounted to an offer to transfer the disputed domain name for valuable consideration in excess of the respondent's documented out-of-pocket costs.
- *Vanounou Clothing Inc v Administrator c/o Domain Admin.*[172] The complainant approached the respondent with an offer to purchase the disputed domain name and the respondent replied with an offer to sell the domain name for US$5,000. On this basis, the panellist held that the respondent had acquired the domain name primarily for the purpose of re-selling it.

[7.17.5] Paragraph 4(b)(i): General Offers to Sell or Domain Name Auctions

Paragraph 4(b)(i) provides that the disputed domain name must be registered primarily for the purpose of selling the domain name 'to the complainant . . . or to a competitor of that complainant'. This suggests that general offers to sell the domain name, such as an auction of the domain name, will not automatically fall within the circumstance set out in paragraph 4(b)(i). As the panellist stated in *Investone Retirement Specialists, Inc v Motohisa Ohno*:[173]

168 NAF Case No FA346763 (17 Dec 2004).
169 WIPO Case No D2005-1109 (13 Dec 2005).
170 NAF Case No FA756894 (4 Sept 2006).
171 WIPO Case No D2006-1263 (16 Dec 2006).
172 NAF Case No FA861280 (26 Jan 2007).
173 WIPO Case No D2005-0643 (2 Aug 2005).

> It is clear that relief is predicated on the fact that the Respondent must be targeting the Complainant, directly or indirectly. The Policy does not contain any per se prohibition against trafficking in domain names. The mere fact that a Respondent may own a large number of domain names for purposes of resale does not constitute an abusive practice. The particular circumstances must be examined for each name in question, to determine whether the registration and use of the name run afoul of the bad faith provisions.

In certain circumstances, however, a general offer to sell a domain name may be no more than an attempt to induce the complainant, or a competitor of the complainant, to purchase the disputed domain name. In such circumstances, the offer to sell the domain name, even though it is not expressly directed to the complainant, may amount to evidence that paragraph 4(b)(i) has been made out. Furthermore, even if a general offer to sell the disputed domain name, such as an offer to sell the domain name at auction, does not fit within paragraph 4(b)(i), as the circumstances set out in paragraph 4(b) are not exhaustive, it may still constitute evidence of bad faith registration and use.

The following are examples of UDRP decisions which have held that a general offer to sell a disputed domain name, such as an auction, is evidence that the respondent registered the domain name primarily for the purpose of selling it to the complainant, or a competitor of the complainant, contrary to paragraph 4(b)(i). In such decisions, a key factor is commonly the extent to which the respondent was aware of the complainant's mark at the time the domain name was registered. Once it is established that the respondent was aware of the complainant's mark, then it is correspondingly easier to infer that a general offer to sell the disputed domain name was directed at the complainant or a competitor of the complainant.

- *CONOCO INC v RDH Computer Solutions.*[174] In response to a 'cease and desist' letter from the complainant, the respondent indicated that it would auction off the domain name, and subsequently offered it from a domain name auction site for a minimum bid of US$1,500. The panellist held that, although the respondent had not attempted to sell the domain name directly to the complainant, the respondent had registered it primarily to transfer it for consideration far in excess of its out-of-pocket expenses.
- *Prisma Presse v BUYDOMAINS.COM.*[175] The respondent offered the disputed domain name, <geomagazine.com>, for sale from its web site for a minimum price of US$650. The panellist held that the continuous unsolicited offer for sale of the domain name for more than the respondent's out-of-pocket expenses was evidence that paragraph 4(b)(i) had been made out.
- *Pfizer Inc v Jason Haft.*[176] The respondent placed the disputed domain name <viagra-fem.com> for auction on eBay with a reserve of US$5,000 on the day that it was registered. The panellist held that, as the respondent clearly had knowledge of the complainant's 'VIAGRA' mark, placing it on eBay was sufficient evidence that the domain name was obtained primarily for the purpose of selling it to the complainant, or the complainant's competitors, for valuable consideration in excess of the respondent's out-of-pocket expenses.
- *Echelon Corporation v RN WebReg.*[177] In response to a 'cease and desist' letter, the respondent offered to sell the disputed domain name, <digitalhome.com>, to the complainant

[174] WIPO Case No D2000-0960 (25 Oct 2000).
[175] WIPO Case No D2001-1073 (22 Oct 2001).
[176] WIPO Case No D2003-0133 (10 Apr 2003).
[177] WIPO Case No D2003-0790 (15 Dec 2003).

for US$20,000, with an indication that a lower amount might be acceptable. When the complainant rejected the offer, the respondent placed the domain name on a domain name auction site with an asking price of US$20,000. The majority of the panel inferred that the respondent knew of the complainant's registered mark, and that, as the respondent had sought a substantial sum of money for a domain name that was identical to the complainant's mark, paragraph 4(b)(i) had been made out. The dissenting panellist held that the domain name was a generic term and that the respondent had a legitimate inter est in reselling a generic domain name.

- *Iconcard SpA v Linecom.*[178] The disputed domain name resolved to a web site that advertised the domain name for sale by auction for a starting price of US$3,500. The panellist held that, as the respondent was most likely aware of the complainant's mark at the time the domain name was registered and the minimum amount set for the auction price was more than the respondent's out-of-pocket expenses, the circumstance set out in paragraph 4(b)(i) had been made out.
- *Cloer Elektrogeräte GmbH v Motohisa Ohno.*[179] The respondent registered the disputed domain name <cloer.com> and, in response to the complainant's offer to purchase the domain name, offered it for sale on a domain name auction site, and invited the complainant to make a bid. Meanwhile, the domain name was 'parked' with a 'parking service' which used the domain name to advertise products that competed with the complainant's products. The panellist found that the respondent's claim that he was unaware of the complainant's mark at the time the domain name was registered was unconvincing. Accordingly, the panellist held that offering the domain name for sale by auction was evidence that the respondent had registered the domain name with the primary purpose of targeting the complainant, with the goal of obtaining a higher price for the domain name than if the complainant were not targeted.

[7.17.6] Paragraph 4(b)(i): Good Faith Offers to Sell the Disputed Domain Name

Offers to sell the disputed domain name, including offers to sell it to the complainant, are not necessarily evidence of bad faith registration and use. In certain circumstances, a respondent may make a legitimate offer to sell the domain name for more than the out-of-pocket costs directly related to the domain name. In *Barlow Lyde & Gilbert v The Business Law Group,*[180] for example, the panellist stated that:

> Standing alone, there is nothing wrong with offering to sell a domain name at a high price. It is a very common business practice.

Similarly, in *Primal Quest, LLC v Gabriel Salas,*[181] the panellist stated that:

> The sale of domain names comprised of common terms, without some indicia of bad faith, is neither unlawful nor evidence of cybersquatting; it is only when domain names are registered for resale with knowledge that the names consist of another's trademark that the conduct fairly may be characterized as cybersquatting.

The circumstances in which a respondent may make a good faith offer to sell the disputed domain name include where the respondent has rights or legitimate interests in the domain

[178] WIPO Case No D2005-1115 (28 Mar 2006).
[179] WIPO Case No D2006-0026 (29 Mar 2006).
[180] WIPO Case No D2005-0493 (24 June 2005).
[181] WIPO Case No D2005-1083 (15 Dec 2005).

name, where the respondent was unaware of the complainant's mark at the time the domain name was registered and where the offer is made as part of good faith negotiations with the complainant. The following decisions illustrate the circumstances in which a respondent may make a legitimate offer to sell a disputed domain name.

- *General Machine Products Company, Inc v Prime Domains.*[182] The respondent registered the disputed domain name, <craftwork.com>, as part of its business of registering common dictionary words and generic terms as domain names for the purpose of resale. In response to an inquiry from the complainant, the respondent offered to lease the domain name to the respondent or to sell it for US$25,000. As the respondent was unaware of the complainant's mark at the time the domain name was registered and had a legitimate interest in the domain name, the panel held that the domain name had not been registered in bad faith.
- *Teradyne, Inc v 4Tel Technology 4.*[183] The respondent initially used the disputed domain name, <4tel.com>, in connection with its computer design business. After the respondent's business was wound up, however, the respondent offered to sell the domain name to the complainant for US$6,000. While the panellist held that the offer to sell the domain name was a bad faith use of the domain name, as the respondent had originally registered the domain name to reflect its business name there was no evidence that the respondent had registered or acquired the domain name for the purpose of selling it. As the domain name had not been registered for an illegitimate purpose, the panellist held that paragraph 4(b)(i) had not been made out.
- *Etam, plc v Alberta Hot Rods.*[184] The respondent used the disputed domain name, <tammy.com>, to direct Internet traffic to an 'adult entertainment' site. After the complainant contacted the respondent, the respondent offered to sell the domain name to the complainant for US $100,000. The panel held that the respondent had been making a legitimate use of the domain name before it had notice of the dispute. Moreover, given that the complainant was an English company, the panellist found that the respondent was unaware of the complainant's mark at the time the domain name was registered. Based on the respondent's legitimate interest in the domain name and its lack of awareness of the complainant's mark, the panel concluded that the offer to sell the domain name did not constitute bad faith.
- *Scholastic Inc v ScholasticAdvising.com.*[185] The respondent, which assisted students with scholarship applications, registered the disputed domain name <scholasticadvising.com> for use in its business. After receiving a 'cease and desist' letter, the respondent offered to sell the domain name to the complainant for US$8,000. Finding that the respondent had undertaken preparations to use the domain name in connection with a bona fide offering of services, the panel concluded that the respondent had not registered the domain name *primarily* for the purpose of selling it to the complainant.
- *The Coca-Cola Company v Svensson.*[186] The respondent registered the disputed domain name, <sprite.nu>, for his son to use to sell 'graphic buttons'. After being approached by the complainant, the respondent initially indicated that he would be willing to sell the domain name for US$1,100, and subsequently suggested that the domain name might be

[182] NAF Case No FA92531 (16 Mar 2000).
[183] WIPO Case No D2000-0026 (9 May 2000).
[184] WIPO Case No D2000-1654 (31 Jan 2001).
[185] WIPO Case No D2001-0946 (4 Nov 2001).
[186] NAF Case No FA103933 (27 Feb 2002).

rented to the complainant. The panellist concluded that the respondent had a legitimate interest in the domain name, which she held was a generic term. On this basis, the panellist held that the respondent had acquired the domain name for his son to use, and not for the purpose of transferring it to the complainant.

- *The Monticello Group, Ltd v Teletravel, Inc.*[187] The respondent initially used the disputed domain name <monticello.com> to provide vanity e-mail addresses for people living at Monticello, Illinois. Subsequently, the respondent offered the domain name for sale from a domain name auction site. The panel held that, as the respondent had a legitimate interest in the domain name and was unaware of the complainant's mark at the time the domain name was registered, the offer to sell the domain name did not constitute bad faith.
- *Cool Cat Fashion BV v Cool Kat.*[188] The respondent used the disputed domain name <cool-kat.com> to advertise its appliance repair business. In reply to a 'cease and desist' letter from the complainant, the respondent offered to sell the domain name for US$250,000. As the panellist held that the respondent had a legitimate interest in the disputed domain name, he held that the offer to sell the domain name in negotiations between the parties did not amount to bad faith.
- *Barlow Lyde & Gilbert v The Business Law Group.*[189] At the time the disputed domain name, <blg.com>, was registered, the complainant, which relied on common law rights, had not registered the 'BLG' service mark. After being contacted by the complainant, the respondent indicated that the current bid for the domain name was US $100,000 and that the complainant would need to better the bid. The panellist held that the evidence did not suggest that the respondent had registered the domain name with the intent to sell it to the complainant, as the respondent was unaware of the complainant's mark at the time the domain name was registered and there was a connection between the respondent's name ('The Business Law Group') and the disputed domain name.
- *Investone Retirement Specialists, Inc v Motohisa Ohno.*[190] The respondent, who resided in Tokyo, registered the domain name <investone.com> and offered it for sale from a domain name auction site. Given that, at the time the domain name was registered, the reputation of the US-based complainant was very limited, the panellist held that the respondent did not register the domain name for the purpose of selling it to the complainant.
- *Primal Quest, LLC v Gabriel Salas.*[191] The respondent claimed that he chose the disputed domain name, <primalquest.com>, to use in conjunction with an artistic web site. Subsequently, the respondent offered the domain name for sale for US$300,000. As the complainant first used the 'PRIMAL QUEST' mark after the domain name was registered, the panellist held that the respondent was unaware of the complainant's mark and that, even if the respondent had registered the domain name for the purpose of resale, in these circumstances that would not amount to bad faith.
- *Jodelle Boorady v Joann Harris Parker.*[192] The respondent registered the disputed domain name, <jodelle.com>, for use with her make-up business and the modelling services of

[187] WIPO Case No D2002-1157 (16 Apr 2003).
[188] WIPO Case No D2005-0385 (2 June 2005).
[189] WIPO Case No D2005-0493 (24 June 2005).
[190] WIPO Case No D2005-0643 (2 Aug 2005).
[191] WIPO Case No D2005-1083 (15 Dec 2005).
[192] NAF Case No FA600883 (14 Feb 2006).

her daughter, 'Jodelle'. In e-mail correspondence initiated by the complainant with a musician known as 'JODELLE', the potential sale of the domain name had been discussed. Finding that the respondent had registered the domain name in connection with a bona fide offering of goods or services, and that the complainant had no rights in the 'JODELLE' mark, the panellist held that there was no evidence that the respondent had registered the domain name with an intent to sell it to the complainant.

Registration to Prevent Use of Mark

[7.18] Registration to Prevent Use of Mark as Domain Name

The second non-exclusive circumstance that is evidence of bad faith registration and use, as set out in paragraph 4(b)(ii), is that:

> the domain name holder has registered the domain name in order to prevent the owner of the trademark or service mark from reflecting the mark in a corresponding domain name, provided that the domain name holder has engaged in a pattern of such conduct.

On its face, there are two parts to paragraph 4(b)(ii): that the respondent has registered the domain name to prevent the trade mark owner from reflecting the mark in a 'corresponding' domain name; and that the domain name holder has engaged in a 'pattern of conduct' of registering domain names to prevent trade mark owners from registering corresponding domain names.

There has been a difference of view among some panellists in relation to the meaning of the first part of paragraph 4(b)(ii). The difference relates to whether the phrase 'reflecting the mark in a corresponding domain name' should be interpreted as meaning 'any' corresponding domain name or as 'a particular' corresponding domain name. As the majority pointed out in *Bruce Springsteen v Jeff Burgar*,[193] adopting the first interpretation would mean that registering a domain name that corresponds to the complainant's mark in the .com gTLD would not prevent the complainant from registering the domain name in other gTLDs, such as .net. In this respect, the majority maintained that:

> In these circumstances what is meant by the word 'corresponding'? Nothing that has been done by Mr Burgar has prevented Bruce Springsteen's official website at <brucespringsteen.net> being registered and used in his direct interests. That is surely a 'corresponding domain name' for these purposes, as the expression 'corresponding domain name' clearly refers back to the words 'trade mark or service mark' rather than the domain name at issue referred to in the first line of paragraph 4(b)(ii).

Adopting the interpretation favoured by the majority in *Bruce Springsteen* would mean that paragraph 4(b)(ii) could rarely be satisfied, as it would almost always be possible for the trade mark owner to reflect its mark in a corresponding domain name in another gTLD or ccTLD. As the panellist explained in *Julie Brown v Julie Brown Club*:[194]

[193] WIPO Case No D2000-1532 (25 Jan 2001).
[194] WIPO Case No D2000-1628 (13 Feb 2001).

> The question is . . . what is meant by the phrase 'reflecting the mark in a corresponding domain name' and above all what is meant by 'a'. Does 'a' mean 'any', or does it mean 'a particular'? If the meaning is the former, as the Respondent asserts, then this form of bad faith can scarcely ever be established. There are many choices of suffix, just as there are of variations in hyphenation, spelling, added words and so on. On the Respondent's construction, there would only be bad faith if the domain squatter had sought to register them all.

In order to give paragraph 4(b)(ii) a degree of effective operation, then, the preferred view is to interpret the phrase 'a corresponding domain name' to mean 'a particular corresponding domain name' in the same gTLD as the domain name registered by the respondent. This interpretation has been adopted by subsequent panel decisions, including the decisions in *Celine Dion v Burgar*[195] and *Bradford & Bingley Plc v Registrant info@fashionID.com 987654321*,[196] which have rejected the view of the majority in *Bruce Springsteen*. Furthermore, the phrase 'corresponding domain name' is not confined to a domain name that is identical to the complainant's mark, but includes domain names that are 'confusingly similar'. As the panellist put it in *NBTY, Inc v LaPorte Holdings*:[197]

> The Panel's view is that the word 'corresponding' in paragraph 4(b)(ii) is a broad term, that encompasses a finding under paragraph 4(a)(i) that the disputed domain name is either identical or confusingly similar to the Complainant's mark.

Given that each domain name is necessarily unique, it could be argued that, adopting the view of the majority of panellists on the interpretation of 'a corresponding domain name' would mean that the registration of any domain name that is identical, or confusingly similar, to a trade mark will prevent the trade mark owner from reflecting the mark in the particular domain name. This approach would, however, effectively confer rights on trade mark owners over all identical or confusingly similar domain names, a result that is clearly not contemplated by the policy. To avoid this outcome, paragraph 4(b)(ii) requires the complainant to establish that the respondent has engaged in a 'pattern of conduct' of blocking the registration of trade marks as domain names. In practice, where the respondent has engaged in a relevant 'pattern of conduct', this will amount to evidence that the respondent has registered the domain name to prevent the trade mark owner from reflecting the mark as a domain name. As the panellist explained in *NBTY, Inc v LaPorte Holdings:*[198]

> The terms 'in order to prevent' in paragraph 4(b)(ii) . . . suggest that the Respondent *intentionally* registered the domain name to prevent the owner reflecting its mark in the domain name. This Panel's view is that the evidence of the Respondent's intent is found in its pattern of conduct. It is unlikely that a pattern of conduct under paragraph 4(b)(ii) would ever arise by accident, in circumstances where there are findings against the Respondent under paragraphs 4(a)(i) and (ii).

The importance of establishing that the respondent has engaged in a 'pattern of conduct' may be illustrated by the decision in *EMAP USA, Inc v Dick Jurgens*.[199] In that dispute, the respondent registered the disputed domain name <motorcyclist.com>, and the complainant, who published the 'MOTORCYCLIST' magazine, claimed that the registration and passive holding of the domain name prevented the complainant from reflecting its mark as a domain name. The complainant pointed to the fact that the respondent had repeatedly

195 WIPO Case No D2000-1838 (13 Feb 2001).
196 WIPO Case No D2002-0499 (16 Aug 2002).
197 WIPO Case No D2005-0835 (30 Sept 2005).
198 WIPO Case No D2005-0835 (30 Sept 2005).
199 WIPO Case No D2001-0311 (26 May 2001).

renewed the domain name after being put on notice of the complainant's rights as evidence that the respondent had engaged in a 'pattern of conduct'. In rejecting the contention that renewal of the one domain name could amount to a 'pattern of conduct', the panel stated that:

> We disagree with Complainant's conclusion. While it is true that Respondent's registration of the Domain Name prevents Complainant from reflecting its alleged mark in the .com domain, we do not agree that Respondent's renewal of its registration constitutes a 'pattern of such conduct' under the Policy. The interpretation urged by Complainant would mean that virtually every domain name holder would satisfy the pattern of conduct requirement, a result that clearly was not the intent of the Policy. Instead, the provision relied upon by Complainant seems to require some showing that the respondent has registered multiple domain names preventing others from using their marks as domain names. Here, there is no pattern of multiple registration. Rather, it appears that Respondent has registered only two domain names—the Domain Name at issue, and his personal name, <dickjurgens.com>.

The key element in establishing the circumstance set out in paragraph 4(b)(ii) is clearly the determination of whether the respondent has engaged in a 'pattern of conduct' which, as explained in *EMAP USA*, must involve the registration of multiple domain names. UDRP decisions have distinguished two forms of multiple registration that are relevant in determining whether there is a 'pattern of conduct': the registration of domain names resulting in multiple UDRP disputes; and a single UDRP dispute involving multiple domain names. The distinction between the two categories of disputes is drawn by the WIPO Consensus View, which provides as follows:

> A pattern of conduct can involve multiple UDRP cases with similar fact situations or a single case where the respondent has registered multiple domain names which are similar to known trademarks, however the registration of two domain names in the same case is not generally sufficient to show a pattern.[200]

In addition to the distinction identified in the Consensus View, a similar distinction has been drawn between multiple registrations of domain names that reflect the marks of different entities, known as 'horizontal' abusive registrations, and multiple registrations of domain names that reflect a single trade mark owner's mark, known as 'vertical' abusive registrations. The distinction between 'horizontal' and 'vertical' abusive registrations was explained by the panellist in *Smoky Mountain Knife Works v Deon Carpenter*,[201] in the following terms:

> First, a domain registrant can operate 'horizontally', targeting multiple entities, perhaps in multiple industries.
>
> Second, a domain registrant can operate 'vertically', targeting a single entity, but registering multiple domains which reflect either different aspects of the target's business, or different alphabetic variations of the target's trademark.

Both 'horizontal' and 'vertical' multiple registrations may be evidence of bad faith under paragraph 4(b)(ii). As the panellist stated in *Telstra Corporation Limited v Ozurls*:[202]

200 WIPO *Overview* para 3.3.
201 eRes Case No AF-230a; 230b (3 July 2000).
202 WIPO Case No D2001-0046 (20 Mar 2001).

A 'pattern of conduct' as required in Paragraph 4.b.(ii) typically involves multiple domain names directed against multiple Complainants, but may involve multiple domain names directed against a single Complainant.

The distinction between the two forms of multiple registration, which may amount to a 'pattern of conduct', are illustrated by the facts in *Smoky Mountain Knife Works* itself. In that dispute, the respondent registered the disputed domain names <smokymountainknives.com> and <smokymountainknife.net>. Subsequently, the respondent registered the domain name <smokymountainknifeworks.com>, which was identical to the complainant's 'SMOKY MOUNTAIN KNIFE WORKS' mark, but was not the subject of the instant dispute. The evidence indicated that the respondent had been a party to a previous UDRP proceeding, involving a different complainant, which had found that the respondent had registered the <blueridgeknife.com> domain name in bad faith.

The panellist first held that, on the basis of the earlier UDRP dispute, the respondent had engaged in a pattern of 'horizontally' abusive registrations. The panellist reached this conclusion even though the respondent had evidently targeted only two trade mark owners. On this point, the panellist stated that:

> A pattern of registering even a very large number of domains which are confusingly similar to registered trademarks may well be innocent, particularly where the trademarks in question are generic words, and the domain registrant is not a participant in the industry in which the generic word has acquired secondary meaning. A domain registrant should be held to a higher standard, however, when it registers multiple domains which target a single industry as to which the domain registrant has a sophisticated understanding. In the instant case, not only does the Respondent presumably have a sophisticated understanding of the cutlery industry, but Respondent has actively targeted several of its most prominent direct competitors. We find this pattern of conduct, even without more, to be persuasive evidence of bad faith.

Secondly, the panellist held that the respondent had also engaged in a pattern of 'vertically' abusive registrations. In determining whether there was a 'pattern of conduct', the panellist considered the registration of a domain name in the .net gTLD to be relevant, as well as the registration of minor alphabetical variations of the complainant's mark. More importantly, however, the evidence indicated that, after receiving actual notice of the complainant's mark, the respondent proceeded to register a domain name that was identical to the complainant's mark. In this respect, the panellist stated:

> It is evident to this Panel that Respondent, not content with registering a series of domains which 'fenced in' Complainant's existing domain, also undertook to preclude Complainant from using the domain as to which Complainant arguably has the strongest rights of all: Namely, a verbatim, letter-for-letter transcription of Complainant's full, four-word trademark. Such a pattern of systematically registering multiple variants of Complainant's trademark is a paradigmatic example of bad-faith registrations intended to foreclose Complainant from reflecting the mark in a corresponding domain name.

The differences in the circumstances involving multiple registrations of domain names where there are multiple UDRP disputes, on the one hand, and multiple domain names in a single dispute, on the other hand, mean that the two different forms of multiple registration are best considered separately.

[7.18.1] Paragraph 4(b)(ii): 'Pattern of Conduct' Arising from Multiple UDRP Disputes Involving Multiple Complainants

A history of UDRP complaints against the respondent, involving multiple complainants and multiple domain names, may amount to a 'pattern of conduct' of registering domain names to prevent trade mark owners from reflecting marks as domain names. If the respondent has been unsuccessful in a number of UDRP decisions involving the registration of famous or well-known marks, then the respondent will almost invariably be found to have engaged in a 'pattern of conduct'. Moreover, the respondent may be found to have engaged in a 'pattern of conduct' even if it has been successful in previous UDRP proceedings.

The following are representative examples of decisions in which panels have found that a history of involvement as a respondent in UDRP disputes is evidence of a 'pattern of conduct'.

- *Celine Dion v Burgar.*[203] The respondent, the notorious cybersquatter Jeff Burgar, registered the domain name <celinedion.com> as part of an extensive pattern of registering celebrity names in the .com gTLD and stockpiling the registrations. The panellist readily found that the respondent had engaged in a 'pattern of conduct' that prevented celebrities from securing the 'straightforward' .com registration of their personal name as a domain name, contrary to paragraph 4(b)(ii).
- *Collections Etc, Inc v Cupcake Patrol.*[204] The respondent registered the domain name <collectionetc.com> and the complainant pointed to the large number of domain name disputes in which the respondent—which was associated with the notorious cybersquatter, John Zuccarini—had been involved, as evidence of bad faith. The panellist held that the respondent's involvement in a 'course of operations' had 'probative value' in inferring that the domain name had been registered in bad faith.
- *PepsiCo, Inc v Phayze Inc.*[205] The respondent registered a number of domain names incorporating the famous 'PEPSI' mark, including <pepsibillions.com>, <pepsimash.com.com> and <pepsi-smash.com>. The complainant referred to seven previous UDRP decisions in which it had been found that the respondent had registered well-known marks in bad faith to argue that the respondent had engaged in a 'pattern of conduct'. The panellist held that the multiple registrations of domain names that were confusingly similar to the complainant's mark and the history of previous UDRP decisions amounted to a 'pattern of conduct'.
- *Arai Helmet Americas, Inc v Goldmark.*[206] The respondent registered the disputed domain name <arai.com>, which was identical to the complainant's 'ARAI' mark, and the complainant claimed that the respondent had engaged in a 'pattern of conduct' of bad faith registration on the basis of a similar, recently decided UDRP dispute. In that dispute, *CBS Broadcasting Inc v Goldmark*,[207] the panellist had held that that the respondent had registered the mark of a major radio station in bad faith. In the instant dispute, the panellist found the respondent's explanation of its intention in registering the domain name as 'implausible' as the explanation given in the previous dispute. The panellist accepted the

203 WIPO Case No D2000-1838 (13 Feb 2001).
204 WIPO Case No D2001-0305 (10 May 2001).
205 WIPO Case No D2003-0693 (2 Dec 2003).
206 WIPO Case No D2004-1028 (22 Jan 2005).
207 WIPO Case No D2004-0330 (24 June 2004).

prior proceedings as evidence that the respondent had engaged in a 'pattern of conduct' of abusive registration.

- *Société BIC v LaPorte Holdings, LLC.*[208] The respondent registered the disputed domain names, <bicgraphics.com> and <bicgraphicsusa.com>, and the complainant claimed that the respondent had engaged in a 'pattern of conduct' of abusive registration. As the respondent had been unsuccessful in at least four previous UDRP disputes, the panellist agreed that the respondent had engaged in such a 'pattern of conduct'. The fact that the respondent had successfully defended itself in one dispute, *Advanced Relational Technology, Inc v Domain Deluxe,*[209] was irrelevant in determining whether there had been a 'pattern of conduct'.
- *NBTY, Inc v LaPorte Holdings.*[210] The respondent registered the disputed domain names <puritianpride.com> and <purtian.com>, which corresponded to the complainant's marks, 'PURITAN' and 'PURITAN'S PRIDE'. Given that the respondent had been an unsuccessful respondent in prior UDRP decisions, including *Société BIC v LaPorte Holdings, LLC,*[211] the panellist held that it had engaged in a 'pattern of conduct'.
- *Mobile Communication Service Inc v WebReg, RN.*[212] The respondent registered the disputed domain name <mobilcom.com> and the complainant claimed that the respondent had engaged in a 'pattern of conduct' on the basis of its involvement in at least two previous UDRP disputes. After conducting a search of UDRP decisions, the panellist found that the respondent had been an unsuccessful respondent in at least five additional disputes. On this basis, the panellist concluded that the respondent was in the business of registering fanciful or expired domain names for the purpose of reselling them. The panellist held that, although it is perfectly legitimate to resell generic domain names, the registration of large numbers of domain names with no attention to whether they correspond to trade marks may amount to a 'pattern of conduct'. In this case, the panellist held that, given that the respondent had been an unsuccessful respondent in a number of UDRP decisions, it was engaging in a 'pattern of conduct' of depriving trade mark owners of the ability to register domain names reflecting their marks.
- *Grundfos A/S v Manila Industries Inc.*[213] The respondent, the registrant of the disputed domain name <grundos.com>, had a 'significant history' of being unsuccessful in UDRP proceedings in circumstances similar to the particular dispute. The panellist held that this was sufficient to establish a 'pattern of conduct' contrary to paragraph 4(b)(ii).
- *Patricia Ann Romance Peterson v Network Operations Center/Alberta Hot Rods.*[214] The respondent, which was associated with Jeff Burgar, registered the domain name <trisharomance.com>, which reflected the professional name of a well-known Canadian artist, as part of a persistent pattern of registering the names of celebrities and well-known individuals as domain names. Given the extensive history of the respondent in UDRP proceedings, the panellist readily concluded that it had engaged in a 'pattern of conduct' of abusive registrations.

208 WIPO Case No D2005-0342 (16 June 2005).
209 WIPO Case No D2003-0567 (13 Oct 2003).
210 WIPO Case No D2005-0835 (30 Sept 2005).
211 WIPO Case No D2005-0342 (16 June 2005).
212 WIPO Case No D2005-1304 (24 Feb 2006).
213 WIPO Case No D2006-1295 (6 Dec 2006).
214 WIPO Case No D2006-1431 (3 Jan 2007).

[7.18.2] Paragraph 4(b)(ii): 'Pattern of Conduct' Arising from Multiple Registrations of Domain Names Reflecting the Complainant's Mark or Marks

Multiple 'vertical' registrations of domain names, meaning the registration of multiple domain names reflecting the mark of a single trade mark owner, may amount to a 'pattern of conduct' of abusive registrations, just as much as a history of registering domain names that reflect the marks of multiple trade mark owners. Furthermore, multiple registrations of domain names that reflect a number of different marks held by a single trade mark owner may also amount to a 'pattern of conduct'. The following are examples of decisions that have held that multiple registrations of domain names that are variations of a single mark, or multiple marks held by a single trade mark owner, constitute a 'pattern of conduct' contrary to paragraph 4(b)(ii).

- *Yahoo! Inc v Eitan Zviely.*[215] The respondent registered 37 domain names that included the complainant's 'YAHOO' mark. Moreover, the evidence indicated that the respondent had registered domain names that included 50 other trade marks. The panellist held that the respondent had engaged in a 'pattern of conduct' of registering hundreds of domain names based on trade marks.
- *Harcourt, Inc v Jeff Fadness.*[216] The respondent registered the disputed domain names <firerescuemagazine.com>, <clinicalpsychiatrynews.com>, <internalmeducinenews.com>, <obgynnews.com>, <pediatricnews.com> and <skinandallergynews.com>, which were identical to the complainant's six marks. The panellist held that one instance of registering multiple domain names that correspond to multiple marks held by the complainant was evidence of a 'pattern of conduct' under paragraph 4(b)(ii).
- *Gamesville.com, Inc v John Zuccarini.*[217] The respondent, a notorious cybersquatter, registered the domain names <gamesvilles.com>, <gamesville.com>, <gamesvill.com>, gamesvlle.com>, <gamevile.com> and <gamevill.com>, which reflected the complainant's 'GAMESVILLE' mark. The registration of the multiple domain names, together with the respondent's previous UDRP history, led the panellist to conclude that the respondent had engaged in a 'pattern of conduct' contrary to paragraph 4(b)(ii).
- *Telstra Corporation Limited v Ozurls.*[218] The respondent registered 15 domain names that incorporated the complainant's 'TELSTRA' mark, including <I-telstra.com>, <telstraaustralia.net>, <telstramobiles.com> and <telstraweb.com>. The panellist held that the registration of 15 domain names that included the complainant's mark together with other generic words obviously involved a 'pattern of conduct' preventing the complainant from reflecting its mark in corresponding domain names.
- *Ourisman Dodge, Inc v Ourisman 'Okie doke' Dodge dot com.*[219] The respondent registered the disputed domain names <ourismandodge.com>, <ourismandodge.org> and <ourismandodge.net>, which reflected the complainant's mark 'OURISMAN DODGE'. The panellist held that the three registrations of the complainant's service mark constituted a 'pattern of conduct' pursuant to paragraph 4(b)(ii).
- *Dell Computer Corporation v Logo Excellence.*[220] The respondent registered the disputed domain names <dellconnect.com>, <delletrade.com>, <delle-trade.com>,

[215] WIPO Case No D2000-0273 (14 June 2000).
[216] NAF Case No FA95247 (9 Aug 2000).
[217] NAF Case No FA95294 (30 Aug 2000).
[218] WIPO Case No D2001-0046 (20 Mar 2001).
[219] WIPO Case No D2001-0108 (22 Mar 2001).
[220] WIPO Case No D2001-0361 (7 May 2001).

<delleway.com>, <delle-way.com>, <dellfirst.com>, <dellmight.com>, <dellpower.com>, <dellwebsite.com> and <dellwork.com>, which reflected the complainant's 'DELL' marks. The evidence indicated that the respondent had intentionally registered a large number of other domain names that incorporated well-known marks for the purpose of resale. The panellist held that the repeated registration of multiple domain names that incorporated well-known marks was a 'pattern of conduct'.

- *Allianz AG And Dresdner Bank AG v Comofer S L.*[221] The respondent registered the disputed domain names <allianz-dresdnerbank.com>, <allianz-dresdner.org> and <allianz-dresdnerbank.info>, which reflected the complainant's 'ALLIANZ' and 'DRESDNER' marks. The first two disputed domain names had been registered prior to a merger of Alliance AG and Dresdner Bank AG, while the third disputed domain name was registered after the merger. The panellist held that, given the registration of the two previous domain names, the registration of the third domain name was part of a 'pattern of conduct' contrary to paragraph 4(b)(ii).
- *Royal Bank of Canada v Henry Chan.*[222] The respondent registered the disputed domain names <rbcroyal.com>, <rbccenturabank.com> and <rbcrewards.com>, which corresponded to the complainant's 'RBC ROYAL BANK', 'RBC CENTURA' and 'RBC REWARDS' marks. The panellist held that the registration of three domain names that were confusingly similar to the complainant's marks was evidence of a 'pattern of conduct' of intentionally selecting domain names to prevent the complainant from reflecting its marks in corresponding domain names.
- *SWATCH AG v Stefano Manfroi.*[223] The respondent registered the disputed domain names <swatchnews.com>, <swatchnews.net>, <swatchnews.org>, <swatchresearch.com>, <swatchresearch.net>, <swatchresearch.org>, <swatchdiscovery.com>, <swatchdiscovery.net> and <swatchdiscovery.org>, which reflected the complainant's 'SWATCH' mark. The evidence indicated that the respondent had also registered the domain names <motorolanews.com> and <motorolaresearch.com>, which reflected the unrelated 'MOTOROLA' mark. The panel held that the registration of the multiple domain names reflecting the 'SWATCH' mark, together with the registration of the domain names relecting the 'MOTOROLA' mark, amounted to a 'pattern of conduct'.
- *Ruth Vending, Inc v Profits On The Web.*[224] The respondent registered the domain names <tuffronts.com> and <tuffronts.net>, which reflected the complainant's 'TUFFRONTS' mark. The panellist held that the respondent had engaged in a 'pattern of conduct' of abusive registration, despite the fact that there was no previous history of involvement in UDRP disputes, and it had registered only two infringing domain names.
- *Felsina SpA v Hannah's Recipes Inc.*[225] The respondent registered the disputed domain names <felsina.biz>, <felsina.com> and <felsina.info>, and the evidence indicated that the respondent had been an unsuccessful respondent in one previous UDRP decision and in two cases under the .info *Sunrise Challenge Policy*. The panellist held that the registration of the 'FELSINA' domain names would have amounted to a 'pattern of conduct', but that the history of unsuccessful involvement in domain name disputes, including those under the *Sunrise Challenge Policy*, reinforced this conclusion.

[221] WIPO Case No D2001-1398 (12 Feb 2002).
[222] WIPO Case No D2003-0031 (5 Mar 2003).
[223] WIPO Case No D2003-0802 (20 Jan 2004).
[224] NAF Case No FA646989 (7 Apr 2006).
[225] WIPO Case No D2006-1444 (1 Feb 2007).

[7.18.3] Paragraph 4(b)(ii): Circumstances in which Multiple Registrations do not Amount to a 'Pattern of Conduct'

In certain circumstances, the registration of multiple domain names will not constitute a 'pattern of conduct' within paragraph 4(b)(ii). As the WIPO Consensus View explains, the registration of two domain names in the one dispute is not generally sufficient to amount to a 'pattern of conduct'. Furthermore, the registration of multiple generic terms as domain names will usually not be an illegitimate 'pattern of conduct' under paragraph 4(b)(ii). The following decisions illustrate the circumstances in which multiple registrations of domain names will not constitute a 'pattern of conduct'.

- *Home Interiors & Gifts, Inc v Home Interiors.*[226] The respondent registered the disputed domain names <homeinteriors.net> and <homeinteriorsandgifts.com>, which corresponded to the complainant's marks. The complainant alleged that the registration of the two domain names constituted a 'pattern of conduct' under paragraph 4(b)(ii). The panellist, however, did not consider that the registration of two domain names 'rises to the level of a pattern of conduct'.
- *Ingersoll-Rand Co v Frank Gully.*[227] The respondent registered the disputed domain names <ingersoll-rand.net>, <ingersoll-rand.org> and <ingersollrand.org>. The panellist held that the registration of only three confusingly similar or identical domain names was not a 'pattern of conduct'.
- *Bellevue Square Managers, Inc v Redmond Web.*[228] The respondent registered the disputed domain name <bellevuesquare.com>, and had also registered the domain names <redmondtowncenter.com>, <studio1design.com> and <ctstbm.com>. The panellist held that as only one of the other domain names, <redmondtowncenter.com>, was sufficiently close to the instant dispute to be relevant, there was no 'pattern of conduct' under paragraph 4(b)(ii).
- *William Hill Organization Limited v Fulfillment Management Services Limited.*[229] The respondent registered the disputed domain name <betwilliamhill.net>, but the evidence indicated that it had also registered the <betwilliamhill.co.uk>. The panellist held that the fact that the respondent had registered one other similar domain name, which it did not intend to use, did not establish a 'pattern of conduct'.
- *Laboratoire Pharmafarm (SAS) v M Sivaramakrishan.*[230] The respondent registered the disputed domain name <pharmafarm.com>, and had also registered the <21st-century-box.com> domain name. The panellist held that the registration of one other domain name, which the respondent did not intend to use, did not establish a 'pattern of conduct'.
- *Hexion Specialty Chemicals GmbH v Pacific Webs Pty Ltd.*[231] The respondent registered the disputed domain name <bakelite.com>, and the evidence indicated that it operated a web site selling generic domain names. The panellist approved the statement of the panellist in *Mobile Communication Service Inc v WebReg, RN*[232] that it is not necessarily 'bad faith to resell domain names that incorporate common dictionary terms if the respondent was unaware of complainant's trademark rights at the time of registration'. In this dispute,

[226] WIPO Case No D2000-0010 (7 Mar 2000).
[227] WIPO Case No D2000-0021 (9 Mar 2000).
[228] WIPO Case No D2000-0056 (27 Mar 2000).
[229] WIPO Case No D2000-0826 (17 Sept 2000).
[230] WIPO Case No D2001-0615 (6 June 2001).
[231] WIPO Case No D2005-1272 (26 Apr 2006).
[232] WIPO Case No D2005-1304 (24 Feb 2006).

unlike in the *WebReg* decision, the respondent had not directly offered the domain name for sale to the complainant, and there was no evidence that the respondent had been involved in previous UDRP proceedings. The panellist therefore held that there was no illegitimate 'pattern of conduct' and that the disputed domain name had not been registered in bad faith.

- *Ferrero SpA v Dante Ferrarini.*[233] The respondent registered the disputed domain name <ilovenutella.com>, and the evidence indicated that he had also registered <ilovenutella.co.uk>. The panellist held that, as the respondent had registered only two domain names, paragraph 4(b)(ii) was not applicable.
- *Walsh Bishop Associates, Inc v Honggi, Honggi Kim.*[234] The respondent registered the disputed domain name <with.com>, and the complainant claimed that this prevented it from reflecting its 'WITH' mark as a domain name. The evidence indicated that the respondent had registered a large number of generic terms as domain names, including <just.com>, <untrue.com>, <like.net>, <when.net> , <dislike.net> and <feel.net>. Although one of the respondents had been an unsuccessful respondent in a previous UDRP dispute, *Prada SA v Mr Juno Kim,*[235] the panellist held that the disputed domain name had not been registered to prevent the complainant from reflecting its mark as a domain name, and that the one previous dispute did not constitute a 'pattern of conduct'

Disrupting Competitor's Business

[7.19] Registration to Disrupt Competitor's Business

The third non-exclusive circumstance that is evidence of bad faith registration and use, as set out in paragraph 4(b)(iii), is that:

> the domain name holder has registered the domain name primarily for the purpose of disrupting the business of a competitor.

As explained by the panel in *RuggedCom, Inc v LANstore, Inc,*[236] paragraph 4(b)(iii) contains the following three elements, which must all be satisfied for this circumstance to be made out:

> - the Respondent knew of the Complainant's business. For the Respondent to have the primary purpose of disrupting its competitor, it is obvious that the Respondent must have known of the Complainant's business. (It is not necessary that the Respondent knew or believed that the Complainant has rights in a mark. It is possible a respondent might seek to disrupt its competitor without a complainant having rights in a mark. But a respondent must at least be aware of the complainant's competing business to have any intention to disrupt it.)
> - the 'primary purpose' of the Respondent's registration of the disputed domain name was to disrupt the Complainant's business. From the terms of paragraph 4(b)(iii), it is not necessary to show that the Complainant's business was actually disrupted—only that the Respondent's primary purpose is disruption. Also, the intent to disrupt must be the 'primary' purpose. No finding

233 WIPO Case No D2006-1163 (23 Nov 2006).
234 NAF Case No FA907521 (6 Mar 2007).
235 WIPO Case No D2003-0757 (13 Nov 2003).
236 WIPO Case No D2005-0760 (15 Nov 2005) para 6.

could be made under paragraph 4(b)(iii) if the Respondent's purpose in registering the domain name is primarily and legitimately connected to its own business, or had some other primarily legitimate purpose, even if the Respondent incidentally intended to disrupt a competitor.
- the Respondent and the Complainant are competitors.

First, as paragraph 4(b)(iii) refers to registration of the domain name for the purpose of disrupting the business of a competitor, it is axiomatic that the domain name cannot be registered for this purpose unless the respondent is aware of the complainant's business. A finding that the respondent had actual or imputed notice of the complainant's mark will often be important in determining whether the disputed domain name has been registered contrary to paragraph 4(b)(iii). The circumstances in which notice of the complainant's mark may be inferred or imputed to the respondent are explained at [7.12]. Nevertheless, as the panel in *RuggedCom* pointed out, the complainant only needs to establish that the respondent was aware of the business of a competitor, not that it was specifically aware of the complainant's mark.

The circumstances in which knowledge of the complainant's business may be inferred is illustrated by the decision in *Two Systems Enterprises Co, Ltd v Sonie's Creations.*[237] In that dispute, the respondent used the disputed domain name <musiclegs.com> for a variety of purposes, including an online catalogue for lingerie, while the complainant used its 'MUSIC LEGS' mark in connection with women's hosiery and stockings. The central issue in the dispute was whether the respondent knew of the complainant's business at the time the domain name was registered. Although there was no direct evidence that the respondent knew of the complainant's mark, the panellist inferred that it knew of the mark from the following circumstances: the respondent was a Canadian business; the complainant's mark was registered in Canada; the complainant sold its 'MUSIC LEGS' hosiery in Canada and elsewhere; the respondent was in the hosiery business in Canada and sold, through its web site, hosiery in competition with the complainant; and the respondent failed to provide any explanation or rationale for its use of the disputed domain name. As the respondent was using the domain name in competition with the complainant, the panellist held that it had been registered primarily to disrupt the complainant's business contrary to paragraph 4(b)(iii).

Secondly, the requirement that the domain name must have been registered 'primarily' for the purpose of disruption means that the respondent must have had the intention of disrupting the business of a competitor at the time the domain name was registered. An intention to disrupt a competitor's business is commonly inferred where the respondent had notice of the complainant's mark at the time the domain name was registered and where the domain name has been actually used to disrupt the business of a competitor. As the panel explained in the *RuggedCom* decision, however, it is not necessary for the domain name to be used to disrupt the complainant's business for paragraph 4(b)(iii) to be made out.

The approach to determining the 'primary purpose' of registration, where a domain name may have been registered for multiple purposes, is illustrated by the *RuggedCom* decision itself. In that dispute, the respondent used the disputed domain name, <rugged-com.net>, to redirect Internet users to the respondent's web site at <lanstore.com>, where the respondent offered products in direct competition with the complainant. After being contacted by the complainant, the respondent used the domain name purely as an 'information' site, but included a link to its <lanstore.com> site and continued to use the com-

[237] eRes Case No AF-0911 (7 Sept 2001).

plainant's domain name, <ruggedcom.com>, as a metatag. In response to the complaint, the respondent claimed that it had registered the domain name solely to provide information relating to industrial Ethernet, and not to disrupt the complainant's business. The panel, however, concluded that, while the provision of information might have been 'a purpose' of registering the domain name, it was not the 'primary purpose'. In determining that the primary purpose was to disrupt the business of the complainant, the panel found the use of the complainant's <ruggedcom.com> as a metatag to be decisive: it was evidence both that the respondent knew of the complainant's business and that the domain name had been registered to disrupt the complainant's business by diverting Internet users from the complainant's site.

As is apparent from the approach adopted in *RuggedCom,* the way in which the respondent has used the disputed domain name is often determinative of whether the domain name has been registered for the purpose of disrupting a competitor's business. The main circumstances in which UDRP panels have found that a disputed domain name has been registered for the primary purpose of disrupting a competitor's business are dealt with at [7.19.1].

Thirdly, the domain name must have been registered for the purpose of disrupting the business of a 'competitor'. The wording of paragraph 4(b)(iii) means that the respondent must be an 'actual competitor' of the complainant, and not merely a 'potential competitor'. In *Cybertill Limited v Darren Casey,*[238] for example, the respondent registered the disputed domain name <cybertill.com> and admitted that the domain name had been registered because the complainant, which owned the 'CYBERTILL' mark, had not registered the domain name. Although the domain names were not actively used, the complainant argued that the respondent was a 'potential competitor' who had registered the domain names with the intention of misleadingly diverting the complainant's consumers to the respondent's web site. Given the derogatory statements made by the respondent in e-mail communications with the complainant, the panellist concluded that the domain names had been registered for the purpose of disrupting the complainant's business. As there was no evidence that the respondent was a competitor of the complainant, however, the panellist held that paragraph 4(b)(iii) had not been made out. At the same time, given that the circumstances set out in paragraph 4(b) are expressed to be 'without limitation', the panellist held that the evidence that the domain name had been registered to disrupt the complainant's business indicated that the domain name had been registered in bad faith, despite the fact that the respondent and complainant were not competitors.

There is an unresolved division of opinion among UDRP panellists about what is meant by the term 'competitor' for the purpose of paragraph 4(b)(iii). On one view, the paragraph only applies where the respondent is in competition with the complainant in the provision of goods or services. On the other view, however, a 'competitor' is interpreted more broadly than a commercial or business competitor, to encompass any person or entity that has an interest in 'opposition' to the complainant. The different views have significant practical consequences. As it is common for panellists who adopt the broader view also to regard the operation of a criticism site as insufficient to establish rights or interests in a disputed domain name, proponents of this view are likely to conclude that a domain name that resolves to a criticism site was registered to disrupt the business of a competitor contrary to paragraph 4(b)(iii). This division of opinion is explained further at [7.19.2].

[238] WIPO Case No D2005-0460 (22 June 2006).

[7.19.1] Paragraph 4(b)(iii): 'Disruption' of the Business of a Competitor

A finding that the disputed domain name has been registered primarily for the purpose of disrupting the business of a competitor will commonly be inferred from the circumstances in which the domain name is used. In practice, the circumstances leading to a conclusion that the domain name has been registered contrary to paragraph 4(b)(iii) will often overlap with those that result in a contravention of paragraphs 4(b)(ii) or 4(b)(iv). The main circumstances in which panels have held that the respondent has registered the domain name with the intention of disrupting a competitor's business are: use of the domain name to promote the respondent's business where the respondent is a competitor of the complainant; use of the domain name to divert Internet users to the sites of business competitors of the respondent; and 'passive holding' of a domain name that is identical, or confusingly similar, to the complainant's mark where the respondent is a competitor of the complainant. In each of these circumstances, evidence that the respondent registered the domain name with actual or imputed notice of the complainant's mark will generally lead to the conclusion that the domain name has been registered in breach of paragraph 4(b)(iii).

The following decisions illustrate the main circumstances in which panels have held that a domain name has been registered primarily to disrupt the business of a competitor.

Use of the disputed domain name to promote the respondent's business which competes with the business of the complainant

While using a domain name for the purpose of legitimate competition in the provision of goods or services is permissible, the use of a competitor's mark as a domain name to attract the competitor's customers to the respondent's commercial site is evidence of bad faith registration for the purpose of disrupting the complainant's business. The following are examples of decisions that have held that the use of a disputed domain name to promote or sell goods or services in competition with the complainant trade mark owner is evidence of bad faith registration under paragraph 4(b)(iii).

- *EthnicGrocer.com, Inc v LatinGrocer.com.*[239] The respondent used the disputed domain names, including <ethnicgrocery.com> and <ethnicgroceries.com>, to divert users to its web site at <latingrocer.com>, which offered Latin foods and products. The panellist held that the use of the domain names, and the minor variation from the complainant's 'ETHNICGROCER.COM' mark, was evidence that the respondent had registered the domain names primarily for the purpose of disrupting the complainant's business.
- *Southern Exposure v Southern Exposure, Inc.*[240] The respondent, which specialised in the production of films and videotapes, used the disputed domain name, <southernexposure.net>, to divert Internet users to its <floridafilmandtape.com> site. The disputed domain name was identical to the complainant's 'SOUTHERN EXPOSURE' mark, which was registered in connection with films and videotapes. The panellist concluded that, as the respondent was also in the film industry, it knew or should have known of the complainant's mark. As the domain name was used to divert Internet users to the respondent's own web site, the panellist held that the domain name had been registered to disrupt the complainant's business contrary to paragraph 4(b)(iii).

[239] NAF Case No FA94384 (7 July 2000).
[240] NAF Case No FA94864 (18 July 2000).

- *Lubbock Radio Paging v Venture Tele-Messaging.*[241] The respondent used the domain names <lubbockradiopagingservice.com> and <lubbockradiopaging.com> to offer sales information for pagers, paging services and communications services in competition with the complainant. As the parties were in competing lines of business in the same market area, the panellist held that the domain names had been registered to disrupt the business of a competitor under paragraph 4(b)(iii).
- *General Media Communications, Inc v Emilio Lopez.*[242] The respondent used the disputed domain name <pennthouse.com> for an active 'adult' entertainment site in competition with the complainant's famous 'PENTHOUSE' sites. The panellist held that an intention to disrupt the complainant's business could be inferred from the respondent's presence in the Internet adult-entertainment industry, his knowledge of the complainant's famous 'PENTHOUSE' mark, and the respondent's creation of an active adult entertainment web site intending to divert consumers looking for the complainant's <penthouse.com> web site to the respondent's <pennthouse.com> site.
- *Dollar Rent A Car Systems, Inc v Patrick Ory.*[243] The respondent used the disputed domain name <dollar-rent-a-car.com> to redirect users to its <onetravel.com> web site, which offered the opportunity to book air fare, hotel accommodation and rental cars in competition with the complainant. The <dollar-car-rentals.net> domain name offered the opportunity to book air fare, hotel accommodation and rental cars directly from the web site. The panellist held that the respondent was disrupting the business of a competitor by offering similar services to those offered by the complainant, which was evidence of bad faith under paragraph 4(b)(iii).
- *Southwest Airlines Co v TRN.*[244] The disputed domain name <southwest-airline.com> was used to provide travel services in competition with the complainant, one of the largest airlines in the United States, including links to discount travel services and 'cheap flights'. As the respondent was depriving the complainant of sales revenue by diverting Internet users from the complainant's web site, the panellist concluded that the domain name had been registered to disrupt the complainant's business in violation of paragraph 4(b)(iii).
- *SA Bendheim Co, Inc v Hollander Glass.*[245] The respondent used the disputed domain name <restorationglass.com> to sell stained glass products in competition with the complainant. The panellist held that the respondent had intentionally used the complainant's 'RESTORATION GLASS' mark as a domain name primarily to disrupt the business of the complainant contrary to paragraph 4(b)(iii).
- *Online Vacation Center, Inc v Domain Active Pty Ltd.*[246] The respondent used the disputed domain name <onlinevactioncenter.com> for a web site that offered travel packages and services in competition with the complainant. The panellist held that redirecting users to a competing travel services web site was evidence of bad faith registration contrary to paragraph 4(b)(iii).
- *Red Lion Hotels Corporation v Damir Kruzicevic.*[247] The disputed domain name <redlionhotels.com> was used to redirect users to the respondent's <cheaprooms.com> web site, which offered reservations for hotels that competed with the complainant, as well

[241] NAF Case No FA96102 (23 Dec 2000).
[242] NAF Case No FA96776 (19 Apr 2001).
[243] NAF Case No FA125229 (14 Nov 2002).
[244] WIPO Case No D2002-0893 (18 Nov 2002).
[245] NAF Case No FA142318 (13 Mar 2003).
[246] NAF Case No FA751635 (6 Sept 2006).
[247] WIPO Case No D2006-1051 (13 Oct 2006).

as for the complainant's hotels. The panellist held that the only credible explanation of the respondent's behaviour was that the domain name had been registered intentionally to disrupt the complainant's business contrary to paragraph 4(b)(iii).

- *Briefing.com, Inc v Marco Dalonzo.*[248] The disputed domain name <otcbriefing.com> resolved to the respondent's web site that offered financial analysis and market commentary in competition with the complainant. The panellist accepted this as evidence of an intention to disrupt the complainant's business for the respondent's benefit contrary to paragraph 4(b)(iii).

Use of the disputed domain name to divert Internet users to the sites of the complainant's competitors

Many panels have held that the use of a disputed domain name to divert Internet users to the sites of competitors of the complainant constitutes evidence of registration for the purpose of disrupting a competitor's business under paragraph 4(b)(iii). Those panels that have held that this practice is evidence of bad faith registration contrary to paragraph 4(b)(iii), even though the respondent is not a direct commercial or business competitor of the complainant, would seem to have adopted the broader view of the meaning of the term 'competitor' in paragraph 4(b)(iii), explained at [7.19.2], although this is not always made explicit in the decision itself. The following are examples of decisions that have held that using a disputed domain name to divert users to the sites of the complainant's competitors is evidence of bad faith registration under paragraph 4(b)(iii).

- *Disney Enterprises, Inc v Nicolas Noel.*[249] The respondent used the disputed domain name <euro-disney.com> to redirect Internet users to the <futurosope.com> site, which related to a theme park operating in competition with the complainant, which owned the famous 'DISNEY' mark. The panellist held that it was reasonable to infer that the respondent had registered the domain name either to disrupt or to create confusion for the complainant's business in bad faith pursuant to paragraph 4(b)(iii).
- *Travant Solutions, Inc v Don Cole.*[250] The respondent used the disputed domain name <travant.com> to redirect Internet users to its web site at <pinnica.com>, which advertised computer software services in competition with the complainant. The panellist held that, as the respondent knew, or should have known, of the complainant's 'TRAVANT' mark at the time the domain name was registered, and as the domain name was used to divert Internet users, the domain name had been registered contrary to paragraph 4(b)(iii).
- *Expedia, Inc v Albert Abramov.*[251] The respondent used the disputed domain name <expediashipping.com> to divert Internet users to the respondent's web site at <pricebreakshipping.com>, which imitated the colour scheme, arrangement of elements and logo of the complainant's <expedia.com> web site. The disputed domain name resolved to a search engine web site that featured links to third-party web sites that offered travel agency services in competition with the complainant's business. The panellist held that the respondent's use of the domain name to operate a search engine site with links to the services of the complainant's competitors was bad faith registration contrary to paragraph 4(b)(iii).

[248] NAF Case No FA888093 (27 Feb 2007).
[249] NAF Case No FA198805 (11 Nov 2003).
[250] NAF Case No FA203177 (6 Dec 2003).
[251] NAF Case No FA817182 (20 Nov 2006).

- *Marriott International, Inc v Hong Kong Names LLC.*[252] The disputed domain name <marriottvacations.com> resolved to a pay-per-clicks web page with links to the complainant's competitors in the hotel and travel industries. The panellist held that redirection of Internet users in this way was evidence that the domain name had been registered in bad faith to disrupt the business of the complainant according to paragraph 4(b)(iii).
- *Ebynum Enterprises, Inc v Tag-Board.com Corporation.*[253] In an unusual dispute, the respondent managed to acquire registration of the disputed domain name, <tag-board.com>, which the complainant used in connection with a web-hosting service. The respondent permitted the domain name to point to the complainant's web site but made a number of threats, including that the domain name would be redirected to competitors' sites unless the complainant complied with its demands, such as the payment of money and inclusion of advertising on the complainant's web site. The panellist held that the respondent was using the domain name intentionally to disrupt the complainant's business, amounting to evidence of a breach of paragraph 4(b)(iii).

'Passive holding' of a domain name that is identical, or confusingly similar, to the complainant's mark where the respondent is a competitor of the complainant and registered the domain name with actual or imputed notice of the complainant's mark.

A number of decisions have held that registration of a domain name that is identical, or confusingly similar, to the complainant's mark, together with 'passive holding' of the domain name, is evidence that the domain name has been registered for the purpose of disrupting the complainant's business under paragraph 4(b)(iii). While it is relatively uncontroversial to infer that 'passive holding' may amount to bad faith registration under paragraph 4(b)(iii) where the respondent is a direct commercial or business competitor of the complainant, the potential for the term 'competitor' to be interpreted as anyone acting in 'opposition', as explained at [7.19.2], would give this circumstance an extremely broad application. The following decisions have held that 'passive holding' is evidence of bad faith registration contrary to paragraph 4(b)(iii).

- *Surface Protection Industries, Inc v The Webposters.*[254] The respondent, a direct competitor of the complainant in the business of selling paints, registered the disputed domain name <mannbrothers.com>, but made no active use of the domain name. Given the competitive relationship between the parties, and their geographical proximity, the panellist inferred that the respondent had registered the domain name with an intent of disrupting, or causing injury of some sort, to the complainant's business contrary to paragraph 4(b)(iii).
- *Wells Fargo and Company v Unasi Management Inc.*[255] The respondent registered 15 domain names that were identical, or confusingly similar, to the complainant's 'WELLS FARGO' mark, including <ewellsfargofinancial.com>, <wellsdfargofinancial.com> and <wellsfargofiinancial.com>, but made no active use of the domain names. The panellist held that registration of confusingly similar domain names, with knowledge of the complainant's rights in the corresponding mark, is evidence of bad faith registration pursuant to paragraph 4(b)(iii).

[252] NAF Case No FA803966 (7 Nov 2006).
[253] NAF Case No FA817104 (15 Dec 2006).
[254] WIPO Case No D2000-1613 (5 Feb 2001).
[255] NAF Case No FA471513 (17 June 2005).

- *The Cheesecake Factory Inc v Say Cheesecake.*[256] The respondent registered the disputed domain name <saycheesecakefactory.com>, and was also the registrant for the <fantasyweddingcakes.com> web site, which advertised 'Wedding Cake Concepts owned by Vickie Sullivan'. No active use was made of the disputed domain name. The panellist held that the retention and non-use of the domain name, together with the respondent's prior knowledge of the complainant's 'CHEESECAKE FACTORY' mark, was passive holding contrary to paragraph 4(b)(iii).
- *The Highland Street Connection v Chris McGrath.*[257] The disputed domain name <highlandstreetfoundation.org> was not actively used but directed users to a site that simply said 'under construction'. Meanwhile, the other disputed domain name, <highlandstreetfoundation.com>, directed users to a site that featured the former name of the complainant. The panel concluded that the respondent's primary motivation in registering the domain names was to frustrate the complainant's efforts to adopt the new 'HIGHLAND STREET FOUNDATION' name and mark, as approved by a majority of the complainant's board. Taking this into account, the panel held that the respondent's passive use of the <highlandstreetfoundation.org> domain name confirmed that the respondent's primary purpose was to deprive the complainant of the use of the domain names.

[7.19.2] Paragraph 4(b)(iii): When is the Respondent a 'Competitor' of the Complainant?

There is an unresolved and persistent division among UDRP panels about the scope of the term 'competitor' in paragraph 4(b)(iii). On one view, the term is confined to commercial or business competitors, while on the other view the term extends to anyone acting in 'opposition' to another, including respondents who use the disputed domain name for a criticism site.

The broader interpretation of the term 'competitor' can be traced to the decision of the panellist in *Mission KwaSizabantu v Benjamin Rost.*[258] In that dispute, the respondent registered the disputed domain names <kwasizabantu.com>, <kwasizabantu.org> and <kwasizabantu.net>, and used them for a web site that contained information and views that were critical of the complainant. As explained at [6.9.6], the panellist held that the use of the domain names to criticise the complainant indicated an intention to tarnish the complainant's mark, and that the respondent had no legitimate interest in the domain name. The panellist held that the use of the domain name to publish highly negative views of the complainant indicated an intention to disrupt the complainant's activities, and that the truth or otherwise of the complainant's views was irrelevant to this finding. In deciding that the respondent had registered the domain names in bad faith contrary to paragraph 4(b)(iii), the panellist concluded that, although the respondent was not a business competitor of the complainant, he was a competitor in the sense that he acted in 'opposition' to the complainant. In this respect, the panellist adopted the following approach to interpreting paragraph 4(b)(iii):

> The Policy does not define who is a 'competitor' for the purpose of the paragraph, i.e. is it a competitor of the domain name owner or a competitor of the Complainant? Neither does the paragraph define what is encompassed by the expression 'business of a competitor'. There is no ref-

[256] WIPO Case No D2005-0766 (12 Sept 2005).
[257] WIPO Case No D2006-0516 (14 Aug 2006).

erence to the business being one where trade is conducted for commercial gain. For this reason and because of the context of paragraph 4(b)(iii) as an example within the general heading of 'bad faith', the Panel holds that the expression 'business' must be liberally construed to cover activities concerning the supply of some goods or services and in respect of which a reputation may be gained. The activities of the Complainant are held to come within the ambit of this term. Regarding the word 'competitor' similarly, the Panel holds that this term naturally should be read as relating to the 'competitor' of the domain name owner. The natural meaning of the word 'competitor' is one who acts in opposition to another and the context does not imply or demand any restricted meaning such as a commercial or business competitor. The Panel holds that the Respondent meets this definition . . .

While it would seem uncontroversial to conclude that the term 'competitor' in paragraph 4(b)(iii) must refer to a competitor of the domain name holder, which will commonly be the complainant, the definition of a 'competitor' as 'one who acts in opposition to another' has been extremely controversial. In adopting the broad interpretation of 'competitor', the panellist was clearly influenced by the differences in wording between paragraphs 4(b)(iii) and 4(b)(iv); paragraph 4(b)(iv) being confined to bad faith use of a domain name 'for commercial gain', while paragraph 4(b)(iii) makes no reference to a commercial purpose.

The narrow approach to the term 'competitor' was first clearly enunciated by the panellist in *Compusa Management Company v Customized Computer Training*.[259] In that dispute, the respondent registered the disputed domain names <bancompusa.com> and <stopcompusa.com> for the purpose of criticising the complainant's products. The panellist held that, although the disputed domain names had been registered for the purpose of disrupting the business of the complainant, there was no bad faith registration under paragraph 4(b)(iii) as the respondent was not a 'competitor' but, if anything, was a 'commentator'. In dismissing the complaint, the panellist considered the attempt to use the policy to silence the respondent was an example of impermissible 'cyber-bullying'.

The broad interpretation of the term 'competitor' in paragraph 4(b)(iii) adopted in *Mission KwaSizabantu* was first rejected by the panellist in *Bosley Medical Group v Michael Kremer*.[260] That dispute concerned the domain name <bosleymedical.com>, which the respondent claimed was registered for the purpose of publishing material that related to the complainants. After finding that the respondent had legitimate interests in the domain name based on free speech and fair use rights, the panellist turned to the complainant's argument that registration of the domain name violated paragraph 4(b)(iii). The panellist denied this contention essentially on the basis that the respondent was a critic, and not a competitor, of the complainant.

In rejecting the *Mission KwaSizabantu* approach, the panellist emphasised the inconsistency of regarding all opponents of a complainant as competitors with the 'fair use' privilege established under paragraph 4(c)(iii) of the UDRP. In this respect, the panellist stated that:

this Panel declines to accept . . . [the] . . . broad definition of 'competitor' to include not only commercial or business competitors, but anyone acting in opposition to another. If that were the rule, all consumer commentary sites whose domain names incorporate the trademark holders' mark would violate the Policy where their criticisms may have an adverse effect on the trademark holder's business, a result inconsistent with the express fair use defense set forth in the Policy. Here, the

[258] WIPO Case No D2000-0279 (7 June 2000).
[259] NAF Case No FA95082 (17 Aug 2000).
[260] WIPO Case No D2000-1647 (28 Feb 2001).

> Respondent has no commercial purpose and is a commentator exercising his fair use and free speech rights, not a competitor.

Examples drawn from panel decisions that have applied the competing approaches is the best way to understand the practical implications of the two views of the meaning of the term 'competitor' in paragraph 4(b)(iii). The following decisions have adopted the broader interpretation of the term 'competitor' as meaning anyone who acts in 'opposition' to another.

- *E & J Gallo Winery v Hanna Law Firm.*[261] The respondents claimed to have registered the disputed domain names <bartlesandjaymes.com> and <bartlesandjaymes.net> as consumer complaint sites. The panellist held that the respondents had no rights or legitimate interests in the domain names, as domain names that were not virtually identical to the complainant's mark could have been chosen for criticism sites. The panellist further held that, given the declared intention of the respondents, the domain names had been registered for the purpose of diverting the complainant's customers to the criticism sites. Referring to the *Mission KwaSizabantu* decision, the panellist stated that the fact that the parties were not commercial competitors was not the end of the inquiry under paragraph 4(b)(iii). In any case, regardless of paragraph 4(b)(iii), the panellist held that the registration of domain names that were virtually identical to the complainant's mark was, in the circumstances, in bad faith.
- *Estée Lauder Inc v estelauder.com.*[262] The respondent claimed that the disputed domain names <estelauder.com> and <estelauder.net> had been registered for use as consumer complaints sites. The panellist held that the respondent had no rights or legitimate interests in the domain names as it could have chosen domain names that were not confusingly similar to the complainant's marks. Given the avowed intention of the respondent in registering the domain names, the panellist held that the respondent had registered the domain names for the purpose of disrupting the complainant's business. Applying the broad definition of the term 'competitor' adopted in *Mission KwaSizabantu,* the panellist held that the respondent was 'competing' with the complainant for the attention of Internet users. In reaching this conclusion, the panellist agreed with the panellist in *Mission KwaSizabantu,* that:

> The fact that the terms of paragraph 4(b)(iii) of the Policy, unlike those of paragraph 4(b)(iv), do not require that a party be motivated by 'commercial gain', further supports the fact that the genre of competition referred to need not be commercial.

- *The Toronto-Dominion Bank v Boris Karpachev.*[263] The respondent had a contractual dispute with an online brokerage service and registered the disputed domain names, <tdwatergouse.com> and <dwaterhouse.com>, for the purpose of 'attacking' the service, which had been acquired by the complainant. The panellist held that the respondent had no rights or interests in the disputed domain names, as there was nothing in the domain names themselves to indicate that they were registered to criticise the complainant. The panellist further held that, as the respondent said it wanted to drive the broking service to bankruptcy, the domain name had been registered for the purpose of disrupting the complainant's business. In deciding that paragraph 4(b)(iii) had been satisfied, the panellist

[261] WIPO Case No D2000-0615 (3 Aug 2000).
[262] WIPO Case No D2000-0869 (25 Sept 2000).
[263] WIPO Case No D2000-1571 (15 Jan 2001).

applied the interpretation of the term 'complainant' adopted in *Mission KwaSizabantu*, to conclude that the respondent, a self-avowed opponent of the complainant, was a competitor.

- *Gardere Wynne Sewell LLP v DefaultData.com.*[264] The respondent used the disputed domain names, <garderewynne.com> and <gardereandwynne.com>, which reflected the name of a law firm, for a general statement criticising the US legal establishment. The panellist held that the respondent had no rights or legitimate interests in the domain names, as the domain names did not identify the sites as protest sites. In determining whether the domain names had been registered contrary to paragraph 4(b)(iii), the panellist acknowledged that the respondent was a 'competitor' of the complainant only in the broad sense that it attempted to convince the complainant's clients to seek alternative services. Adopting this broad interpretation, the panellist held that the respondent's activities fell within the term 'competitor', and that paragraph 4(b)(iii) had been satisfied.
- *Adams and Reese LLP v American Distribution Systems Inc.*[265] The respondent used the disputed domain name <adamsandreese.com>, which reflected the name of the complainant's law firm, to display graphics of dollar signs, sharks circling a diver and fish being eaten by larger fish. The panellist held that the complainant had no rights or legitimate interests in the domain name as the domain name itself did not indicate that it was used for the purpose of criticism or parody. The panellist followed the *Mission KwaSizabantu* approach to find that 'competitor' in the context of paragraph 4(b)(iii) 'does not imply or demand any restricted meaning such as a commercial or business competitor'. On that basis, the panellist held that the domain name had been registered contrary to paragraph 4(b)(iii).
- *Jenner & Block LLC v Defaultdata.com.*[266] As in *Adams and Reese*, the respondent used the disputed domain names <jennerblock.com> and <jennerandblock.com>, which corresponded to the name of the complainant's law firm, mainly to display graphics of dollar signs, sharks circling a diver and fish being eaten by larger fish. The respondent claimed that 'messing with the frail egos of paranoid lawyers is pure sport to me'. The panellist held that a respondent that uses a domain name for a criticism or parody site will not have rights or legitimate interests in the domain name if 'it does not on its face indicate that any site to which it leads is not that of the trademark or service mark holder but is instead a site for criticism or parody'. Referring to *Mission KwaSizabantu*, the panellist held that a 'competitor' under paragraph 4(b)(iii) is 'one who acts in opposition to another.' As, under this definition, the respondent was a 'competitor' of the complainant, and given that the respondent's purpose was to disrupt the business of the complainant, the panellist held that paragraph 4(b)(iii) had been satisfied.
- *Darl McBride v Peter Prior.*[267] The complainant brought an action against a company that produced open source software and the respondent, an advocate for open source software, used the disputed domain name <darlmcbride.com> to redirect users to a web site that contained material that criticised the complainant. The panellist held that redirecting users to a site that was ideologically opposed to the complainant did not confer rights or legitimate interests on the respondent. Concluding that the respondent was, in the broad sense, a competitor of the complainant, the panellist held that using a domain name

[264] WIPO Case No D2001-1093 (16 Nov 2001).
[265] NAF Case No FA102860 (6 Feb 2002).
[266] NAF Case No FA117310 (27 Sept 2002).
[267] NAF Case No FA201643 (17 Nov 2003).

identical to another's mark to criticise the owner of the mark demonstrates bad faith registration and use under paragraph 4(b)(iii).

- *Greater Houston Convention and Visitors Bureau v John Alden.*[268] The respondent, who was not a business competitor of the complainant, used the disputed domain name <houstonfilmcommission.com> to redirect Internet users to a commercial web site that hosted unrelated links, a search engine and advertisements. Following *Mission KwaSizabantu,* the panellist held that the respondent was a 'competitor' for the purpose of paragraph 4(b)(iii), and that the domain name had been registered to disrupt the business of the complainant.
- *CryptoLogic, Inc v Internet Billions Domains Inc.*[269] The respondent used the disputed domain names, including <cryptologic.net>, <ecryptologic.com> and <e-cryptologic.com>, for critical and highly inflammatory comments regarding the complainant, a software gaming company. The panel held that the respondent had no rights or legitimate interests in the domain names and that use of the domain names for a criticism site was bad faith registration under paragraph 4(b)(iii).
- *C & W Motors v Carlos Alvarez.*[270] The respondent registered the disputed domain names <arizonamotorsports.com> and <tucsonmotorsports.com>, but was not a business competitor of the complainant. The panellist held that the respondent could have no rights or legitimate interests in the domain names, even if they were used for criticism sites, as the domain names were identical to the complainant's mark and would create initial interest confusion. Although the respondent was not a business competitor of the complainant, the panellist followed *Mission KwaSizabantu* to conclude that the respondent was 'acting in opposition' to the complainant, and that the domain names had been registered to disrupt the complainant's business contrary to paragraph 4(b)(iii).
- *Gerber Products Company v LaPorte Holdings.*[271] The disputed domain names <gerberinsurance.com>, <gerberfood.com> and <gerberproducts.com> at one point contained links to the sites of the complainant's competitors, but the links were later disabled. The respondent was not, however, a business competitor of the complainant. Nevertheless, adopting the *Mission KawSizabantu* approach, the panellist held that the term 'competitor' should encompass 'a person who acts in opposition to another'. As the respondent had acted in opposition to the complainant, the panellist held that the domain names had been registered in bad faith.
- *MySpace, Inc v Myspace Bot.*[272] The respondent used the disputed domain name, <myspacebot.com>, to sell software that circumvented the security system for the complainant's well-known <myspace.com> web site. The respondent's software would, for example, allow purchasers to send unsolicited e-mails to 'MYSPACE' users. The panellist found that the use of the disputed domain name to redirect Internet users for commercial gain was neither a bona fide offering of goods or services pursuant to paragraph 4(c)(i) nor a legitimate non-commercial fair use of the disputed domain name pursuant to paragraph 4(c)(iii). As the respondent used the domain name to offer software that interfered with the legitimate use of the complainant's web site, the panellist held that the respon-

[268] NAF Case No FA298292 (6 Sept 2004).
[269] NAF Case No FA318925 (5 Nov 2004).
[270] WIPO Case No D2005-0270 (2 May 2005).
[271] WIPO Case No D2005-1277 (8 Feb 2006).
[272] NAF Case No FA672161 (19 May 2006).

dent had registered the domain name in bad faith under paragraph 4(b)(iii), even though the respondent was not a direct business competitor of the complainant.

- *St Lawrence University v Nextnet Tech.*[273] The respondent used the disputed domain name <stlawu.com> for a web site that included a directory of links to third party web sites, including sites offering educational services in competition with the complainant. The panellist held that redirecting Internet users seeking information on the complainant to competing web sites was evidence of bad faith registration contrary to paragraph 4(b)(iii). Although the respondent was not a business competitor of the complainant, the panellist held that it was a 'competitor' for the purpose of paragraph 4(b)(iii) as it 'acted in opposition' to the complainant.

The following are examples of decisions that have adopted the narrow interpretation of the term 'competitor' as meaning a direct commercial or business competitor, and as not extending to all opponents of the complainant.

- *Tribeca Film Center, Inc v Lorenzo Brusasco-Mackenzie.*[274] The respondent, who had been an intern of the complainant, claimed that he intended to use the disputed domain name <tribecafilmcenter.com> as a 'fan' site for the complainant. As the panellist held that mere intentions are insufficient to establish rights or interests based on fair use under paragraph 4(c)(iii), he concluded that the complainant had established that the respondent had no rights or legitimate interests in the domain name. In relation to the complainant's claim under paragraph 4(b)(iii), however, the panellist held that bad faith registration had not been established, as the respondent was not a competitor of the complainant. Acknowledging the split between UDRP panellists on this issue, the panellist stated that:

  > a respondent can 'disrupt [ ] the business of a competitor' only if it offers goods or services that can compete with or rival the goods or services offered by the trademark owner. If Paragraph 4(b)(iii) were given the meaning advanced by Complainant, respondents would be found to have disrupted the business of competitors in far too many cases, and the Policy's bad faith requirement would be diluted beyond recognition.

- *Vishwa Nirmala Dharma v Sahaja Yoga Ex-Members Network.*[275] The respondent used the disputed domain name <sahaja-yoga.org> as a criticism site for disaffected former members of the 'SAHAJA YOGA' group. The majority of the panel held that the respondent had a legitimate interest in using the web site to provide information about 'SAHAJA YOGA'. Referring to the *Bosley Medical Group* decision, the majority held that the domain name had not been registered in bad faith, in part because the ex-members of the complainant who used the domain name for the purpose of criticism were not 'competitors' of the complainant under paragraph 4(b)(iii). The dissenting panellist agreed that the respondent had not registered the domain name in breach of paragraph 4(b)(iii), but concluded that, as paragraph 4(b) is not exhaustive, use of a domain name that is identical to the complainant's mark for a criticism site amounts to bad faith registration and use.
- *Britannia Building Society v Britannia Fraud Prevention.*[276] The respondent used the <britanniabuildingsociety.org> domain name for a web site that was highly critical of the complainant building society. As explained at [6.9.9], adopting the 'complaints site'

[273] NAF Case No FA881234 (21 Feb 2007).
[274] WIPO Case No D2000-1772 (10 Apr 2001).
[275] WIPO Case No D2001-0467 (16 June 2001).
[276] WIPO Case No D2001-0505 (6 July 2001).

approach, the panellist held that the respondent had rights or legitimate interests in the disputed domain name. Following his decision in *Tribeca Film Center,* the panellist held that the respondent was not in competition with the complainant for the provision of goods or services and that, consequently, paragraph 4(b)(iii) was not satisfied.

- *Savin Corporation v savinsucks.com.*[277] The respondent registered the <savinsucks.com> domain name but made no active use of it. The panel held that, as there was no evidence that the primary purpose of the registration was to disrupt the business of the complainant, and as the complainant was not a 'competitor' of the respondent within the meaning of the policy, there was no bad faith registration under paragraph 4(b)(iii).
- *Satchidananada Ashram v Domain Administrator.*[278] The respondent used the disputed domain names, including <yogaville.com>, <yogaville.net> and <integralyoga.com>, to direct users to a site that was critical of the complainant. The panel held that the respondent had rights or legitimate interests in the domain names on the basis that they were being used for a legitimate non-commercial or fair use. Concluding that the respondent was not a competitor of the complainant, but was merely using the disputed domain names to host a web site that offers a public space to discuss the complainant, the panel held that the domain names had not been registered in bad faith.
- *Covance, Inc v The Covance Campaign.*[279] As explained at [6.9.10], the panellist adopted the 'totality of circumstances' approach to conclude that the respondent animal rights group had rights or legitimate interests in the <covancecampaign.com> web site, which was used to criticise the complainant. The panellist also held that the domain name had not been registered for the purpose of disrupting the complainant's business under paragraph 4(b)(iii) as the respondent was not a competitor of the complainant. In this respect, the panellist observed that an organisation that protests against the activities of a complainant cannot be a 'competitor' for the purposes of paragraph 4(b)(iii).
- *Advanced Internet Technologies, Inc v Mark Bianchi.*[280] The respondent, who operated a web design business, registered the disputed domain names <aitsucks.com> and <aitsucks.net>, which reflected the complainant's 'AIT' marks, which were used in connection with the complainant's web-design and web-hosting business. The domain names were used solely for the purposes of criticising the complainant and as a forum for complaints about the complainant. After a review of the respondent's web site, the panellist held that, as it was a legitimate criticism site, the complainant had not established that the respondent had no rights or legitimate interests in the domain names. In concluding that the domain names had not been registered contrary to paragraph 4(b)(iii), the panellist stated:

> It appears that Respondent has registered the domain name primarily for the purpose of disrupting Complainant's business through criticism of Complainant's services. The vast majority of the comments on the Respondent's website are critical of the Complainant. This is no doubt due to the fact that those customers who are satisfied with Complainant's services are not looking for a website entitled <aitsucks.com> on which to note their satisfaction. However, the disruption to Complainant's business is taking place through protected criticism of the Complainant's services. That type of disruption is not what is contemplated by UDRP paragraph 4(b)(iii). The type of disruption contemplated by the UDRP is diversion of consumers from the Complainant's website

[277] NAF Case No FA103982 (5 Mar 2002).
[278] NAF Case No FA125228 (13 Dec 2002).
[279] WIPO Case No D2004-0206 (30 Apr 2004).
[280] NAF Case No FA245965 (11 May 2004).

to the Respondent's website for commercial purposes. Even though both Respondent and Complainant design websites and may therefore potentially compete on a very small scale, Respondent is not at this time using the <aitsucks.com> or <aitsucks.net> website for that purpose.

- *Ryanair Limited v Michael Coulston.*[281] The respondent used the disputed domain name <ryanaircampaign.org> for a site that was critical of the complainant's business practices. As explained at [6.9.10], the panellist applied the *Covance* 'totality of circumstances' approach to find that the respondent had a legitimate interest in the domain name. Adopting the narrow view of paragraph 4(b)(iii), that a 'competitor' is an entity that is in competition with a complainant for the provision of goods or services, the panellist followed his previous decisions in *Tribeca Film Center* and *Britannia Building Society* to conclude that, as the respondent was not a competitor of the complainant, the domain name was not registered in bad faith under paragraph 4(b)(iii).

Commercial Gain from Confusing Use

[7.20] Commercial Gain from Confusing Use

The fourth non-exclusive circumstance that is evidence of bad faith registration and use, as set out in paragraph 4(b)(iv), is that:

> by using the domain name, the domain name holder has intentionally attempted to attract, for commercial gain, Internet users to its web site or other on-line location, by creating a likelihood of confusion with the complainant's mark as to the source, sponsorship, affiliation, or endorsement of its web site or location or of a product or service on its web site or location.

The following five elements must be established for the circumstance set out in paragraph 4(b)(iv) to be satisfied:

- the domain name must be used;
- intentionally to attempt to;
- divert Internet users to the domain name holder's web site or other online location;
- for commercial gain;
- by creating a likelihood of confusion with the complainant's mark as to the source, sponsorship, affiliation or endorsement of the domain name holder's web site or location, or of a product or service on its web site or location.

First, paragraph 4(b)(iv) differs from the other circumstances set out in paragraph 4(b) in that it refers to 'use', and not registration, of the disputed domain name. As explained at [7.5], there is a tension between the wording of paragraph 4(a)(iii), which requires the disputed domain name to have been registered and used in bad faith, and paragraph 4(b)(iv), which only refers to bad faith 'use'. As further explained at [7.5], the preferred approach to determining whether the third element of the UDRP has been satisfied in the circumstances set out in paragraph 4(b)(iv) is that once the complainant has established the circumstances set out in paragraph 4(b)(iv), then an inference can be drawn that the domain name was

[281] WIPO Case No D2006-1194 (12 Dec 2006).

registered in bad faith, unless the respondent has produced evidence that the domain name has been registered in good faith. As explained at [7.6], the domain name does not need to resolve to an active web site in order for it to be 'used'. Moreover, as explained at [7.10], it has been established that, in certain circumstances, passive holding or passive warehousing can amount to bad faith use of the disputed domain name.

Secondly, the respondent must have used the domain name intentionally to attract Internet users to the domain name holder's web site or other online location. As explained at [7.20.1], the intention of the respondent is to be determined objectively from the circumstances surrounding the registration and use of the disputed domain name. Moreover, as paragraph 4(b)(iv) does not require the attraction of Internet users to be the primary intention of the respondent, the paragraph will be satisfied provided that this is one of the respondent's intentions in using the domain name.

Thirdly, as the majority pointed out in *Park Place Entertainment Corporation v Anything.com Ltd*,[282] using the domain name to attract Internet users essentially means 'diverting' Internet users to the respondent's web site or to other sites, such as the sites of competitors of the complainant.

Fourthly, as explained at [7.20.2], a domain name may be used for 'commercial gain' even if all of the earnings from the domain name are used for a 'non-commercial' purpose. Moreover, as further explained, the requirement that the domain name be used for commercial gain means that a non-commercial use, such as a use of the domain name for a non-commercial criticism site, will not fall within paragraph 4(b)(iv).

Fifthly, as explained at [7.20.3], the use of a disputed domain name will not amount to bad faith use under paragraph 4(b)(iv) if there is no 'likelihood of confusion' with the complainant's mark. For example, the use of a generic domain name to redirect Internet users may not contravene paragraph 4(b)(iv), as there is unlikely to be a 'likelihood of confusion' with the complainant's site. The main forms of redirection that have been held to create a 'likelihood of confusion' include: use of the domain name to attract users to the respondent's web site for commercial gain; use of the domain name to redirect users to other sites in return for commercial gain, such as 'click-through' revenue; use of the domain name to redirect users to pornographic sites; and 'parking' the domain name with a registrar, or third party advertiser, that uses the domain name for advertising or links to other web sites.

[7.20.1] Paragraph 4(b)(iv): 'Intentionally' Attempting to Attract Internet Users to the Domain Name Holder's Web Site or Other Online Location

Paragraph 4(b)(iv) requires the complainant to establish that the respondent *intentionally* attempted to attract Internet users to its web site or other online location. As the panellist stated in *Memorydealers.com, Inc v Dave Talebi*:[283]

> Any intentional use of a competitor's mark, even a 'weak' mark, to attract, divert, or mislead the competitor's customers is not legitimate use.

The intention of the respondent is usually inferred from the way in which the disputed domain name is used. In other words, the 'intention' of the respondent referred to in paragraph 4(b)(iv) is not the respondent's 'subjective' intention in using the domain name to divert Internet users, but is to be objectively determined from the respondent's conduct.

[282] WIPO Case No D2002-0530 (16 Sept 2002).
[283] WIPO Case No D2004-0409 (20 July 2004).

The reasons for adopting an objective approach to determining the respondent's intention in using the domain name under paragraph 4(b)(iv) were explained by the panellist in *Paule Ka v Paula Korenek*.[284] In that dispute, the respondent registered the <paulekacreations.com> domain name, which reflected the complainant's 'PAULE KA' mark, and used it on a web site for her gift basket business. The respondent claimed that, after a visit to Paris where she was impressed by the complainant's shop, she had herself become known by the 'PAULE KA' nick-name.

On the facts of the dispute, there was no evidence that any of the circumstances set out in paragraph 4(b)(i)–(iii) had been made out. Nevertheless, the respondent clearly knew of the complainant's business and had used the domain name for a commercial purpose in a way that was likely to create confusion. Finding that the respondent had intentionally attempted to attract Internet users, the panellist refused to take into account the respondent's claimed subjective intention in registering the domain name. In doing so, the panellist analysed the requirement for the domain name be 'intentionally' used in the following terms:

> how is the 'intentional' element required under paragraph 4(b)(iv) to be defined? The Panel is guided in its consideration of the issue by the fact that a subjective test of intent (thus considered more or less as a *mens rea* element) would be difficult if not impossible to apply given that credibility must be assessed only on the basis of documentary evidence. It is difficult to enter the minds of the parties to determine their subjective intent. The proper test in this Panel's view is whether the objective consequence or effect of the Respondent's conduct is to free-ride on the Complainant's goodwill, whether or not that was the primary (subjective) intent of the Respondent.

As the panellist pointed out, paragraph 4(b)(iv) does not require the respondent to have the 'primary' intention of using the domain name to attract Internet users, but only that this is one of the respondent's intentions. Applying the objective approach to determining 'intention' under paragraph 4(b)(iv), the respondent formulated the following test for deciding whether the domain name has been intentionally used to attract Internet users in the circumstances of this case:

> knowledge of the mark at the time of registration followed by use in commerce without a legitimate interest and likely to create confusion (and thus allow the Respondent to free-ride on the Complainant's goodwill) . . .

Finding that the respondent knew of the complainant's mark at the time of registration, that she had no legitimate interest in the domain name, and that use of the domain name was likely to create confusion, the panellist held that the domain name was being used in bad faith contrary to paragraph 4(b)(iv).

The objective test for establishing the intention of using a disputed domain name under paragraph 4(b)(iv) has been applied in a number of subsequent panel decisions. The following are examples of decisions that have applied the *Paule Ka* objective test.

- *Glaxo Group Limited, SmithKlineBeecham Corporation v Michale Kelly*.[285] The disputed domain name <discount-imitrex-pharmacy.com> resolved to a web site from which the respondent offered the complainant's 'IMITREX' brand medication for sale along with the products of the complainant's competitors. In the circumstances, the panellist inferred that the respondent knew of the complainant's mark at the time the domain name was registered and concluded that the primary intent in using the domain name was to trade

[284] WIPO Case No D2003-0453 (24 July 2003).
[285] WIPO Case No D2004-0262 (17 May 2004).

on the complainant's goodwill by creating a likelihood of confusion with the complainant's mark. Referring to the *Paule Ka* decision, the panellist held that the domain name was being used in bad faith contrary to paragraph 4(b)(iv).

- *Lifetime Products, Inc v IQ Management Corporation.*[286] The disputed domain name <lifetimeproducts.com> was used to divert Internet users to a search engine web site, which listed numerous commercial products. Applying the *Paule Ka* test, the panellist found that the domain name was used contrary to paragraph 4(b)(iv) as the respondent knew of the complainant's mark at the time the domain name was registered, was being used in commerce without a legitimate interest and was likely to create confusion with the complainant's mark.
- *Julie & Jason, Inc v Faye Scher.*[287] The respondent initially used the disputed domain name <themahjonggmaven.com> to direct users to her <wherethewindsblow.com> site, where she offered mah jong games and accessories in competition with the complainant. In response to the complaint, the respondent claimed that she was commonly known as a 'mah jong maven'. Applying the *Paule Ka* test, the panellist held that, as the respondent knew of the mark at the time the domain name was registered, had no legitimate interest in the domain name, and that use of the domain name was likely to create confusion, the domain name was used in bad faith under paragraph 4(b)(iv).
- *Permatex, Inc v Mounir Mati.*[288] The disputed domain name <permatex-me.com> provided a link to a travel service unrelated to the complainant and the <permetex-me.com> web site advertised the respondent's own web-hosting services. Applying the *Paule Ka* test, the panellist inferred that the respondent knew of the complainant's mark at the time the domain names were registered, the domain names were used in commerce without a legitimate interest and were likely to cause confusion. Consequently, the panellist concluded that the domain names were used in bad faith contrary to paragraph 4(b)(iv).
- *Lo Monaco Hogar, S L v MailPlanet.com, Inc.*[289] The respondent claimed that it was using the disputed domain name <lomonaco.com> as part of a business of using expired domain names for a vanity e-mail service. The domain name was parked with a third party Internet advertising company, that placed active links on the web site, such as 'Latex', which linked to web-pages including the complainant's and to competitors of the complainant. The panel held that the domain name was being used to divert Internet users and, applying the objective test of intention, concluded that the respondent had intended to divert Internet users.
- *Elite Model Management Corporation v Wesley Perkins.*[290] The disputed domain name <elitemodels.com> was used for automatically generated advertisements and articles, and the domain name <elitemodels.net> simply contained links to other sites. Adopting the *Paule Ka* objective test, the panellist held that use of the domain name to divert Internet users was evidence of a bad faith intention under paragraph 4(b)(iv).

The conduct of third parties may be taken into account in determining whether the respondent is intentionally using the domain name to divert Internet users contrary to paragraph 4(b)(iv). The relevance of third party conduct in determining whether or not

[286] WIPO Case No D2004-0719 (12 Nov 2004).
[287] WIPO Case No D2005-0073 (6 Mar 2005).
[288] WIPO Case No D2005-0954 (16 Nov 2005).
[289] WIPO Case No D2005-0896 (2 Dec 2005).
[290] WIPO Case No D2006-0297 (15 May 2006).

there is a bad faith intention under paragraph 4(b)(iv) divided the panel in *Park Place Entertainment Corporation v Anything.com Ltd.*[291]

In that dispute, the respondent, who had registered the disputed domain name <flamingo.com>, had an agreement with a third party service provider under which the respondent was paid for links and advertisements that the third party placed on the respondent's web site. Pursuant to this agreement, the respondent's web site included a list of 'topics' with links to other web sites, including Las Vegas hotels that competed with the complainant's 'FLAMINGO' hotel and casino. The majority held that the diversion of Internet users to third party web sites was evidence of the respondent's intention to use the domain name to divert Internet users for commercial gain under paragraph 4(b)(iv), and that the fact that a third party was effectively operating the respondent's web site did not protect the respondent from this finding. The dissenting panellist, however, was more inclined to believe the respondent's claim that it had registered the domain name because 'flamingo' is a common word with wide appeal, and that the conduct of the third party service provider should not be the sole or primary evidence of the state of mind of the respondent.

The majority of panels that have considered this issue have agreed with the majority that an intention to divert Internet users may be inferred from the conduct of third parties, an approach that is supported by the application of the objective test to determine the respondent's intention under paragraph 4(b)(iv).

[7.20.2] Paragraph 4(b)(iv): Diverting Internet Users for 'Commercial Gain'

Paragraph 4(b)(iv) requires the complainant to establish that the respondent is using the domain name to divert Internet users for 'commercial gain'. This mirrors the wording of paragraph 4(c)(iii) of the UDRP, which provides that a respondent has rights or legitimate interests in a disputed domain name if it makes a legitimate non-commercial or fair use of the domain name 'without intent for commercial gain'. What amounts to 'commercial gain' under paragraph 4(c)(iii) is dealt with at [6.9.5] which explains, for example, that the fact that the respondent donates all of its gains to charity does not convert a 'commercial' use to a 'non-commercial' use.

This principle is equally applicable in determining whether the domain name has been used for 'commercial gain' under paragraph 4(b)(iv). In *The Professional Golfers' Association of America (PGA) v Provisions, LLC,*[292] for example, the respondent, a charitable organisation that donated money to help children with developmental disabilities and autism, used the disputed domain name <ryder2004.com> to resell tickets and hospitality packages to the complainant's golfing events. The panellist held that the respondent was using the domain name to attract Internet users for commercial gain contrary to paragraph 4(b)(iv), and that it was irrelevant whether the respondent later gave some or all of its earnings to charity.

The requirement that the domain name be used for 'commercial gain' means that non-commercial criticism sites do not fall within paragraph 4(b)(iv). The importance of this requirement is illustrated by the decision in *Estée Lauder Inc v estelauder.com, estelauder.net and Jeff Hanna.*[293] In that dispute, the respondent, an American law firm specialising in personal injury and product liability, used the disputed domain names, <estelauder.com> and <estelauder.net>, for consumer criticism sites. The complainant argued that the respondent

[291] WIPO Case No D2002-0530 (16 Sept 2002).
[292] WIPO Case No D2004-0576 (13 Sept 2004).
[293] WIPO Case No D2000-0869 (25 Sept 2000).

was using the domain names to attract Internet users to solicit potential claimants for its personal injury practice, and therefore was using the domain names for 'commercial gain'. In this respect, the panellist considered the fact that the law firm did not advertise its services on the relevant web sites to be equivocal, as it might be consistent with a desire to conceal a commercial intent under the guise of a criticism site. In the absence of evidence of any actual 'commercial gain' from the use of the domain names, however, the panellist held that the complainant had failed to meet its burden of proof in establishing that the sites were 'commercial' under paragraph 4(b)(iv).

[7.20.3] Paragraph 4(b)(iv): 'Diverting' Internet Users by Creating a 'Likelihood of Confusion'

Paragraph 4(b)(iv) specifies that the disputed domain name must be used intentionally to attract Internet users to the respondent's web site or other online location by creating 'a likelihood of confusion'. 'Attracting' Internet users has been interpreted to mean 'diverting' or 'redirecting' users to the respondent's web site or another online location. Not all forms of 'redirection' of Internet users will, however, fall within paragraph 4(b)(iv), as the redirection must create a 'likelihood of confusion' with the complainant's mark, as to the source of the respondent's web site or other location, or the source of a product on the respondent's web site or other location. Consequently, a redirection of users that does not create a 'likelihood of confusion' with the complainant's mark will not fall within paragraph 4(b)(iv).

The need for a redirection to create a 'likelihood of confusion' is illustrated by the decision in *Shirmax Retail Ltd v CES Marketing, Inc.*[294] In that dispute, the respondent used the disputed domain name <thyme.com> to redirect users to the respondent's site that advertised generic domain names for sale. Given the generic nature of the 'THYME' mark, the panellist held that, even though the <thyme.com> domain name was used to attract Internet users for commercial gain, it had not done so by creating a likelihood of confusion with the complainant's 'THYME' mark.

An important issue in determining whether the disputed domain name is used to attract users by creating a 'likelihood of confusion' is whether the presence of a disclaimer on the respondent's web site is sufficient to dispel any consumer confusion. The approach that panels have adopted to this issue is dealt with at [7.15] above.

In most cases where there is an attempt to redirect users for commercial gain, where the domain name is identical or confusingly similar to the complainant's mark and where the complainant has actual or imputed notice of the complainant's mark, the redirection will be considered likely to create confusion with the complainant's mark. The following are the main forms of redirection that have been held to fall within paragraph 4(b)(iv) of the policy.

Use of the domain name to attract users to the respondent's web site for commercial gain

Use of the disputed domain name to attract Internet users to the respondent's web site for commercial gain is a common example of a bad faith use of the domain name in a way that illegitimately 'free rides' on the complainant's goodwill or reputation contrary to paragraph 4(b)(iv). The following are examples of decisions that have held that diversion of users to the respondent's site for commercial gain amounts to bad faith use under paragraph 4(b)(iv).

[294] eRes Case No AF-0104 (20 Mar 2000).

- *Busy Body, Inc v Fitness Outlet Inc.*[295] The disputed domain name <efitnesswarehouse.com> was used to sell exercise equipment in competition with the complainant. The panellist held that the respondent was using the domain name intentionally to attempt to attract the complainant's customers to its web site contrary to paragraph 4(b)(iv). Although there was no evidence that the respondent had ever successfully attracted users to the web site, the panellist pointed out that paragraph 4(b)(iv) only requires the respondent to have 'attempted' to attract users, and not that it has 'successfully' attracted users.
- *MathForum.com, LLC v Weiguang Huang.*[296] The disputed domain name <drmath.com> was used by the respondent to provide services relating to the teaching of mathematics in competition with the complainant. The panellist held that the respondent was using the domain name to attract users to its web site by creating a likelihood of confusion with the complainant's mark contrary to paragraph 4(b)(iv).
- *Arthur Guinness Son & Co (Dublin) Limited v Dejan Macesic.*[297] The disputed domain name <guiness.com> was used for a web site that contained content relating to home brewing and sports, and later expanded to become a 'discussion board' for various related topics. The site generated revenue by means of advertising banners. The panellist held that the domain name was being used both for a discussion group and for the commercial gain of generating advertising revenue. On this basis,, the panellist held that the domain name was being used misleadingly to divert Internet users contrary to paragraph 4(b)(iv). The panellist added that, even if the use of the domain name did not fall within paragraph 4(b)(iv) on the basis that the web site was non-commercial, the domain name was being used in bad faith because it was being used intentionally to mislead Internet users into visiting the respondent's web site in the belief that they were visiting the complainant's web site
- *International Organization for Standardization v International Supplier Operations Audit Services.*[298] The disputed domain names <isoquality.com>, <isoaudit.com> and <isoconsultant.net> were used for a web site that advertised the respondent's services, which related to quality standards. The panellist held that the respondent had deliberately used the domain names to attempt to attract Internet users, who would believe they were accessing a site of the complainant, to its web site for commercial gain. Although the services offered from the respondent's web site were 'free', the panellist held that the sites were being used for 'commercial gain', as the web site would present the respondent with business opportunities that otherwise might not exist.
- *Mamas & Papas (Holdings) Limited v QTK Internet.*[299] The disputed domain name <mamasandpapas.com> was used for a web site that contained a variety of material, including advertisements for 'singles' and spyware removal services. The panel held that this was bad faith use contrary to paragraph 4(b)(iv).
- *Wellspring Products, LLC v Meridian LifeForce Inc.*[300] The disputed domain names, including <ionspacanada.com>, <ionspadetoxcleanse.com>, and <ionspafootbath.com>, resolved to the respondent's <ionicspahealth.com> web site, which offered products in competition with the complainant. The panellist held that this constituted an intentional effort to attract, for commercial gain, Internet users to the respondent's web site by creating a likelihood of confusion with complainant's mark under paragraph 4(b)(iv).

[295] WIPO Case No D2000-0127 (22 Apr 2000).
[296] WIPO Case No D2000-0743 (17 Aug 2000).
[297] WIPO Case No D2000-1698 (25 Jan 2001).
[298] WIPO Case No D2002-0460 (9 July 2002).
[299] WIPO Case No D2004-0496 (14 Aug 2004).
[300] NAF Case No FA812927 (17 Nov 2006).

Use of the domain name to redirect users to other sites in return for commercial gain, such as 'click-through' revenue

The redirection of users to a site, or sites, that are unrelated to the complainant for commercial gain, such as obtaining 'click-through' revenue, is a form of redirection that has been held to create a 'likelihood of confusion' in many UDRP decisions. Where users are redirected to commercial sites, UDRP panels will infer that the diversion is for a 'commercial gain'. The following are examples of decisions that have held that redirection of users to sites that are unrelated to the complainant amounts to bad faith use contrary to paragraph 4(b)(iv).

- *Nokia Corporation v Nokiagirls.com.*[301] The <nokiagirls.com> web site included a banner that contained links to sites including <shopnow.com> and <altavista.com>, presumably for 'click-through' revenue. The panellist held that the use of the domain names was aimed at generating revenue through 'click-through' traffic, thereby taking a 'free ride' on the goodwill of the complainant's NOKIA trade mark, contrary to paragraph 4(b)(iv).
- *Reuters Limited v Global Net 2000, Inc.*[302] The disputed domain name <ruters.com> was used to redirect users to the respondent's web site at <global2000.com>, which was used for the respondent's web-hosting services. The panellist held that this was an intentional attempt to redirect Internet users for commercial gain under paragraph 4(b)(iv).
- *AltaVista Company v Andrew Krotov.*[303] The disputed domain name <altavisga.com> was used to provide links to the <webprovider.com> web site, which included a directory of links under subject headings, including services that competed with the complainant's services. The panellist held that the link to the third party web site amounted to a redirection of users that created a likelihood of confusion under paragraph 4(b)(iv).
- *Nokia Corporation v Private.*[304] The disputed domain name <wwwnokia.com> was used to redirect users to the <bestoftheweb.com> site, which contained links to web sites offering a wide range of products, including electronic products that competed with the complainant's products. The panellist held that the respondent had intentionally attempted to attract, for commercial gain, Internet users to the web site to which the domain name pointed, by creating a likelihood of confusion with the complainant's trade mark under paragraph 4(b)(iv).
- *Ticketmaster Corporation v DiscoverNet, Inc.*[305] The disputed domain name <ticketmaster.net> was used to divert Internet users to the respondent's <ticketpro.com> web site, which advertised theatre, concert and sporting event tickets in competition with the complainant's ticketing service. The panellist held that this use manifested an intention to benefit commercially from the confusion of Internet users who are redirected to the respondent's site as to source, sponsorship, affiliation or endorsement, between the complainant and the services offered by the respondent, contrary to paragraph 4(b)(iv).
- *Nintendo of America, Inc v Pokemonplanet.net.*[306] The disputed domain name <pokemonplanet.net> was used to sell the complainant's merchandise and to direct users via a link to another site, presumably for 'click-through' revenue. Based on the format of

[301] WIPO Case No D2000-0102 (18 Apr 2000).
[302] WIPO Case No D2000-0441 (13 July 2000).
[303] WIPO Case No D2000-0091 (25 Oct 2000).
[304] WIPO Case No D2000-1271 (3 Nov 2000).
[305] WIPO Case No D2001-0252 (9 Apr 2001).
[306] WIPO Case No D2001-1020 (25 Sept 2001).

the respondent's web site, the panellist held that the domain name was being used in bad faith under paragraph 4(b)(iv).

- *Alpha Xi Delta, Inc v Befriend Internet Services.*[307] The disputed domain name <alphaxidelta.com> resolved to a web site that included links to web sites at <apartments.com>, <springstreet.com> and <travelocity.com>, as well as to the respondent's own web site at <befriend.com>. The panellist held that the use of the domain name was a misappropriation of the complainant's goodwill in its mark contrary to paragraph 4(b)(iv).
- *Ltd Commodities, Inc v CostNet.*[308] The disputed domain name <ltdcomodities.com> directed users to the <linkster.com> web site, which included a directory of links grouped in general categories, including services that competed with the complainant's. The respondent earned 'click-through' revenue from the domain name in return for redirecting users to the third party web site. The panellist held that the respondent's use of the domain name was an intentional attempt to attract users to another online location offering services similar to the complainant's contrary to paragraph 4(b)(iv).
- *Best Western International, Inc v QTK Internet.*[309] The disputed domain name <bestweston.com> was used to redirect users to <roomsusa.com>, a travel reservation site that allowed consumers to book hotel and motel rooms worldwide, in competition with the complainant. The panellist held that this was an intentional attempt to divert Internet users contrary to paragraph 4(b)(iv).
- *Corbis Corporation v RegisterFly.com.*[310] The respondent used the disputed domain names, including <corbisclicks.com> and <corbisexchange.com>, to direct Internet users to the sites of the complainant, which is a well-known provider of digital images and footage. The panellist held that diverting users to the complainant's competitors for 'click-through' revenue was bad faith under paragraph 4(b)(iv) as it 'free rides' on the complainant's mark and goodwill.
- *Nicholas V Perricone, M D v Compana LLC.*[311] The disputed domain name <perricone.com> resolved to a web site that included links to skin care products in competition with the complainant's, as well as links for other products. From the nature of the respondent's web site, the panellist inferred that the respondent was receiving 'click-through' revenue from the links. Finding a likelihood of consumer confusion, the panellist held that the domain name was being used in bad faith contrary to paragraph 4(b)(iv).

Use of the domain name to redirect users to pornographic sites

The use of the disputed domain name to redirect Internet users to pornographic web sites is almost always evidence that the domain name is being used contrary to paragraph 4(b)(iv). The following are examples of decisions that have found that redirection of users to pornographic web sites amounts to an intentional attempt to divert Internet users under paragraph 4(b)(iv).

- *National Football League Properties, Inc v One Sex Entertainment Co.*[312] The respondent used the <chargergirls.net> domain name to redirect Internet users to its pornographic

[307] NAF Case No FA100681 (28 Nov 2001).
[308] WIPO Case No D2002-0031 (3 May 2002).
[309] NAF Case No FA124994 (17 Oct 2002).
[310] WIPO Case No D2006-0546 (14 July 2006).
[311] NAF Case No FA825260 (13 Dec 2006).
[312] WIPO Case No D2000-0118 (17 Apr 2000).

web site <onesex.com>. The panellist held that the domain name was being used to attract users to its web site by creating a likelihood of confusion regarding whether the complainant was the source, sponsor, affiliate or endorser of its web site. The panellist added that there was a likelihood of confusion, even though any confusion of an association would be dispelled once users visited the respondent's site.

- *Ty, Inc v OZ Names.*[313] The respondent linked the disputed domain names, <beaniesbabies.com>, <beanybabies.com>, <beanybaby.com>, and <ty.org>, to its web sites at <37.com> and <megaco.com>, both of which contained graphic, 'adult' oriented material. The panellist concluded that this amounted to bad faith registration and use.
- *Land O' Lakes Inc v Offbeat Media Inc.*[314] The disputed domain name <land-o-lez.com> was used for a pornographic web site. The home-page for the web site featured a confusingly similar pornographic depiction of the respondent's 'LAND O' LAKES' design mark. The panellist held that this was a bad faith attempt to attract Internet users contrary to paragraph 4(b)(iv).
- *America Online, Inc v East Coast Exotics.*[315] The disputed domain name <aolwomen.com> was used for a pornographic web site and, after the complaint was lodged, contained a link to another pornographic site. The panellist held that the domain name was being used in bad faith under paragraph 4(b)(iv).
- *Microsoft Corporation v Paul Horner.*[316] The disputed domain name <microsoftsite.com> was used for a web site that contained pornographic material and links to pornographic web sites. The panellist held that the respondent had contravened paragraph 4(b)(iv) as the domain name was confusingly similar to the complainant's mark, was being used to direct users to its web site by creating confusion as to sponsorship or affiliation, and was being used for commercial purposes.
- *Perlier SpA v Mr. Stanley Filoramo.*[317] The disputed domain name <perlier.net> was used for a pornographic web site and visitors to the site were 'mousetrapped'. The panellist held that this use of the domain name created a likelihood of confusion with the complainant's mark, attracting users who were not looking for a pornographic web site but for the complainant's products. In reaching this conclusion, the panellist stated that 'the mere diversion of domain names to a pornographic web site, whatever the motivation of Respondent, is in and of itself certainly consistent with the finding that the disputed domain name was registered and is being used in bad faith'.
- *Metro-Goldwyn-Mayer Studios Inc v Michael McGreevy.*[318] The disputed domain name <rocky.com> was used to redirect Internet users to a commercial web site. The web site included links to third-party web sites that competed with the complainant's business, as well as links to adult-oriented content. Inferring that the respondent received revenue from 'click-through' traffic, the panellist concluded that the domain name was being used in bad faith pursuant to paragraph 4(b)(iv).
- *Expedia, Inc v Vinyl Grooves, Inc.*[319] The disputed domain name <sexpedia.com> resolved to a web site that featured pornographic photographs and links to 'adult'-oriented web sites. The panellist held that the respondent's use of the confusingly similar

[313] WIPO Case No D2000-0370 (27 June 2000).
[314] NAF Case No FA96451 (23 Feb 2001).
[315] WIPO Case No D2001-0661 (10 July 2001).
[316] WIPO Case No D2002-0029 (27 Feb 2002).
[317] WIPO Case No D2003-0848 (17 Dec 2003).
[318] NAF Case No FA709144 (29 June 2006).
[319] NAF Case No FA781917 (3 Oct 2006).

<sexpedia.com> domain name was likely to cause confusion among Internet users as to the complainant's sponsorship of, or affiliation with, the disputed domain name contrary to paragraph 4(b)(iv).

'Parking' the domain name with a registrar, or third party advertiser, that uses the domain name for advertising or links to other web sites

Domain name 'parking' is a practice used by registrars or third party Internet advertisers, whereby a domain name that is not actively used by the registrant is used by the registrar or advertiser to display advertising or links to other web sites. In such cases, the domain name holder will usually receive 'click-through' revenue based on Internet traffic generated by the site. Where the domain name is used to earn revenue from 'parking', this will generally amount to bad faith redirection of Internet users for commercial gain under paragraph 4(b)(iv), even though the registrant is not responsible for determining the content that appears on the web site. Another form of domain name 'parking', which usually does not involve payments to the registrant, involves posting a placeholder, such as an 'under construction' notice, on the web site, with no commercial content. Although this form of 'parking' is unlikely to fall under paragraph 4(b)(iv), as it involves no 'commercial gain', it may otherwise amount to bad faith registration or use as a form of 'passive holding'. The following are examples of decisions that have held that 'parking' a domain name in return for commercial gain constitutes bad faith use contrary to paragraph 4(b)(iv).

- *Park Place Entertainment Corporation v Anything.com Ltd.*[320] The disputed domain name <flamingo.com> was parked with an Internet search service provider that placed links to third party web sites. The links that were placed on the web site were determined by the advertiser by means of a program that determined the links that were associated with the domain name. The majority of the panel inferred that the respondent knew of the complainant's mark at the time the domain name was registered, and that 'parking' the domain name indicated an intention to use it to redirect Internet users contrary to paragraph 4(b)(iv). In reaching this conclusion, the majority held that the conduct of the Internet search service provider could be used to infer bad faith use under paragraph 4(b)(iv).
- *Associated Newspapers Limited v Domain Manager.*[321] The disputed domain name <mailonsunday.com> was used by a third party service provider for a web site that contained links to the sites of newspapers, including the complainant's site and the sites of competitors to the complainant. The panellist inferred that the respondent was commercially benefiting from the misleading domain name by earning 'click-through' fees, and that the domain name was being used contrary to paragraph 4(b)(iv).
- *B and J García, SL, Arnedo, Spain v Gorila.*[322] The disputed domain name <gorila.com> was parked with a domain name 'parking service' that placed automatically generated links on the web site. The panellist held that use of the domain name for a generic directory of commercial links was bad faith use under paragraph 4(b)(iv).
- *Sanofi-Aventis v 1N4 Web Services.*[323] The disputed domain name <amaryllid.com> was parked with a third party advertiser that placed 'contextual' links on the web site and made 'click-through' payments to the respondent. The panellist held that the domain name was

[320] WIPO Case No D2002-0530 (16 Sept 2002).
[321] NAF Case No FA201976 (19 Nov 2003).
[322] WIPO Case No D2004-1071 (4 Mar 2005).
[323] WIPO Case No D2005-0938 (24 Nov 2005).

being used to attempt to attract, for commercial gain, Internet users to the respondent's web site by creating a likelihood of confusion with the complainant's mark. In reaching this conclusion, the panellist stated that 'the Respondent either passively condoned or had a blameworthy indifference to the material posted to its domains by another'.

- *Allen-Edmonds Shoe Corporation v joseph.*[324] The disputed domain name <allenedmondshoes.com> was parked with the registrar, which used it for links to several shoe retailers who were competitors of the complainant. Although there was no evidence that the respondent had registered the domain name with the purpose of generating advertising revenue, the panellist held that this did not prevent a finding that the domain name was being used to redirect users for commercial gain contrary to paragraph 4(b)(iv).
- *Diners Club International Ltd v O P Monga.*[325] The disputed domain name <wwwdiners-clubus.com> was parked with a third party 'parking service' ('Sedo') that used the web site for 'targeted links' and paid the respondent 'click-through' fees. The panellist held that, by agreeing to Sedo's 'parking policy' and not having 'un-parked' the domain name, the respondent was using the domain name intentionally to attract Internet users for commercial gain contrary to paragraph 4(b)(iv). In reaching this conclusion, the panellist stated that:

> The fact that such links were provided by Sedo does not excuse Respondent. It was Respondent herself who caused the links to appear, since she freely decided to 'park' the disputed domain name with Sedo, and to obtain revenue per click.

- *Prestige Brands Holdings, Inc v The domain is not for sale.*[326] The disputed domain name <prell.com> was parked with a third party 'parking service' that placed a variety of material on the web site, and paid the respondent 'click-through' revenue. Inferring that the respondent must have known of the complainant's mark at the time the domain name was registered, the panellist concluded that the respondent intended to earn 'click-through' revenue by diverting Internet users that were looking for the complainant's products. In deciding that the domain name was used in bad faith under paragraph 4(b)(iv), the panellist took into account the fact that the respondent would have been aware that the 'parking service', by its very nature, would put links on the page that related to the complainant's products.

[324] NAF Case No FA624511 (28 Feb 2006).
[325] NAF Case No FA670049 (22 May 2006).
[326] WIPO Case No D2006-0608 (17 July 2006).

Appendix 1

ICANN's Structure

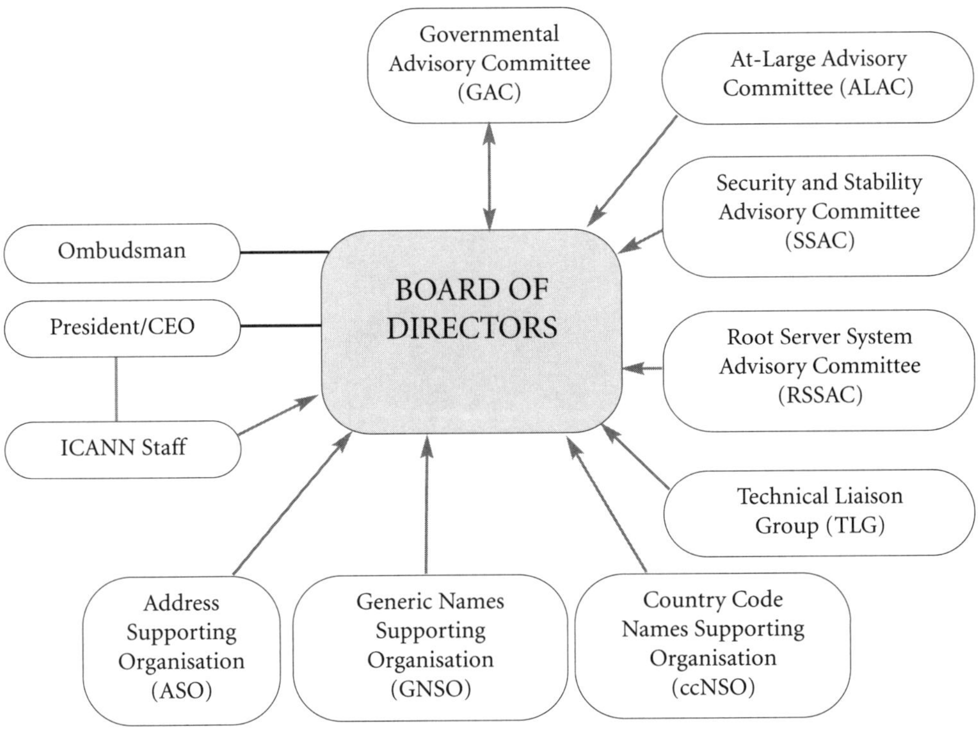

Committees of the Board of Directors

- Audit Committee
- Board Governance Committee
- Committee on Conflicts of Interest
- Committee on Reconsideration
- Compensation Committee
- Executive Committee
- Finance Committee
- Meetings Committee

Other Committees and Task Forces

- President's IANA Consultation Committee
- President's Strategy Committee
- President's Standing Committee on Privacy
- President's Advisory Committee for IDNs
- Nominating Committee
- ICANN Board and ICANN Governmental Advisory Committee Working Group

Appendix 2

Uniform Domain Name Dispute Resolution Policy

Policy Adopted: August 26, 1999
Implementation Documents Approved: October 24, 1999

ICANN

Notes:

1. This policy is now in effect. See www.icann.org/udrp/udrp-schedule.htm for the implementation schedule.

2. This policy has been adopted by all accredited domain-name registrars for domain names ending in .com, .net, and .org. It has also been adopted by certain managers of country-code top-level domains (e.g., .nu, .tv, .ws).

3. The policy is between the registrar (or other registration authority in the case of a country-code top-level domain) and its customer (the domain-name holder or registrant). Thus, the policy uses "we" and "our" to refer to the registrar and it uses "you" and "your" to refer to the domain-name holder.

Uniform Domain Name Dispute Resolution Policy

(As Approved by ICANN on October 24, 1999)

1. Purpose. This Uniform Domain Name Dispute Resolution Policy (the "Policy") has been adopted by the Internet Corporation for Assigned Names and Numbers ("ICANN"), is incorporated by reference into your Registration Agreement, and sets forth the terms and conditions in connection with a dispute between you and any party other than us (the registrar) over the registration and use of an Internet domain name registered by you. Proceedings under Paragraph 4 of this Policy will be conducted according to the Rules for Uniform Domain Name Dispute Resolution Policy (the "Rules of Procedure"), which are available at www.icann.org/udrp/udrp-rules-24oct99.htm, and the selected administrative-dispute-resolution service provider's supplemental rules.

2. Your Representations. By applying to register a domain name, or by asking us to maintain or renew a domain name registration, you hereby represent and warrant to us that (a) the statements that you made in your Registration Agreement are complete and accurate;

(b) to your knowledge, the registration of the domain name will not infringe upon or otherwise violate the rights of any third party; (c) you are not registering the domain name for an unlawful purpose; and (d) you will not knowingly use the domain name in violation of any applicable laws or regulations. It is your responsibility to determine whether your domain name registration infringes or violates someone else's rights.

3. Cancellations, Transfers, and Changes. We will cancel, transfer or otherwise make changes to domain name registrations under the following circumstances:

a. subject to the provisions of Paragraph 8, our receipt of written or appropriate electronic instructions from you or your authorized agent to take such action;

b. our receipt of an order from a court or arbitral tribunal, in each case of competent jurisdiction, requiring such action; and/or

c. our receipt of a decision of an Administrative Panel requiring such action in any administrative proceeding to which you were a party and which was conducted under this Policy or a later version of this Policy adopted by ICANN. (See Paragraph 4(i) and (k) below.)

We may also cancel, transfer or otherwise make changes to a domain name registration in accordance with the terms of your Registration Agreement or other legal requirements.

4. Mandatory Administrative Proceeding.

This Paragraph sets forth the type of disputes for which you are required to submit to a mandatory administrative proceeding. These proceedings will be conducted before one of the administrative-dispute-resolution service providers listed at www.icann.org/udrp/approved-providers.htm (each, a "Provider").

a. Applicable Disputes. You are required to submit to a mandatory administrative proceeding in the event that a third party (a "complainant") asserts to the applicable Provider, in compliance with the Rules of Procedure, that

(i) your domain name is identical or confusingly similar to a trademark or service mark in which the complainant has rights; and

(ii) you have no rights or legitimate interests in respect of the domain name; and

(iii) your domain name has been registered and is being used in bad faith.

In the administrative proceeding, the complainant must prove that each of these three elements are present.

b. Evidence of Registration and Use in Bad Faith. For the purposes of Paragraph 4(a)(iii), the following circumstances, in particular but without limitation, if found by the Panel to be present, shall be evidence of the registration and use of a domain name in bad faith:

(i) circumstances indicating that you have registered or you have acquired the domain name primarily for the purpose of selling, renting, or otherwise transferring the domain name registration to the complainant who is the owner of the trademark or service mark or to a competitor of that complainant, for valuable consideration in excess of your documented out-of-pocket costs directly related to the domain name; or

(ii) you have registered the domain name in order to prevent the owner of the trademark or service mark from reflecting the mark in a corresponding domain name, provided that you have engaged in a pattern of such conduct; or

(iii) you have registered the domain name primarily for the purpose of disrupting the business of a competitor; or

(iv) by using the domain name, you have intentionally attempted to attract, for commercial gain, Internet users to your web site or other on-line location, by creating a likelihood of confusion with the complainant's mark as to the source, sponsorship, affiliation, or endorsement of your web site or location or of a product or service on your web site or location.

c. How to Demonstrate Your Rights to and Legitimate Interests in the Domain Name in Responding to a Complaint. When you receive a complaint, you should refer to Paragraph 5 of the Rules of Procedure in determining how your response should be prepared. Any of the following circumstances, in particular but without limitation, if found by the Panel to be proved based on its evaluation of all evidence presented, shall demonstrate your rights or legitimate interests to the domain name for purposes of Paragraph 4(a)(ii):

(i) before any notice to you of the dispute, your use of, or demonstrable preparations to use, the domain name or a name corresponding to the domain name in connection with a bona fide offering of goods or services; or

(ii) you (as an individual, business, or other organization) have been commonly known by the domain name, even if you have acquired no trademark or service mark rights; or

(iii) you are making a legitimate noncommercial or fair use of the domain name, without intent for commercial gain to misleadingly divert consumers or to tarnish the trademark or service mark at issue.

d. Selection of Provider. The complainant shall select the Provider from among those approved by ICANN by submitting the complaint to that Provider. The selected Provider will administer the proceeding, except in cases of consolidation as described in Paragraph 4(f).

e. Initiation of Proceeding and Process and Appointment of Administrative Panel. The Rules of Procedure state the process for initiating and conducting a proceeding and for appointing the panel that will decide the dispute (the "Administrative Panel").

f. Consolidation. In the event of multiple disputes between you and a complainant, either you or the complainant may petition to consolidate the disputes before a single Administrative Panel. This petition shall be made to the first Administrative Panel appointed to hear a pending dispute between the parties. This Administrative Panel may consolidate before it any or all such disputes in its sole discretion, provided that the disputes being consolidated are governed by this Policy or a later version of this Policy adopted by ICANN.

g. Fees. All fees charged by a Provider in connection with any dispute before an Administrative Panel pursuant to this Policy shall be paid by the complainant, except in cases where you elect to expand the Administrative Panel from one to three panelists as provided in Paragraph 5(b)(iv) of the Rules of Procedure, in which case all fees will be split evenly by you and the complainant.

h. Our Involvement in Administrative Proceedings. We do not, and will not, participate in the administration or conduct of any proceeding before an Administrative Panel. In addition, we will not be liable as a result of any decisions rendered by the Administrative Panel.

i. Remedies. The remedies available to a complainant pursuant to any proceeding before an Administrative Panel shall be limited to requiring the cancellation of your domain name or the transfer of your domain name registration to the complainant.

j. Notification and Publication. The Provider shall notify us of any decision made by an Administrative Panel with respect to a domain name you have registered with us. All decisions under this Policy will be published in full over the Internet, except when an Administrative Panel determines in an exceptional case to redact portions of its decision.

k. Availability of Court Proceedings. The mandatory administrative proceeding requirements set forth in Paragraph 4 shall not prevent either you or the complainant from submitting the dispute to a court of competent jurisdiction for independent resolution before such mandatory administrative proceeding is commenced or after such proceeding is concluded. If an Administrative Panel decides that your domain name registration should be canceled or transferred, we will wait ten (10) business days (as observed in the location of our principal office) after we are informed by the applicable Provider of the Administrative Panel's decision before implementing that decision. We will then implement the decision unless we have received from you during that ten (10) business day period official documentation (such as a copy of a complaint, file-stamped by the clerk of the court) that you have commenced a lawsuit against the complainant in a jurisdiction to which the complainant has submitted under Paragraph 3(b)(xiii) of the Rules of Procedure. (In general, that jurisdiction is either the location of our principal office or of your address as shown in our Whois database. See Paragraphs 1 and 3(b)(xiii) of the Rules of Procedure for details.) If we receive such documentation within the ten (10) business day period, we will not implement the Administrative Panel's decision, and we will take no further action, until we receive (i) evidence satisfactory to us of a resolution between the parties; (ii) evidence satisfactory to us that your lawsuit has been dismissed or withdrawn; or (iii) a copy of an order from such court dismissing your lawsuit or ordering that you do not have the right to continue to use your domain name.

5. <u>All Other Disputes and Litigation</u>. All other disputes between you and any party other than us regarding your domain name registration that are not brought pursuant to the mandatory administrative proceeding provisions of Paragraph 4 shall be resolved between you and such other party through any court, arbitration or other proceeding that may be available.

6. <u>Our Involvement in Disputes</u>. We will not participate in any way in any dispute between you and any party other than us regarding the registration and use of your domain name. You shall not name us as a party or otherwise include us in any such proceeding. In the event that we are named as a party in any such proceeding, we reserve the right to raise any and all defenses deemed appropriate, and to take any other action necessary to defend ourselves.

7. <u>Maintaining the Status Quo</u>. We will not cancel, transfer, activate, deactivate, or otherwise change the status of any domain name registration under this Policy except as provided in Paragraph 3 above.

8. <u>Transfers During a Dispute</u>.

a. Transfers of a Domain Name to a New Holder. You may not transfer your domain name registration to another holder (i) during a pending administrative proceeding brought pursuant to Paragraph 4 or for a period of fifteen (15) business days (as observed in the location of our principal place of business) after such proceeding is concluded; or (ii) during a pending court proceeding or arbitration commenced regarding your domain name unless the party to whom the domain name registration is being transferred agrees, in writing, to be bound by the decision of the court or arbitrator. We reserve the right to cancel any transfer of a domain name registration to another holder that is made in violation of this subparagraph.

b. Changing Registrars. You may not transfer your domain name registration to another registrar during a pending administrative proceeding brought pursuant to Paragraph 4 or for a period of fifteen (15) business days (as observed in the location of our principal place of business) after such proceeding is concluded. You may transfer administration of your domain name registration to another registrar during a pending court action or arbitration, provided that the domain name you have registered with us shall continue to be subject to the proceedings commenced against you in accordance with the terms of this Policy. In the event that you transfer a domain name registration to us during the pendency of a court action or arbitration, such dispute shall remain subject to the domain name dispute policy of the registrar from which the domain name registration was transferred.

9. Policy Modifications. We reserve the right to modify this Policy at any time with the permission of ICANN. We will post our revised Policy at least thirty (30) calendar days before it becomes effective. Unless this Policy has already been invoked by the submission of a complaint to a Provider, in which event the version of the Policy in effect at the time it was invoked will apply to you until the dispute is over, all such changes will be binding upon you with respect to any domain name registration dispute, whether the dispute arose before, on or after the effective date of our change. In the event that you object to a change in this Policy, your sole remedy is to cancel your domain name registration with us, provided that you will not be entitled to a refund of any fees you paid to us. The revised Policy will apply to you until you cancel your domain name registration.

Appendix 3

ICANN

Rules for Uniform Domain Name Dispute Resolution Policy

Policy Adopted: August 26, 1999
Implementation Documents Approved: October 24, 1999

Note: These rules are now in effect. See www.icann.org/udrp/udrp-schedule.htm for the implementation schedule.

Rules for Uniform Domain Name Dispute Resolution Policy (the "Rules")

(As Approved by ICANN on October 24, 1999)

Administrative proceedings for the resolution of disputes under the Uniform Dispute Resolution Policy adopted by ICANN shall be governed by these Rules and also the Supplemental Rules of the Provider administering the proceedings, as posted on its web site.

1. Definitions

In these Rules:

Complainant means the party initiating a complaint concerning a domain-name registration.

ICANN refers to the Internet Corporation for Assigned Names and Numbers.

Mutual Jurisdiction means a court jurisdiction at the location of either (a) the principal office of the Registrar (provided the domain-name holder has submitted in its Registration Agreement to that jurisdiction for court adjudication of disputes concerning or arising from the use of the domain name) or (b) the domain-name holder's address as shown for the registration of the domain name in Registrar's Whois database at the time the complaint is submitted to the Provider.

Panel means an administrative panel appointed by a Provider to decide a complaint concerning a domain-name registration.

Panelist means an individual appointed by a Provider to be a member of a Panel.

Party means a Complainant or a Respondent.

Policy means the Uniform Domain Name Dispute Resolution Policy that is incorporated by reference and made a part of the Registration Agreement.

Provider means a dispute-resolution service provider approved by ICANN. A list of such Providers appears at www.icann.org/udrp/approved-providers.htm.

Registrar means the entity with which the Respondent has registered a domain name that is the subject of a complaint.

Registration Agreement means the agreement between a Registrar and a domain-name holder.

Respondent means the holder of a domain-name registration against which a complaint is initiated.

Reverse Domain Name Hijacking means using the Policy in bad faith to attempt to deprive a registered domain-name holder of a domain name.

Supplemental Rules means the rules adopted by the Provider administering a proceeding to supplement these Rules. Supplemental Rules shall not be inconsistent with the Policy or these Rules and shall cover such topics as fees, word and page limits and guidelines, the means for communicating with the Provider and the Panel, and the form of cover sheets.

2. Communications

(a) When forwarding a complaint to the Respondent, it shall be the Provider's responsibility to employ reasonably available means calculated to achieve actual notice to Respondent. Achieving actual notice, or employing the following measures to do so, shall discharge this responsibility:

(i) sending the complaint to all postal-mail and facsimile addresses (A) shown in the domain name's registration data in Registrar's Whois database for the registered domain-name holder, the technical contact, and the administrative contact and (B) supplied by Registrar to the Provider for the registration's billing contact; and

(ii) sending the complaint in electronic form (including annexes to the extent available in that form) by e-mail to:

(A) the e-mail addresses for those technical, administrative, and billing contacts;

(B) postmaster@<the contested domain name>; and

(C) if the domain name (or "www." followed by the domain name) resolves to an active web page (other than a generic page the Provider concludes is maintained by a registrar or ISP for parking domain-names registered by multiple domain-name holders), any e-mail address shown or e-mail links on that web page; and

(iii) sending the complaint to any address the Respondent has notified the Provider it prefers and, to the extent practicable, to all other addresses provided to the Provider by Complainant under Paragraph 3(b)(v).

(b) Except as provided in Paragraph 2(a), any written communication to Complainant or Respondent provided for under these Rules shall be made by the preferred means stated by the Complainant or Respondent, respectively (see Paragraphs 3(b)(iii) and 5(b)(iii)), or in the absence of such specification

(i) by telecopy or facsimile transmission, with a confirmation of transmission; or

(ii) by postal or courier service, postage pre-paid and return receipt requested; or

(iii) electronically via the Internet, provided a record of its transmission is available.

(c) Any communication to the Provider or the Panel shall be made by the means and in the manner (including number of copies) stated in the Provider's Supplemental Rules.

(d) Communications shall be made in the language prescribed in Paragraph 11. E-mail communications should, if practicable, be sent in plaintext.

(e) Either Party may update its contact details by notifying the Provider and the Registrar.

(f) Except as otherwise provided in these Rules, or decided by a Panel, all communications provided for under these Rules shall be deemed to have been made:

(i) if delivered by telecopy or facsimile transmission, on the date shown on the confirmation of transmission; or

(ii) if by postal or courier service, on the date marked on the receipt; or

(iii) if via the Internet, on the date that the communication was transmitted, provided that the date of transmission is verifiable.

(g) Except as otherwise provided in these Rules, all time periods calculated under these Rules to begin when a communication is made shall begin to run on the earliest date that the communication is deemed to have been made in accordance with Paragraph 2(f).

(h) Any communication by

(i) a Panel to any Party shall be copied to the Provider and to the other Party;

(ii) the Provider to any Party shall be copied to the other Party; and

(iii) a Party shall be copied to the other Party, the Panel and the Provider, as the case may be.

(i) It shall be the responsibility of the sender to retain records of the fact and circumstances of sending, which shall be available for inspection by affected parties and for reporting purposes.

(j) In the event a Party sending a communication receives notification of non-delivery of the communication, the Party shall promptly notify the Panel (or, if no Panel is yet appointed, the Provider) of the circumstances of the notification. Further proceedings concerning the communication and any response shall be as directed by the Panel (or the Provider).

3. The Complaint

(a) Any person or entity may initiate an administrative proceeding by submitting a complaint in accordance with the Policy and these Rules to any Provider approved by ICANN. (Due to capacity constraints or for other reasons, a Provider's ability to accept complaints may be suspended at times. In that event, the Provider shall refuse the submission. The person or entity may submit the complaint to another Provider.)

(b) The complaint shall be submitted in hard copy and (except to the extent not available for annexes) in electronic form and shall:

(i) Request that the complaint be submitted for decision in accordance with the Policy and these Rules;

(ii) Provide the name, postal and e-mail addresses, and the telephone and telefax numbers of the Complainant and of any representative authorized to act for the Complainant in the administrative proceeding;

(iii) Specify a preferred method for communications directed to the Complainant in the administrative proceeding (including person to be contacted, medium, and address information) for each of (A) electronic-only material and (B) material including hard copy;

(iv) Designate whether Complainant elects to have the dispute decided by a single-member or a three-member Panel and, in the event Complainant elects a three-member Panel, provide the names and contact details of three candidates to serve as one of the Panelists (these candidates may be drawn from any ICANN-approved Provider's list of panelists);

(v) Provide the name of the Respondent (domain-name holder) and all information (including any postal and e-mail addresses and telephone and telefax numbers) known to Complainant regarding how to contact Respondent or any representative of Respondent, including contact information based on pre-complaint dealings, in sufficient detail to allow the Provider to send the complaint as described in Paragraph 2(a);

(vi) Specify the domain name(s) that is/are the subject of the complaint;

(vii) Identify the Registrar(s) with whom the domain name(s) is/are registered at the time the complaint is filed;

(viii) Specify the trademark(s) or service mark(s) on which the complaint is based and, for each mark, describe the goods or services, if any, with which the mark is used (Complainant may also separately describe other goods and services with which it intends, at the time the complaint is submitted, to use the mark in the future.);

(ix) Describe, in accordance with the Policy, the grounds on which the complaint is made including, in particular,

(1) the manner in which the domain name(s) is/are identical or confusingly similar to a trademark or service mark in which the Complainant has rights; and

(2) why the Respondent (domain-name holder) should be considered as having no rights or legitimate interests in respect of the domain name(s) that is/are the subject of the complaint; and

(3) why the domain name(s) should be considered as having been registered and being used in bad faith

(The description should, for elements (2) and (3), discuss any aspects of Paragraphs 4(b) and 4(c) of the Policy that are applicable. The description shall comply with any word or page limit set forth in the Provider's Supplemental Rules.);

(x) Specify, in accordance with the Policy, the remedies sought;

(xi) Identify any other legal proceedings that have been commenced or terminated in connection with or relating to any of the domain name(s) that are the subject of the complaint;

(xii) State that a copy of the complaint, together with the cover sheet as prescribed by the Provider's Supplemental Rules, has been sent or transmitted to the Respondent (domain-name holder), in accordance with Paragraph 2(b);

(xiii) State that Complainant will submit, with respect to any challenges to a decision in the administrative proceeding canceling or transferring the domain name, to the jurisdiction of the courts in at least one specified Mutual Jurisdiction;

(xiv) Conclude with the following statement followed by the signature of the Complainant or its authorized representative:

"Complainant agrees that its claims and remedies concerning the registration of the domain name, the dispute, or the dispute's resolution shall be solely against the domain-name holder and waives all such claims and remedies against (a) the dispute-resolution provider and panelists, except in the case of deliberate wrongdoing, (b) the registrar, (c) the registry administrator, and (d) the Internet Corporation for Assigned Names and Numbers, as well as their directors, officers, employees, and agents."

"Complainant certifies that the information contained in this Complaint is to the best of Complainant's knowledge complete and accurate, that this Complaint is not being presented for any improper purpose, such as to harass, and that the assertions in this Complaint are warranted under these Rules and under applicable law, as it now exists or as it may be extended by a good-faith and reasonable argument."; and

(xv) Annex any documentary or other evidence, including a copy of the Policy applicable to the domain name(s) in dispute and any trademark or service mark registration upon which the complaint relies, together with a schedule indexing such evidence.

(c) The complaint may relate to more than one domain name, provided that the domain names are registered by the same domain-name holder.

4. Notification of Complaint

(a) The Provider shall review the complaint for administrative compliance with the Policy and these Rules and, if in compliance, shall forward the complaint (together with the explanatory cover sheet prescribed by the Provider's Supplemental Rules) to the Respondent, in the manner prescribed by Paragraph 2(a), within three (3) calendar days following receipt of the fees to be paid by the Complainant in accordance with Paragraph 19.

(b) If the Provider finds the complaint to be administratively deficient, it shall promptly notify the Complainant and the Respondent of the nature of the deficiencies identified. The Complainant shall have five (5) calendar days within which to correct any such deficiencies, after which the administrative proceeding will be deemed withdrawn without prejudice to submission of a different complaint by Complainant.

(c) The date of commencement of the administrative proceeding shall be the date on which the Provider completes its responsibilities under Paragraph 2(a) in connection with forwarding the Complaint to the Respondent.

(d) The Provider shall immediately notify the Complainant, the Respondent, the concerned Registrar(s), and ICANN of the date of commencement of the administrative proceeding.

5. The Response

(a) Within twenty (20) days of the date of commencement of the administrative proceeding the Respondent shall submit a response to the Provider.

(b) The response shall be submitted in hard copy and (except to the extent not available for annexes) in electronic form and shall:

(i) Respond specifically to the statements and allegations contained in the complaint and include any and all bases for the Respondent (domain-name holder) to retain registration and use of the disputed domain name (This portion of the response shall comply with any word or page limit set forth in the Provider's Supplemental Rules.);

(ii) Provide the name, postal and e-mail addresses, and the telephone and telefax numbers of the Respondent (domain-name holder) and of any representative authorized to act for the Respondent in the administrative proceeding;

(iii) Specify a preferred method for communications directed to the Respondent in the administrative proceeding (including person to be contacted, medium, and address information) for each of (A) electronic-only material and (B) material including hard copy;

(iv) If Complainant has elected a single-member panel in the Complaint (see Paragraph 3(b)(iv)), state whether Respondent elects instead to have the dispute decided by a three-member panel;

(v) If either Complainant or Respondent elects a three-member Panel, provide the names and contact details of three candidates to serve as one of the Panelists (these candidates may be drawn from any ICANN-approved Provider's list of panelists);

(vi) Identify any other legal proceedings that have been commenced or terminated in connection with or relating to any of the domain name(s) that are the subject of the complaint;

(vii) State that a copy of the response has been sent or transmitted to the Complainant, in accordance with Paragraph 2(b); and

(viii) Conclude with the following statement followed by the signature of the Respondent or its authorized representative:

"Respondent certifies that the information contained in this Response is to the best of Respondent's knowledge complete and accurate, that this Response is not being presented for any improper purpose, such as to harass, and that the assertions in this Response are warranted under these Rules and under applicable law, as it now exists or as it may be extended by a good-faith and reasonable argument."; and

(ix) Annex any documentary or other evidence upon which the Respondent relies, together with a schedule indexing such documents.

(c) If Complainant has elected to have the dispute decided by a single-member Panel and Respondent elects a three-member Panel, Respondent shall be required to pay one-half of the applicable fee for a three-member Panel as set forth in the Provider's Supplemental Rules. This payment shall be made together with the submission of the response to the Provider. In the event that the required payment is not made, the dispute shall be decided by a single-member Panel.

(d) At the request of the Respondent, the Provider may, in exceptional cases, extend the period of time for the filing of the response. The period may also be extended by written stipulation between the Parties, provided the stipulation is approved by the Provider.

(e) If a Respondent does not submit a response, in the absence of exceptional circumstances, the Panel shall decide the dispute based upon the complaint.

6. Appointment of the Panel and Timing of Decision

(a) Each Provider shall maintain and publish a publicly available list of panelists and their qualifications.

(b) If neither the Complainant nor the Respondent has elected a three-member Panel (Paragraphs 3(b)(iv) and 5(b)(iv)), the Provider shall appoint, within five (5) calendar days following receipt of the response by the Provider, or the lapse of the time period for the submission thereof, a single Panelist from its list of panelists. The fees for a single-member Panel shall be paid entirely by the Complainant.

(c) If either the Complainant or the Respondent elects to have the dispute decided by a three-member Panel, the Provider shall appoint three Panelists in accordance with the procedures identified in Paragraph 6(e). The fees for a three-member Panel shall be paid in their entirety by the Complainant, except where the election for a three-member Panel was made by the Respondent, in which case the applicable fees shall be shared equally between the Parties.

(d) Unless it has already elected a three-member Panel, the Complainant shall submit to the Provider, within five (5) calendar days of communication of a response in which the Respondent elects a three-member Panel, the names and contact details of three candidates to serve as one of the Panelists. These candidates may be drawn from any ICANN-approved Provider's list of panelists.

(e) In the event that either the Complainant or the Respondent elects a three-member Panel, the Provider shall endeavor to appoint one Panelist from the list of candidates provided by each of the Complainant and the Respondent. In the event the Provider is unable within five (5) calendar days to secure the appointment of a Panelist on its customary terms from either Party's list of candidates, the Provider shall make that appointment from its list of panelists. The third Panelist shall be appointed by the Provider from a list of five candidates submitted by the Provider to the Parties, the Provider's selection from among the five being made in a manner that reasonably balances the preferences of both Parties, as they may specify to the Provider within five (5) calendar days of the Provider's submission of the five-candidate list to the Parties.

(f) Once the entire Panel is appointed, the Provider shall notify the Parties of the Panelists appointed and the date by which, absent exceptional circumstances, the Panel shall forward its decision on the complaint to the Provider.

7. Impartiality and Independence

A Panelist shall be impartial and independent and shall have, before accepting appointment, disclosed to the Provider any circumstances giving rise to justifiable doubt as to the Panelist's impartiality or independence. If, at any stage during the administrative proceeding, new circumstances arise that could give rise to justifiable doubt as to the impartiality or independence of the Panelist, that Panelist shall promptly disclose such circumstances to the Provider. In such event, the Provider shall have the discretion to appoint a substitute Panelist.

8. Communication Between Parties and the Panel

No Party or anyone acting on its behalf may have any unilateral communication with the Panel. All communications between a Party and the Panel or the Provider shall be made to a case administrator appointed by the Provider in the manner prescribed in the Provider's Supplemental Rules.

9. Transmission of the File to the Panel

The Provider shall forward the file to the Panel as soon as the Panelist is appointed in the case of a Panel consisting of a single member, or as soon as the last Panelist is appointed in the case of a three-member Panel.

10. General Powers of the Panel

(a) The Panel shall conduct the administrative proceeding in such manner as it considers appropriate in accordance with the Policy and these Rules.

(b) In all cases, the Panel shall ensure that the Parties are treated with equality and that each Party is given a fair opportunity to present its case.

(c) The Panel shall ensure that the administrative proceeding takes place with due expedition. It may, at the request of a Party or on its own motion, extend, in exceptional cases, a period of time fixed by these Rules or by the Panel.

(d) The Panel shall determine the admissibility, relevance, materiality and weight of the evidence.

(e) A Panel shall decide a request by a Party to consolidate multiple domain name disputes in accordance with the Policy and these Rules.

11. Language of Proceedings

(a) Unless otherwise agreed by the Parties, or specified otherwise in the Registration Agreement, the language of the administrative proceeding shall be the language of the Registration Agreement, subject to the authority of the Panel to determine otherwise, having regard to the circumstances of the administrative proceeding.

(b) The Panel may order that any documents submitted in languages other than the language of the administrative proceeding be accompanied by a translation in whole or in part into the language of the administrative proceeding.

12. Further Statements

In addition to the complaint and the response, the Panel may request, in its sole discretion, further statements or documents from either of the Parties.

13. In-Person Hearings

There shall be no in-person hearings (including hearings by teleconference, videoconference, and web conference), unless the Panel determines, in its sole discretion and as an exceptional matter, that such a hearing is necessary for deciding the complaint.

14. Default

(a) In the event that a Party, in the absence of exceptional circumstances, does not comply with any of the time periods established by these Rules or the Panel, the Panel shall proceed to a decision on the complaint.

(b) If a Party, in the absence of exceptional circumstances, does not comply with any provision of, or requirement under, these Rules or any request from the Panel, the Panel shall draw such inferences therefrom as it considers appropriate.

15. Panel Decisions

(a) A Panel shall decide a complaint on the basis of the statements and documents submitted and in accordance with the Policy, these Rules and any rules and principles of law that it deems applicable.

(b) In the absence of exceptional circumstances, the Panel shall forward its decision on the complaint to the Provider within fourteen (14) days of its appointment pursuant to Paragraph 6.

(c) In the case of a three-member Panel, the Panel's decision shall be made by a majority.

(d) The Panel's decision shall be in writing, provide the reasons on which it is based, indicate the date on which it was rendered and identify the name(s) of the Panelist(s).

(e) Panel decisions and dissenting opinions shall normally comply with the guidelines as to length set forth in the Provider's Supplemental Rules. Any dissenting opinion shall accompany the majority decision. If the Panel concludes that the dispute is not within the scope of Paragraph 4(a) of the Policy, it shall so state. If after considering the submissions the Panel finds that the complaint was brought in bad faith, for example in an attempt at Reverse Domain Name Hijacking or was brought primarily to harass the domain-name holder, the Panel shall declare in its decision that the complaint was brought in bad faith and constitutes an abuse of the administrative proceeding.

16. Communication of Decision to Parties

(a) Within three (3) calendar days after receiving the decision from the Panel, the Provider shall communicate the full text of the decision to each Party, the concerned Registrar(s), and ICANN. The concerned Registrar(s) shall immediately communicate to each Party, the Provider, and ICANN the date for the implementation of the decision in accordance with the Policy.

(b) Except if the Panel determines otherwise (see Paragraph 4(j) of the Policy), the Provider shall publish the full decision and the date of its implementation on a publicly accessible web site. In any event, the portion of any decision determining a complaint to have been brought in bad faith (see Paragraph 15(e) of these Rules) shall be published.

17. Settlement or Other Grounds for Termination

(a) If, before the Panel's decision, the Parties agree on a settlement, the Panel shall terminate the administrative proceeding.

(b) If, before the Panel's decision is made, it becomes unnecessary or impossible to continue the administrative proceeding for any reason, the Panel shall terminate the administrative proceeding, unless a Party raises justifiable grounds for objection within a period of time to be determined by the Panel.

18. Effect of Court Proceedings

(a) In the event of any legal proceedings initiated prior to or during an administrative proceeding in respect of a domain-name dispute that is the subject of the complaint, the Panel shall have the discretion to decide whether to suspend or terminate the administrative proceeding, or to proceed to a decision.

(b) In the event that a Party initiates any legal proceedings during the pendency of an administrative proceeding in respect of a domain-name dispute that is the subject of the complaint, it shall promptly notify the Panel and the Provider. See Paragraph 8 above.

19. Fees

(a) The Complainant shall pay to the Provider an initial fixed fee, in accordance with the Provider's Supplemental Rules, within the time and in the amount required. A Respondent electing under Paragraph 5(b)(iv) to have the dispute decided by a three-member Panel, rather than the single-member Panel elected by the Complainant, shall pay the Provider one-half the fixed fee for a three-member Panel. See Paragraph 5(c). In all other cases, the Complainant shall bear all of the Provider's fees, except as prescribed under Paragraph 19(d). Upon appointment of the Panel, the Provider shall refund the appropriate portion, if any, of the initial fee to the Complainant, as specified in the Provider's Supplemental Rules.

(b) No action shall be taken by the Provider on a complaint until it has received from Complainant the initial fee in accordance with Paragraph 19(a).

(c) If the Provider has not received the fee within ten (10) calendar days of receiving the complaint, the complaint shall be deemed withdrawn and the administrative proceeding terminated.

(d) In exceptional circumstances, for example in the event an in-person hearing is held, the Provider shall request the Parties for the payment of additional fees, which shall be established in agreement with the Parties and the Panel.

20. Exclusion of Liability

Except in the case of deliberate wrongdoing, neither the Provider nor a Panelist shall be liable to a Party for any act or omission in connection with any administrative proceeding under these Rules.

21. Amendments

The version of these Rules in effect at the time of the submission of the complaint to the Provider shall apply to the administrative proceeding commenced thereby. These Rules may not be amended without the express written approval of ICANN.

Appendix 4

GNSO Final Report, *Introduction of New Generic Top-Level Domains* (8 August 2007)

Principles, Recommendations and Implementation Guidelines

	PRINCIPLES	MISSION & CORE VALUES
A	New generic top-level domains (gTLDs) must be introduced in an orderly, timely and predictable way.	M1 & CV1 & 2, 4–10
B	Some new generic top-level domains should be internationalised domain names (IDNs) subject to the approval of IDNs being available in the root.	M1–3 & CV 1, 4 & 6
C	The reasons for introducing new top-level domains include that there is demand from potential applicants for new top-level domains in both ASCII and IDN formats. In addition the introduction of new top-level domain application process has the potential to promote competition in the provision of registry services, to add to consumer choice, market differentiation and geographical and service-provider diversity.	M3 & CV 4–10
D	A set of technical criteria must be used for assessing a new gTLD registry applicant to minimise the risk of harming the operational stability, security and global interoperability of the Internet.	M1–3 & CV 1
E	A set of capability criteria for a new gTLD registry applicant must be used to provide an assurance that an applicant has the capability to meets its obligations under the terms of ICANN's registry agreement.	M1–3 & CV 1
F	A set of operational criteria must be set out in contractual conditions in the registry agreement to ensure compliance with ICANN policies.	M1–3 & CV 1
G	The string evaluation process must not infringe the applicant's freedom of expression rights that are protected under internationally recognized principles of law.	

	PRINCIPLES	MISSION & CORE VALUES
1	ICANN must implement a process that allows the introduction of new top-level domains. The evaluation and selection procedure for new gTLD registries should respect the principles of fairness, transparency and non-discrimination.	
	All applicants for a new gTLD registry should therefore be evaluated against transparent and predictable criteria, fully available to the applicants prior to the initiation of the process. Normally, therefore, no subsequent additional selection criteria should be used in the selection process.	M1-3 & CV1-11
2	Strings must not be confusingly similar to an existing top-level domain or a Reserved Name.	M1-3 & C1-6-11
3	Strings must not infringe the existing legal rights of others that are recognized or enforceable under generally accepted and internationally recognized principles of law.	
	Examples of these legal rights that are internationally recognized include, but are not limited to, rights defined in the Paris Convention for the Protection of Industry Property (in particular trademark rights), the Universal Declaration of Human Rights (UDHR) and the International Covenant on Civil and Political Rights (ICCPR) (in particular freedom of expression rights).	CV3
4	Strings must not cause any technical instability.	M1-3 & CV 1
5	Strings must not be a Reserved Word.	M1-3 & CV 1 & 3
6*	Strings must not be contrary to generally accepted legal norms relating to morality and public order that are recognized under international principles of law. Examples of such principles of law include, but are not limited to, the Universal Declaration of Human Rights (UDHR), the International Covenant on Civil and Political Rights (ICCPR), the Convention on the Elimination of All Forms of Discrimination Against Women (CEDAW) and the International Convention on the Elimination of All Forms of Racial Discrimination, intellectual property treaties administered by the World Intellectual Property Organisation (WIPO) and the WTO Agreement on Trade-Related Aspects of Intellectual Property (TRIPS).	M3 & CV 4

* The NCUC submitted Minority Statements on Recommendations 6 and 20. The remainder of the Recommendations have support from all GNSO Constituencies.

	PRINCIPLES	MISSION & CORE VALUES
7	Applicants must be able to demonstrate their technical capability to run a registry operation for the purpose that the applicant sets out.	M1-3 & CV1
8	Applicants must be able to demonstrate their financial and organisational operational capability.	M1-3 & CV1
9	There must be a clear and pre-published application process using objective and measurable criteria.	M3 & CV6-9
10	There must be a base contract provided to applicants at the beginning of the application process.	CV7-9
11	[Replaced with Recommendation 20 and Implementation Guideline P and inserted into Term of Reference 3 Allocation Methods section]	
12	Dispute resolution and challenge processes must be established prior to the start of the process.	CV7-9
13	Applications must initially be assessed in rounds until the scale of demand is clear.	CV7-9
14	The initial registry agreement term must be of a commercially reasonable length.	CV5-9
15	There must be renewal expectancy.	CV5-9
16	Registries must apply existing Consensus Policies and adopt new Consensus Policies as they are approved.	CV5-9
17	A clear compliance and sanctions process must be set out in the base contract which could lead to contract termination.	M1 & CV1
18	If an applicant offers an IDN service, then ICANN's IDN guidelines must be followed.	M1 & CV1
19	Registries must use only ICANN accredited registrars in registering domain names and may not discriminate among such accredited registrars.	M1 & CV1
20*	An application will be rejected if an expert panel determines that there is substantial opposition to it from a significant portion of the community to which the string may be explicitly or implicitly targeted.	

* The NCUC submitted Minority Statements on Recommendations 6 and 20. The remainder of the Recommendations have support from all GNSO Constituencies.

	PRINCIPLES	MISSION & CORE VALUES
IG A	The application process will provide a pre-defined roadmap for applicants that encourages the submission of applications for new top-level domains.	CV 2, 5, 6, 8 & 9
IG B	Application fees will be designed to ensure that adequate resources exist to cover the total cost to administer the new gTLD process. Application fees may differ for applicants.	CV 5, 6, 8 & 9
IG C	ICANN will provide frequent communications with applicants and the public including comment forums.	CV 9 & 10
IG D	A first come first served processing schedule within the application round will be implemented and will continue for an ongoing process, if necessary. Applications will be time and date stamped on receipt.	CV 8-10
IG E	The application submission date will be at least four months after the issue of the Request for Proposal and ICANN will promote the opening of the application round.	CV 9 & 10
IG F*	If there is contention for strings, applicants may: i) resolve contention between them within a pre-established timeframe ii) if there is no mutual agreement, a claim to support a community by one party will be a reason to award priority to that application. If there is no such claim, and no mutual agreement a process will be put in place to enable efficient resolution of contention and; iii) the ICANN Board may be used to make a final decision, using advice from staff and expert panels.	CV 7-10
IG H*	Where an applicant lays any claim that the TLD is intended to support a particular community such as a sponsored TLD, or any other TLD intended for a specified community, that claim will be taken on trust with the following exceptions: (i) the claim relates to a string that is also subject to another application and the claim to support a community is being used to gain priority for the application; and (ii) a formal objection process is initiated.	CV 7 - 10

* The NCUC submitted Minority Statements on Implementation Guidelines F, H & P. The remainder of the Implementation Guidelines have support from all GNSO Constituencies.

	PRINCIPLES	MISSION & CORE VALUES
	Under these exceptions, Staff Evaluators will devise criteria and procedures to investigate the claim. Under exception (ii), an expert panel will apply the process, guidelines, and definitions set forth in IG P.	
IG H	External dispute providers will give decisions on objections.	CV 10
IG I	An applicant granted a TLD string must use it within a fixed timeframe which will be specified in the application process.	CV 10
IG J	The base contract should balance market certainty and flexibility for ICANN to accommodate a rapidly changing market place.	CV 4-10
IG K	ICANN should take a consistent approach to the establishment of registry fees.	CV 5
IG L	The use of personal data must be limited to the purpose for which it is collected.	CV 8
IG M	ICANN may establish a capacity building and support mechanism aiming at facilitating effective communication on important and technical Internet governance functions in a way that no longer requires all participants in the conversation to be able to read and write English.	CV 3–7
IG N	ICANN may put in place a fee reduction scheme for gTLD applicants from economies classified by the UN as least developed.	CV 3–7
IG O	ICANN may put in place systems that could provide information about the gTLD process in major languages other than English, for example, in the six working languages of the United Nations.	CV 8–10
IG P*	The following process, definitions and guidelines refer to Recommendation 20. **Process** Opposition must be objection based. Determination will be made by a dispute resolution panel constituted for the purpose.	

* The NCUC submitted Minority Statements on Recommendations 6 and 20. The remainder of the Recommendations have support from all GNSO Constituencies.

PRINCIPLES	MISSION & CORE VALUES
The objector must provide verifiable evidence that it is an established institution of the community (perhaps like the RSTEP pool of panelists from which a small panel would be constituted for each objection).	

Guidelines

The task of the panel is the determination of substantial opposition.

a) **substantial**—in determining substantial the panel will assess the following: signification portion, community, explicitly targeting, implicitly targeting, established institution, formal existence, detriment

b) **significant portion**—in determining significant portion the panel will assess the balance between the level of objection submitted by one or more established institutions and the level of support provided in the application from one or more established institutions. The panel will assess significance proportionate to the explicit or implicit targeting.

c) **community**—community should be interpreted broadly and will include, for example, an economic sector, a cultural community, or a linguistic community. It may be a closely related community which believes it is impacted.

d) **explicitly targeting**—explicitly targeting means there is a description of the intended use of the TLD in the application.

e) **implicitly targeting**—implicitly targeting means that the objector makes an assumption of targeting or that the objector believes there may be confusion by users over its intended use.

f) **established institution**—an institution that has been in formal existence for at least 5 years. In exceptional cases, standing may be granted to an institution that has been in existence for fewer than 5 years.

   Exceptional circumstances include but are not limited to a re-organization, merger or an inherently younger community.

   The following ICANN organizations are defined as established institutions: GAC, ALAC, GNSO, ccNSO, ASO.

g) **formal existence**—formal existence may be demonstrated by appropriate public registration, public

	PRINCIPLES	MISSION & CORE VALUES
	historical evidence, validation by a government, intergovernmental organization, international treaty organization or similar. h) **detriment**—the objector must provide sufficient evidence to allow the panel to determine that there would be a likelihood of detriment to the rights or legitimate interests of the community or to users more widely.	
IG Q	ICANN staff will provide an automatic reply to all those who submit public comments that will explain the objection procedure.	
IG R	Once formal objections or disputes are accepted for review there will be a cooling off period to allow parties to resolve the dispute or objection before review by the panel is initiated.	

Bibliography

1. Reports and Policy Statements

Clinton, WJ and Gore, A, Jr, *Framework for Global Electronic Commerce* (1 July 1997), available at http://www.technology.gov/digeconomy/framewrk.htm.

Committee on ICANN Evolution and Reform (CIER), *ICANN: A Blueprint for Reform* (20 June 2002), available at http://www.icann.org/committees/evol-reform/blueprint-20jun02.htm.

Department of Commerce (US), *Statement Regarding Extension of Memorandum of Understanding with ICANN* (20 September 2002), available at http://www.ntia.doc.gov/ntiahome/domainname/agreements/docstatement-09192002.htm.

—— *Statement Regarding Extension of Memorandum of Understanding with ICANN* (16 September 2003), available at http://www.ntia.doc.gov/ntiahome/domainname/agreements/sepstatement-09162003.htm.

DNSO Working Group-A (ICANN), *WG-A Final Report to the Names Council* (29 July 1999), available at http://www.dnso.org/dnso/notes/19990729.WGA-report.html.

—— *WG-A Final Report to the ICANN Board* (3 August 1999), available at http://www.icann.org/dnso/wga-final-report.htm.

European Commission, *World Summit on the Information Society in Tunis, Internet Governance: Frequently asked questions,* Memo/05/428 (Brussels, 15 November 2005)., available at http://ec.europa.eu/information_society/activities/internationalrel/docs/wsis/m06-172.en.pdf.

Geist, M, *Fair.com? An Analysis of the Allegations of Systematic Unfairness in the ICANN UDRP* (August 2001), available at http://aix1.uottawa.ca/~geist/geistudrp.pdf.

ICANN, Staff Report, *Uniform Dispute Resolution Policy for gTLD Registrars* (24 August 1999), available at http://www.icann.org/santiago/udrp-staff-report.htm.

—— *Staff Report on Implementation Documents for the Uniform Dispute Resolution Policy* (29 September 1999), available at http://www.icann.org/udrp/staff-report-29sept99.htm.

—— *Second Staff Report on Implementation Documents for the Uniform Dispute Resolution Policy* (24 October 1999), available at http://www.icann.org/udrp/udrp-second-staff-report-24oct99.htm.

—— *A Unique, Authoritative Root for the DNS,* Internet Coordination Policy 3 (ICP-3, 9 July 2001), available at http://www.icann.org/icp/icp-3.htm.

—— *Strategy: Introduction of New Generic Top-Level Domains* (30 September 2004), available at http://www.icann.org/tlds/new-gtld-strategy.pdf.

—— *Status Report on the sTLD Evaluation Process* (28 November 2005; updated 3 December 2005), available at http://www.icann.org/tlds/stld-apps-19mar04/stld-status-report.pdf

—— Generic Names Supporting Organisation (GNSO), *Final Report on the Introduction of New Generic Top-Level Domains* (8 August 2007). Part A, available at http://gnso.icann.org/issues/new-gtlds/pdp-dec05-fr-parta-08aug07.htm.

—— Generic Names Supporting Organisation (GNSO), *Final Report on the Introduction of New Generic Top-Level Domains* (8 August 2007). Part B, available at http://gnso.icann.org/issues/new-gtlds/pdp-dec05-fr-partb-08aug07.htm.

Lynn, S, *President's Report—ICANN: The Case for Reform* (24 February 2002), available at http://www.icann.org/general/lynn-reform-proposal-24feb02.htm.

National Telecommunications and Information Administration (NTIA) (US), *A Proposal to Improve the Technical Management of Internet Names and Addresses,* 63 Fed. Reg 8,825 (1998), available at http://www.ntia.doc.gov/ntiahome/domainname/022098fedreg.htm.

—— *Management of Internet Names and Addresses* 63 Fed Reg 31,741 (1998), available at http://www.ntia.doc.gov/ntiahome/domainname/6_5_98dns.htm.

—— *U.S. Principles on the Internet's Domain Name Addressing System* (30 June 2005), available at http://www.ntia.doc.gov/ntiahome/domainname/USDNprinciples_06302005.htm

—— *The Continued Transition of the Technical Coordination and Management of the Internet Domain Name and Addressing System* 71 Fed Reg 30,388 (2006).

New TLD Evaluation Planning Task Force (NTEPPTF), *Final Report* (31 July 2002), available at http://www.icann.org/committees/ntepptf/final-report-31jul02.htm.

Palfrey, J, Chen, C, Hwang, S and Eisenkraft, N, *Public Participation in ICANN: A Preliminary Study* (December 2003), available at http://cyber.law.harvard.edu/icann/publicparticipation/.

Summit Strategies International, *Evaluation of the New gTLDs: Policy and Legal Issues*, Prepared for ICANN by Summit Strategies International (10 July 2004), available at http://icann.org/tlds/new-gtld-eval-31aug04.pdf.

World Intellectual Property Organization (WIPO), *The Management of Internet Names and Addresses: Intellectual Property Issues*, Interim Report of the WIPO Internet Domain Name Process (23 December 1998), available at http://www.wipo.int/amc/en/processes/process/rfc/3/index.html.

—— *The Management of Internet Names and Addresses: Intellectual Property Issues*, Report of the WIPO Internet Domain Name Process (30 April 1999), available at http://www.icann.org/amc/en/processes/process1/report/.

—— *The Recognition of Rights and the Use of Names in the Internet Domain Name System*, Second Process Interim Report (12 April 2001), available at http://www.wipo.int/export/sites/www/amc/en/docs/report-rfc3.pdf.

—— *The Recognition of Rights and the Use of Names in the Internet Domain Name System*, Report of the Second WIPO Internet Domain Name Process (3 September 2001), available at http://www.wipo.int/export/sites/www/amc/en/docs/report-final2.pdf.

World Summit on the Information Society (WSIS), *Declaration of Principles. Building the Information Society: a Global Challenge in the New Millennium*, Document WSIS-03/GENEVA/DOC/4-E (12 December 2003), available at http://www.itu.int/wsis/docs/geneva/official/dop.html.

——*Plan of Action*, Document WSIS-03/GENEVA/DOC/5-E (12 December 2003), http://www.itu.int/wsis/docs/geneva/official/poa.html.

——*Tunis Agenda for the Information Society*, Document WSIS-05/TUNIS/DOC/6(Rev.1)-E (18 November 2005), available at http://www.itu.int/wsis/docs2/tunis/off/6rev1.html.

—— *Tunis Commitment*, Document WSIS-05/TUNIS/DOC/7-E (18 November 2005), available at http://www.itu.int/wsis/docs2/tunis/off/7.html.

Working Group on Internet Governance (WGIG), *Report on Internet Governance* (June 2005), available at http://www.wgig.org/docs/WGIGREPORT.pdf.

—— *Background Report* (June 2005), available at http://www.wgig.org/docs/Background-Report.htm.

2. Books

Abbate, J, *Inventing the Internet* (Cambridge, Mass, The MIT Press, 1999).

Albitz, P, and Liu, C, *DNS and BIND*, 4th edn (Sebastopol, Cal, O'Reilly & Associates, 2001).

Black, U, *Internet Architecture: An Introduction to IP Protocols* (Upper Saddle River, NJ, Prentice Hall PTR, 2000).

Cross, R, and Harris, JW, *Precedent in English Law*, 4th edn (Oxford, Oxford University Press, 1991).

Davison, M, Johnstone, K, and Kennedy, P, *Shanahan's Australian Law of Trade Marks & Passing Off*, 3rd edn (Sydney, Law Book Co, 2003).

Greenblatt, B, *Internet Directories* (Upper Saddle River, NJ, Prentice Hall PTR, 2001).

Hafner, K, and Lyon, M, *Where Wizards Stay Up Late* (New York, Simon & Schuster, 1996).
Held, D, and McGrew, A (eds), *The Global Transformations Reader* (Cambridge, Polity Press, 2000).
Kitchin, D, Llewelyn, D, Mellor, J, Meade, R, Moody-Stuart, T, and Keeling, D, *Kerly's Law of Trade Marks and Trade Names,* 14th edn (London, Sweet & Maxwell, 2005).
Ladas, SP, *Patents, Trademarks, and Related Rights: National and International Protection* (Cambridge, Mass, Harvard University Press, 1975).
McCarthy, JT, *Trademarks and Unfair Competition,* 4th edn (Eagan, Minn, Thompson/West 2006).
Mueller, M, *Ruling the Root* (Cambridge, Mass, The MIT Press, 2002).
Naughton, J, *A Brief History of the Future* (London, Phoenix, 1999).
Rhodes, RAW, *Understanding Governance* (Buckingham, Open University Press, 1997).
Wadlow, C, *The Law of Passing Off,* 3rd edn (London, Sweet & Maxwell, 2004).

3. Journal Articles

Coase, R, 'The 1987 McCorkle Lecture: Blackmail' (1988) 74 *Virginia Law Review* 655.
Froomkin, AM, 'Wrong Turn in Cyberspace: Using ICANN to Route Around the APA and the Constitution' (2000) 50 *Duke Law Journal* 17.
—— 'ICANN's "Uniform Dispute Resolution Policy"—Causes and (Partial) Cures' (2002) 67 *Brooklyn Law Review* 605.
Klein, H, 'Understanding WSIS: An Institutional Analysis of the UN World Summit on the Information Society' (2004) 1 *Information Technologies and International Development* 3.
Morison, WL, 'Unfair Competition and Passing Off' (1956) 3 *Sydney Law Review* 50.
Mueller, M, 'ICANN and Internet Governance' (1999) 1 *info* 497.
Oppedahl, C, 'How Is a Domain Name Like a Cow?' (1997) 15 *John Marshall Journal of Computer and Information Law* 437.
Shavell, S, 'An Economic Analysis of Threats and their Illegality: Blackmail, Extortion and Bribery' (1993) 141 *University of Pennsylvania Law Review* 1877.
Weinberg, J, 'ICANN and the Problem of Legitimacy' (2000) 50 *Duke Law Journal* 187.
White, G, 'ICANN's Uniform Domain Name Dispute Resolution Policy in Action' (2001) 16 *Berkeley Technology Law Journal* 229.

Index